The Present Status of the Quantum Theory of Light

Fundamental Theories of Physics

*An International Book Series on The Fundamental Theories of Physics:
Their Clarification, Development and Application*

Editor: ALWYN VAN DER MERWE
University of Denver, U.S.A.

Editorial Advisory Board:

LAWRENCE P. HORWITZ, *Tel-Aviv University, Israel*
BRIAN D. JOSEPHSON, *University of Cambridge, U.K.*
CLIVE KILMISTER, *University of London, U.K.*
PEKKA J. LAHTI, *University of Turku, Finland*
GÜNTER LUDWIG, *Philipps-Universität, Marburg, Germany*
ASHER PERES, *Israel Institute of Technology, Israel*
NATHAN ROSEN, *Israel Institute of Technology, Israel*
EDUARD PROGOVECKI, *University of Toronto, Canada*
MENDEL SACHS, *State University of New York at Buffalo, U.S.A.*
ABDUS SALAM, *International Centre for Theoretical Physics, Trieste, Italy*
HANS-JÜRGEN TREDER, *Zentralinstitut für Astrophysik der Akademie der
    Wissenschaften, Germany*

Volume 80

The Present Status of the Quantum Theory of Light

*Proceedings of a Symposium
in Honour of Jean-Pierre Vigier*

edited by

Stanley Jeffers
*Department of Physics and Astronomy,
York University,
North York, Toronto, Canada*

Sisir Roy
*Physics and Applied Mathematics Unit,
Indian Statistical Institute,
Calcutta, India*

Jean-Pierre Vigier
*Department of Physics,
Université Pierre et Marie Curie,
Paris, France*

and

Geoffrey Hunter
*Department of Chemistry,
York University,
North York, Toronto, Canada*

KLUWER ACADEMIC PUBLISHERS
DORDRECHT / BOSTON / LONDON

A C.I.P. Catalogue record for this book is available from the Library of Congress.

ISBN 0-7923-4337-9

Published by Kluwer Academic Publishers,
P.O. Box 17, 3300 AA Dordrecht, The Netherlands.

Kluwer Academic Publishers incorporates
the publishing programmes of
D. Reidel, Martinus Nijhoff, Dr W. Junk and MTP Press.

Sold and distributed in the U.S.A. and Canada
by Kluwer Academic Publishers,
101 Philip Drive, Norwell, MA 02061, U.S.A.

In all other countries, sold and distributed
by Kluwer Academic Publishers Group,
P.O. Box 322, 3300 AH Dordrecht, The Netherlands.

Printed on acid-free paper

All Rights Reserved
© 1997 Kluwer Academic Publishers
No part of the material protected by this copyright notice may be reproduced or
utilized in any form or by any means, electronic or mechanical,
including photocopying, recording or by any information storage and
retrieval system, without written permission from the copyright owner.

Printed in the Netherlands

Professor Jean-Pierre Vigier

These proceedings of the symposium

"The Present Status of the Quantum Theory of Light"

are dedicated to the memory of our colleague

Richard Deane Prosser

born December 1936, deceased February 1996.

TABLE OF CONTENTS

Preface	xiii
Introduction to the Conference	xv
A Quantum of Light Shed on a Classical Problem .... Cynthia Kolb Whitney	1
A Classical Photon Model .... R. J. Beil	9
Stationary Vacuum-Polarization "P Fields": The Missing Element in Electromagnetism and Quantum Mechanics .... Malcolm H. MacGregor	17
Electrons and Photons as Soliton-Waves .... Geoffrey Hunter	37
Topological Solitons in Non-Linear Electrodynamics .... German Kälbermann	45
A Particle or Photon in a Field .... J. Kajamaa	57
The Myth of the Photon ...... Trevor W. Marshall and Emilio Santos	67
Could the Photon Be Superfluous? .... Mendel Sachs	79
Fluctuation-Dissipation Quantum Electrodynamics G.F. Efremov, A. Y. Chekhov, L.G. Mourokh and M.A. Novikov	97
Statistical Theory of Photon in the Nonlinear Media and in Vacuum .... G.F. Efremov, L.G. Mourokh and M.A. Novikov	103
Einstein-de Broglie-Proca Theory of Light and Simultaneous Existence of Transverse and Longitudinal Photons Sisir Roy and Malabika Roy	107

Origin, Observation and Consequences of the $B^{(3)}$ Field .. M.W. Evans 117

Opto-Magnetic Effects in Non Absorbing Media: Problems of Measurement and Interpretation .. S. Jeffers, M. Novikov and G. Hathaway 127

Generalized Equations of Electrodynamics of Continuous Media
.... V.M. Dubovik, B. Saha and M.A. Martsenuyk 141

Maxwellian Analysis of Reflection and Refraction
.... R. D. Prosser, S. Jeffers and J. Desroches 151

Solutions of Maxwell Equations for a Hollow Curved Wave Conductor
.... V. Bashkov and A. Tchernomorov 159

Maxwell's Equations Directly from the Dynamics of Point Particles
.... G. N. Ord 165

Obtaining the Schrödinger and Dirac Equations from the Einstein/KAC Model of Brownian Motion by Projection .... G. N. Ord 169

Derivation of the Schrödinger Equation .... Thomas B. Andrews 181

An Appraisal of Stochastic Interpretations of Quantum Mechanics
.... Millard Baublitz, Jr. 193

Compatible Statistical Interpretation of Quantum Beats . Mirjana Božić 205

Stochastic Non-Markov Model of Quantum Mechanical Behaviour
.... A. T. Gavrilin 217

Essentially Pure Particle Formulation of Quantum Mechanics
.... S. R. Vatsya 223

A Fundamental Force as the Deterministic Explanation of Quantum Mechanics .... Billie Jack Dalton 235

TABLE OF CONTENTS

A Geometric Approach to the Quantum Mechanics of de Broglie and Vigier
.... W. R. Wood and G. Papini   247

Stable Orbits as Limit Cycles .... S. Bergia   259

The Correspondence Principle: Periodic Orbits from Quantum Mechanics
.... Daniel Provost   269

Neutron Interferometric Experiments on Quantum Mechanics
.... G. Badurek   281

Enigmatic Neutrons .... V. K. Ignatovich   293

Experiments to Test the Reality of de Broglie Waves .... J. R. Croca   305

Preselected Quantum Optical Correlations .... Mladen Pavičić   311

Apparent Contradiction in EPR Correlations .... Luiz Carlos Ryff   323

The Wave-Particle Duality .... Eftichios Bitsakis   333

Quantum Mechanical Tunneling in a Causal Interpretation
.... Michael Clarkson   349

Quantum Uncertainty, Wave-Particle Duality and Fundamental Symmetries
.... Peter Rowlands   361

On the Contradiction between Quantum Mechanics and Relativity:
a Superluminal Quantum Morse Telegraph .... A. Garuccio   373

Incompatibility between Einstein's Relativity and Lorentz's Equations
.... Paul Marmet   383

On Self-Interaction and (Non-)Perturbative Quantum GRT
.... H.-H. v. Borzeszkowski   395

Classical Physics Foundations for Quantum Physics .... Lloyd Motz and David W. Kraft	405
Inertial Transformations: a Review .... F. Selleri	413
The Lorentz Invariance Revisited .... M. Surdin	437
A New Type of Massive Spin-One Boson: and its Relation with Maxwell Equations .... D. V. Ahluwalia	443
Implications of Extended Models of the Electron for Particle Theory and Cosmology .... E. J. Sternglass	459
The Takabayasi Moving Frame, from the A Potential to the Z Boson .... Roger Boudet	471
The Nature of the Cosmological Redshift .... Mariano Moles	483
A Quantum Digital Theory of Light .... N. V. Pope	495
Toward a Comprehensible Physical Theory: Gravity and Quantum Mechanics .... Hüseyin Yilmaz	503
Hidden Background Field and Quantum Non-Locality . R. Ramanathan	527
Einstein-Podolsky-Rosen Correlations within Bell Region R. Horodecki	535
Electromagnetic Gauge as Integration Condition and Selection of the Source Adhering Gauge .... O. Costa de Beauregard	541
Author Index	547
Subject Index	549

PREFACE

THE PRESENT STATUS OF THE QUANTUM THEORY OF LIGHT

In August of 1995, a group of over 70 physicists met at York University for a three-day symposium in honour of Professor Jean-Pierre Vigier. The attendance included theoretical and experimental physicists, mathematicians, astronomers and colleagues concerned with issues in the philosophy of science. The symposium was entitled "The Present Status of the Quantum Theory of Light" in accordance with Professor Vigier's wishes but in fact encompassed many of the areas to which Professor Vigier has contributed over his long and distinguished career. These include stochastic interpretations of quantum mechanics, particle physics, and electromagnetic theory. The papers presented at the symposium have been arranged in this proceedings in the following approximate order: ideas about the nature of light and photons, electrodynamics, the formulation and interpretation of quantum mechanics, and aspects of relativity theory.

Some of the papers presented deal with alternate interpretations of quantum phenomena in the tradition of Vigier, Bohm et al. These interpretations reject the account given in purely probabilistic terms and which deems individual quantum events to be acausal and not amenable to any analysis in space-time terms. As is well known, Einstein and others also rejected the purely statistical account of quantum mechanics. As stressed by Professor Vigier at the symposium, the current experimental situation now allows for the first time for individual quantum events to be studied, e.g. using parametric down converters, since a coincidence detection ensures that each detector registered the arrival of only one photon. This situation is very exciting since it may well be that experimental challenges to the orthodox interpretation may be realized. Some very interesting experiments making use of such techniques were reported at the symposium.

The local organizing committee would like to acknowledge the support received for this symposium from the Office of the Dean of Pure and Applied Science, Dean R. Prince and the Master of Bethune College, Paul Delaney. We would also like to acknowledge the efforts of Mrs. M. Filseth, B. de Sousa and C. Mariaselvanayagam, who contributed much to the success of this event.

S. Jeffers, Department of Physics and Astronomy
G. Hunter, Department of Chemistry
S. Joshi, Department of Physics and Astronomy
York University, Toronto

INTRODUCTION TO THE CONFERENCE

The fact that we are sitting here in Toronto is proof that we still don't really understand what light is. In fact, perhaps the point of this conference (and no doubt succeeding conferences) is to ask questions, but what are the right questions? New experiments are shaking the field. We live at an exciting time where one can detect photons one by one, so that we may discover new answers to old questions.

The first set of questions revolve around Maxwell's equations. Question one is: Are Maxwell's waves real? That is to say, do they carry energy and momentum, and what is their relation to particles? If following the Copenhagen school they are only probability waves, then how is it that they seem to coexist with particles as shown in the Otake experiment?

Do photons travel in space and time? When you observe light from some distant galaxy, has it traveled along a given path? When a photon collides with an electron, we know that energy momentum is locally conserved: there is directionality in corresponding directional recoil. Now if the photon traveled along a given path, then energy cannot be exactly conserved if it only goes through one slit in the double-slit experiment.

What is the connection between photons and Maxwell's waves? One of the ways of connecting particles with Maxwell's waves is to assume that the particles beat in phase with the real electromagnetic field. This was the starting point for de Broglie in his thesis. If this is true for photons, how do they describe correctly probabilistic events?

We know that the photon interferes with itself, but does it only interfere with itself? The experiments of Pfleegor and Mandel seem to indicate that coherent light from independent sources can interfere (Pfleegor and Mandel, 1967), even when there is only one photon present in the apparatus at a time. Can it be influenced by other waves, and if so, under what conditions?

If the Maxwell waves are real, what medium are they carried by? Is it the stochastic zero-point medium proposed by Dirac? From Dirac's point of view, "we are rather forced to have an aether" (Dirac, 1951). But if there is an aether, then we would predict that after a certain distance, interference in the double slit experiment should decrease, or correlations in EPR type of experiments should begin to fade. Some experiments bearing on these questions have been done (Jeffers and Sloan, 1994) but more need to be done.

Is light just photons, or is it a field plus photons? The Casimir effect would lead us to believe that it is both, because there are no photons exchanged in the Casimir effect. The same is true for the Bartlett and Cole experiment (Bartlett and Cole, 1985), showing the reality of the Maxwellian displacement current.

Is there a quantum potential? The quantum potential allows us to discuss certain naturally occurring concepts that cannot even be discussed in the Copenhagen interpretation of quantum mechanics, such as the time it takes for a particle to tunnel. Can it be detected, as has been proposed by various authors? If photons are "piloted" by Maxwellian fields, then are they piloted by the quantum potential? This quantum potential also leads to predictions that are not able to be made in the Copenhagen interpretation of quantum mechanics. For example:

- The quantum potential interpretation of the double-slit experiment leads us to predict that the photon passes through one slit and never crosses the central axis of symmetry (see Prosser and Jeffers, these proceedings). We can propose the following experiment to confirm this: put a strip of film that is mainly transparent along the axis of symmetry. Spray a photographic emulsion onto the film, such that it is still mainly transparent, but has on average some light sensitivity. Turn on the double slit for a significant period of time. If the film subsequently shows no absorbtion of photons along the central axis, then a prediction of the quantum potential that cannot be made in the Copenhagen interpretation is confirmed.
- The quantum potential interpretation predicts that the forward components of the wave-packet in either physical or momentum spaces are preferentially diffracted by the double slits. This leads us to predict that the fringes that are farthest away from the slits will have their frequency shifted slightly towards higher energies. This should be experimentally testable.
- The work function of the photoelectric effect should be modulated by the quantum potential. Panarella did some experimental tests on this a decade ago (Panarella, 1986), but the work needs to be redone in the single-photon regime.

Then what about the sources of the magnetic field? Even if photons are particles, there still remains the question of particles as sources of the electromagnetic field. If you assume, as did the founders of quantum mechanics, that particles are dimensionless points, you run into the many unsolved problems of the Lorentz-Dirac self-energy of the particle. Is the renormalisation of the mass and the self-interaction of the field on the electron just a mathematical trick or a real phenomenon? Not even the founders of quantum electrodynamics were satisfied with the theory; Dirac didn't believe in it (Hora and Shepanski, 1978).

So one is tempted to drop the point aspect of particles for a more complex model like Schwinger and others (Schwinger, 1983), of an extended structure for the photon or electron. A classical model of the spinning electron can be constructed that has no runaway solutions, and does not radiate in the Bohr orbits (Gutowski et al., 1977). The center of charge precesses about the center of mass with the de Broglie frequency, and the extended structure can be constructed in a covariant fashion. But Lorentz also proposed this, and his ideas ran into a number of problems; perhaps it is the appropriate moment to re-examine these objections. Yet if we don't adopt a point of view such as this, how else can you justify the non-radiation of the accelerating Bohr orbits?

The extended electron model gives us insight into the question: what is spin? We still don't have a definitive intepretation of Planck's constant. It was invented in a thermodynamical context, in the solution of the divergences of the Maxwellian theory of the black-body spectrum, then arises apparently independently as the unit of quantum mechanical spin, or in the diffusion constant of hydrodynamical models such as Madelung and Nelson (Nelson, 1966), or in the Feynman paths (Feynman and Hibbs, 1965).

Obviously the quantum theory of light is inherently related to the special theory of relativity. We know that the EPR type of interactions take place faster than the speed of light (Aspect et al.). What carries such interactions? Are they compatible with relativity (see Garuccio, these proceedings)? Do they break causality or not?

There are longitudinal solutions to Maxwell's equations that arise if the photon has a non-zero rest mass (see Evans, these proceedings). These solutions could provide a mechanism by which phase information between correlated particles is propagated faster than light. If an interaction is faster than light in one frame, then there exists a frame in which it is instantaneous, presumably in the center of mass frame of the correlated particles. Presumably it corresponds locally with the frame in which the 2.7°K background radiation frame is isotropic. This can, and should, now be tested.

The existence of a massive photon, with phase interactions that are instantaneous in the center of mass rest frame of interacting charged particles also implies the possible existence of an absolute rest frame. Perhaps it is time to repeat the Michelson-Gayle (Michelson and Gayle, 1925) and Sagnac experiments: do the Sagnac type fringes move when one aligns the experiments with respect to the 2.7°K microwave background?

It would appear from some astronomical observations that gravitational interactions take place instantaneously (Beckmann, 1987), (von Flandern, 1993, p. 49). This points towards a possible similarity between the Newton and Coulomb forces.

We live at an exciting time, especially with the new technology of parametric down-converters, which are providing some of the first really single photon experiments. The Alley (these proceedings), Otake, Chiao (Steinberg et al., 1993) experiments are producing results that show in the clearest of possible ways the implicit non-locality of quantum mechanics. Single photon tunneling can be probed for the first time and seems to conform to one of the predictions of the Bohm/de Broglie theory, namely that it is the forward components of the wave packet that preferentially tunnel (see Clarkson, these proceedings).

The quantum theoory of light is still in a shaky situation. Are Maxwell's equations the ultimate stage of the game? Should we add non-linear terms to the theory, or work on soliton solutions? Do Coulomb forces propagate faster than light?

We are certain about much less than we physicists have lead people to believe. It is not even clear what are the basic equations, or the interpretation of quantum mechanics. Prigogine has shown the non-equivalence of the Hamiltonian (Schrödinger) and Liouvillian (density matrix) formulations of quantum mechanics (Prigogine et al., 1973), contrary to what is taught in most textbooks (von Neuman). Beckmann has derived the Schrödinger equation from Maxwell's equations with only classical considerations (Beckmann, 1987). Ord shows the connection between the Feynman formalism, and the Dirac, Schrödinger and Maxwell equations (see Ord, these proceedings), and we have elaborated the connection between the Feynman paths and the Bohm trajectories (Vigier, 1989).

As the results of NASA's Hubble telescope begin to arrive, we see that the Hubble constant is not what we thought it was, and is perhaps not even constant. If so, then there is big trouble for the Big Bang (Arp and van Flandern, 1992)! We can say that everything is on the table now. It is as if we are in as much of a crisis in physics at the turn of the millennium as we were at the turn of the last century. We have to go right back to Newton: many of Newton's questions are still there. Action-at-a-distance is a fantastic example of the resurrection of old problems in modern physics. The magnetic interactions between charged particles might lead in certain situations to a new quantum chemistry, and explain new phenomena in plasma discharges.

We can now ask new questions, or demand precise answers to old questions. But perhaps the most important is to not rest content with old answers when they are clearly unsatisfactory. Really, nobody knows what light is.

To quote Newton, it is as if "we are children on the beach, and the ocean of truth lies still unexplored before us." Let us hope that this conference gives us all a swimming lesson!

J.-P. Vigier
M. Clarkson

References

Arp, H.C. and van Flandern, T. (1992) The case against the big bang, *Physics Letters A* 164, 263.
Bartlett, D.F. and Cole, T.R. (1985) *Physical Review Letters* 55, 59.
Beckman, P. (1987) *Einstein Plus Two*, Golem Press, Boulder, Colorado.
Dirac, P.A.M. (1951) Is there an aether? *Nature* 168, 906.
Feynman, R.P. and Hibbs, A.R. (1965) *Quantum Mechanics and Path Integrals*, McGraw Hill, New York.
Gutowski, D., Moles, M., and Vigier, J.-P. (1977) Hidden-parameter theory of the extended Dirac electron, *Nuovo Cimento B* 39, 193.
Hora, H. and Shepanski, J. editors (1978) *Directions in Physics*, Wiley, New York.
Jeffers, S. and Sloan, J. (1994) An experiment to detect "empty" waves, *Found.Phys. Lett.* 7, 4, 333.
Michelson, A.A. and Gayle, H.G. (1925) The effect of the earth's rotation on the velocity of light, *Astrophysical Journal* 61, 137.
Nelson, E. (1966). *Physical Review* 150, 1079.
Panarella, E. (1986) Nonlinear behaviour of light at very low intensities, in Honig, W.H., Kraft, D.M., and Panarella, E. (eds), *Quantum Uncertainties*, Plenum, New York, p. 105.
Pfleegor, R. and Mandel, L. (1967) Interference of independent photon beams, *Physical Review* 159, 1084.
Prigogine, I., George, C., Henin, F. and Rosenfeld, L. (1973) A unified formulation of dynamics and thermodynamics, *Chemica Scripta* 4, 5.
Schwinger, J. (1983) *Foundations of Physics* 13, 373.
Steinberg, A.M., Kwiat, P.W. and Chiao, R.Y. (1993) Measurement of the single-photon tunneling time, *Physical Review Letters* 73, 708.
Vigier, J.-P. (1989) Real physical paths in quantum mechanics, in *Proceedings of the 3rd Int. Symposium on the Foundations of Quantum Mechanics*, Tokyo, p. 140.
von Flandern, T. (1993) *Dark matter, missing planets and new comets*, North Atlantic Books.

A QUANTUM OF LIGHT SHED ON A CLASSICAL PROBLEM

CYNTHIA KOLB WHITNEY
Tufts University Electro Optics Technology Center
Medford, MA 02155, U.S.A.

Abstract

The quantum concept of the light photon, elaborated with possibly non-zero rest mass, is used to re-examine the classical problem of retarded electromagnetic potentials. A conflict with the classical Lienard-Wiechert formulation is displayed. An alternative formulation is recommended.

1. Introduction

The classical theory of retarded electromagnetic potentials is usually attributed to Lienard[1] and Wiechert[2], who developed a methematical model based on Euclidean geometry. Their work predated Einstein's special relativity theory (SRT), but has since been incorporated into that theory by means of various arguments, all claiming to represent the propagation of the classical light waves that are obtained from Maxwell's equations.

A new argument based on modern light quanta ought to support the same conclusions in regard to retarded electromagnetic potentials. But the present paper argues that in fact it does not. We are therefore forced to reconsider, and possibly alter, both the classical arguments and the potentials that result from them.

2. The Classical Formulation

The classical Lienard-Wiechert potentials are

$$\Phi, \mathbf{A} = Q[(1,\beta) / \kappa R]_{ret} \qquad (1)$$

where Φ and $\mathbf{A}$ are, respectively, scalar and vector potentials, Q is source charge, β is $\mathbf{v}/c$ with $\mathbf{v}$ being source velocity, κ is $1 - \mathbf{n} \cdot \beta$ with $\mathbf{n}$ being $\mathbf{R}/R$, $\mathbf{R}$ is source-to-observer spatial vector, and "ret" means source parameters are evaluated at retarded time, i.e. causally connected time, i.e. $t_{ret} = t - R_{ret}/c$.

S. Jeffers et al. (eds.), The Present Status of the Quantum Theory of Light, 1–8.
© *1997 Kluwer Academic Publishers. Printed in the Netherlands.*

Interesting features of the Lienard-Wiechert potentials include the following:

1) Source coordinates are defined implicitly. That is, t_{ret} is a function of R_{ret}, and R_{ret} is a function of t_{ret}. This feature may serve to warn the cautious user that constructing an analysis from the point of view of the observer may be tricky.

2) The scalar potential Φ is not a symmetric function of the source speed v, and the vector potential **A** is not an antisymmetric function of v. This behavior seems to conflict with behavior that might be expected for Φ,**A** if they were obtained by Lorentz transformation of the four-vector $(\Phi_0, \mathbf{0})$ from the coordinate frame of the source.

The classical Lienard-Wiechert potentials are nevertheless incorporated into SRT. The first step is rewriting them as

$$\Phi, \mathbf{A} = Q[\, \gamma(1, \beta) / \gamma \kappa R \,]_{ret} \qquad (2)$$

Then the numerator $[\gamma(1,\beta)]_{ret}$ is recognized as the unit-normalized source-velocity four-vector $V^\mu{}_{ret}$. Assuming uniform motion, the subscript "ret" has no effect on V^μ. The denominator $[\gamma \kappa R]_{ret}$ is recognized as the frame-invariant scalar inner product $[V_\mu R^\mu]_{ret}$, where the four-vector $R^\mu{}_{ret} = [R, \mathbf{R}]_{ret}$ is the zero-length source-to-observer coordinate difference. Thus

$$\Phi, \mathbf{A} = Q[\, V^\mu / V_\mu R^\mu \,]_{ret} \qquad (3)$$

A noteworthy feature of this formulation is that the denominator $[V_\mu R^\mu]_{ret}$ is not a frame-invariant scalar *number*, but rather a frame-invariant scalar *function* that varies with some appropriate time argument. This behavior makes this inner product quite different from some of the other inner products familiar in SRT, inner products that yield truly constant numbers, like rest mass or charge. We shall return to this point later.

3. The Quantum Approach

If we were to address the problem of retarded potentials within the context of quantum photons instead of classical waves, we would change many things. For example, istead of infinite wave extent, we would have a localized blip, and instead of phase velocity, we would have centroid velocity. In place of an energy density $(E^2+B^2)/2$ defined by fields, we would have a total energy hv defined by frequency, and in place of momentum density ExB/c defined by fields, we would have total momentum hv/c defined by frequency. Indeed, we would not have fields at all; we would have only photons, real and virtual, to mediate forces.

It is clear that the quantum photons are very different from classical fields. Polarization is agood example: quantum electrodynamics admits not two but four polarization states.[3] It is also clear that quantum photons are very similar to quantum particles. Interference is a good example: both kinds of entities experience interference.

Photons would become even more particle-like if we were to consider possibly non-zero photon rest mass. This would add something really decisive to direct the analysis.

Note that if photons have mass, then the population of photons shed from a source has a *center* of mass. Center of mass was always a useful concept in classical mechanics, and it will be an equally useful concept in a quantum-based review of retarded electromagnetic potentials.

In the observer's coordinate frame, this center of mass moves as time elapses. But in the source's coordinate frame, it is permanently positioned right at the source. This suggests that the analysis should definitely begin in the source coordinate frame.

4. In the Source Coordinate Frame

In the source coordinate frame, let time and space coordinates be t_0 and x_0, y_0, z_0. Let the observer move at some uniform velocity **v** past the source. Let $t_0 = 0$ be the time of closest approach between source and observer. Let a photon leave the source at $t_0 = t_{emit0}$ and arrive at the observer at $t_0 = t_{abs0}$. The distance the photon travels is

$$R_0 = c(t_{abs0} - t_{emit0}) \tag{4a}$$

while the observer travels distance

$$v(t_{abs0} - t_{emit0}) = \beta R_0 \tag{4b}$$

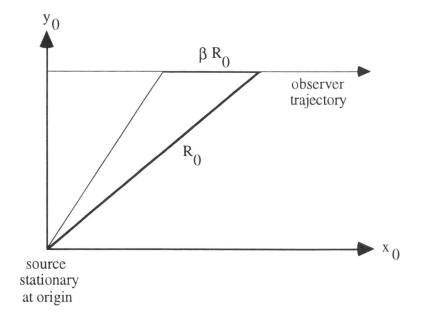

Figure 1. Situation display in source coordinate frame

Without loss of generality, we can make the direction of **v** be along the x_0 axis, and we can make the observer path lie in the x_0,y_0 plane. The situation in the source coordinate frame is then as illustrated by Fig. 1.

As time evolves, R_0 evolves with it. It is possible to track R_0 as a function of t_{abs0} with v as a parameter. The R_0 is a symmetric function of both t_{abs0} and v. It has asymptotes with slopes $\pm v$, and it looks like Fig. 2.

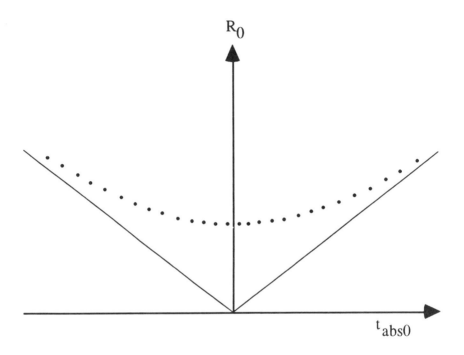

Figure 2. Evolution of R_0 with t_{abs0} in the source coordinate frame.

It is also possible to track R_0 as a function of t_{emit0}, again with v as a parameter. The R_0 is then not a symmetric function of either t_{emit0} or v. Its asymptotes have slopes $-v/(1+\beta)$ and $+v/(1-\beta)$, and it looks like Fig. 3.

Observe that every numerical value that occurs in $R_0(t_{abs0})$ (Fig. 2) also occurs in $R_0(t_{emit0})$ (Fig. 3), only for a different value of time argument. If we were to suppress the distinguishing subscripts "abs0" and "emit0" on the time arguments, then the function depicted on Fig. 3 would appear to anticipate the function depicted on Fig. 2. The magnitude of the time shift would be $t_{abs0}-t_{emit0}$, the time required for light propagation over the distance R_0. This observation forshadows a similar but more troubling one later on.

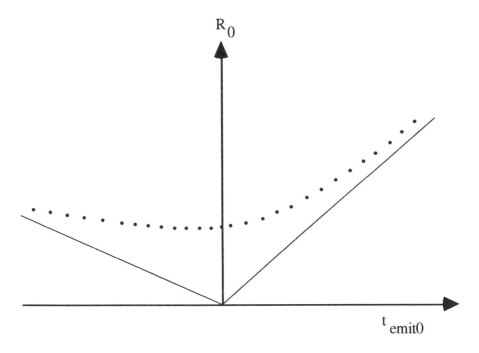

Figure 3. Evolution of R_0 with t_{emit0} in the source coordinate frame.

5. In the Observer Coordinate Frame

Clearly, R_0 provides the denominator for the scalar potential Φ_0 in the source coordinate frame, and it is R_0 that the denominator $[\gamma \kappa R]_{ret}$ in (2) or $[V_\mu R^\mu]_{ret}$ in (3) promises to recover while using only variables evaluated in the observer coordinate frame. So consider plots of R_0 analogous to Figs. 2 and 3, but the observer coordinate frame. For this we require new time variables t_{abs} and t_{emit} determined by Lorentz transformations.

First we have

$$t_{abs} = \gamma (t_{abs0} - v\, x_{abs0} / c^2) \tag{5a}$$

with $x_{abs0} = v\, t_{abs0}$, so that

$$t_{abs} = \gamma (t_{abs0} - v^2\, t_{abs0} / c^2) \equiv t_{abs0} / \gamma \tag{5b}$$

So when transformed to the observer coordinate frame, Figure 2 just shrinks in width. The essential symmetry of the curve does not change. The only change is that the slopes of it's asymptotes change from $\pm v$ to $\pm \gamma v$.

Similarly we have

$$t_{emit} = \gamma(t_{emit0} - v\, x_{emit0}/c^2) \qquad (6a)$$

with $x_{emit0} = 0$, so that

$$t_{emit} = \gamma\, t_{emit0} \qquad (6b)$$

So when transformed to the observer coordinate frame, Figure 3 just stretches in width. It remains asymmetrical. The slopes of its asymptotes change from $-v/(1+\beta)$ and $+v/(1-\beta)$ to

$$-v/\gamma(1+\beta) = -v\,[(1-\beta)/(1+\beta)]^{1/2} \qquad (7a)$$

and

$$+v/\gamma(1-\beta) = +v\,[(1+\beta)/(1-\beta)]^{1/2} \qquad (7b)$$

6. Back to the Classical Result

Now consider the classical $[\gamma\kappa R]_{ret}$ denominator for the potentials in (2). Plotted versus t_{abs}, the $[\gamma\kappa R]_{ret}$ is an asymmetrical function with asymptotes

$$-v\,\gamma(1-\beta) = -v\,[(1-\beta)/(1+\beta)]^{1/2} \qquad (8a)$$

and

$$+v\,\gamma(1+\beta) = -v\,[(1+\beta)/(1-\beta)]^{1/2} \qquad (8b)$$

Observe that there is a *perfect match* between the asymptotes (7a) and (8a), and between the asymptotes (7b) and (8b). Indeed, upon careful examination there is a perfect match between the *entire function* R_0 plotted versus t_{emit} and the function $[\gamma\kappa R]_{ret}$ plotted versus t_{abs}.

Conversely, there is *no match at all* between $[\gamma\kappa R]_{ret}$ plotted versus t_{abs} and R_0 also plotted versus t_{abs}. Where R_0 is a symmetric function of both t_{abs} and v, $[\gamma\kappa R]_{ret}$ is not a symmetrical function of either variable. Althogh every numerical value that occurs in $R_0(t_{abs})$ also occurs in $[\gamma\kappa R]_{ret}(t_{abs})$, it occurs earlier there. That is, the function $[\gamma\kappa R]_{ret}(t_{abs})$ *anticipates* the function $R_0(t_{abs})$. The magnitude of the time shift is just $t_{abs}-t_{emit}$, the time to propagate over R_{ret}.

In short, $[\gamma\kappa R]_{ret}$ does *not* recover R_0 as promised. The $[\gamma\kappa R]_{ret}$ anticipates R_0 by the amount of time required for propagation over R_{ret}. The fact that $[\gamma\kappa R]_{ret}$ is equivalent to an inner product, i.e. $[V_\mu R^\mu]_{ret}$, means only that it is an invariant; it does does *not* mean that it *the* invariant that corresponds to the correct time argument.

7. Remedy

Clearly, the needed R_0 is some *other* function of **R**$_{ret}$ and **v**. What this other function is can be determined by trigonometry. Figure 4 shows R_0 and $[\gamma \kappa R]_{ret}$ as hypotenuses of right triangles, along with the perpendicular sides needed to satisfy the Pythagorean theorem.

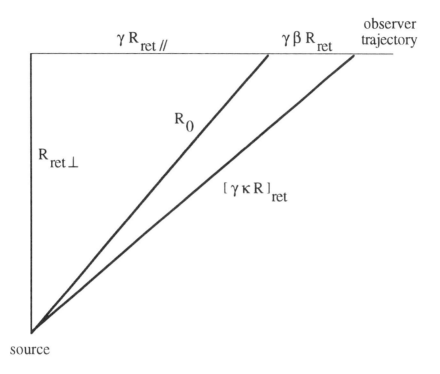

Figure 4. Determination of R_0 from R_{ret} and v.

Because coordinates perpendicular to motion are unaffected by coordinate-frame change, we have

$$R_{0\perp} = R_{ret\perp} \tag{9a}$$

It then follows that

$$R_{0//} = \gamma R_{ret//} \tag{9b}$$

and

$$R_0 = (R_{ret\perp}^2 + \gamma^2 R_{ret//}^2)^{1/2} \tag{9c}$$

It is *this* evaluation of R_0 that belongs in the potentials. In place of (2) or (3), we need

$$\Phi, A = Q \ [\ \gamma(1,\beta) \ ]_{ret} / R_0 = Q \ V^\mu_{ret} / R_0 \qquad (10)$$

We must *abandon* the usual assumption that the needed R_0 is the classical $[\gamma \kappa R]_{ret}$. or its relativistic equivalent $[V_\mu R^\mu]_{ret}$.

8. Discussion

The standard Lienard-Wiechert potentials have stood unchallenged for a very long time. But they do lead to unacceptable problems,[4] and a new formulation is definitely needed. The present paprer offers a new formulation that is consistent with the concept of photons that are similar to quantum particles.

The proposed new formulation does have at least one characteristic reminiscent of the classical formulation; namely, a "shrinkage" phenomenon. Standard theory says that, in the observer coordinate frame, the straight-line pattern of E field lines just shrinks by γ in the direction of source motion.[5] Somewhat similarly, the present theory says that, in the observer coordinate frame, the photon propagation path just shrinks by γ in the direction of source motion.

Acknowledgment

The author is grateful to Prof. J. P. Vigier for having suggested that concept of non-zero photon rest mass might shed light on the classical problem of retarded potentials.

References

1. A. Lienard, "Champ electrique et magnetique produit par une chjarge electrique concentree en un point et animee d'un movement quelconque," Eclairage Electrique **16**, 5-14, 53-59, 106-112 (1898).

2. E. Wiechert, "Elektrodynamische Elementargesetze," Archives Neerlandes **5**, 549-573 (1900).

3. R. P. Feynman, QED, The Strange Theory of Light and Matter, p.120, (Princeton University Press, Princeton N. J., 1988).

4. C. K. Whitney, "A Gedanken Experiment with Relativistic Fields," Galilean Electrodynamics **2**, 28-29 (1991).

5. J. D. Jackson, *Classical Electrodynamics*, Second Edition, Fig. 11.9 (John Wiley and Sons, New York, Chichester, Brisbane, Toronto, Singapore, 1975).

A CLASSICAL PHOTON MODEL

R. G. Beil
313 S. Washington
Marshall, TX 75670

ABSTRACT. Photon-like objects are constructed from the Brittingham solutions of the four-dimensional wave equation. The objects have one-dimensional, localized, but non-singular extension in the null direction. Their properties computed using classical electrodynamics correspond to the experimental values for photons.

In most of physics literature, both classical and quantum, the photon is depicted as a plane wave. This model is chosen perhaps more because of its mathematical convenience than its physical suitability.

The plane wave is actually sometimes difficult to reconcile with physical intuition because it implies that the photon can be extended over the whole universe. Also, the plane wave has infinite energy. One would think that there should be a better model for something as fundamental as a photon.

It is common now to argue that photons are not real particles at all and should not be subject to visualizable models. This is the attitude of most interpretations of quantum electrodynamics. Photons are supposed to be only the quantized excitations of the normal modes of the system. This may be satisfactory for boxes, but somehow does not seem appropriate for the light from a star or even the localized scintillations on a phosphor screen.

This difference in viewpoint goes back at least to the discussion by Einstein [1] of *Kugelstrahlung*, spherical radiation, versus *Nadelstrahlung*, needlelike radiation. Einstein concluded that only the radiation along a single null path, *Nadelstrahlung*, could satisfy the requirements of relativistic dynamics. A recent discussion of these arguments has been given by Cormier-Delanoue [2].

Occasionally, other physicists have expressed their discomfort with the plane wave model [3]. Infrequent attempts have been made to represent the photon as a finite distribution or wave packet [4]. It is "common knowledge" that such representations may be possible, but specific realizations are rare. Perhaps the most interesting older work in this line is that of Synge [5]. A fairly recent review of photon models is given by Kidd, Ardini, and Anton [6], with an extensive list of references.

One example of an alternative to plane waves, which goes to the other extreme, are highly singular delta function solutions. These solutions were studied recently by Barut [7].

Another alternative, advocated by Hillion [8], is the Brittingham solutions to the wave equation. These are quite remarkable because they can be localized, but dispersionless bundles of energy. The existence of these solutions was unknown until 1983, but their properties are already well established and an extensive literature exists, mostly in electrical engineering journals.

What the Brittingham solutions offer is a specific mathematical realization of the *Nadelstrahlung* concept. They give a well-defined blueprint for constructing an object which is a localized bundle of energy propagating in a null direction [8]. The actual construction of such an object and the calculation of its properties is presented here.

The procedure will be to first set up a system of coordinates in the frame of the particle itself. General Brittingham solutions of the wave equation for the electromagnetic four-potential are found in these coordinates. A certain linear combination of these solutions produces a potential vector which is identified as the electromagnetic potential of the photon. Calculations are then made using classical electromagnetic theory of the four-momentum, spin, helicity, and current due to this potential. These results agree with experimental values.

The rectilinear frame coordinates are labeled $X^\alpha = (T; X, Y, Z)$ and are related at some point (which is taken to be the instantaneous origin of the frame) to the laboratory coordinates x^μ by a Lorentz transformation. The coordinate T is in a timelike direction but is taken for convenience to have the dimension of a length. The transformation matrix forms an orthonormal tetrad $\partial X^\alpha / \partial x^\mu = X^\alpha_{,\mu}$ where

$$\eta_{\alpha\beta} X^\alpha_{,\mu} X^\beta_{,\nu} = \eta_{\mu\nu} \qquad (1)$$

at the point.

The coordinates X^α have been used in prior work [9, 10]. These references should be consulted for details of the geometry.

Null coordinates can also be defined: $U = X + T$, $V = X - T$ with

$$U_{,\mu} U^{,\mu} = 0, \ V_{,\mu} V^{,\mu} = 0, \ U_{,\mu} V^{,\mu} = -2 \qquad (2)$$

where $U^{,\mu} = U_{,\nu} \eta^{\mu\nu}$.

A CLASSICAL PHOTON MODEL

The particle frame is taken to be transporting in the U-direction along the line $V = 0$. This is a null path and U can be identified as the parameter of the path. So the space part of the propagation is in the $+X$ direction with Y and Z in the transverse plane. The transport of the tetrad determines the transport of the photon itself. The photon is assumed to reside on a moving frame which propagates in a null direction. This is a special case of transport by the Poincaré group.

Notice that this tetrad scheme makes it very easy to include polarization. For example a transverse polarization would be given by a linear combination of $Y^{,\mu}$ and $Z^{,\mu}$.

It is assumed, in general, that the electromagnetic potential vector can be expressed as

$$A^\mu = X_\beta^{,\mu} F^\beta(X^\alpha) \tag{3}$$

where the polarization could have components in any of the four coordinate directions. For simplicity, one direction ϵ^μ could be chosen so that $A^\mu = \epsilon^\mu F(X^\alpha)$. F is a solution of the wave equation in the frame of the particle, i.e.

$$\Box F = \eta^{\alpha\beta}(\partial^2 F/\partial X^\alpha \partial X^\beta) = 0 \tag{4}$$

It is useful to introduce polar coordinates σ and θ in the $Y - Z$ plane: $\sigma^2 = Y^2 + Z^2$, $\theta = \arctan(Z/Y)$, such that the wave equation (using also U and V) becomes:

$$4\frac{\partial^2 F}{\partial U \partial V} + \frac{1}{\sigma}\frac{\partial F}{\partial \sigma} + \frac{\partial^2 F}{\partial \sigma^2} + \frac{1}{\sigma^2}\frac{\partial^2 F}{\partial \theta^2} = 0 \tag{5}$$

Now, instead of the usual separated cylindrical solutions, the Brittingham solutions are taken:

$$F = \frac{f(P)}{V}, \quad P = \frac{\sigma^2}{V} + U \tag{6}$$

where f can be any twice differentiable function of P. For a survey of the literature of these solutions one could start with [11] and [12]. The fact that (6) is a solution of (5) can easily be verified by direct substitution. Note that P is not a function of θ.

These waves have the "built-in" property of being nondispersive. Particular forms of these waves can have a transverse structure which does not change as the wave is propagated in the U direction [8] . These are soliton-like solutions of the *linear* wave equation. The only problem with these solutions is that they have infinite energy, as do plane waves. There has been considerable discussion as to whether finite energy solutions of this type exist [8]. As a matter of fact, such solutions do exist. They can be obtained by assuming certain forms for $f(P)$ and taking a superposition of waves in the transverse plane. This is now demonstrated:

The four-volume element $dT\,dX\,dY\,dZ$ in these coordinates is

$$d^4\Omega = \frac{1}{2}dU\,dV\,\sigma\,d\sigma\,d\theta \tag{7}$$

Integration over each of the coordinates should be accompanied by a factor $k/(2\pi)^{\frac{1}{2}}$ where k is a wave number. This is common usage in quantum electrodynamics and goes with the application of symmetric Fourier transforms. The factor also preserves the dimensionality of the result.

So a two-dimensional potential is formed by integrating over the transverse plane:

$$A^\mu(U,V) = \frac{A_0 Y^{,\mu} k^2}{2\pi} \int_0^\infty \int_0^{2\pi} \frac{1}{V} f(P) d\theta \sigma d\sigma \tag{8}$$

The transverse polarization direction, $Y^{,\mu}$, is chosen. The constant A_0 is to be determined.

From the definition of P, one has $\sigma d\sigma = V dP/2$, so the integral is

$$A^\mu(U,V) = \frac{1}{2} A_0 Y^{,\mu} k^2 \int_U^\infty f(P) dP \tag{9}$$

The θ integration just gives 2π.

At this juncture a particular form for $f(P)$ could be chosen. There are numerous possibitities, but one which gives favorable results is

$$f(P) = -\left[2\sin(kP) + 4(m^2/k)P\cos(kP)\right] \exp(-m^2 P^2) \tag{10}$$

This produces

$$A^\mu(U) = A_0 Y^{,\mu} k \cos(kU) \exp(-m^2 U^2) \tag{11}$$

Note that the V-dependence fortuitously disappears so the resulting potential has only a one-dimensional extension on the U axis and is localized around $U = 0$. Eq.(11) is a reasonable form for a localized wave. The constant m determines the extent of the Gaussian envelope of the wave. It has the dimensions of a wave number and eventually will be shown to be equatable to k. The expression (11) also has the correct dimensionality of an electromagnetic potential vector if A_0 is proportional to $(\hbar c)^{\frac{1}{2}}$.

The potential is easily verified to be itself a solution of the wave equation (5) so the discussion could have begun at (11). It was instructive, however, and perhaps also physically significant to show how the potential can be constructed from the solutions (6).

The potential (11) is linearly polarized. For a photon potential a circularly polarized wave is taken:

$$A^\mu(U) = A_0 \left[Y^{,\mu} g_Y(U) + Z^{,\mu} g_Z(U)\right] \tag{12}$$

with

$$\begin{aligned} g_Y(U) &= ik\cos(kU)\exp(-m^2 U^2) \\ g_Z(U) &= ik\sin(kU)\exp(-m^2 U^2) \end{aligned} \tag{13}$$

This is similar to some previous classical discussions of photons [13]. The imaginary factor in (13) is inserted to make the norm of the potential positive.

The potential (12) can be normalized in order to fix the constant A_0. Since A^μ is presumed to be the elementary state of one photon it is reasonable to normalize it to the level of a single quantum of radiation:

$$\int_{-\infty}^{\infty} A_\mu A^\mu \, dU = \hbar\omega \qquad (14)$$

This condition is the only "quantum" assumption used here; the rest is purely classical.

For the potential (12) with (13),

$$\int_{-\infty}^{\infty} A_\mu A^\mu \, dU = -A_0^2 \int_{-\infty}^{\infty} \left[(g_Y)^2 + (g_Z)^2\right] dU$$

$$= A_0^2 k^2 \int_{-\infty}^{\infty} \exp(-2m^2 U^2) \, dU = \frac{A_0^2 k^2}{m} \left(\frac{\pi}{2}\right)^{\frac{1}{2}} \qquad (15)$$

Comparison with (14) gives

$$A_0^2 = \left(\frac{2}{\pi}\right)^{\frac{1}{2}} \frac{m\hbar\omega}{k^2} \qquad (16)$$

The potential (12) is now taken as a starting point for the computation of various physical properties. The particular solutions (13) with (16) will eventually be used, but the discussion will be done in terms of general functions $g_Y(U)$ and $g_Z(U)$ since other forms than (13) are possible.

First,

$$\frac{\partial A^\mu}{\partial x^\nu} = A_0 \left(g'_Y Y^{,\mu} U_{,\nu} + g'_Z Z^{,\mu} U_{,\nu}\right) \qquad (17)$$

The prime denotes differentiation with respect to the argument U of the functions. This potential obviously satisfies the Lorentz gauge $\partial A^\mu / \partial x^\mu = 0$.

The field tensor for this potential is

$$F^{\mu\nu} = A_0 \left[g'_Y \left(U^{,\mu} Y^{,\nu} - U^{,\nu} Y^{,\mu}\right) + g'_Z \left(U^{,\mu} Z^{,\nu} - U^{,\nu} Z^{,\mu}\right)\right] \qquad (18)$$

It is readily shown that this is a null field.

The physical properties of the object are given in terms of three standard tensor densities:

$$\Theta^{\mu\nu} = \frac{1}{4\pi} \left(F^{\mu\rho} F_\rho{}^\nu + \frac{1}{4} \eta^{\mu\nu} F_{\kappa\lambda} F^{\kappa\lambda}\right)$$

$$S^{\nu\kappa\lambda} = -\frac{1}{4\pi c} \left(F^{\nu\kappa} A^\lambda - F^{\nu\lambda} A^\kappa\right)$$

$$J^\nu = \frac{c}{4\pi} \frac{\partial F^{\mu\nu}}{\partial x^\mu} \qquad (19)$$

These densities are, respectively, the energy-momentum, the intrinsic angular momentum of the field, and the current.

The results for the potential (12) are

$$\Theta^{\mu\nu} = \frac{A_0^2}{4\pi}\left[(g'_Y)^2 + (g'_Z)^2\right] U^{,\mu}U^{,\nu}$$

$$S^{\nu\kappa\lambda} = -\frac{A_0^2}{4\pi c} R^{\nu\kappa\lambda}$$

$$\begin{aligned}R^{\nu\kappa\lambda} &= g_Y g'_Y \left[Y^{,\nu}\left(U^{,\lambda}Y^{,\kappa} - U^{,\kappa}Y^{,\lambda}\right)\right] \\ &+ g_Z g'_Y \left[U^{,\nu}\left(Y^{,\kappa}Z^{,\lambda} - Y^{,\lambda}Z^{,\kappa}\right) + Y^{,\nu}\left(U^{,\lambda}Z^{,\kappa} - U^{,\kappa}Z^{,\lambda}\right)\right] \\ &+ g_Y g'_Z \left[U^{,\nu}\left(Z^{,\kappa}Y^{,\lambda} - Z^{,\lambda}Y^{,\kappa}\right) + Z^{,\nu}\left(U^{,\lambda}Y^{,\kappa} - U^{,\kappa}Y^{,\lambda}\right)\right] \\ &+ g_Z g'_Z \left[Z^{,\nu}\left(U^{,\lambda}Z^{\kappa} - U^{,\kappa}Z^{,\lambda}\right)\right]\end{aligned}$$

$$J^\nu = 0 \tag{20}$$

The density $\Theta^{\mu\nu}$ is well known to be conserved and gauge invariant for a general free field. Since $U^{,\mu}$ is in the direction of propagation the identification $U^{,\mu} = k^\mu/k$ can be made. So the tensor is proportional to a standard classical form. It also has the correct dimensionality of energy/three-volume since the g's have the dimensionality of k or $length^{-1}$.

The density $S^{\nu\kappa\lambda}$ is not conserved in general, but is, in fact, conserved for the type of potential given by (12). This density is also not gauge invariant, but since, in effect, a particular gauge (Lorentz) is determined by (12), gauge invariance is not critical.

The vanishing current is a result which, of course, should be expected.

In order to get four-momentum and spin, the corresponding expressions in (20) should be integrated over a three-surface. In this case, the surface can be derived from the four-volume (7):

$$dS_\nu = \frac{1}{2} V_{,\nu} dU \sigma d\sigma d\theta \tag{21}$$

But the object being dealt with is already two-dimensional (in the U, V plane) since the integration over the transverse (Y, Z) plane has already been performed in (8). So integration should be done only over the null coordinate U. The four-momentum is thus,

$$P^\mu = \frac{2\pi}{ck^2} \int_{-\infty}^{\infty} \frac{1}{2} \Theta^{\mu\nu} V_{,\nu} dU \tag{22}$$

The factor $2\pi/k^2$ is consistent with the usage discussed in relation to (8) and produces correct dimensionality.

Similarly, the integral for the spin is

$$S^{\kappa\lambda} = \frac{2\pi}{k^2} \int_{-\infty}^{\infty} \frac{1}{2} S^{\nu\kappa\lambda} V_{,\nu} dU \tag{23}$$

The results using (20) along with (1) and (2) are

$$P^\mu = -\frac{A_0^2}{2ck^2} \int_{-\infty}^{\infty} \left[(g'_Y)^2 + (g'_Z)^2\right] U^{,\mu} dU$$

$$S^{\kappa\lambda} = \frac{A_0^2}{2k^2 c} \int_{-\infty}^{\infty} [g_Z g'_Y - g_Y g'_Z] \left(Y^{,\kappa} Z^{,\lambda} - Y^{,\kappa} Z^{,\lambda}\right) dU \qquad (24)$$

Now, with the particular functions (13),

$$P^\mu = \frac{A_0^2}{2c} \int_{-\infty}^{\infty} \left(k^2 + 4m^4 U^2\right) \exp(-2m^2 U^2) U^{,\mu} dU$$

$$= \frac{A_0^2}{2c} \left(\frac{k^2}{m} + m\right) \left(\frac{\pi}{2}\right)^{\frac{1}{2}} U^{,\mu} \qquad (25)$$

For $k = m = \omega/c$ and recalling (16),

$$P^\mu = \hbar\omega U^{,\mu}/c = \hbar k^\mu \qquad (26)$$

Also,

$$S^{\kappa\lambda} = \frac{A_0^2}{2c} \int_{-\infty}^{\infty} k \exp(-2m^2 U^2) \left(Y^{,\kappa} Z^{,\lambda} - Y^{,\lambda} Z^{,\kappa}\right) dU$$

$$= \frac{A_0^2 k}{2mc} \left(\frac{\pi}{2}\right)^{\frac{1}{2}} \left(Y^{,\kappa} Z^{,\lambda} - Y^{,\lambda} Z^{,\kappa}\right) = \frac{\hbar}{2} \left(Y^{,\kappa} Z^{,\lambda} - Y^{,\lambda} Z^{,\kappa}\right) \qquad (27)$$

From this antisymmetric tensor a spin vector

$$S_\mu = \frac{1}{2} \epsilon_{\mu\nu\kappa\lambda} S^{\nu\kappa} U^{,\lambda} \qquad (28)$$

is constructed. The vector $U^{,\lambda}$ must be used here since it defines the direction of propagation. S_μ is computed as

$$S_\mu = \frac{\hbar}{4} \epsilon_{\mu\nu\kappa\lambda} \left(Y^{,\nu} Z^{,\kappa} - Y^{,\kappa} Z^{,\nu}\right) \left(X^{,\lambda} + T^{,\lambda}\right) = -\hbar \left(T_{,\mu} + X_{,\mu}\right) = -\hbar U_{,\mu} \qquad (29)$$

So the helicity is

$$-S_\mu V^{,\mu}/U_{,\mu} V^{,\mu} = \hbar \qquad (30)$$

A value of $-\hbar$ would be obtained for reverse polarization.

The results (26) and (30) show that the constructed objects thus have the known properties of photons.

So it might be said that what has been developed here is a sort of "classical quantum". This is an extended, but localized wave bundle formed of classical solutions. A key difference is that the solutions are not the usual plane waves, but the Brittingham waves. These are a specific example of Einstein's *Nadelstrahlung* and offer a compromise between point particles and infinite plane waves.

These objects are waves which behave like particles or particles which behave like waves. They might offer a resolution of the age-old duality problem.

References

[1] Einstein, A. (1917) *Phys. Z.* **18**,121.

[2] Cormier-Delanoue (1989) *Found. Phys.* **19**,1171.

[3] Jaynes, E. (1978) in L. Mandel and E. Wolf (eds.), *Coherence and Quantum Optics*, Plenum, N.Y.

[4] Geppert, D. V. (1964) *Phys. Rev. B* **134**,1407.

[5] Synge, J. L. (1972) *Relativity: the Special Theory*, North-Holland, Amsterdam.

[6] Kidd, R., Ardini, J., and Anton, A. (1989) *Am. J. Phys.* **57**,27.

[7] Barut, A. O. (1990) *Phys. Lett. A* **143**,349.

[8] Hillion, P. (1992) *Phys. Lett. A* **172**,1.

[9] Beil, R. G. (1993) *Found. Phys.* **23**,1587.

[10] Beil, R. G. (1995) *Found. Phys.* **25**,717.

[11] Donnelly, R. and Ziolkowski, R. (1993) *Proc. Roy. Soc. London A* **440**,541.

[12] Hillion, P. (1991) *Int. J. Theor. Phys.* **30**,197.

[13] Rohrlich, F. (1965) *Classical Charged Particles*, Addison-Wesley, Reading, Mass.

STATIONARY VACUUM-POLARIZATION "P FIELDS": THE MISSING ELEMENT IN ELECTROMAGNETISM AND QUANTUM MECHANICS

MALCOLM H. MAC GREGOR
Lawrence Livermore National Laboratory
University of California
P. O. Box 808, Livermore, CA 94550, USA

Abstract

A spectroscopic analysis of electrons and photons shows that, with an extension of our usual concept of basis states, these particles admit essentially classical representations. By assigning mass and charge "lattices" to the vacuum state, we can define basic "particle-hole" and "antiparticle-antihole" vacuum-polarization excitation pairs, and we can deduce their equations of motion. We denote these (non-rotating) P-H and $\bar{P}$-$\bar{H}$ pairs as "polarons" and "antipolarons." With the aid of the polaron basis states, we can reproduce not only the photon, but also the "zeron," which is the quantum of the wave packet that accompanies the photon. The photon is spectroscopically reproduced as a linked polaron-antipolaron pair in a rotational mode. The zeron is a rotating polaron that closely mimics the photon electromagnetically, but which carries only infinitesimal amounts of energy and momentum. In its motion through space, a wave train of rotating zerons induces a stationary polaron electric field pattern (a "P field") of opposite polarity in the vacuum state. This static P field performs a number of crucial functions: *(1)* the P field "self-synchronizes" individual zerons into accurately correlated ψ-wave packets, both for plane waves and for ψ-wave interference patterns; *(2)* the electric field strength of the P field in overlapping interference regions is proportional to the *square* of the corresponding ψ-wave amplitude, and in its steering mechanism this interference P field converts ψ-wave amplitudes into quantum mechanical probability distributions; *(3)* the requirement that the P field must close on itself furnishes the quantization of atomic electron orbitals in accurate multiples of the de Broglie wavelength λ; *(4)* a pair of time-reversed (counter-rotating and spin-reversed) electrons, protons, or neutrons can share a common mutually-enhanced P field, thus creating the "pairing force" mechanism that dominates both atomic and nuclear structure, and which also leads to the Cooper pairs of superconductivity; *(5)* we can provide a physical basis for the ψ-wave orthogonality relations that dictate atomic shell structure by imposing the requirement that the superposed P fields of two electrons in different orbitals do not (on the average) "steer" each other's electrons. Classical electric E-field lines are reproduced as chains of linked polarons, and magnetic H-field lines are reproduced as linked-polaron loops, with the Lorentz force acting to prevent the collapse of the loops.

1. Enigmatic Electrons

This Symposium is devoted to The Present Status of the Quantum Theory of Light. Thus it is primarily focused on the properties of photons and electromagnetic waves. However, we know from the successes of quantum electrodynamics that electrons and photons are closely related objects, and as such should be studied together. In particular, formalisms that can be shown to apply to one may also apply to the other. It is with this rationale that we start with a brief discussion of the electron.

The *enigma* of the electron has long been recognized. In the context of classical formalisms, its observed mass and spin angular momentum combine to mandate a Compton-sized particle ($R \sim 4 \times 10^{-11}$ cm), and its observed charge and magnetic moment lead to this same conclusion. But actual measurements of the "size" of the electron lead to a vastly smaller upper limit ($R < 10^{-16}$ cm). Apart from these spectroscopic paradoxes, the enigma of the electron is compounded by the difficulty of deciding whether it should be regarded as a discrete localized particle (in Compton and Møller scattering), or as a wave (in the Schrödinger equation, which makes no reference to particles), or as sometimes one and sometimes the other (in electron virtual-double-slit experiments). From the viewpoint of *physical realism (PR)*, which we pursue here, real physical effects are necessarily produced by real physical entities. Since electron particle and wave phenomena can be observed in the same experiment, localized electrons and extended electron waves must simultaneously exist (as must localized photons and extended electromagnetic waves). Although not in the mainstream of present-day scientific thought, this *PR* viewpoint has had a long and continuing development [1].

The postulated existence of the electron as a localized entity leads to the intellectual challenge of producing a spectroscopic model for the electron. This is more than just an idle curiosity. Since we can't actually "see" an electron, we have to represent its properties mathematically. The question is: *what mathematics are we allowed to use?* Classical physics provides us with equations for calculating spins and magnetic moments, but do these equations apply to an object the size of the electron? Furthermore, what *is* the size of the electron? These questions have been dealt with in a recent monograph by the present author [2], and here we merely summarize the key results.

The interactions of the electron are known to be purely electromagnetic, and to be point-like in nature. Thus the charge on the electron must be point-like. However, if its magnetic moment arises from a rotating charge, the radius of the current loop must be Compton-sized. Similarly, the spin angular momentum of the electron requires a rotating mass of Compton-like dimensions. Since the electron mass does not contribute to its interactions, the mass must be non-interacting (neutrino-like). These properties of the electron lead to a model which consists of a non-interacting spherical mechanical mass with a charge on the equator. The sphere is rotating at the relativistic limit in which its equator is moving at or just below the velocity of light, c. If we calculate the inertial properties of the rotating mass relativistically [2], then this more-or-less classical model gives the correct gyromagnetic ratio between the magnetic moment and spin angular momentum of the electron—which is a result that has sometimes been regarded as impossible to achieve. This model for the electron is shown in Figure 1, where the spin axis is tilted with respect to the z-axis of quantization at the quantum-mechanically

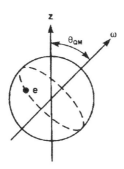

Figure 1. A spectroscopic model for the electron [2].

prescribed angle θ_{QM} for a spin 1/2 particle, which is also the angle at which the electric quadrupole moment of the rotating current loop vanishes. The standard spectroscopic features of the electron are reproduced, and also some intriguing quantum features. A comprehensive discussion of this model in given in ref. [2].

2. Enigmatic Photons

If we can devise a representational model for the electron, can we do the same for the photon? The answer is *yes*, if we are willing to extend our usual definition of basis states. The electron model shown in Fig. 1 features an ordinary point electric charge e, but it requires the inclusion of a new basis state—a *non-interacting mechanical mass*. The photon model, in which photons carry angular momentum but have no net rest mass, leads to an additional basis state—a *hole* state, which must be regarded as a "hole" in the underlying vacuum manifold. With the addition of this state, we can accurately reproduce the standard properties of the photon, and we can also devise a quantum to represent the "empty waves" that surround and accompany the photon and guide it through space. Before we develop this model mathematically, we first discuss the properties of the vacuum manifold and of "hole" states, including their motion in an applied external electrostatic field. "Hole" states differ in an essential respect from the well-studied "negative-mass" states of special relativity. Many of the ideas we discuss here with respect to photons and empty waves have been described in detail in a recent publication [3], which gives pertinent references to the literature.

3. Non-Rotating Vacuum-State "Polaron" Particle-Hole Excitation Pairs

The formalism of lattice gauge theory was developed by assigning a computational lattice to the vacuum state. In rough analogy to this, we conceptually picture the vacuum state as simultaneously containing two lattices—a "particle" mass manifold L that is studded with negative electric charges $-e$, and an "antiparticle" mass manifold $\bar{L}$ that is studded with positive electric charges $+e$. The excitations of these manifolds correspond to the *vacuum fluctuations* that give rise, for example, to the Casimir effect [2,3]. A basic "particle-hole" excitation corresponds to the displacement of a "particle" mass m and an associated charge $-e$ from its place in the lattice L, with the concomitant production of a vacancy—a "hole"—that appears as a (virtual) mass $-m$ with a (virtual) charge $+e$. A corresponding "antiparticle-antihole" pair is produced by exciting an antiparticle mass $\bar{m}$ and an associated charge $+e$ from its place in the antilattice $\bar{L}$. These are the fundamental "polaron" and "antipolaron" vacuum polarization pairs.

To study the behavior of these polaron pairs, we must ascertain the equations of motion for the "particle" and "hole" excitation states. The "particle" masses and charges are the ones we customarily deal with, and these states follow the conventional Newtonian equations of motion. The "hole" effective masses and charges, however, are different. The significance of a hole state in a mass manifold is that a *new degree of freedom* is now available which was not there before. A neighbor can move into the hole and fill it, thereby creating a new hole at the former position of the neighbor. If a mechanical "push" is applied from (say) the left, the left neighbor is moved into the hole, so that the hole itself moves to the *left*, in the direction *opposite* to the push. Thus, mechanically, a hole state behaves formally like a "negative mass." It appears in Newton's equations with a negative sign for the mass [3]. But the behavior of the hole under an external electrostatic field is different. The hole state seems to have a positive charge, since the negative charge which used to be at that location in the lattice is missing. However, in actuality the hole has no charge. An external electrostatic field will not act directly on the hole, but instead will act on the neighbors of the hole (just as the mechanical push did). Consider an applied electrostatic field that is directed so as to move a positive charge to (say) the right. It actually acts on a negatively-charged right-hand neighbor of the hole and pulls it to the left and into the hole, thus moving the hole to the *right*. Hence the "hole" moves under the influence of an electric field in the same manner as its associated "particle" does. It does *not* behave electrostatically like a negative mass. This may appear to be a trivial result, but it does not seem to have been noted before in the literature (until ref. [3]), and it is has a crucial effect on vacuum-state excitations. The Newtonian equations of motion for the particle and hole states are:

$$\vec{P} = +m\vec{v}, \qquad d\vec{P}/dt = +\vec{F} = +m\vec{a}, \qquad \text{particle state} \qquad (1)$$

$$\vec{P} = -m\vec{v}, \qquad d\vec{P}/dt = +\vec{F}_{mech} = -m\vec{a}, \qquad \text{hole state} \qquad (2)$$

$$\vec{P} = -m\vec{v}, \qquad -d\vec{P}/dt = +\vec{F}_{elec} = +m\vec{a}, \qquad \text{hole state} \qquad (3)$$

With respect to possible vacuum-state excitations, consider first a hypothetical *positive-mass, negative-mass* pair that carry opposite electric charges. Such a pair, once excited, cannot de-excite: the positive-mass member is drawn electrostatically toward the negative-mass member, which, however, runs away from the positive-mass member. Thus this cannot be a model for a vacuum-state excitation. But now consider instead a *positive-mass, hole* pair ($P^- - H^+$), a *polaron*. Once excited, the polaron soon de-excites, with its two members being drawn together electrostatically. Thus the polaron can serve as a suitable basis for transient vacuum-state excitations. Moreover, the polaron can be prevented from de-exciting by setting it into rotation (so that it becomes a "zeron"). If an appropriate frequency is assigned, the mechanical centrifugal forces and the Coriolis force balance out the electrostatic attraction, and a stable configuration is reached. This is essentially the model that we arrive at for both the photon (a linked rotating polaron-antipolaron pair) and its associated "empty wave," the "zeron" (a single rotating polaron or antipolaron), as we demonstrate below. (It should be noted that stable rotation is not possible for a positive-mass, negative-mass pair.)

Polarons are stable only under the action of external forces, which can arise from stationary or moving electric charges. As we demonstrate in Sec. 12, polarons linked end-to-end form field lines that require a "ground border" for their stability. These constitute the E field lines of classical electrostatics. Polarons linked end-to-end in complete loops do not need a ground border, but they require the Lorentz force to keep the loops from collapsing. These constitute the H field lines of classical magnetism.

One feature of the $P-H$ polarons which seems to be phenomenologically suggested is that the particle state P remains bound to its associated hole state H, in the sense that the two eventually reunite. The electrostatic field lines of force which exist between them are seemingly unbreakable. Another feature is that negative electric charges excite only the lattice L to produce polarons, whereas positive electric charges excite only the lattice $\bar{L}$ to produce antipolarons. This has important consequences for the production of magnetic field lines, and also for the differences in behavior of electron, positron, and photon wave packets in the presence of external electric fields.

We assume that in the creation of a $P-H$ polaron pair, the mass m which is excited out of its position in the lattice L is spinless. An alternative formulation would be to assume that it has a spin of $1/2$ $\hbar$, as in the case of the Dirac sea. Similarly, we assume that the charge carried by the mass m is the electronic charge e, and that the forces between the internal polaron charges are the ordinary coulomb forces. A different force law could be used without changing any of the essential results of the paper. What we are presenting here is a *scenario* rather than a detailed theory. The scenario can be correct even if some of the assumptions used in its implementation must later be changed.

4. Stable Rotating "Zeron" Particle-Hole Pairs: the Fundamental Wave Quanta

One of the main tasks in constructing a model for a photon or an electromagnetic wave is to reproduce its *frequency*. From the *PR* viewpoint, this model must exhibit a periodic motion, which is most logically a rotation. We denote *rotating* polaron and antipolaron particle-hole pairs as "zerons" and "antizerons," respectively. Figure 2 shows a rotating $P-H$ zeron as seen in both a stationary co-moving frame of reference (left) and a rotating co-moving frame (right). The left figure demonstrates the manner in which the particle and hole both follow the same circular orbit, even though the momentum vectors for the hole state are formally reversed. The right figure shows how the centrifugal force and Coriolis force act to counterbalance the inward electrostatic force [3].

Consider the orbit equations for the *particle* state P. These are the force equation

$$m\omega^2 r = e^2 / 4r^2 \tag{4}$$

and the angular momentum equation

$$mr^2\omega = \hbar / 2, \tag{5}$$

where this member of the rotating zeron has been assigned an angular momentum of $\hbar/2$ (half of the angular momentum of the photon). Eliminating the mass m, we obtain

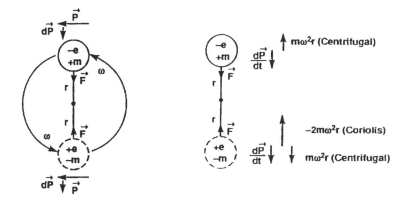

Figure 2. A rotating particle-hole "zeron" pair, as viewed in stationary (left) and rotating (right) co-moving frames. The particle states (top) follow the standard Newtonian equations of motion. The hole states (bottom) have reversed momentum vectors, but behave electrostatically in the same manner as the particle states.

$$\hbar\omega = e^2 / 2r. \qquad (6)$$

The energy quantum $\hbar\omega$ is the Planck energy for a photon that has frequency ω, and the term $e^2/2r$ is the electrostatic energy of the rotating pair. Similar orbit equations for the *hole* state H can be written down, with a Coriolis force added, and these show that the orbit of the hole H matches that of the particle P [3].

The angular momentum of a rotating $P-H$ pair is formally equal to zero, so it cannot serve as a representation of the photon. But it can serve as the quantum for the photon wave packet. This wave quantum necessarily carries essentially zero energy in addition to its zero angular momentum, which is why we denote it as a "zeron". Electromagnetically, the zeron closely resembles the photon. In particular, it carries the same rotational frequency as the photon. We have the freedom in this model to adjust the "negative hole energy" so as to cancel out the positive energy components of the zeron, including its electrostatic energy [3].

5. A Rotating Particle-Hole, Antiparticle-Antihole Quartet: the Photon

Photons, unlike polarons and zerons, have particle-antiparticle symmetry. In order to reproduce the photon, it is necessary to simultaneously create a particle-hole, antiparticle-antihole quartet, and to merge them together, as shown in Figure 3. The oppositely-charged hole states coalesce at the center, and the particle and antiparticle states rotate around one another (Fig. 3 left, the co-moving frame). Equations (4) - (6) also apply to this configuration, which reproduces the spectroscopic features of the photon [3], including its angular momentum $\hbar$, longitudinal circular polarization, and total energy $\hbar\omega$ in the co-moving frame (the negative energies of the hole states cancel out the positive mechanical energies of the particle states). Electromagnetically, the photon is a traveling and slowly rotating electric dipole (Fig. 3 right, the laboratory frame), a configuration which has been demonstrated by Bateman [4] to be consistent with Maxwell's equations. Equations (4) and (5) have three variables, and thus are underconstrained. These equations admit solutions for all values of ω, which is a result we require

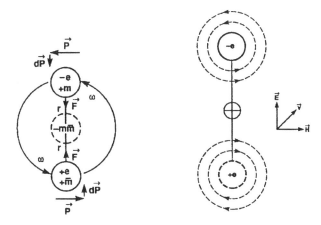

Figure 3. A model for the photon. The left diagram is in a non-rotating frame which is co-moving with the photon, and it shows the orbital momentum vectors. The right diagram is in the laboratory frame, with the electric dipole moving forward into the page, and it shows the transverse and orthogonal E and H electromagnetic fields, which are in accordance with Maxwell's equations [4].

phenomenologically, since photons occur with a continuous range of energies. The problems associated with relativistic transformations are briefly considered in Sec. 12.

The models for the photon and zeron that we have presented here are "generic" in the sense that they illustrate one kind of model which might be constructed. The use of spin 1/2 masses (Dirac states) instead of the spinless masses employed here would possibly lead to representations of photons and zerons as spin 1 and spin 0 combinations of these states. The essential point we are trying to establish is that it is possible to devise models for the photon and the zeron (the "empty wave quantum") in which they each appear as rotating electric dipoles, and yet at the same time reproduce the radically different mechanical properties of these states.

6. Representation of a Circularly-Polarized Electric Wave as a Train of Zerons

The photon is clearly an electromagnetic entity. The wave that surrounds, accompanies, and guides the photon must also be electromagnetic, since it *is* the electromagnetic wave. In the present model, we picture the electromagnetic wave as being composed of a synchronized cloud of rotating zerons, each of which carries the velocity and frequency of the wave. Each zeron has effective electric dipole charges, and these charges are responsible for the electromagnetic characteristics of the wave. If the photon is traveling in a straight line, the wave is a plane wave. When the wave packet impacts on a narrow slit, the part of the wave that passes through the slit is "jostled," so that the zerons all across the slit region receive lateral impulses from the slit edges. This causes the wave to fan out in a spherical or cylindrical manner, as observed experimentally in diffraction experiments. Since the speed and rotational frequency of the zerons are unaltered, the initial-state wavelength is maintained in the diverging final-state wave.

A single photon has circular polarization. A single rotating zeron also has circular polarization. The photon and zeron each represent a transversely-rotating, forward-traveling electric dipole. If we represent a circularly-polarized electromagnetic wave as

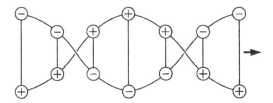

Figure 4. A succession of phase-shifted electric dipoles that can be used to represent a transverse circularly-polarized electromagnetic wave. This assembly of dipoles travels forward (to the right) in the direction of the wave propagation. The dipoles can be formed either as (a) non-rotating *polarons*, or (b) rotating *zerons*. Both of these configurations reproduce the observed Young and Fraunhofer interference amplitudes. In case (a), an observer at a fixed location in space sees a rotating electric vector. This is the usual assumption that is made about electromagnetic waves [5]. In case (b), the forward-moving and rotating zerons trace out the positions shown here, and thus appear at a fixed location in space with a fixed orientation of the electric field vector. Since polarons are unstable, whereas zerons are intrinsically stable, case (b) is the correct choice from the present viewpoint.

a collection of traveling electric dipoles, then, in order to reproduce the observed Young and Fraunhofer interference amplitudes, a linear train of these dipoles must be successively phase-shifted so as to reproduce the observed wavelength of the wave. However, given this configuration, which is displayed in Figure 4, we still have two choices. We can create wave motion with this ensemble of dipoles either by (a) moving the entire configuration forward (to the right), or else (b) moving the configuration to the right and at the same time letting the dipoles rotate. That is, we can use either (a) non-rotating polarons or (b) rotating zerons as the vacuum-polarization dipoles in the wave train. From the vantage point of a fixed observer, these two choices are quite different. In case (a), points of constant phase in the wave (the polarons) travel to the right, and the fixed observer sees a rotating electric field. This is the way that circularly-polarized electromagnetic fields are conventionally pictured [5], and the moving zero-phase points of the wave train are used as markers to track the phase relationships. In case (b), the traveling and rotating zerons trace out the successive path positions shown in Figure 4, so that the zeron wave train motion becomes time-independent, and the fixed observer sees a stationary electric field. Cases (a) and (b) correspond to the wave amplitude equations $e^{i(kx-\omega t)}$ and e^{ikx}, respectively, where these equations give the location of constant-phase points on the wave amplitude. Although case (a) constitutes the usual picture of a traveling wave [5], it is ruled out here because the stationary (non-rotating) polaron dipoles are unstable. From the standpoint of the present phenomenology, this is a crucial result. If all of the zerons that pass a point in space have the same electric dipole orientation, they will excite stationary particle-hole (or antiparticle-antihole) polaron pairs of the opposite polarity. Ordinarily, such *P-H* polaron pairs would quickly de-excite, but as long as the zeron train keeps passing by, they will be held apart. The spatial ensemble of these zeron-induced vacuum-polarization polaron pairs constitutes a vacuum state "*P* field", and it accurately mirrors (with reverse polarity) the phase relationships of the zeron ψ field. As the trailing edge of the zeron wave packet is reached, and the zeron intensity tapers off, the *P* field starts to de-excite, and it feeds whatever energy it contains back into the receding wave packet. Thus, even though it is a stationary polaron field, the *P* field constantly adjusts its boundaries to match those of the "overlying" zeron ψ wave packet.

7. A Self-Synchronizing Zeron Wave Packet: the "P field" Generator

As we demonstrated in Figure 4, a linear train of rotating and forward-moving phase-shifted zerons all assume the same orientation as they pass a fixed region of space. They thus *linearly polarize* that region of space. This linear polarization—the production of stationary particle-hole or antiparticle-antihole polaron or antipolaron pairs of opposite polarity in the spatial manifold—persists as long as the zeron cloud is passing through, and then de-excites. The interesting point about this process is that after the leading edge of the zeron wave train has polarized the vacuum state, thus creating a P field, this P field in turn acts back on the remaining portion of the zeron wave train and serves to accurately synchronize the individual zerons: it requires each succeeding zeron in the train to "follow the leader" and pass through a region of space with precisely the same electric orientation as the preceding zerons. An "electric filter" is thus set up at each spatial point, and it rigidly controls the phases of the ensuing zerons. As long as each zeron maintains its original *velocity* and *rotational frequency*, this filter ensures that the spatial orientation of the successive zeron phases in a wave train will be such as to accurately reproduce the *wavelength* that corresponds to this velocity and frequency. Even if the zerons are initially produced with random phases, this self-synchronization mechanism quickly forces them into the phase orientation shown in Fig. 4. Conceptually, it is useful to think of a linear train of self-synchronized zerons as forming a "zeron tunnel" in the vacuum state, and to denote the P-field polarization structure that corresponds to this zeron tunnel as a "P-field streamline." *Empirically, the streamlines echo the trajectory of the particle contained in the wave packet.* In the case of photons, the streamlines are unaffected by external electromagnetic fields, and hence are straight lines. In the case of electrons or positrons, the streamlines are deflected by external electromagnetic fields, and they correspond to similar trajectories followed by an electron or positron, respectively. This difference between wave packets is accounted-for by constructing electron and positron wave packets of zerons and anti-zerons, respectively, and constructing photon wave packets of equal numbers of zerons and antizerons. Electrons contain negative charges and thus excite the lattice L, whereas positrons contain positive charges and thus excite the lattice $\bar{L}$. Photons contain negative and positive charges in equal numbers, and thus excite both lattices equally.

In a linearly-moving photon wave packet, the streamlines are all parallel. As a result of the P-field interaction, the zerons are organized into linear wave trains which combine together to form a tightly-knit electromagnetic wave packet. As has been observed in astronomical telescopes, these accurately-correlated photon wave packets can extend laterally over distances of many meters.

In a photon interference field, such as that produced in single-slit or double-slit Fraunhofer or Young experiments, the final-state streamlines overlap and interpenetrate to form a web-like structure, while still maintaining their linear character.

When coherent co-moving photon wave packets overlap, they create a common P field in the region of overlap, and this serves to link the packets together. This linkage is *bosonic*, and it is manifested as the photon clumping observed in the Hanbury-Brown Twiss effect. Individual photon wave packets, like individual photons, are circularly polarized. A plane-polarized electromagnetic wave consists of combinations of right- and left-circularly-polarized photons and wave packets. The P-field linkage synchronizes combinations of overlapping right-handed and left-handed wave packets into plane-polarized fields, with the P fields in the overlap regions exhibiting a plane-polarized structure. Electron wave packets are *fermionic*, and electrons (except in time-reversed atomic or superconducting orbits, as discussed in Sec. 10), do not share a common P field.

As an alternate scenario, we can envisage this same kind of P-field vacuum-state polarization being produced with conventional circularly-polarized electromagnetic waves (Sec. 6). In this case, the electric vector at a fixed point in space is *rotating* [5], so the P field consists of a collection of rotating but non-moving P—H pairs; that is, of stationary *zerons* rather than stationary *polarons*. One difficulty with this picture is that zerons are more stable than polarons, which could lead to problems with energy conservation if the zerons do not die away immediately after the passage of the wave packet.

8. Photon Single-Slit Diffraction: Amplitudes and Probabilities

Although zerons produce a different electrostatic pattern at a fixed region of space than do conventionally-pictured electromagnetic waves (Sec. 6), they both lead to the same ψ-wave amplitudes in single-slit and double-slit diffraction-interference experiments. Computer calculations of zeron phase fields for cylindrical outgoing waves past the plane of the slits were carried out that accurately reproduce both single-slit diffraction amplitudes and double-slit wave interference oscillations at the image plane. The plan behind these calculations is as follows.

A linearly-moving collection of zerons is organized by the underlying P-field spatial polarization into a collinear ensemble of accurately-synchronized P-field streamlines, as described in Sec. 7. When this plane wave is incident on a narrow slit, a lateral disturbance is created in the portion of the wave packet that passes through the slit, and the zerons emerge on the far side of the slit plane with a distribution of lateral velocity components *at each local region in space* (Huygens' principle), thus destroying the original P-field pattern. The zerons in each segment of the slit opening then fan out in a cylindrical manner, forming a new interpenetrating web of P-field streamlines and a new overall P-field diffraction pattern. Each zeron keeps its original absolute velocity and original wave frequency, so the wavelength of each outgoing cylindrical wave component matches that of the incoming plane wave. After leaving the slit region, the zerons move through the vacuum state in linear trajectories, and they freely pass through one another without interacting (as mandated by the linear superposition principle for ψ-wave fields). The leading zerons in the expanding wave packet polarize the underlying vacuum state, which in turn acts back on the ensuing zerons to lock them into this new pattern, just as in the formation of a plane wave. Hence a new and different P field is established, in which the different components of the wave interact to produce the interference pattern. In the case of single-slit diffraction, for example, wave components from all across the slit combine together with slightly different phases to form the characteristic diffraction pattern at the image plane. The induced polarization of the "underlying" vacuum state by the "overlying" ψ wave results in an electric standing-wave P-field pattern—with opposite polarity—that is created as sort of a *negative template* for the ψ-wave amplitude distribution.

By following the zeron wave quanta in detail, we can accurately reproduce the standard ψ-wave *amplitudes* that occur in the various photon interference experiments. But the problem then emerges of reproducing the quantum mechanical *probability* distributions, which depend on the absolute *squares* of the amplitudes. We need to find a physical quantity that is relevant in steering photons through the zeron wave field, and which itself depends on the square of the zeron amplitude. Such a quantity is provided by the *underlying vacuum-state polarization*—the P field itself! When P-field streamlines overlap, as they do in an interference region, the zerons they channel are superimposed. The polarization intensity that the superimposed zerons create in the underlying vacuum state is proportional to the *square* of the zeron density. For example, *two* superposed in-phase zeron electric dipoles have *four* times the internal electric field of a single zeron dipole, and hence produce four times the corresponding polaron vacuum-state polarization. Thus the P-field electric polarization created by a zeron interference wave field is a static mapping of the *square* of the zeron ψ-wave amplitude at each point in space, with the orientation of the vacuum-state electric P-field vectors opposite in direction to the orientation of the zeron electric fields which produced them. Although, empirically, the individual overlapping zerons do not interact with one another (do not disturb each others' streamlines), they do act together in the production of P-field polarons.

We arrive at the following scenario for the way in which photons are steered in a single-slit diffraction experiment. An incident photon that is traveling in a straight line through a bundle of stationary P-field streamlines is locked into the overall phase pattern of the plane wave packet. In fact, more than one photon can lock into the same wave packet, since photons are bosons. The P-field fibers serve as streamlines for the photon motion. When the photon passes through a slit and enters the area where a new outgoing diffraction P-field has been created by the forward components of the wave packet, it loses its synchronization with the original linear P field and must establish a new phase relationship with the reconstituted P field. This loss of synchronization is partly because of the change in direction of the outgoing streamlines, and partly because of an interference-field zeron phase shift that occurs in the region just behind the slit due to the longer pathlengths of the diverging and overlapping streamlines. This phase shift typically amounts to about 45 degrees, and it causes the photon to resynchronize with a new streamline. No matter what lateral position it has as it passes through the slit, the photon faces an ensemble of new P-field streamlines that lead radially outward in all forward directions, in agreement with Huygens' principle. (The radius of an optical zeron, and hence of an optical P-field streamline, can be estimated from Eqs. (4 - 6) to be on the order of Angstroms (but see Sec. 12), so that a slit a fraction of a mm in width can accommodate a large number of streamlines.) Empirically, we know that the photon selects among these possible new paths in such a way as to satisfy the probability distribution of quantum mechanics. In terms of P fields, this means that the path selection is proportional to the attractive *electric field intensities* of the streamlines (which in turn are proportional to the absolute squares of the vector sums of the overlapping zeron amplitudes that correspond to the streamlines). The stationary electrostatic fields of a streamline guide the emerging photon into a trajectory along the streamline. These radial streamlines are of course the paths in which the photon can asymptotically lock back into phase with the expanding P field. The summation over photon trajectories along these streamlines reproduces the standard probability distribution. Double-slit interference-diffraction experiments also follow this same general scenario.

9. The Electron Double-Slit Experiment

Double-slit experiments are carried out with electrons as well as photons. A charged metallicized filament in an electron microscope breaks an electron wave packet into two parts, which are electrostatically deflected around the filament and then recombined at the far side to form a double-slit interference pattern. The fact that this kind of experiment closely duplicates the optical double-slit experiment (except for a large scaling in the distances involved) demonstrates that the formation of a P field by a moving electron wave packet must be closely analogous to the photon case discussed above. De Broglie showed that the frequency ω of a zeron created by a moving electron is given by the equation

$$\hbar\omega = mc^2 . \qquad (7)$$

The *frequency* of the zeron, which is uniquely related to the mass element m of the rotating P—H pair (Eqs.4 - 5), thus depends *only on the relativistic mass of the electron.* This "mass linkage" between the moving electron and the vacuum state evidently determines the frequency of the vacuum-state excitations *independently of anything else.* Electrons, protons, neutrons, muons, kaons, and helium atoms accurately obey Eq. (7), so this zeron frequency ω does not seem to be correlated with any "internal frequency" in the particle. If we identify the right-hand side of Eq. (7) with the *total energy* of the particle, then this equation turns into the Planck equation for the frequency of a photon, which again demonstrates the close analogy between electron and photon wave packets.

The superluminal forward *wave velocity* of the zeron, assuming it has an infinitesimal (but nonzero) total energy, is a straightforward consequence of special relativity: it is given by the equations of perturbative special relativity [7]. Hence the *de Broglie wavelength* λ of the zeron, which is the ratio of its velocity and frequency, is uniquely determined for each electron energy. There are three important conclusions we can infer from these results: *(1)* zeron wave packets are deflected in electrostatic fields in the same manner as are the electrons which produced them; *(2)* superluminally-moving (phase velocity) zerons create the electron P field, since it is the de Broglie wavelength λ which governs the interference effects that are produced; *(3)* the interference P field that is produced by a moving electron persists until the electron impacts at the detection plane, even though the portion of the zeron wave that passes on the far side of the metallicized filament in the electron microscope is cut off from the rest of the zeron wave and from the moving electron for a distance of typically 400,000 λ. The fact that electron-produced zerons follow the same orbits in an electrostatic field as do electrons is of crucial importance for atomic and molecular orbitals, as we discuss below. This fact was accounted-for by de Broglie when he showed that the wave equation $\lambda = h/p$ ties together Maupertuis' variational principle for particle propagation with Fermat's principle for wave propagation. The effect of external electric and magnetic fields on electron wave packets can be observed in experiments involving Rydberg orbitals [8].

In the present formalism, the fact that electron and positron wave packets are deflected by external electric fields whereas photon wave packets are not is accounted-for by postulating that positive charges excite only the lattice L, and negative charges excite only the lattice $\bar{L}$ (Sec. 7).

10. The Quantization of Atomic and Molecular Orbitals

We have sketched the manner in which the polarization of the vacuum state by a zeron ψ-wave packet creates a P field that stabilizes the wave packet. In the case of overlapping diffraction-interference regions, the P field also converts ψ-wave amplitudes into a probability distribution for the accompanying particle. One of the most spectacular consequences of these vacuum-state P fields may be their effect on atomic orbitals. As de Broglie demonstrated, the quantized Bohr orbits in an atom have path lengths which are integral multiples of the electron de Broglie wavelength λ. From the present viewpoint, this quantization is a consequence of the P field that is created by an orbiting electron: the P field must accurately close on itself. This can only occur when the path is an integral number of wavelengths λ. The quantization of atomic orbits is frequently spoken of in terms of the ψ-wave being single-valued (closing on itself), even though the ψ-wave is often regarded as nothing more than a mathematical construct. The physical reality behind this quantization is the necessity for creating a continuous circular or elliptical P field to lock in the orbiting electron. In fact, this P field, once formed, is so stable, and the electron is locked in so tightly, that it will not admit another electron into the same orbit. It will, however, admit a (time-reversed) counter-rotating electron with opposite spin orientation, which traces out and reinforces this same spatial P-field polarization pattern. The two electrons *share a common P field*, which electrostatically is more intense than the P field that corresponds to a single orbiting electron, since we now have overlapping in-phase zerons. The "pairing force" thus produced in these matching electron orbitals, or in matching proton or neutron orbitals in the nuclear shell model, is very strong. These paired-off particles do not participate easily in interactions: it is the unpaired particles that dominate atomic and nuclear reactions. Interleaved P fields that correspond to the same energy and angular momentum link together to form the "closed shells" which characterize atomic and nuclear structure. Low temperature superconductivity occurs when electrons bind together in (time-reversed) opposite-linear-momentum spin-up, spin-down (Cooper) pairs, even though they may be spatially well-separated from one another. This suggests that the Cooper electron pairs share a common P field, in analogy to the atomic electron pairs.

The orthogonality relation that exists among overlapping electron orbits in an atom is the condition that, on the average, the motion of an electron in one orbit is not perturbed by the underlying P fields of the other orbits.

Quantized atomic electrons interact with other quantized atomic electrons only via their P fields. This is why the Schrödinger equation yields accurate results even though it takes no account of direct electron-electron interactions.

An electron that is bound into a stationary P-field orbit does not radiate. We can think of the electron as being in a sort of "superconducting" state. When an electron moving linearly in space is decelerated or is accelerated laterally, a part of its accompanying wave packet is ripped away and appears as *bremsstrahlung* (braking radiation) photons. In a somewhat analogous situation, a Cooper pair of superconducting electrons can be accelerated across a Josephson junction by a potential V, where it acquires an energy of $2eV$. Since the pair cannot store this energy, and since it moves without resistance, it radiates a photon of energy $\hbar\omega = 2eV$. This is the ac Josephson effect.

When two atoms are combined together into a molecule [9], the mutually-overlapping P fields of the initial electron orbitals in the region between the atomic centers can be superposed either "in-phase" or "out-of-phase." For the "in-phase" case, electrons tend to concentrate in the overlapping region; whereas for the "out-of-phase" case, which contains a wave function node, the electron density in the central region between the atoms is low, as is the associated P field intensity. Molecular reactions are dominated by "in-phase" pairings. The strongest molecular reactions are those which require minimum P-field reorganization as individual atomic electron orbitals are converted into corresponding molecular orbitals. The conversion of the electron atomic orbitals into molecular orbitals seems to occur simultaneously rather than sequentially for the affected orbitals. This same kind of simultaneous conversion of P fields can also be observed in atomic nuclei. For example, a ^{40}Ca nucleus that splits into two ^{20}Ne nuclei yields gamma rays which correspond predominantly to low-lying levels in ^{20}Ne, thus indicating a rapid shell model organization of the final-state ^{20}Ne nuclei.

11. Electrostatic and Magnetic Fields as Linked Polaron Excitation

The vacuum-polarization P field, as we have defined it here, is a stationary spatial manifold of polaron and/or antipolaron particle-hole pairs that is created by a coherent zeron and/or antizeron wave packet. The P-field polaron excitations persist as long as the wave packet is present. This is not the only type of polaron excitation that can take place. Polaron excitations also occur as the electrostatic and magnetic E and H fields of classical electromagnetism, which thus emerge as close analogs of the wave-packet P field. External static or moving electric charges create the electromagnetic E and H fields, and the stability of the E-field and H-field polaron excitations is provided by these same external charges.

Let us first discuss the electrostatic E field. If an *external* negative electric charge e^- is introduced into a region of space, it will produce particle-hole polaron excitations of the lattice L. (A positive charge e^+ produces excitations of the lattice $\bar{L}$.) Consider an individual polaron, which we assume to contain effective internal charges e^- and e^+. The external charge e^- serves to separate the positively charged "hole" from the negatively charged "particle" in the polaron, and to orient the polaron along a line radiating out from the charge center. This is not a stable configuration for a single polaron. But if the polaron can excite a second polaron along the same radial line and then link to it electrostatically, this will stabilize the first polaron. A third polaron excitation stabilizes

Figure 5. A notation for polaron and antipolaron particle-hole pairs.

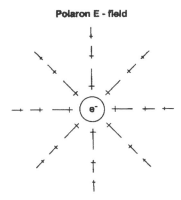

Figure 6. Polaron E-field excitation chains surrounding a stationary negative electric charge. (A stationary positive electric charge produces antipolaron excitations.) Each E-field polaron chain must for stability terminate at an oppositely-charged "grounding border."

the second polaron. This process can be extended as a chain of excitations all along the radial line, which must ultimately terminate in a "grounding border" in order to stabilize the last polaron. Thus, for stability, the chain of linked polarons must be "anchored" electrostatically at both ends. Figure 5 shows a schematic notation that can be used to denote polaron and antipolaron excitations. Figure 6 depicts a polaron electrostatic field surrounding an external electric charge e^-. The polaron excitations map out the electrostatic E-field lines. "Tension" in the field lines is produced by the tendency of the internal particle and hole states in the polarons to draw together electrostatically and de-excite.

Now consider the magnetic H field. Magnetic flux lines are created by moving electric charges, which constitute an electric current. In the case of a linear current, the magnetic flux lines form circles around the current. The electric current produces the Lorentz force, which acts so as to keep the circular flux lines from collapsing. These magnetic flux lines are logically formed from polaron excitations linked end-to-end, just like the polaron E-field lines described above, but in the form of complete circles. The stability of these flux lines is due to a combination of the polaron linkages around the circle, which keep the individual polarons from collapsing, together with the Lorentz force, which keeps the entire polaron loop from collapsing. It is obvious that these polaron magnetic field lines are stable only if they constitute topologically-complete loops. *Negative* electric currents produce *polaron* loops that circle the current in one direction, and *positive* electric currents produce *antipolaron* loops that circle the current in the other direction. Figure 7 shows a schematic representation of a polaron

Polaron H - field

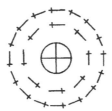

Figure 7. Polaron H-field excitation chains surrounding a moving (into the page) negative electric charge. (A moving positive charge produces antipolaron excitations with the opposite charge orientation.) Each H-field polaron chain must for stability form a complete loop.

E - field repulsion

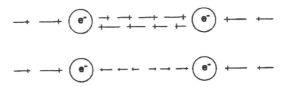

Figure 8. The top diagram shows a schematic superposition of two like-sign electrostatic polaron E fields, with only the horizontal polaron chains being displayed (see Fig. 6). The bottom diagram illustrates the manner in which the polaron excitations cancel out in the region between the two charges. This cancellation leads to an overall net *repulsion* between the charges, even though the individual polaron forces are all *attractive*.

H - field attraction

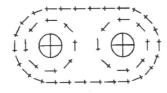

Figure 9. The polaron magnetic H-field lines that result from two co-moving (into the picture plane) like-sign electrical charges. In the region where oppositely-directed polaron lines from the two charges come together, they join to form an overall loop that leads to an attractive force between the moving charges.

magnetic field line. Mathematically, the Lorentz force that sustains the current loop is calculated by making the momentum replacement

$$m\vec{v} \;\to\; m\vec{v} + e/c\,\vec{A},$$

for the moving charged particles, where A is the vector potential.

When several external electric charges are present, the polaron E-field lines add up linearly. Figure 8 illustrates this superposition schematically for the case of two negative electric charges. The superposition of polaron H-field lines for multiple currents is nonlinear. Figure 9 shows the addition of polaron magnetic flux lines for two adjacent like-sign currents.

12. Questions about Special Relativity and Lorentz Transformations

The equations of special relativity were originally devised with respect to electromagnetic fields, which move at the velocity c. However, they apply equally well to electrons and other systems which move at much slower velocities. A difficulty with these equations is that we can use them to transform between various subluminal reference frames, but the frame moving at $v = c$ is a special case, and the Lorentz transformations become singular at this velocity. Hence we have no mathematical way of transforming out of the frame co-moving with the photon and into the laboratory frame.

The electron model that is shown in Figure 1 is relativistically correct [2]. Hence we might expect that the corresponding photon and zeron models shown in Figures 2 and 3, which are based on the same (classical) equations for angular momentum and electromagnetic effects, are also relativistically correct. Without the complete transformation equations, we cannot draw this conclusion. But we can make a few inferences. We know that masses scale relativistically as $m = \gamma \, m_o$ and frequencies scale as $\omega = \omega_o / \gamma$, where m_o and ω_o are the values in some reference frame, and $\gamma = 1 / \sqrt{(1 - v^2/c^2)}$ is the transformation factor for a velocity v relative to the reference frame. In Eqs. (4) and (5), the radius r is relativistically invariant, because the motion of the system is at right angles to r. Thus the centrifugal force term $m\omega^2 r$ in Eq. (4) scales as $1 / \gamma$. The right hand side of Eq. (4), which is valid in the co-moving frame (Figure 3, left), must be modified in a moving frame (Figure 3, right) to allow for both the relativistic flattening of the electrostatic field and also the addition of the magnetic field that arises from the forward-moving charges. When these factors are inserted, it turns out that the right-hand side of Eq. (4) also varies as $1 / \gamma$ [10]. Thus Eq. (4) exhibits the proper relativistic behavior. In Eq. (5), both sides are independent of γ, so it is also relativistically correct. But since γ goes to infinity at $v = c$, we cannot transform in a continuous manner from the frame co-moving with the photon (Figure 3, left) to the stationary laboratory frame (Figure 3, right). What we *can* say is that the photon has an experimentally-determinable frequency ω in the laboratory frame, and that it must also have a finite (but unknown) frequency ω_o in the co-moving frame, so there is an effective (but unknown) transformation parameter γ_{EFF} that in some manner represents the proper way of going to the limit $v = c$ in the Lorentz transformation equations.

The use of a mechanical mass to reproduce the angular momentum of the photon, as shown in Figure 3, raises a problem for a system moving at the velocity c, since the mass formally becomes infinite at that energy. However, the introduction of a matching "hole" state (with a "virtual negative mass") may act to cancel this energy divergence, so that *particle-hole pairs* become admissible at luminal velocities. In fact, the existence of a luminal velocity for a particle may perhaps be taken as an indication that hole states are present in the structure of the particle. The angular momentum of the photon causes macroscopically-observable effects, so it is a real physical quantity.

Electromagnetically, the important feature of the electric dipole model of the photon is that it is in accordance with Maxwell's equations, as Bateman demonstrated many years ago [4]. Relativistically, the particle-hole vacuum-state polaron pairs that appear in photons and zerons, if treated as single entities, may have transformation properties which are tractable at all velocities—subluminal, luminal, and superluminal.

13. Summary and Final Remarks

The models we have presented here for the electron, photon, and zeron are mathematical *particle* models which accurately reproduce the known spectroscopic properties of these states. The discussion of *fields*, however, is largely qualitative. What we have tried to do is develop a plausible scenario for the manner in which these fields function. In particular, we have introduced a new field, a stationary vacuum-polarization P field, which is generated by a moving electromagnetic or ψ wave packet (these two types of wave packet are closely related). The P field is, in the opinion of the author, the key ingredient that has been missing from quantum mechanical formulations. The plausible existence of the P field was deduced by introducing a specific electromagnetic representation for the ψ wave. The main points with respect to the generation of the P field by ψ waves are the use of zerons as wave quanta, the recognition that a train of forward-moving and rotating zerons has a fixed electric phase orientation at a stationary point in space, the consequent polarization of the underlying vacuum state (with opposite polarity) by the leading edge of the zeron wave train, and the manner in which this polarized vacuum state reacts back on the train of individual zeron quanta to stabilize them and rigidly phase-lock them into the wave train, thereby creating "zeron tunnels" and "P-field streamlines" in the wave packet and underlying vacuum state. In diffraction-interference experiments, where ψ-wave fields are superimposed, the intensity of the P-field polarization at each point in space is proportional to the *square* of the magnitude of the ψ wave at that point in space. (The ψ wave is the vector sum of the zeron quanta which are simultaneously present at that point in space.) In a single-slit or double-slit interference-diffraction experiment, a photon, after passing anywhere through a slit, is faced with an ensemble of P-field streamlines that extend forward in all radial directions. It also experiences a relative phase shift that desynchronizes it from the wave packet and forces it to choose a new final-state wave-packet streamline. The photon selects among these final-state channels on the basis of their relative (attractive) electric polarization intensities, and thus reproduces the standard probability distribution of quantum mechanics.

In stable atomic orbits, the P-field polarization pattern must close on itself, thus quantizing these orbits in units of the de Broglie wavelength λ that characterizes the generating Schrödinger ψ waves, and hence also the P field. Two time-reversed (counter-rotating and spin-reversed) electrons can share a common P field, thereby enhancing its strength. This "pairing force" dominates shell structure in both atoms and atomic nuclei, with this same mechanism also playing a role in the time-reversed Cooper pairs of superconductivity.

The P field is composed of stationary $P-H$ and $\overline{P}-\overline{H}$ vacuum-polarization pairs, denoted as polarons and antipolarons, which correspond to excitations of the particle and antiparticle lattices L and $\overline{L}$, respectively. Zerons are rotating polarons. The classical electric E fields and magnetic H fields are reproduced as polaron excitations. Particle wave packets are reproduced as zeron excitations. Negative electric charges excite only polarons and zerons, whereas positive electric charges excite only antipolarons and antizerons. Photons, which contains internal charges of both signs, excite equal numbers of zerons and antizerons. There are two direct consequences of this excitation mechanism: *(1)* photon wave packets are not deflected by electrostatic fields, whereas electron and positron wave packets are; *(2)* negatively charged and positively charged electric currents induce magnetic field lines that are aimed in opposite directions.

The P field that underlies a wave packet is an *opposite-polarity* vacuum-polarization manifold which appears as an "enhanced negative template" of the ψ wave that created it. In this context, it is interesting to consider the muon deep-inelastic experiments that have been carried out on polarized protons [11]. These experiments probe the vacuum state around the proton, and they have led to the conclusion that the quark-antiquark pairs which surround the proton are polarized in the direction *opposite* to the proton's net spin.

The neutrino, which has no internal electric charges, cannot produce ψ waves or P fields, which could account for its very small interaction cross section.

References

1. Selleri, F.: *Quantum Paradoxes and Physical Reality*, Kluwer Academic Publishers, Dordrecht (1990).
2. Mac Gregor, M. H..: *The Enigmatic Electron*, Kluwer Academic Publishers, Dordrecht (1992).
3. Mac Gregor, M. H.: *Bull. A.P.S.* **39** (1994), 1153; "Model basis states for photons and 'empty waves'," *Foundations of Physics Letters* **8** (1995), 135-160.
4. Bateman, H.: *Phil. Mag.* **46** (1923), 977; also see Jackson, J. D.: *Classical Electrodynamics*, Second Edition, Wiley, New York (1975), pp. 552 - 556, for somewhat similar ideas.
5. See, for example, Jackson, J. D.: *ibid.*, pp. 273 - 275.
6. Strekalov, D.V., Sergienko, A.V., Klyshko, D.N., and Shih, Y.H.: *Phys. Rev. Lett.* **74** (1995) 3600.
7. Mac Gregor, M. H.: *Lett. Nuovo Cimento* **44** (1985), 697.
8. For a review of Rydberg atoms, see Alber, G. and Zoller, P.: *Physics Reports* **199**, (1991) 231 - 280.
9. See Woodward, R. B. and Hoffmann, R.: *The Conservation of Orbital Symmetry*, Verlag Chemie, Academic Press, Weinheim/Bergstr. (1970).
10. French, A. P.: *Special Relativity*, Chapman and Hall, New York (1968), pp. 234 - 250.
11. Jaffe, R. L.: *Scientific American* **48** (9) (1995), 24.

This work was performed under the auspices of the U. S. Department of Energy by the Lawrence Livermore National Laboratory under contract No. W-7405-ENG-48.

ELECTRONS AND PHOTONS AS SOLITON-WAVES

The DE BROGLIE and SCHRÖDINGER WAVES of an ELECTRON,
The ELECTROMAGNETIC SOLITON WAVE of a PHOTON

GEOFFREY HUNTER
Department of Chemistry, York University,
Toronto, Ontario, Canada M3J 1P3
Tel: 416-736-5306 Fax: 416-736-5936
Email: ghunter@yorku.ca

1. The Wave and Particle Paradigms

Whether the physical world is composed of indivisible **atoms**, or alternatively is an infinitely divisible continuum, is a question that has concerned natural philosophers for hundreds of years; specifically Newton speculated about whether light is composed of a stream of particles or is a wave motion within some underlying medium (the "ether").

The question became a paradox in the early part of this century (1905-30) through the discoveries that:

- light exhibits both wave properties (diffraction and interference) and particle properties (Compton and photo-electric effects),
- electrons exhibit the wave property of diffraction as well as behaving as (charged) particles in electric and magnetic fields and in producing localized fluorescence (emission of light) on striking certain solid ("phosphor") coatings on the inside of cathode ray tubes.

Orthodox ("Copenhagen") quantum theory accepts the paradox without attempting to explain it. That both light and electrons behave as particles in some experiments and as waves in others is accepted as an inexplicable phenomenon of quantum mechanics; the particular experimental arrangement (the process of "measurement") determines whether the light or electron exhibits wave or particle-like behaviour. This orthodox philosophy asserts that it is not possible to know the nature of light or an electron prior to the act of measurement.

The apparently indivisible particles (electrons, protons, and some unstable particles) have become regarded by many scientists as the *elementary constituents of the physical world.* The currently popular paradigm of theoretical physics regards the particle as an elementary ingredient of the mathematical theory; particle-particle interactions are produced ("mediated") by "exchange" of other (in some cases hypothetical) particles (including the particle of light - the photon).

This currently popular paradigm of physical theory has been inherited from the laws of Newtonian mechanics derived from the motions of planets and such objects as cannonballs. These motions are quantitatively described by regarding each moving body (planet/cannonball) as a specific mass located "at a point" - the centre of gravity of the body, the finite size of the body being neglected in calculating the motion.

These concepts and equations of classical mechanics have been transferred to the microscopic world of atoms, electrons and protons via the concept of **particle**, albeit within the formalism and mathematics of quantum, as distinct from classical mechanics.

The general thesis of this article is that both light and electrons are localized ("soliton")[1] waves; their wave properties are derived from their intrinsic wave nature, and their particle properties are derived from the localized nature of the wave. Thus the paradox of the wave-particle duality is, in principle resolved because the electron or photon is a localized wave whose centre behaves like a particle.

2. The Non-Elementary Nature of Particles

The concept of a particle is of a point-like object, the size being either infinitesimal (truly "at a point") or alternatively finite but small. For the particle concept to be elementary (i.e. without internal structure that would have to be described in more elementary terms), the particle would have to exist at a point, or it would have to be a rigid body - probably a rigid sphere.[2]

If the elementary particle existed at a point, and if it is something physical (as distinct from mathematical), then the elementary physical object would have to have an infinite energy density, which is physically untenable; **there is no physics at a mathematical point**.

The other possibility is that the particle is a *rigid* body. This would require the transfer of momentum across the particle's finite size during a collision to require either infinite forces within it (which is physically

---

[1] I use the term "soliton" in the same sense as Coleman [1], which (as he explains) is not the same as the meaning in the established mathematical theory of solitons based upon solutions of non-linear differential equations [2].

[2] The concept of something being **elementary** is that it is conceptually self-sufficient; thus an *elementary* finite-sized particle would have to be a **rigid** body, otherwise we would need some more elementary property to describe its internal elasticity.

untenable), or alternatively an infinite speed of transmission of the impact across its finite size, which is inconsistent with the principle that the velocity of light is the upper limit for the transmission of all material interactions. These arguments originate from Lucas and Hodgson [3].

Hence the concept of **Particle** is not *Elementary*: the nature and structure of particles (such as electrons, protons, photons etc) must be described in terms other than "particle". This article posits that particles are waves – localized (soliton) waves. On this basis the paradox of the wave-particle duality paradox will be resolved in favour of the wave – the other principle paradigm of theoretical physics.

3. The Elementary Nature of the Wave Concept

A wave always has a finite extent - the energy of the wave is distributed throughout its volume; the energy density is finite everywhere. When a wave collides with another physical object it deforms: the impact spreads across the extent of the wave at a finite speed (often equal to, and always no larger than the velocity of light). The forces involved are finite.

When a wave collides with a "solid" object its direction of propagation will be bent (i.e. diffraction). Since the angle of bending (around an edge or through a slit) will depend upon the impact parameter of the wave with the edge, this bending can also lead to "interference"; i.e. preferential bending (higher intensity) in certain directions. The classical concept of interference (á la Huygen's principle) is probably artifactual as in the classical, Maxwellian theory developed by Richard Prosser [4].

4. Mass as Localized Wave Motion

Einstein (1905) quantified the idea that mass is simply a form of energy in his famous equation:

$$E = mc^2 \qquad (1)$$

Some scientists regard particles (especially the electron and proton) as elementary constituents of nature; they also regard particles as the sources of fields, a gravitational field arising from the particle's mass and an electromagnetic field arising from its charge and/or magnetic moment.

The idea that a particle is a localized wave was introduced by de Broglie [5]. By combining Einstein's equation (1) with Planck's equation:

$$E = h\nu \qquad (2)$$

relating the energy, E of a quantum "oscillator" to its frequency ν, he postulated that an elementary particle such as an electron is some kind of

localized (soliton) wave whose frequency of oscillation is related to its mass by:
$$h\nu = mc^2 \qquad (3)$$
i.e. what we regard as the (inertial) mass of the particle is, according to de Broglie's proposal, simply the vibrational energy (divided by c^2) of a localized, oscillating field (most likely the electromagnetic field). From this standpoint inertial mass is not an elementary property of a particle, but rather a property derived from the localized oscillation of the (electromagnetic) field. De Broglie described this equivalence between mass and the energy of oscillational motion (quantitatively expressed by (3)) as "*une grande loi de la Nature*" (a great law of nature) [5].

For an electron, the computed frequency of the wave is $\nu = 1.2356 \times 10^{20}\,\text{sec}^{-1}$ (Hertz), corresponding to a wavelength of $\lambda = c/\nu = 2.4263 \times 10^{-12}$ metre (the electron's Compton radius/2π).

5. The Particle's Internal, De Broglie Wave

De Broglie derived the fundamental relationship between the momentum and wavelength of a moving particle by considering a Lorentz transformation of the particle's internal frequency (given by (3)) to the reference frame of an observer (or apparatus) within which the particle is moving at constant velocity.

An alternative procedure is to postulate that the time-dependence of the particle's internal wave motion in its rest-frame has the form:
$$\psi = exp(i2\pi\nu t) = exp(i\omega t) \qquad (\omega = 2\pi\nu) \qquad (4)$$

The Lorentz transformation from the particle's rest frame to the frame of a apparatus or observer moving at velocity v (along the x axis) relative to the particle is:
$$t = \gamma(t' + x'v/c^2) \qquad x = \gamma(x' + t'v) \qquad (5)$$
where $\gamma = 1/\sqrt{1-(v/c)^2}$ and the primed variables, (x', t') are the coordinates in the observer's frame of reference.

Hence the wave (4) becomes in the observer's frame (x', t'):
$$\psi = exp(i2\pi\nu\gamma x'v/c^2) \times exp(i2\pi\nu\gamma t') = exp(i2\pi/\lambda') \times exp(i2\pi\nu') \qquad (6)$$

The time-dependence of this wave is (for small velocities – $\gamma \approx 1$, $\nu' \approx \nu$) virtually the same as the orginal internal particle-wave, but its space dependence has a wavelength λ' that is simply the well-known de Broglie wavelength of the particle as manifest in electron diffraction experiments:
$$\lambda' = c^2/(\nu\gamma v) = h/(\gamma m v) \qquad (7)$$

– the de Broglie wavelength of a particle of mass m moving with relativistic momentum $\gamma m v$.[3]

The space-factor of the transformed wave (6) is the Schroödinger wavefunction of the particle of mass m moving (relative to the observer or apparatus) with velocity v.[4] This space-factor satisfies the time-independent Schrödinger equation:

$$\frac{\partial^2}{\partial x'^2} exp(ikx') = -k^2 exp(ikx') \qquad k = \gamma mv/\hbar = p'_x/\hbar \qquad (8)$$

and since the non-relativistic kinetic energy, T of the particle is $T = p'^2_x/2m$,[5] this Schrödinger equation may be written in the familar general form:

$$-\frac{\hbar^2}{2m}\frac{\partial^2 \psi}{\partial x'^2} = T\psi = E - V\psi \qquad (9)$$

where E is the (constant) total non-relativistic energy and V is the (usually variable) potential energy; ψ is only equal to $exp(ikx')$ when the potential energy V is constant. A very similar "derivation" of the Schrödinger equation has been given by Andrews [6].

Schrödinger's wave is a kind of "beat frequency" (much lower frequency and longer wavelength than de Broglie's internal electron wave) produced by the relative motion of the particle and the observer.[6] This idea has the potential to resolve the long-standing problem of the wave-particle duality paradox; how can the electron sometimes behave like a particle and at other times as a (Schrödinger) wave ? The answer is that the particle is always a (de Broglie) wave localized around its centre of energy (mass) that we tend to regard as a point particle. Its Schrödinger wave is a manifestation

---

[3] Some scientists interpret γm as the dynamic mass being larger than the rest mass, but this interpretation is not strictly in accord with the Lorentz transformation of the energy-momentum vector.

[4] This overall result (shown to me by Roger Jennison in Canterbury, in September 1994) is remarkable; from De Broglie's postulate that the electron is (internally) some kind of wave motion, one can derive (using simply a Lorentz transformation) the Schrödinger wavefunction for the particle as seen by an "observer" such as a diffraction grating.

[5] The relativistic momentum of the moving particle is precisely γmv from the Lorentz Transformation of the energy-momentum vector [3]. The non-relativistic assumptions in the Schrödinger equation (9) are the equation of the exact relativistic momentum with $2mT$ and further the equation of T with $E-V$ especially when V is not constant, in which case the Lorentz transformation (5) should in principle be an infinitesimal transformation that is a function of the varying velocity v. This procedure for deriving a relativistic wave equation has (as far as I know) never been carried out.

[6] The frequency and velocity of the Schrödinger wave is *not* necessarily obtained from the time-dependence of (6). De Broglie [5] considered the two waves having the "group velocity" v and the "phase velocity" c^2/v; the latter is a speed faster than the velocity of light c.

(an artifact ?) of its internal wave motion as experienced by a relatively moving observer (or apparatus).

6. The Shape of the Particle's Internal Wave

A reasonable conjecture is that the particle's internal wave, Ψ satisfies the classical, D'Alembert wave equation (for electromagnetic waves - that propagate at the speed of light c):

$$\left\{\frac{\partial^2}{\partial x^2} + \frac{\partial^2}{\partial y^2} + \frac{\partial^2}{\partial z^2} - \frac{1}{c^2}\frac{\partial^2}{\partial t^2}\right\}\Psi = 0 \qquad (10)$$

Assuming that the wave Ψ can be written as a product of the time factor (4) and a space-dependent factor $f(x, y, z)$:

$$\Psi(x, y, z, t) = exp(i\omega t) \times f(x, y, z) \qquad (11)$$

substitution of this product form leads to the result that the space factor $f(x, y, z)$ is a solution of the Helmholtz equation [7]:

$$\left(\frac{\partial^2}{\partial x^2} + \frac{\partial^2}{\partial y^2} + \frac{\partial^2}{\partial z^2}\right)f(x, y, z) + \left(\frac{mc}{\hbar}\right)^2 f(x, y, z) = 0 \qquad (12)$$

This initial result about the space-dependence of the particle's internal wave is not pursued here except to note that the internal wave may be a tensor wave (such as the 6-component electromagnetic field) rather than a scalar field as has been implicitly assumed in the above algebra.

7. Photons as Soliton Waves

The Photon has been modeled as a soliton wave [8, 9, 10]. Our model [8] is a non-plane wave solution of Maxwell's equations that has the correct spin-angular momentum of the photon ($\pm\hbar$), the two values corresponding to right and left circularly polarized photons. Although the mathematical form of the electric and magnetic fields is obtained by solution of the linear Maxwell equations, they are constrained by the relativistic principle of causality to lie within a circular ellipsoid whose long axis is the axis of propagation. The length of the ellipsoid is one wavelength and its cross-sectional circumference is also one wavelength; hence it is a prolate ellipsoid whose diameter is approximately one-third (accurately $1/\pi$) of its length.

This causally constrained soliton model of the Photon predicts some of the experimental properties of monochromatic light:

- the optimal resolving power of a monochromatic microscope is stated in textbooks of elementary physics to be "a little less than a third of a wavelength"; $1/\pi$ is "a little less than a third",

- the minimum slit-width (or orifice diameter) required for transmission of circularly polarized light is λ/π; this effective photon diameter was acurately confirmed by our own experiments with microwaves [8] within the experimental error of a half a percent,
- the threshold intensity for the production of multiphoton-absorption in focussed laser beams is predicted by this photon diameter of λ/π; the threshold beam intensity corresponds to adjacent photons in the beam touching each other. We have called this intensity the photon's *intrinsic* intensity, because it is the average light intensity within the photon's cross-sectional circle of diameter λ/π perpendicular to its direction of propagation. Its value is given by:

$$I_p = 4\pi hc^2/\lambda^4 \qquad (13)$$

For example at 523 nm (middle of the visible spectrum) this intrinsic intensity is 1 megawatt per square centimetre. At beam intensities higher than the intrinsic intensity photons necessarily overlap; i.e. there are two or more photons in the same place at the same time, which is the requisite condition for multiphoton absorption to occur [8].

This prediction of the threshold beam intensity for multiphoton absorption to occur correctly predicts the requisite experimental intensities (typically produced in focussed laser beams). The predicted threshold intensity has also been confirmed by multi-photon absorption experiments with radiation from a CW CO_2 laser ($\lambda = 10.5$ micrometres, $I_p = 6$ watts per square centimetre); thus the expression (13) scales with wavelength (as λ^4) correctly.

8. Resolution of the Wave-Particle Duality Paradox

Recognition that the concept of particle is not elementary, and that the elementary concept of wave can be used to construct "elementary" particles (photons, electrons, etc) that have both particle-like (locality and momentum) and wave-like (diffraction and interference) properties, may be the way to resolve the Wave-Particle Duality Paradox [11].

The concept of particle should be abandoned as an elementary ingredient of physical theory (as distinct from a pragmatically useful, approximate concept). Our further understanding of the nature of the "particles" of physics is likely to involve soliton waves.

This conclusion suggests that the currently-in-vogue paradigm in theoretical physics (in which interactions between particles are represented by exchange of other particles) should be replaced by a paradigm based upon waves in order to describe the physical world in terms of truly elementary concepts.

References

1. S.Coleman, *Aspects of Symmetry*, Cambridge University Press, 1985, p.186.
2. Longren and Scott, *Solitons in Action*, Academic Press, New York, 1978.
3. J.R.Lucas and P.E.Hodgson, *Spacetime and Electromagnetism*, Clarendon Press, Oxford, 1990.
4. R.D. Prosser, *The Interpretation of Diffraction and Interference in terms of Energy Flow*, International Journal of Theoretical Physics, **Vol.15**, pp.169-180 (1976).
5. Thèses présenées par Louis de Broglie, premier thèse "Researches sur La Théorie des Quanta", Masson et Companie, Editeurs, Libraires de l'Academie de Médecine, PARIS, 1924. Reprinted in *Annales de la Fondation Louis De Broglie*, Vol.17, p.1 (1992).

    A reconsideration and extension (in english) has been given by Mioara Mugur-Schächter (*Quantum Mechanics and Relativity: Attempt as a New Start*, Foundations of Physics Letters. Vol.2, No.3, pp.261-286 (1989).

    A translation of part of de Broglie's thesis entitled *The Relationship Between the Quantum and Relativity* has been published in the *American Journal of Physics*, Vol.40, pp.1315-1319 Sept.1972.
6. Thomas B. Andrews, article *Derivation of the Schrödinger Equation/*, in these Proceedings.
7. P.M.Morse and H.Feshbach, *Methods of Theoretical Physics*, Part I, p.271, McGraw-Hill, New York, 1953.
8. Geoffrey Hunter and Robert L.P. Wadlinger *Photons and Neutrinos as Electromagnetic Solitons*, Physics Essays, **Vol.2** pp.158-172 (1989).
9. R. G. Beil, article *A Classical Photon Model* in these Proceedings.
10. J.P. Vigier, reprint seen during the symposium, circa 1992.
11. S.Diner, D.Fargue, G.Lochak and F.Selleri (Editors), *The Wave-Particle Dualism*, D.Reidel, Dordrecht, (1984).

TOPOLOGICAL SOLITONS IN NONLINEAR ELECTRODYNAMICS

German Kälbermann
Racah Institute of Physics
Hebrew University of Jerusalem
Jerusalem 91904, Israel

Abstract

Massive topological solitons of integer spin arise in Maxwell's electrodynamics with the nonlinear gauge condition of Dirac and Nambu. The absolutely conserved topological charge of the solitons can be identified to the electric charge. The states exist only when the electromagnetic field is endowed with quantized rotational collective degrees of freedom.

1. Introduction

Maxwell's electrodynamics is based on the separation of variables into the electromagnetic field and matter. Matter is described classically by means of the dynamics of point charges. Quantum mechanically matter is described by means of charged matter fields. Maxwell's electrodynamics is the basis for quantum electrodynamics, the most successful field theory that we have today. The unification of electromagnetism and weak interaction in the Glashow-Salam-Weinberg theory is extremely successful too. The distinction between matter fields and gauge fields remains in this theory too. Moreover, both matter and gauge fields are treated as structureless, a fact supported by the the experiments at the present time.
Since the beginning of the century the have been numerous attempts to describe the electron as a lump of localized energy in the electromagnetic field [1-4]. More recently there has been a revival of the old ideas supplemented by topological considerations that could explain the absolute conservation of electric charge [5, 6]. A simple way to unify the picture of gauge and matter fields in electrodynamics, was provided long ago by Dirac [7] and Nambu [8]. The key ingredient in their approach was the imposition of a nonlinear constraint on the electromagnetic potentials. In this gauge the electromagnetic matter current originates from the gauge field itself, without explicit matter degrees of freedom. Many classical phenomena of charges interacting with the electromagnetic field could then be reproduced from the fields themselves. Although the use of this gauge could potentially violate Lorentz invariance, Nambu has already stated that: 'the fact that the vector potential is unobservable makes the breakdown of Lorentz invariance only superficial...' [8]. In perturbation theory with the photon field regarded as a fluctuation one can retrieve a Lorentz invariant and gauge invariant formulation provided the scale of the nonlinear constraint is much bigger than the relevant energy of the photons. We will see below that this scale will be approximately 1 TeV. The

nonlinear constraint will start to affect the predictions of electrodynamics only for photons above this energy scale. Maxwell's electrodynamics in the nonlinear gauge is very similar to a nonlinear σ model. In such models it is possible to define a topologically conserved current as a homotopy application between fields and spacetime. An absolutely conserved topological current was the key ingredient in the pioneering work of Skyrme [9]. He suggested that topological solitons in a meson field theory may be identified as baryons, the topological charge playing the role of baryon number. So starting from a meson field theory it is possible to describe baryons too. A similar attitude can be taken in electrodynamics in the nonlinear gauge. The actual value of the electromagnetic potentials will determine the topological charge. In fact, this is the only place where the potentials will enter. All the other observables will depend, as usual, on the derivatives of the fields.

Righi and Venturi took advantage of this idea and tried to describe the electron as a topological soliton solution of Maxwell's electrodynamics in the nonlinear gauge [5]. They were able to find spherically symmetric solutions whose radius was the classical electron radius of the order of a few fermi, a fact unacceptable by present day experiments. They generated the spin of the object expanding the electromagnetic field into a static part and a quantum fluctuation. However, this procedure breaks the nonlinear gauge condition and consequently destroys the topological nature of the states. More recently Rodrigues et. al. [6] attempted to circumvent the spin problem by introducing a pseudovector classical field to Maxwell's action. For this case too, the radius of the object turns out to be much to big. Despite the only partial success of the nonlinear gauge approach in describing charged particles as solitons, there appears to be a clear indication that this theory is quite rich and deserves further attention. The nonlinear gauge approach can be implemented in other classical extensions of electromagnetism also, such as the electromagnetic theory of Born and Infeld [3], or the higher order theory of Podolsky [4]. These theories are nonlinear even without the gauge condition, and were designed to obtain classical solitonic solutions of electromagnetism. The theory of Born and Infeld has classical solutions with the classical electron radius and Podolsky's electrodynamics has solutions whose size depend on a new mass parameter. (For a discussion of the quantization of Podolsky's electrodynamics see ref. [10].)

In the present work we will find topological soliton solutions of Maxwell's electrodynamics in the nonlinear gauge. Massive and charged vector bosons of integer spin will emerge when we endow the electromagnetic field with quantized rotational collective coordinates. If we identify the ground state of massive spin one bosons with the electroweak charged gauge bosons W^+ and W^-, then the model predicts the existence of a spin zero partner degenerate with it. The radius of these objects turns to be of the order of 10^{-5} fermi, essentially pointlike for present day accelerator capabilities. The states that we will find differ from the classical solutions by that they are inherently quantum mechanical and owe their existence to the introduction of rotational degrees of freedom.

2. Topological solitons in nonlinear electrodynamics

We start from the standard Maxwell lagrangian for the electromagnetic field without sources,

TOPOLOGICAL SOLITONS IN NON-LINEAR ELECTRODYNAMICS

$$L = \frac{-1}{16\pi} \int d^3x \, F_{\mu\nu} F^{\mu\nu} \tag{1}$$

where

$$F_\mu^\nu = \partial_\mu A^\nu - \partial^\nu A_\mu \tag{2}$$

in the nonlinear gauge

$$A^\mu A_\mu = A_0^2 - \vec{A}^2 = -\alpha^2 \tag{3}$$

with α a real constant.

In Eq. (3) we have chosen the negative sign in order to obtain a positive static energy and a positive rotational energy (see below). A positive sign was preferred in previous works although there is no *a-priori* reason to do so. Nambu considered both sign variants and showed the viability of the nonlinear gauge in perturbative QED.

Due to the nonlinear constraint of Eq. (3) there are only three independent degrees of freedom. We can choose them to be the three space components of the electromagnetic potential. The equations of motion of the fields turn out to have automatically source terms that are absent in the linear case [5, 7]. These source terms depend solely on the fields. Without these terms there will not be nontrivial solutions to the equations of motion. The electromagnetic sources are the fields themselves.

The constraint of Eq. (3) has to be obeyed also asymptotically far from the sources in the vacuum. If we do not want to break rotational invariance, then, the only way to implement it, is by demanding the boundary condition $\vec{A} = 0$ and $A_0 = \pm i \alpha$ at spatial infinity. The electromagnetic potentials of Eq. (2) are then forced to be complex. (We will see below that the appearance of complex potentials does not pose a real problem.) We can rewrite the constraint of Eq. (3) by extracting the complex factor from the time component of the electromagnetic potential to obtain a constraint identical to that of the nonlinear σ model. The transformed fields now span the manifold SU(2). At the same time the above boundary condition at infinity makes all the points at infinity equivalent, effectively transforming the space manifold from R^3 to S^3.

As in the case of the Skyrme model one can define a homotopy application between space S^3 and field manifold SU(2) = S^3. The winding number of the application will then define the topological charge of the model.

The topological current whose time component yields the absolutely conserved charge is here defined [5, 9].

$$N^\mu = \frac{i}{12\pi^2 \alpha^4} \varepsilon^{\mu\nu\kappa\beta} \varepsilon_{\lambda\rho\sigma\tau} A^\lambda \partial_\nu A^\rho \partial_\kappa A^\sigma \partial_\beta A^\tau \tag{4}$$

It is a trivial exercise to convince ourselves that this current is absolutely conserved regardless of the dynamics provided the constraint of Eq. (3) is obeyed.

We will generate our solutions from the spherically symmetric ansatz

$$A_0 = \pm \alpha \, i \cos(\theta(r))$$
$$A^i = \alpha \, \hat{n} \sin(\theta(r)) \tag{5}$$

the + (-) sign corresponds to positively (negatively) charged solitons of integer topological charge, and

$$\hat{n} = \hat{r}\,\Theta(R-r) + \tilde{n}\,\Theta(r-R)$$
$$\tilde{n} = (a_0^2 - \vec{a}^2)\,\hat{r} + 2\,a_0\,\vec{a}\times\hat{r} + 2\,\vec{a}\cdot\hat{r}\,\vec{a} \tag{6}$$

A_0 and $\vec{a}$ are collective coordinate variables depending on time restricted by the condition [11]

$$a_0^2 + \vec{a}^2 = 1 \tag{7}$$

implying

$$\hat{n}^2 = \tilde{n}^2 = 1 \tag{8}$$

We have split the unit vector of Eq. (6) into an inner and an outer region because we want solutions that yield the normal electrostatic potential outside the particle's core. A more appropriate solution must surely be simultaneously time dependent and radial dependent. However, we want to deal with the most simple possible ansatz still solvable with elementary methods and consistent with the desired properties.

The soliton will then consist of a rigidly rotating core of electromagnetic field energy up to a radius R - to be determined below - and an asymptotic long range electrostatic tail. Inside the core of the soliton there will also be a magnetic field.

The rotational collective coordinates of Eq.(6) will produce topological solitons of nonvanishing angular momentum. The ansatz provides an exact solution when the collective coordinates are absent and an approximate solution when they are included, this is no different than the procedure one uses to quantize solitons and in particular Skyrmeons [11].

The electromagnetic fields for the ansatz of Eq. (5) are

$$\vec{E} = \alpha\left(\mp i\,\theta'\,\hat{r} - \dot{\hat{n}}\,\Theta(r-R)\right)\sin(\theta) \tag{9}$$

$$\vec{B} = \alpha\,\Theta(r-R)\left(\vec{\nabla}\times\hat{n}\,\sin(\theta) + \hat{r}\times\hat{n}\,\theta'\cos(\theta)\right) \tag{10}$$

where an overdot denotes time derivatives and a prime space derivatives.

A well defined static electric field at the origin fixes the boundary condition for the angle θ of Eq. (6) to be $\theta(0) = n\pi$, with n an integer. At the same time, this boundary condition determines the charge of the solitons to be n.

The nonvanishing part of the magnetic field at the origin is purely time dependent and averages to zero, therefore the magnetic field is well defined at the origin too. The electrostatic field of Eq. (9) is purely imaginary. This may seem to contradict basic physics. However, this contradiction is only apparent. The interaction between charged particles is real, as can be seen from a simple additive ansatz for the two body problem appropriate at large separations between the charges. The imaginary factor i drops out in the electrostatic interaction. Consider a two-body ansatz consistent with the nonlinear constraint. As for the Skyrme model a simple such ansatz consistent with the topology is given by:

$$A_0 = \pm i\, \alpha(\pm \cos(\theta_1)\cos(\theta_2) - \hat{n}_1 \cdot \hat{n}_2 \sin(\theta_1)\sin(\theta_2))$$

$$\vec{A} = \alpha(\pm \cos(\theta_1)\sin(\theta_2)\,\hat{n}_2 + \cos(\theta_2)\sin(\theta_1)\,\hat{n}_1 - \hat{n}_1 \times \hat{n}_2 \sin(\theta_1)\sin(\theta_2)) \quad (11)$$

where

$$\hat{n}_{1,2} = \left(\hat{r}\,\Theta(r-R) + \tilde{n}\,\Theta(R-r)\right)_{1,2}$$

$$\tilde{n}_{1,2} = \left((a_0^2 - \vec{a}^2)\,\hat{r} + 2\,a_0\,\vec{a} \times \hat{r} + 2\,\vec{a} \cdot \hat{r}\,\vec{a}\right)_{1,2}$$

$$\vec{r}_{1,2} = \vec{r} \pm \frac{\vec{D}}{2} \quad (12)$$

The ansatz of Eq. (11) summarizes the four possible charge configurations of particles 1 and 2, each one represented by its own angle $\theta(r_{1,2})$ and unit vector $n_{1,2}$ centered at the position of the charge. The vector D represents the separation between the charges in a Born-Oppenheimer type of approach. The ansatz is exact at large separations D and obeys the constraint of Eq. (3) as can be checked by explicit calculation. It represents a topological charge of +2, -2 or 0 (two cases) depending on the choice of signs. Substituting this ansatz into Eq. (1) it is easy to see that the electrostatic potential determined by $\left(\vec{\nabla} A_0\right)^2$ is real. Inside the particles cores there exist spin dependent interactions that are real too after evaluation of the corresponding collective coordinate matrix elements.

The attractive or repulsive character of the electrostatic interaction is governed by the + or - signs in Eq. (11). The same sign determines the topological charge. This is the reason for the identification of the topological charge with the electric charge.

In the framework of nonlinear electrodynamics, any charged object will behave accordingly to the above ansatz (11) and therefore the interactions will be real. In this approach there will not be pointlike charges, all charged particles have to be topological solitons, whose charge is defined in terms of the absolutely conserved topological charge of the extended object.

Returning now to the one-body problem it is straightforward to verify that the ansatz of Eq. (6) has vanishing linear momentum P. The Poynting vector

$$\vec{S} = \frac{\vec{E} \times \vec{B}}{4\pi} \quad (13)$$

has a zero value time averaged and its integral defining the linear momentum of the object vanishes upon angular integration. The latter assertion is quite easy to check because the expression (13) is essentially proportional to the radial versor $\hat{r}$. Therefore not only there is no momentum flow on the average, but also the overall linear momentum vanishes as well as its square. The object is at rest. One may wonder how would we give the charge a boost. This is not a trivial question in the case of solitons and a partial answer, at least in the nonrelativistic realm, can be given by introducing a collective parameter D as in Eq. (12) depending on time and quantizing it. The time derivative of D would represent the center of mass velocity. However, this procedure is valid only at low speeds and reflect a Galilean nonrelativistic drag. The parameter D can then be promoted to a quantum variable and a Schroedinger equation determining the motion of the object will follow. We then see that the complex character of the one-body ansatz is consistent with observations.

To further convince ourselves of the reality of the momentum operator, let us perform a classical boost on the electromagnetic fields [12]. Consider a Lorentz boost in the x direction

$$E'_1 = E_1 \qquad B'_1 = B_1 \qquad (14)$$

$$E'_2 = \gamma(E_2 - \beta B_3) \qquad B'_2 = \gamma(B_2 + \beta E_3) \qquad (15a)$$
$$E'_3 = \gamma(E_3 + \beta B_2) \qquad B'_3 = \gamma(B_3 - \beta E_2) \qquad (15b)$$

Inserting Eqs. (14-15) in Eq. (13) we find that all the crossed terms in the electric and magnetic fields vanish due to averaging over collective coordinates (linearity in the magnetic field) and integration over solid angle. The only remaining terms for the linear momentum in the x direction are quadratic in the electric and magnetic fields that are real as the energy itself (see below). A more accurate treatment would demand the distortion of the soliton, however, this is beyond the scope of the present work.

Below we will see that after quantization the angular momentum is well defined too and its matrix element is real.

Inserting Eqs. (9-10) into Eq. (1) and integrating over angles we obtain

$$L = (\dot{a}_0^2 + \dot{\vec{a}}^2) I \pm 2 i \, a_0 \, \dot{a}_0 \, K - a_0^2 \, \vec{a}^2 \, U - N \qquad (16)$$

where

$$I = \frac{4\alpha^2}{3} \int_0^R r^2 \, dr \, \sin^2(\theta) \qquad (17)$$

$$K = \frac{4\alpha^2}{3} \int_0^R r^2 \, dr \, \sin^2(\theta) \, \theta' \qquad (18)$$

$$U = \frac{4\alpha^2}{3} \int_0^R r^2 \, dr \left\{ \cos^2(\theta) \, \theta'^2 + \frac{\sin(2\theta) \, \theta'}{2 \, r} + 3 \frac{\sin^2(\theta)}{r^2} \right\} \qquad (19)$$

$$N = \frac{\alpha^2}{2} \int_0^\infty r^2 \, dr \sin^2(\theta) \, \theta'^2 \qquad (20)$$

In Eq. (19) we have neglected small terms that appear because our ansatz is not an exact solution to the Euler-Lagrange equations.
We quantize the collective degrees of freedom in the standard manner [11]. Symmetrizing the term linear in the time derivatives and applying the transformation on the wave functions

$$\psi \to e^{\frac{K\cos(2\lambda)}{2}} \psi \qquad (21)$$

where

$$a_0 = \cos(\lambda(t)) \qquad (22)$$

we find the hamiltonian of the model

$$H = \frac{C_2(SU(2))}{4I} + a_0^2 \vec{a}^2 \, U + N = T_{rot} + V_{rot} + E_{static} \qquad (23)$$

where $C_2(SU(2))$ is the quadratic Casimir of SU(2). Not surprisingly we find the hamiltonian of a rigidly rotating top with a potential term depending on the collective coordinates. It is evident from Eq. (23) that our choice of a complex ansatz yields the positive definite energy. This would not have been the case had we taken a real ansatz.

The angular momentum operator - depending on time derivatives of the collective coordinates - reads

$$\vec{J} = \vec{L} + \vec{S} = \int d^3x \left(-E_1 \vec{r} \times \vec{\nabla} A^1 + \vec{E} \times \vec{A} \right) \qquad (24)$$

In order to find the eigenstates of Eq. (24), we first consider the solutions for R>>r. In that region there is no contribution from the collective coordinates, both statically and dynamically. The Euler-Lagrange equation for the soliton profile becomes there

$$\theta'' + \frac{2\theta'}{r} + \theta'^2 \cot(\theta) = 0 \qquad (25)$$

Continuity of the angle θ and its derivative requires the radius R to be at the point where θ goes through $\pi/2$. At that point the equations for the inner part and the outer part coincide and a smooth matching is achieved. The solution of Eq. (25) satisfying the boundary conditions at R and infinity is

$$\cos(\theta) = 1 - \frac{R}{r} \qquad (26)$$

Eq. (26) can be used to find the electrostatic potential at large distances. Asymptotically far from the origin, $r \gg R$ we must have the electric field of a charged particle

$$-\alpha \theta' \sin(\theta) = \frac{e}{r^2} \tag{27}$$

where e is the proton charge ($e^2 = 1/137$). Eqs. (26-27) imply the remarkable relation

$$\alpha R = e \tag{28}$$

Now we can estimate the importance of the terms in Eq. (23)

$$T_{rot} \approx \alpha R^{-3} = \frac{\alpha}{e^3}$$

$$V_{rot} \approx \alpha^2 R = \alpha e \tag{29}$$

It is clear from Eqs. (29) that the most important contribution to the energy comes from the kinetic rotational part T_{rot}, while the potential part depending on the collective coordinates is much smaller, and can be treated as a perturbation. T_{rot} can be diagonalized exactly.

The solutions of T_{rot} are the well known eigenstates of the Casimir of SU(2). These states are labeled by a parameter k. that takes integer and half-integer values.

The energy of the solitons becomes

$$E_k = \frac{k(k+1)}{I} + \left(a_0^2 \vec{a}^2\right)_{av} U + N \tag{30}$$

where the middle term of Eq. (30) is evaluated as a matrix element in each eigenstate of the quadratic Casimir. The ground state corresponds to $k = 1/2$. For this solution we also have $L^2 = S^2 = 3/4$ with $L_z, S_z = \pm 1/2$. It is quite surprising that both the spin and 'orbital' angular momentum are quantized in half-integer values. We can construct eigenstates of total angular momentum J using standard Clebsh-Gordan coefficients. The ground state consists then of a spin one object and a spin zero state both degenerate in energy. For a generic value of k we will have 2 (2 k+1) eigenstates of L_z and S_z, which can be rearranged into eigenstates of total angular momentum J. For each k value we will have two states one of spin $2k$ with (2 (2k)+1) projections and a spin zero state. This is the very peculiar signature of the model, namely, for each higher spin state there will be a twin state of spin zero.

Evaluating the matrix element of the collective coordinate potential term in Eq. (23) in the ground state we find

$$E_k = \frac{3}{4I} + \frac{U}{4} + N \tag{31}$$

Let us now consider the coupling to vibrational excitations of the electromagnetic field. Such couplings are very important in topological soliton models, like the Skyrme model [10, 11], where they induce substantial energy subtractions [13]. The same effect arises in electrodynamics, where we can get positive or negative contributions depending on the details of the electromagnetic field configuration [14]. The subtraction is even more marked when the rotational kinetic energy is the most important term. Following the procedure of ref. [13] we can copy step by step what was done for the rigid quantization of the Skyrme model to the present case. This is due to the fact that our collective coordinate quantization is identical to that of the Skyrmeon. The isospin fluctuation modes are here replaced by the the different polarizations of the electromagnetic field. The analogy is exact with regard to the symmetrization procedure that leads to a subtraction to the soliton energy equal to the rotational energy of the ground state. Inserting this result into the ground state energy we find the mass of the ground state soliton:

$$M = \frac{U}{4} + N \tag{32}$$

The integrodifferential Euler-Lagrange equation for the soliton profile obtained from Eq. (32) for r>R becomes

$$\left(\theta'' + \frac{2\theta'}{r}\right)\left(\sin^2(\theta) + \frac{2}{3}\cos^2(\theta)\right) + \frac{\theta'^2 \sin(2\theta)}{6} - \frac{\sin(2\theta)}{r^2} = 0 \tag{33}$$

Solutions to Eq. (33) are functions of the cutoff radius R. Inserting these numerical solutions in Eq. (30) we find

$$M = \frac{e^2}{R} \tag{34}$$

half of the mass comes from the collective coordinate potential energy and half from the electrostatic energy. The only scale available in the problem is α, or equivalently R. Once this number is fixed all the properties of the states follow.

The spin of the lowest energy state suggests the identification with the gauge vector bosons W^+ and W^-. In the electroweak unified theory, the gauge bosons acquire their mass through the Higgs mechanism. Here we speculate that the massive gauge bosons are topological solitons in the nonlinear gauge of electrodynamics. In order to make a more tangible connection with the W bosons, we should be able to describe the weak interactions in the present language. A major stumbling block for this purpose is the fact we have not been able to generate spin 1/2 objects yet. However, other possible identifications like with the ρ meson, seem less appropriate still, although they cannot be ruled out completely.

With such identification, we can pin down the value of the radius and scale of the model to be $R = 1.8 \times 10^{-5}$ fermi and $\alpha = 936$ GeV. The value of the scale α, predicts that the effect of the nonlinear constraint will start to show up for photons above this energy. Higher spin states lie extremely high in energy, as can be seen from the scaling properties of Eqs. (29). The first excited state consisting of a spin two and a spin zero boson will have an energy of approximately approximately 1500 TeV.

Let us now consider the gyromagnetic factor of the spin one bosons. Using the standard definition of the magnetic moment of a distribution of charge and the gyromagnetic factor g,

$$\vec{\mu} = \int d^3x \, \vec{r} \times \vec{N} = \frac{g\,e}{2\,M}\,\vec{J} \qquad (35)$$

where N is the spatial component of 'anomalous' current of Eq. (4), M is the mass of the soliton of Eq. (2) and J is the total angular momentum, we find

$$g = \frac{-4\,M\,K}{\pi\,I} \qquad (36)$$

The gyromagnetic ratio of Eq. (36) is scale invariant and does not depend on the actual values of R and α. For any chosen value we find $g = 1.6$ well within the experimental limits [15]. Although $g = 2$ for a point relativistic particle, an extended object can violate this restriction. The deviation from $g = 2$ may spoil the renormalizability of the standard model of the weak interactions. However, at the present stage of development of the model, this seems like a premature conclusion especially because we do not know how to include spin 1/2 objects consistently in the picture - as extended topological solitons - and less we know how to deal with the weak interaction within the same framework.

In summary, the model predicts a tower of integer spin states degenerate in energy with spin zero charged states separated by a large energy gap from the ground state. The states owe their existence and spin to the rotational degrees of freedom. The experimental situation concerning the existence of a scalar particle degenerate in energy with the gauge bosons is not clear [16], while higher states are in any event beyond the reach of experiment at the present time.

Acknowledgements

I thank Prof. Shmuel Elitzur for theoretical input and Prof. Guiora Mikkenberg for his remarks concerning the experimental situation.

References

1. Abraham, M., *Ann. Phys. (Leipzig)* **10**, 105 (1903).
2. Lorentz, H. A., *Theory of Electrons*, Dover, New York (1952).
3. Born, M. and Infeld, L., *Proc. Roy. Soc.* **A144**, 425 (1934).
4. Podolsky, B., *Phys. Rev.* **62**, 68 (1942). See also, F. Bopp, Ann. der Physik (Leipzig) **38**, 345 (1940).

5. Righi, R. and Venturi, G., *Int. Jour. of Theor. Phys.* **21**, 63 (1982); and references therein.
6. Rodrigues, W. A., Vaz, J. and Recami, E., *Found. of Phys.* **23**, 469 (1993).
7. Dirac, P.A.M., *Proc. Roy. Soc.* **A209**, 292 (1951) ; **A212**, 330 (1952), **A223**, 438 (1954); *Nature* **168**, 906 (1961). See also Schroedinger, E., *Nature* **169**, 538 (1952).
8. Nambu, Y., *Prog. Theor. Phys.*, Suppl., Extra number, **190** (1968).
9. Skyrme, T. H. R., *Proc. Roy. Soc. London*, **A260**, 127 (1961).
10. Barcelos-Neto, J., Galvão, C.A.P. and Natividade, C.P., *Z. Phys.* **C52**, 559 (1991).
11. Adkins, G. S., Nappi , C. R. and Witten, E., *Nucl. Phys.* **B228**, 552 (1983).
12. Jackson, J.D., *Classical electrodynamics*, Chapter 11, John Wiley & Sons.
13. Verschelde, H. and Verbecke, H., *Nucl. Phys.* **A500**, 573 (1989). For a calculation of the Casimir energy in the Skyrme model see Moussalam, M. and Kalafatis, D., *Phys. Lett.* **B272**,196 (1991).
14. Milton, K. A., *Ann. of Physics* **127**,49 (1980).
15. Particle Data Group, *Phys. Rev.* **D50**,1173 (1994).
16. Mikkenberg, G., private communication.

A PARTICLE OR PHOTON IN A FIELD

J. KAJAMAA
Kajamaa Engineering Oy
Tukholmankatu 2, FIN-00250
Helsinki, Finland

1. Introduction

Classical mechanics is based upon the motion of rigid bodies in force fields. The rigid body has a mass m and the field has an acceleration a. This was the starting point for the scientific considerations that led to an intense technological development. For the first time a physical object and a field, interacting with each other, were considered simultaneously. Important physical laws connected with this basic phenomenon could be derived and expressed in an exact mathematical way. The force acting on the body was equal to the mass times the acceleration.

Later this straightforward and deterministic picture was changed for two reasons: firstly it was realized that if the velocity of the object is very large, this approach is insufficient (the mass of the particle apparently increases at large velocities, leading to the special theory of relativity); secondly if the object is very small (either a particle or a light quantum) then new features appear (leading to quantum mechanics). Specifically the quantum of electromagnetic radiation has no rest mass; this phenomenon cannot be explained by the laws of classical mechanics, because these laws are based upon observations of a rigid body in a force field in which the concept of mass has a central role.

A particle or a radiation quantum (photon) are different physical objects. Though the general laws are the same in both the macroworld and the microworld, the physical models of particles and photons differ from each other. The philosophical question is: what is the mass ? Is it a point-wave or something else.

Mathematically we can describe, for example, the motion of planet earth by assuming that its mass is located at a point - its centre of gravity. Ironically, the motion of an elementary particle, which is certainly tiny and point-like, cannot be described as that of a classical mass-point. In addition to its varying (with velocity) mass, the particle has a wave character (its de Broglie wavelength). It is important to emphasize that we cannot comprehend the mechanical properties of a particle by direct observations; the concept that a particle is a mass-point is an extrapolation from observations on macroscopic bodies.

The light-quantum (photon) is even less appropriately described as a mass-point in a force field. It does have momentum (and hence by inference a mass), but only when it is moving; it simply cannot exist at rest. On the other hand its velocity is constant. Also the radiation of quanta can sometimes be described as the properties of a field, rather than by a collection of individual particle-like quanta.

2. The Differences between a Photon and a Particle

A light-quantum (photon) is the simplest kind of physical object. In a gravitational field it behaves as a mass-point moving at the velocity of light. Since it reacts with the gravitational field in this way we conclude that from the standpoint of classical mechanics, the photon has a mass. Alternatively, from a microworld viewpoint, the concept of photon-mass is that it is just (localized) energy. The distinction between these two viewpoints is very important.

The velocity of a light-quantum is constant and is the same constant for all light-quanta - a universal constant (the velocity of light) - regardless of the velocity of the source. From the viewpoint of classical mechanics this invariance of velocity is astonishing, for if the momentum of a mass-point is increased, its velocity increases. In the microworld of light-quanta and particles this classical, macroworld mechanical model does not pertain.

An important question is: are the physical laws really the same in both systems ? The above example shows that there is certainly no such law as the velocity conservation law. Instead the energy and momentum conservation laws are valid. In the case of the light-quantum, it reacts to the pushing force, but instead of accelerating, its frequency changes. In this case the change of momentum changes the frequency rather than the velocity. This is due to the photon having a different physical structure from a point-mass; the physical phenomenon is, in principle, the same.

In the original form of the important mechanical law, the derivative of the momentum with respect to time is equal to the force. With rigid-bodies (acting as point-masses) the time derivative of the velocity alone is sufficent, because the mass does not change. However, as was shown above, there is another way to change the momentum. The frequency of the quantum is directly proportional to its momentum and to its energy. This means that the energy of the quantum changes in the same way as its momentum. This behaviour is different from that of a rigid body; in the latter the kinetic energy is proportional to the square of its velocity, while its momentum is directly proportional to its velocity.

What are the essential differences between these objects as far as their interaction with a force is concerned ? Primarily, the velocity of a rigid body increases when a force acts upon it. Molecules resemble rigid bodies, but light-quanta do not; the latter's frequency changes, not its velocity, under the same physical influence.

How does a single elementary particle behave then ? Usually it is considered as a mass-point - a quasi rigid-body. However, at high velocities its mass increases, so that it is not a mass-point.

According to the relativistic model of an elementary particle, it is a localized field formed from two quanta. It is a quantum structure with rest mass and its external velocity is always less than the velocity of light c. It fulfils the energy and momentum conservation laws, whereas the relativistic mass point does not.

Consequently the relativistic model, which is based upon energy rather than mass, is physically applicable.

Is the concept of mass essential ? Should we use it in the microworld at all ? The following equations describe a collision between two particles before (index $_1$) and after (index $_2$) the collision:

$$E = \frac{m_0 c^2}{\sqrt{1-(v/c)^2}} \qquad p = \frac{m_0 v}{\sqrt{1-(v/c)^2}} \qquad \Sigma E_1 = \Sigma E_2 \qquad \Sigma p_1 = \Sigma p_2 \tag{1}$$

$$E = E_0 \sqrt{\frac{c+v}{c-v}} \qquad p = \frac{E_0}{c}\left(\sqrt{\frac{c+v}{c-v}} - 1\right) \qquad \Sigma E_1 = \Sigma E_2 \qquad \Sigma p_1 = \Sigma p_2 \tag{2}$$

The first set of equations (refeqn1) contain mass and have no solution, while the second set (2) are based upon energy rather than mass, and they can always be solved.

3. An Elementary Particle in a Force Field

How does a particle really behave ? Is its behaviour a combination of the behaviours of a quantum and a rigid body ? Usually this problem is treated by assuming that the particle is a rigid body with a mass that varies according to the equations of special relativity and hence its momentum is determined by the second of equations (1). It is also possible to define the total energy, E, of the particle from this standpoint of a velocity-dependent mass, and the kinetic energy can be determined in the same way. In addition, an approximation to this energy turns out to be equal to the above mentioned kinetic energy of a rigid body. This approximation is accurate when the velocity of the particle is much smaller than the velocity of light, c.

This usual way of treating the motion of a particle is favoured because it is closely connected with the macroscopic mechanical world. This makes it easy to accept even though the variable mass is not understandable in classical mechanical terms. Some physicists avoid the concept of a velocity-dependent mass by simply emphasizing the energy and momentum conversation laws (1,2). However, for a collision between 2 particles, or between a particle and a light-quantum, the equations based upon the energy and momentum conservation laws cannot be solved at all.

This insolvability arises from the fact that the energy and momentum are not directly proportional to each other; their relationship is more complicated. For example in the case of a light-quantum, $E = pc$. The changes of energy and momentum when the collision partners do not obey the same physical laws, is not solvable from the basic laws of energy and momentum conservation.

In the case where the collision partners are two rigid bodies at low velocities (when their kinetic energies are directly proportional to their momenta), the equations can be solved; the difference of velocities can be cancelled and the set of equations have a general solution.

Under these circumstances we have to choose another, more physical, way to solve the problem. The concept of quantum, which certainly represents the microworld, is taken as a starting point. A relativistic model of an elementary particle has been derived [1, 2, 3, 4, 5].

When a particle is constructed by resonating and reflecting quanta, it inherits new kinds of important mechanical properties. The concept of mass has no central position. Instead of mass, everything is dealt with, defined, measured against, and confirmed by means of the energy, which is described by the frequency of the oscillations within the particle.

The essential difference between this particle model and the rigid body model, is that the energy and momentum are directly proportional to each other. This arises because the momentum and the total energy of the particle include a contribution from its internal motion as well as from it motion relative to the observer's frame of reference.

Thus a particle can store energy in two ways; one resembles the energy of a rigid-body and the other that of a quantum. When this kind of particle accelerates, not only does its velocity increase, but its internal structure changes as well; it becomes flatter due to the Doppler effect. The wavelength decreases, the frequency increases, and consequently both the

A PARTICLE OR PHOTON IN A FIELD

energy and the momentum increase. The essential point is that it is not only the increase of mass which contributes to the change in the energy and the momentum of the particle. The generalization of the concept of mass is not enough in this respect; the overall interaction between the object and the field has to be considered.

4. Objects Consisting of a Collection of Particles

An object such as an atom or a molecule can be regarded as a rigid body in all processes in which the object does not lose or gain any of its constituent particles (electrons and atomic nuclei), and in which the internal (relative) state of motion of the constituent particles does not change.

The total rest mass of such a composite object is equal to the sum of the masses of its constituent particles plus the energy of their relative (internal) motion (divided by c^2).

Similarly while the individual constituents may have substantial momenta, the vector sum of all the constituent particle momenta must be zero in the object's rest frame. This cancellation will prevail even when the object is moving relative to the observer's frame of reference.

In general, when such a composite object interacts with a field or collides with another object, not all of the constituent particles will be affected in the same way. For this reason in some processes the internal state of the object will be disturbed; e.g. when the internal state is changed by absorption of a photon or by collision with high-speed electrons. However, in all interaction processes in which the final internal state of the object is the same as its initial state, the dynamics of the interaction (conservation of energy and momentum) can be treated as though the object were a rigid body with its total mass located at its centre of mass.

5. The Mass-Point and Relativistic Particle Models

The essential difference between the mass-point model and the relativistic model, is that the former is based upon the particle having a rest-mass, m_0, while the latter is based upon its rest-frequency, ν_0, of the resonating electromagnetic radiation. The relation between m_0 and ν_0 is:

$$m_0 c^2 = h\nu_0 \qquad (3)$$

The corresponding varying (velocity dependent) mass and frequency are related to the rest values by:

$$m = \frac{m_0}{\sqrt{1 - (v/c)^2}} \qquad (4)$$

$$\nu = \nu_0 \sqrt{\frac{c+v}{c-v}} = \frac{\nu_0(1+v/c)}{\sqrt{1-(v/c)^2}} \qquad (5)$$

These equations are Lorentz invariant.

While mass is a scalar, a frequency is associated with a directed oscillation. The structure of the relativistic model is dualistic: firstly it is a localized field which behaves like a (scalar) mass, and secondly it is an oscillating field resembling a vector, which is parallel to the propagation direction of the particle. The formula for the dynamic frequency, ν, can be divided into two parts, one from the scalar and the other from the vector contribution. Consequently, this makes both the momentum and the total energy dualistic in the same way.

A comparison of the two approaches (mass point and relativistic) can be made by writing the new E and p as functions of the present energy and momentum, denoted by E_m and p_m respectively. It follows from (1,2) that the varying quantities are given by:

$$p = (m - m_0)c + p_m \qquad E = E_m(1 + v/c) \qquad (6)$$

The original equations are:

$$p = \frac{h}{c}(\nu - \nu_0) = \frac{h}{\lambda_{dB}} \qquad E = h\nu = h\nu_0\sqrt{\frac{c+v}{c-v}} \qquad (7)$$

Clearly the momentum, p, has an internal part which is not directly related to v, but only through m. In addition, it has an external part which is equal to the present momentum, mv. Similarly the energy, E, has two parts; the first part is due to the increase in mass, and the other part to the movement itself. The relationship between λ_{dB} and λ_{dB_m} is more complicated; however, λ_{dB} becomes equal to λ_{dB_m} when $v \ll c$.

It is concluded that a rigid body and the quantum are two extreme cases as far as the change of momentum is concerned. The present formulas for E and p of a single particle, describe, in fact, the behaviour of a rigid body at relativistic velocities. They have been obtained by generalizing the concept of mass.

The relativistic model of an elementary particle is, instead, some kind of a combination of a rigid body and a quantum. It takes into account both the scalar and the vector character of the essential entities, the momentum and the energy.

6. Summary

In classical mechanics the mass is a ratio between the force and the acceleration. It can be determined as a result of an interaction between an object and the force field.

A PARTICLE OR PHOTON IN A FIELD

In relativity theory, the rest mass of a particle, m_0, is assumed to be that of a mass point; the varying (velocity dependent) mass is proportional to it. The total and kinetic energies, and the momentum, are derived in terms of mass.

On the other hand, the de Broglie wavelength is not derived in this way; the interaction between the object and the force field is not the starting point. Instead, the classical concept of mass is applied in the beginning. Therefore the formulas for E, p and T describe the (theoretical) relativistic behaviour of a rigid body, which is a collection of molecules or a mass point physically similar to a rigid body. Types of mass points other than a rigid body do not exist in reality. In addition, their energy, E, and momentum, p, do not fulfil the conservation laws.

From the point of view of the mass point model itself, this means that the increase in mass due to an increase in velocity is taken into account, but the total increase of the energy and momentum is not. This is because the particles move also as parts of the atoms in the object; the vector sum of their internal momenta is always zero.

The equations are inconsistent with each other and the interrelations between different quantities are complicated: e.g.

$$E^2 = (pc)^2 + (m_0 c^2)^2 \tag{8}$$

with the energy, E, a complicated function of the de Broglie wavelength, λ_{dB}.

The relativistic model of an elementary particle is constructed by localizing an electromagnetic radiation field (e.g. by resonance and reflection). The rest frequency, ν_0, of the resulting standing wave is determined from the rest energy, E_0. Then the corresponding frequency, ν, of the moving particle is derived on the basis of the Doppler effect. Thus all the important quantities are defined: the total energy, the momentum and the de Broglie wavelength.

The mass is defined classically by differentiating momentum with respect to time; it is a result of an interaction between the object and the field. In this case E and p fulfil the conservation laws.

From the point of view of the relativistic model the full contribution of the interaction between the object (an elementary particle), and the force field, is taken into account.

The equations are consistent with each other and the interrelations between different quantities are simple: e.g.

$$E = pc + E_0 = hc\left(\frac{1}{\lambda_{dB}} + \frac{1}{\lambda_0}\right) \tag{9}$$

Appendix: Properties and Relations of Particles and Photons

	rigid–body	photon	relativistic–model	mass–point
m	F/a	E/c^2	$\dfrac{h\nu_0}{c\sqrt{c^2-v^2}}$	$\dfrac{m_0 c}{\sqrt{c^2-v^2}}$
m_0	m	0	E_0/c^2	E_0/c^2
p	mv	h/λ	h/λ_{dB}	$\dfrac{m_0 v}{\sqrt{1-(v/c)^2}}$
E		$h\nu = pc$	$h\nu = h\nu_0 \sqrt{\dfrac{c+v}{c-v}}$	$\dfrac{m_0 c^2}{\sqrt{1-(v/c)^2}}$
T	$\tfrac{1}{2}mv^2$	pc	$pc = h(\nu - \nu_0)$	$(m - m_0)c^2$
λ		λ	$\dfrac{\lambda_0}{\sqrt{\dfrac{c+v}{c-v}}-1}$	$\dfrac{h}{m_0 v}$
E		pc	$pc + E_0$	$\sqrt{(pc)^2 + (m_0 c^2)^2}$
T	$\dfrac{p^2}{2m}$	pc	pc	$\dfrac{c^2}{v}p - m_0 c^2$

Notes:

1. The entries in the first column on the table indicate the physical quantity tabulated in each row:

m = mass

m_0 = rest-mass

p = momentum

E = Total Energy

T = Kinetic Energy

λ = de Broglie wavelength

E = Total Energy

T = Kinetic Energy

2. The formula for the varying mass (row 1) is applicable, but the structural description of the object is insufficient or missing; e.g. it does not explain how a particle radiates or emits a photon so that the velocity of light, c, remains constant.
3. In row 3 the momentum, p, is derived mechanically, not by means of the de Broglie wavelength. It represents only part of the momentum. Its functional form is incorrect.
4. In row 4 the total energy, E, describes only that part of the energy which is due to the increase in mass with velocity. Its functional form is not correct.
5. Similarly the kinetic energy, T, in row 5 is incorrect.
6. The entries in row 6 are for the <u>non-relativistic</u> de Broglie wavelength.
7. In row 7 the formulas for the total energy, E, as a function of the momentum, p, are incorrect.
8. The formulas for E and p do not fulfil the energy and momentum conservation laws.
9. The present formulas for E, T and p, describe the (theoretical) relativistic behaviour of a group of molecules; i.e. a rigid body.
10. The de Broglie wavelength of a particle is measured directly by means of diffraction methods. The mass of a single particle is measued by means of a force field. Consequently one has to assume the type of interaction which prevails between the object and the field. This study does not deal with the equation of motion itself; it is taken according to the world view. this study is mainly concerned with only one scalar quantity, the mass.

References

1. Annales de la Fondation Louis de Broglie, **17**, No.3, 275-293.
2. *Atomitason tekniikka*, WSOY (1991).
3. *Laatu ratkaisee modernissa fysiikassakin*, Otatieto No.896 (1993).
4. Abstracts of the conference: *Physical Interpretation of Relativity Theory*, held at Imperial College, London, September 1994.
5. A Finnish TV Program in which a computer animation of the model was depicted (1995).

THE MYTH OF THE PHOTON

TREVOR W. MARSHALL
Department of Mathematics,
University of Manchester,
Manchester M13 9PL, U.K.

AND

EMILIO SANTOS
Departamento de Física Moderna,
Universidad de Cantabria,
39005 Santander, Spain

We have shown that all "single-photon" and "photon-pair" states, produced in atomic transitions, and in parametric down conversion by nonlinear optical crystals, may be represented by positive Wigner densities of the relevant sets of mode amplitudes. The light fields of all such states are represented as a real probability ensemble (not a pseudoensemble) of solutions of the unquantized Maxwell equation.

The local realist analysis of light-detection events in spatially separated detectors requires a theory of detection which goes beyond the currently fashionable single-mode "photon" theory. It also requires us to recognize that there is a payoff between detector efficiency and signal-noise discrimination. Using such a theory, we have demonstrated that all experimental data, both in atomic cascades and in parametric down conversions, have a consistent local realist explanation based on the unquantized Maxwell field.

Finally we discuss current attempts to demonstrate Schroedinger-cat-like behaviour of microwave cavities interacting with Rydberg atoms. Here also we demonstrate that there is no experimental evidence which cannot be described by the unquantized Maxwell field.

We conclude that the "photon" is an obsolete concept, and that its misuse has resulted in a mistaken recognition of "nonlocal" phenomena.

1. Introduction

Since the beginning of modern physics, that is since the seventeenth century, there have been two views as to the nature of light. The corpuscular view has, traditionally, been supported by those most strongly attracted to formal elegance, whilst the undulatory view has been supported by those who insist on the necessity for science to give explanations of phenomena [1]. It is no coincidence that one of the earliest and strongest statements against trying to explain field phenomena is in Newton's *Opticks*, and that a similar statement was made, nine years later, in the Preface to his *Principia*.

Formal elegance, combined with Newton's authority, dominated the eighteenth and the first part of the nineteenth century. Light corpuscles, going near to a sharp edge, experienced, according to their ideas, instantaneous (in today's terminology *nonlocal*) interactions with the edge, and that caused a phenomenon they called *inflexion*. Today that phenomenon is called diffraction, and the name has changed because Young, Fresnel, Faraday and Maxwell taught us that nonlocal "explanations" are not explanations at all. What they gave us instead was a consistently *wave* explanation of diffraction and interference, and theirs remains the only explanation of those phenomena right up to the present day.

So how does it come to pass that the strongest claims to have observed "quantum nonlocality" now come from certain opticians? We suggest that it is because certain opticians have allowed themselves to be seduced by formal elegance, just like Newton's immediate successors. Indeed, just like their intellectual ancestors, they have been carried away by a formally elegant *mechanical* theory. Yes, quantum *mechanics* is elegant, but only as long as it applies to systems with a few degrees of freedom. Light fields have infinite degrees of freedom, and a mature treatment of them requires the considerably less elegant apparatus of *quantum field theory* - not only less elegant, but bristling with all sorts of problems associated with divergences and renormalizations.

Will it be necessary to abandon the quantum formalism in order to obtain a local description of "multiphoton" processes? We cannot yet give a complete answer to this question, but we do assert that, with a certain natural extension of the term "classical", all of the light fields, including those currently classified as "nonclassical", which have so far been produced in the laboratory are, in fact, entirely classical; they are adequately described by the *unquantized* Maxwell equations. We shall see that, in order to extend our notion of "classical", it will be necessary first to escape from Hilbert space and place ourselves in classical phase space; this enables us to adopt the point of view, which has been anathematized by the Copenhagen school, that electromagnetic waves are real waves, in ordinary space and time, having both amplitude and phase. Whereas it may seem natural, as long as we are imprisoned in Hilbert space, to think of "photons", created at one point and absorbed at another, the phase-space description we advocate keeps us entirely within the confines of classical electromagnetic theory. We believe that this step has already been taken, but not fully acknowledged, by a substantial part of the quantum-optics community. For example, three review articles [2-4] on light squeezing make extensive use of phase-space diagrams, and one of them[3] states explicitly that the photon description of the light field is not helpful in the understanding of the phenomenon. We now propose to extend this judgement to the light emitted in atomic-cascade and parametric-down-conversion processes, as well as the microwave radiation contained in cavities.

2. What is a classical light field?

At the moment the accepted convention is to define as "classical", or more precisely "Glauber-classical", a field which is a mixture of pure coherent states. For a single mode of the field, the density matrix of such a state is [5]

$$\hat{\rho} = \int |\alpha\rangle P(\alpha) \langle \alpha | d^2\alpha, \quad (1)$$

where P is nonnegative and

$$|\alpha\rangle = \exp(\alpha \hat{a}^+ - \alpha^* \hat{a}) | 0\rangle. \quad (2)$$

As we have argued elsewhere [6-9], there is a strong case for extending the notion of "classical" to the set of states whose Wigner density is nonnegative. All Glauber-classical states are classical in this wider sense. Indeed the Wigner density of the above single-mode state is

$$W(\alpha) = 2\pi^{-1} \int P(\alpha') \exp(-2 |\alpha - \alpha'|^2) d^2\alpha'. \quad (3)$$

Any quantum state, defined by a density matrix, has a well defined Wigner density, but in general there is no guarantee that this density will be a nonnegative function. The vacuum state is Glauber-classical with

$$P(\alpha) = \delta^2(\alpha), \quad (4)$$

and therefore with

$$W(\alpha) = 2\pi^{-1} \exp(-2 |\alpha|^2). \quad (5)$$

According to some experts[10,11], the difference between the P- and W-representations is entirely formal, but we disagree[9]. Equation (4) suggests that the vacuum is truly empty, whereas equation (5) suggests that there is a real nontrivial distribution of mode amplitudes and phases in the vacuum. Indeed such a view suggests that the *word* "vacuum" is (like "inflexion") obsolete, and that we should call it something else (like "plenum"). This point of view is implicitly supported in that part of the quantum-optics community which takes phase-space diagrams seriously[2-4]. More explicitly, the concept of the real zeropoint field has been central to *stochastic electrodynamics* since the early 1960s[12-14], and its role has been acknowledged more recently in quantum electrodynamics[15].

We consider Max Planck[16] to be the originator, in 1911, of the real zeropoint field, so we shall call light fields with nonnegative Wigner densities "Planck-classical", and will consider a field to be nonclassical only if it is not Planck-classical. There are some Planck-classical fields which are not Glauber-classical. An outstanding example of such a field is the squeezed vacuum state

$$|\zeta\rangle = exp(\zeta \hat{a}^{+2} - \zeta^* \hat{a}^2)|0\rangle. \tag{6}$$

So, using our new definition, are there *any* nonclassical states of the light field? The simple answer is, of course, Yes. Indeed most states of the Hilbert space are nonclassical, because, from a theorem of Hudson[17], generalized by Soto and Claverie[18], no pure states other than the Gaussian subset have nonnegative Wigner density. In particular, the single-mode one-photon Fock state

$$|1\rangle = \hat{a}^+ |0\rangle \tag{7}$$

has Wigner density

$$W_1(\alpha) = 2\pi^{-1}(4|\alpha|^2 - 1)\exp(-2|\alpha|^2). \tag{8}$$

With respect to the "nonclassical" states of the light field currently widely reported as having been observed, our response is that something approximating the squeezed vacuum, as described by equation (6), *has* been observed; this, however, according to our new classification, is a *classical* state, though not Glauber-classical. As for Fock states, represented, for example, by equation (7), we consider that the claims to have observed them are incorrect, and that discussions on such exotic properties (quantum nonlocality, entanglement etc.), which such states would have, if they were to exist, are misguided.

3. Is the "one-photon" state classical?

We have just seen that the *single-mode* one-photon state, represented by equation (7), is not Planck-classical. But we find it amazing that anyone[19] should try to discuss such questions as locality and causality on the basis of waves which fill the whole of space and time! The single-mode representation of a real atomic signal is clearly inadequate.

If we wish to represent the output of a *real* atomic source, we must take account not only of the fact that each atomic light signal occupies a finite time interval (typically about 5ns), but also that neither the time nor the direction of the emitted radiation can be controlled. (We are advocating a return to an unambiguously *wave* description of light, so any signal is emitted into a range of directions. Nevertheless, the spatial distribution of the signal will be influenced, for example, by the atom's charge distribution at the time of emission; this cannot be controlled.)

The first of these requirements leads us to a multimode description of the light signal, while the second forces us to abandon its description as a pure state (that is a vector in Hilbert space) and use, instead, a density matrix. We have shown[20] that, after incorporating these two features, the density matrix is that of a chaotic state, that is its Wigner density is Gaussian and the state is classical. If, however, the signal is part of an atomic-cascade process, it is possible to use one signal in the cascade to monitor the observation time of its partner, as in the experiment of Grangier, Roger and Aspect[21]. In that case[20] we must use a different Wigner density - again positive, however - for the subensemble of monitored signals, and this, as we shall show in a later section, allows a purely wave explanation of what Grangier, Roger and Aspect thought was corpuscular behaviour.

4. The role of the zeropoint field

The real zeropoint field plays a crucial role in explaining how purely wave phenomena may be misinterpreted as evidence of corpuscular behaviour. Recognition of its role would be a convincing vindication of Max Planck[16], because he introduced the concept of the zeropoint field precisely in order to oppose Einstein's *Lichtquanten*, which were the forerunners of photons.

We consider first the way in which a theory with a real zeropoint field views the action of a beam splitter. Such a device was used by Grangier, Roger and Aspect[21] to demonstrate a phenomenon they called *anticorrelation* in the outgoing channels. If we consider the "vacuum" to be empty, then it seems almost unavoidable to assume that the intensity of any incoming classical signal is equal to the sum of the intensities in the outgoing channels, and also that the detection probability in both channels is proportional to the signal intensities in those channels. With such a description it is not possible to explain the anticorrelation data; these were interpreted by Grangier, Roger and Aspect as evidence that the whole "photon" goes into one or other of the outgoing channels. It is easy to see, qualitatively, how the explanatory power of a purely wave theory is increased by the recognition of a real zeropoint field. A beam splitter does not simply split an incoming wave into two parts; it *mixes* together *two* incoming waves, one of them from the "vacuum", to give the two outgoing waves (see Figure 1).

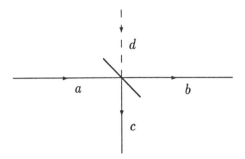

Figure 1: The beam splitter mixes the incoming signal (a) with the relevant modes of the zeropoint field (d) to give the signals (b and c) in the two outgoing channels.

Something similar occurs in a nonlinear optical crystal. An intense coherent input causes two modes of the zeropoint field, initially uncorrelated, to become both enhanced in their amplitudes and correlated (see Figure 2). This in turn causes correlated "photon" counts in the outgoing channels. The current name for what occurs in the crystal is *parametric down conversion* but this is yet another example (like "inflexion") of a bad concept - the "photon" - giving rise to a misleading name and description; it describes an incoming photon of the coherent beam as converting into two completely new photons. But all modes of the field are *already present* before the intervention of the coherent beam and the nonlinear crystal. The *down conversion* is, more accurately, a *correlated amplification* of certain modes of the zeropoint field.

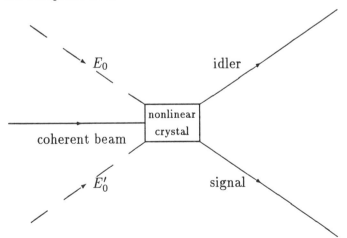

Figure 2: The coherent beam modifies the two initially uncorrelated zeropoint amplitudes E_0 and E_0' to produce the two correlated ("signal" and "idler") outputs.

The two outgoing signals from the beam splitter in Figure 1 are given by[22]

$$E_b = TE_a + iRE_d, \tag{9}$$
$$E_c = TE_d + iRE_a, \tag{10}$$

where R and T are real coefficients satisfying

$$R^2 + T^2 = 1. \tag{11}$$

The two outgoing signals from the nonlinear crystal in Figure 2 are given by[22]

$$E_{\text{signal}} = E_0 + gVE_0^{'*}, \tag{12}$$
$$E_{\text{idler}} = E_0' + gVE_0^*, \tag{13}$$

where g is a coupling constant and V is the analytic signal of the coherent beam. Because these outputs are linearly related to the inputs, we have been able to deduce that the joint Wigner density of the outgoing beams is positive, both when the inputs are all Gaussian, as in Figure 2, and when one of the inputs is "single-photon", as would be the case in the Grangier-Roger-Aspect experiment. We remark that the linearity property holds always in the beam-splitter case, but that it holds only to first-order perturbation approximation in the nonlinear-crystal case. In the full quantum formalism, higher-order effects could, possibly, give an outgoing Wigner density taking negative values[9,10], but such effects are, at present, not experimentally observable.

5. The theory of detection

We have just seen that taking account of the zeropoint field leads us to a different understanding of certain optical devices. In particular, the recognition of the previously "missing" inputs, as in Figure 1 and Figure 2, means that the sum of the intensities of the outgoing signals is not equal to the intensity of just one incoming signal. This new feature of a zeropoint field theory is sufficient to take away the mystery of *enhancement* at a beam splitter. Applying this idea to a *polarizing* beam splitter, we have been able to show[23,24] that all the "nonlocal" data for polarizations of light signals from atomic cascades[25] have a local explanation. However, it is now necessary to modify the theory of detection. All previous semiclassical theories[26] have omitted the zeropoint field, and it has been assumed that the detection probability is proportional to the signal intensity. Since the total intensity in all the zeropoint modes is enormously greater than any signals, it must follow that all detectors are "blind", or nearly so, to the zeropoint intensity. This must be so even when the "signal" is the light from the Sun and the detectors are our own eyes!

The subtraction of the zeropoint noise is, we claim, already a feature of the standard theory, in which *light detectors are considered to be normal-ordering devices*[10]. The probability of joint detection in the two outgoing channels of Figure 1 is given by[5]

$$\Pr[\text{joint detection}] = \eta_b \eta_c \int_0^\tau dt \int_0^\tau dt' \langle N[\hat{I}_b(t)\hat{I}_c(t')]\rangle, \qquad (14)$$

where N denotes that the field amplitudes in $\hat{I}_b$ and $\hat{I}_c$ are normally ordered, and η_b, η_c are the detector efficiencies. We have shown that an equivalent expression is

$$\Pr[\text{joint detection}] = \eta_b \eta_c \int_0^\tau dt \int_0^\tau dt' \langle S[\{\hat{I}_b(t) - I_0\}\{\hat{I}_c(t') - I_0\}]\rangle, \qquad (15)$$

where S denotes a symmetric ordering of the field amplitudes, and I_0 is the zeropoint intensity in the relevant modes. This enabled us to replace the whole expression by an integral[5], over the classical phase space, involving the (positive) Wigner density, and hence give a purely wave explanation of the anticorrelation of the photoelectron counts in these channels. We have made a similar analysis[22] of the correlated signals in Figure 2, and hence have been able to give a purely wave explanation of the experiments (some of which the authors describe as "mind boggling") described in Reference[19].

It remains a problem to explain how this formal subtraction of the zeropoint is actually achieved by the detectors. There must exist some positive functional of the field amplitudes whose average value, weighted by the Wigner density, is very small when only the zeropoint is present. Note that real detectors all have a finite dark rate, so the zeropoint will always give *some* detection events; hence the theory of detection we are demanding will actually explain more than the current theory.

In lieu of such a theory we constructed a simpler model theory[23,24] in order to illustrate how the noise subtraction, in combination with the enhancement mechanism described in the previous section, gave rise to certain "particle-like" counting statistics in the two channels.

The explanation of how detectors are able to extract signals from the very large zeropoint background is a very difficult problem which we have not yet managed to solve. The day that theoretical physicists begin seriously to confront it is when they will begin, perhaps, to recover the respect of the rest of the scientific community. Despite our failure to resolve it, we state our conclusions, namely that the photon is obsolete, that light is nothing but waves, and that all wave fields fluctuate (see the opening sentence of Reference[4]). The next section is nothing but a postscript to this conclusion.

6. The microwave field in a cavity

Proposals have been made for the construction of experimental situations resembling the Schroedinger cat[27,28], in which two quite large objects, namely a Rydberg atom and a microwave cavity, are put in a superposition state, so that certain of their properties are "entangled".

Since it is not, at present, possible to observe directly the state of the cavity, this entanglement (if it really existed!) could not be demonstrated so readily as would be the case, with "perfect" detectors, for entangled light signals. Hence, any serious attempt to construct a Schroedinger cat must either seek to entangle two successive Rydberg atoms passing through the same cavity[29] or make some plausible additional assumption about the single Rydberg atom. We have taken the latter course[30], since we are sure that it leads to experimental requirements which are easily achievable with current technology. The additional hypothesis we propose is that the two-state stochastic process represented by the Rydberg atom be *stationary*. With this condition the Rydberg atom may be treated as a macroscopic system - it is bigger than a protein molecule - and the inequalities for such systems, deduced by Leggett and Garg[31] (also called temporal Bell inequalities) should apply. With the very high Q-values claimed by experimenters for the cavity, it should be possible, according to current theory, to demonstrate a violation of the Leggett-Garg inequality, but our analysis of the data[32] so far available shows no such violation. In the light of our experience with atomic cascades, one should be modest about the conclusions one draws. Hypotheses which seem plausible before doing an experiment should, properly, often be rejected in the light of the new evidence. This was our experience with Clauser and Horne's[33] hypothesis of no enhancement at a beam splitter. For the moment our inclination is to persist with the stationarity hypothesis for the Rydberg atom. It seems to us highly probable that the Q-values currently claimed for supercooled cavities may not take fully into account all the possible relaxation mechanisms for the radiation in the cavity, and it would not take much relaxation to convert the "ideal" quantum electrodynamic autocorrelation of the process into one satisfying the Leggett-Garg inequalities.

We wish Jean-Pierre Vigier a very happy birthday.

Acknowledgement

We acknowledge financial assistance of DGICYT Project No. PB-92-0507 (Spain).

References

1. T. W. Marshall, Found. Phys. **22**, 363 (1992).
2. R. E. Slusher and B. Yurke, Sci. Amer. **258**, 32 (1988).
3. E. Giacobino, C. Fabre, A. Heidmann and S. Reynaud, La Recherche **21**, 170 (1990).
4. R. Loudon and P. L. Knight, J. Mod. Opt. **34**, 709 (1987).
5. J. Peřina, *Quantum Statistics of Linear and Nonlinear Optical Phenomena* (Reidel, Dordrecht,1984).
6. T. W. Marshall and E. Santos, Found. Phys. Lett. **5** ,573 (1992).
7. T. W. Marshall, E. Santos and A. Vidiella-Barranco, *Proc. 3rd Int. Workshop on Squeezed States and Uncertainty Relations* (D. Han, et al eds.) (NASA Conf. Series, No. 3270, 1994) page 581.
8. T. W. Marshall and E. Santos, Phys. Rev. A **41**, 1582 (1990).
9. T. W. Marshall, Phys. Rev. A **44**, 7854 (1991).
10. P. Kinsler and P. D. Drummond, Phys. Rev. A **44**, 7848 (1991).
11. P. Milonni, *The Quantum Vacuum* (Academic, San Diego,1993) page 142.
12. T. H. Boyer, in *Foundations of Radiation Theory and Quantum Electrodynamics*
(A. O. Barut, ed.) (Plenum, New York, 1980).
13. L. de la Peña, in *Stochastic Processes Applied to Physics and Other Related Fields*
(B. Gomez, S. M. Moore, A. M. Rodriguez-Vargas and A. Rueda, eds.) (World Scientific, Singapore, 1983).
14. T. H. Boyer, Sci. Amer. August 1985.
15. J. Dalibard, J. Dupont-Roc and C. Cohen-Tannoudji, J. Phys. (Paris) **43**, 1617 (1982).
16. M. Planck, *Theory of Heat Radiation* (Dover, New York, 1959).
17. R. L. Hudson, Rep. Math. Phys. **6**, 249 (1974).
18. F. Soto and P. Claverie, J. Math. Phys. **24**, 97 (1983).
19. D. M. Greenberger, M. A. Horne and A. Zeilinger, Phys. Today **46**, 22 (1993).
20. T. W. Marshall, in *Fundamental Problems in Quantum Physics* (M. Ferrero and A. van der Merwe, eds.) (Kluwer, Dordrecht, 1995).
21. P. Grangier, G. Roger and A. Aspect, Europhys. Lett. **1**, 173 (1986).
22. A. Casado, T. W. Marshall and E. Santos, preprint Univ. de Cantabria FMESC 3 (1995).

23. T. W. Marshall and E. Santos, Found. Phys. **18**, 185 (1988).
24. T. W. Marshall and E. Santos, Phys. Rev. A **39**, 6271 (1989).
25. A. J. Duncan and H. Kleinpoppen, in *Quantum Mechanics versus Local Realism*
(F. Selleri, ed.) (Plenum, New York, 1988).
26. L. Mandel, Prog. in Optics **13**, 27 (1976).
27. E. Schroedinger, Proc. Camb. Phil. Soc. **31**, 555 (1935).
28. S. Haroche, M. Brune, J. M. Raimond and L. Davidovich, in *Fundamentals in Quantum Optics* (F. Ehlotzki, ed.) (Springer, Berlin, 1993).
29. S. J. D. Phoenix and S. M. Barnett, J. Mod. Opt. **40**, 979 (1993).
30. S. F. Huelga, T. W. Marshall and E. Santos, Phys. Rev. A, Rapid Communications, in press.
31. A. J. Leggett and A. Garg, Phys. Rev. Lett. **54**, 587 (1985).
32. G. Rempe, H. Walther and N. Klein, Phys. Rev. Lett. **58**, 353 (1987).
33. J. F. Clauser and M. A. Horne, Phys. Rev. D **10**, 526 (1974).

COULD THE PHOTON BE SUPERFLUOUS?

MENDEL SACHS
Department of Physics, State University of New York at Buffalo
Buffalo, N.Y. 14260

1. Introduction

Max Planck's discovery of the quantum of electromagnetic radiation - the *photon* - and its use to explain phenomena according to the 'old quantum theory' led to a revolutionary stance in physics - the 'new quantum theory' - led primarily by Niels Bohr and Werner Heisenberg.

In the early decades of the twentieth century Einstein discovered that the photoelectric effect could be described with Planck's quanta of monochromatic electromagnetic radiation, with energy $E = h\nu$, with seemingly localized particle-like properties, while under different sorts of experimentation, they appeared to be a strictly wave-like phenomenon. This led Einstein to pronounce the concept of *wave-particle dualism* for radiation. Two decades later Louis de Broglie extended this dualistic concept to bodies with inertial mass, such as electrons. The wave-like properties of electrons were subsequently observed in electron diffraction studies.

The concept of *wave-particle dualism* was then extended to be the basis for a new revolution in physics - quantum mechanics - for the microdomain of matter. The philosophical basis of this theory was expounded by Bohr, in terms of his *principle of complementarity* - a pluralistic view in which logically exclusive propositions are accepted as mutually true, provided that the consequences of each are verified under different sorts of circumstances of measurement. This view in physics, in turn, is compatible with the epistemological stand of *logical positivism*.

The focus of this essay is on the role of the *photon* in modern physics. Even though its introduction in the early years of this century, first from Planck's discovery and its role of the explanation of 'blackbody radiation', then to its role in terms of the quantized energy levels of atoms, in Bohr's model, and the explanation for the emission and absorption of quantized radiation by atomic matter, then to the discovery of Bose and Einstein (independently) of the statistics of such a gas of photons, at an equilibrium temperature, and finally to the role of photons in relativistic quantum field theory applied to electrodynamics (*quantum electrodynamics*), were eminently successful from an empirical point of view, would it be possible to exorcise the photon altogether from explanations of these data? In the remainder of this chapter, I will discuss reasons for the advantages of doing this, though at the expense of abandoning many currently held views.

2. The History of Quantum Theory

Let us recall the significance of the sequence of developments in the history of the present day quantum theory. First, nonrelativistic quantum mechanics - Schrödinger's 'wave mechanics' or Heisenberg's 'matrix mechanics' - was outstandingly successful in its predictions of the empirical facts concerning micromatter, at sufficiently low energy - the nature of molecules, atoms, nuclei and elementary particles. But it was recognized (historically, before nonrelativistic quantum mechanics was formulated) that such a theory of micromatter must *necessarily* be expressed in a way that would be compatible with the rules of Einstein's theory of special relativity.

The reason for this necessity is, in my view, that the role played by the quantum jump in quantum mechanics is foundational. A basic ingredient of the quantum theory is an *unbreakable* triad: 'emitter-signal-absorber'. The emitter and absorber components of this triad have a nonrelativistic limit in their mathematical description. That is to say, one can always find a frame of reference that is at rest with respect to them. This has to do with the fact that these components of the fundamental triad have inertial mass associated with them. On the other hand, the 'signal' component of the triad (the 'photon') does not have any nonrelativistic limit in its description - there is no frame of reference that is at rest with respect to the 'signal'! This was an important feature of Einstein's theory of relativity that he discovered in its early development. That is, it was found that light always propagates at the speed c, with respect to any frame of reference, if the laws of nature (initially, Maxwell's equations for electromagnetism) are to remain covariant with respect to transformations to any arbitrary inertial frame.

It follows from the foregoing reasoning that the quantum theory must be described at the outset in a way that is compatible with the rules of the probability calculus that defines this theory, and the rules of special relativity (and eventually general relativity) theory. The unbreakable triad, 'emitter-signal-absorber' must be represented this way initially in the general form of the theory. After this would be done, one may then take the nonrelativistic limit of those components of the triad associated with 'emitter' and 'absorber', to smoothly recover the form of nonrelativistic quantum mechanics, as $v/c \to 0$. The latter formalism would then be a mathematical approximation for the truly relativistic theory that unifies the matter components of the triad with its signal component (the 'photon'). The latter theory is *quantum electrodynamics*.

The first step in relativizing nonrelativistic quantum mechanics was taken by Paul Dirac, who found that a necessary generalization would be to proceed from Schrödinger's formalism for wave mechanics, in which the solutions are scalar, complex functions of the space and time coordinates, to complex, two-component spinor functions of the space and time coordinates, in order to satisfy the requirement of relativistic covariance. Dirac then extended this form, in terms of two coupled two-component wave equations (sometimes referred to as 'Majorana's equation') to the single four-component 'bispinor' equation for wave mechanics. The coupled two-component spinor equations reduce to the single bispinor equation be removing the reflection asymmetry from the elements of the underlying Poincaré group of special relativity.

Dirac's later generalization of Schrödinger's wave mechanics to the bispinor form led to an improvement regarding the predictive power of wave mechanics, both for bound state systems and systems that entail scattering. However, this was not yet a complete relativistic generalization because the electromagnetic potential in Dirac's bispinor equation was still treated classically, rather than in terms of quantized photons, that are supposedly created (by emitting matter when it loses energy) and annihilated (by absorbing matter when it gains energy). Thus, Dirac's version of wave mechanics in terms of spinor variables is indeed a 'semi-classical' theory.

The problem of completing the relativistic generalization of quantum mechanics was then taken up by Dirac, in his formulation of *quantum field theory*. This theory entails the introduction of photons to represent the electromagnetic coupling between the 'emitter' and the 'absorber', wherein the latter two components of the 'emitter-signal-absorber' triad, as well as the 'signal'(photon) are expressed as linear operators that create and annihilate matter and radiation - operators that are simulated from the classical form of the electrodynamic interaction.

3. The Formulation of Quantum Electrodynamics

How does one formulate the final expression of relativistic quantum field theory in a closed form? This has never been done! Instead, what one does is to express the solutions for the coupling of matter to radiation, electrodynamically, in the form of a perturbation series, wherein the first term of the series is the unperturbed 'free field'. The second term of the series represents the matter fields coupled to a single photon, the third terms represents the coupling of the matter fields to two photons wherein all possible couplings are taken into account, including those that entail a particle of matter acting on itself via the photon fields, the latter are the 'self-energy' terms.

In this way, a 'solution' of the theory is expressed in terms of an infinite series of terms, with each term expressing a higher order in the perturbing electrodynamic interaction with matter. Thus, when the photons are involved in the coupling of matter fields, the nth term of the series solution depends on α^n, where $\alpha = e^2/\hbar c$ is the 'fine structure constant', whose magnitude is the order of 1/137. We see, then that these coefficients in the perturbation series become increasingly small as n becomes increasingly large. The trouble with this mathematical scheme is that each of the terms in the series entails the product of a small coefficient multiplied by a factor that becomes greater quicker than α^n becomes small, with increasing n. Thus, the series of terms that are to represent the coupling of the emitter to the absorber diverge. The implication is that any predicted quantity from this solution, such as the charge and the mass of an electron, diverge to infinity. One must then conclude that the infinite series of terms considered do not in fact represent a real 'solution' of the quantum electrodynamical problem.

It must then be concluded that at this particular stage of the development of the quantum theory, it failed! This is because 1) there is no closed form expression for the relativistic quantum field theory of 'absorber-signal-emitter', and 2) there were no finite solutions demonstrated, even if such a closed form of the theory should exist. At this stage, then, one may not say that nonrelativistic quantum mechanics is a mathematical approximation for a relativistic quantum theory, since the latter has not been shown to exist!

About twenty years after the discovery of this 'failure' of the quantum theory, 'renormalization' techniques were discovered that provide a mathematical recipe for subtracting the infinities of the divergent series away (with other infinities), so as to yield finite answers for the predictions of physical properties of elementary matter, such as the electron. Nevertheless, there is a trouble here in that such a set of rules for the determination of finite values of the properties of elementary matter is not demonstrably mathematically consistent. That is to say, while some of these numbers may (and did!) lead to remarkably close agreement with the empirical data, it should be possible, in principle, to predict other, different values for the same physical situations. The latter could follow from cutting off the divergent series in different ways, and then subtracting the remainders anew - thereby yielding different predictions for the same physical phenomena that had been successfully obtained from the first recipe!

Even if the latter 'renormalization' could be carried out in a mathematically consistent way, there still remains the more serious problem in my judgement, that there is not yet a closed form solution for the coupling of the emitter and absorber to the signal (photon) that mediates them in the electrodynamic interaction. The latter closed form should go smoothly into the forms of the solutions of the equations of nonrelativistic quantum mechanics, as $v/c \to 0$.

The foregoing argument then says that at the present stage of physics, nonrelativistic quantum mechanics, which in itself is mathematically sound, as a nonrelativistic approximation, is likely an approximation for *some* relativistic theory, but there is no reason to believe that it is based on any of the concepts of the Copenhagen/Göttengen schools - concepts of indeterminism, linearity and a probability calculus as foundational, as well as the irreducible subjectivity in the acausal interpretation of the measurement process.

The two major numerical successes of quantum electrodynamics were its predictions, to high accuracy, of the Lamb shift and the anomalous magnetic moment of the electron. Both of these effects are not even predicted qualitatively from Dirac's semiclassical spinor theory. In the case of the bound states of hydrogen, Dirac's theory predicts a certain degeneracy in its energy levels, that quantum electrodynamics in contrast predicts to be lifted. This prediction of quantum electrodynamics led Willis Lamb and his coworkers to an experimental verification of the theory. In theory, the effect results from the coupling of the hydrogenic electron to the 'physical vacuum' - the infinite sea of photons and electron-positron pairs, acausally transmuting into each other and, in turn, acting on the bound hydrogenic electron. The theory predicts that these 'radiative effects' shift the Dirac energy levels of hydrogen that were degenerate, by different amounts, thus leading to the observation of extra energy levels. The numerical prediction for this effect follows from the expression of the infinite perturbation series for the hydrogenic electron, properly renormalized, giving close agreement with the experimental data.

The second successful numerical agreement with experimentation is an extra contribution to the magnetic moment of the free electron, beyond its Dirac value. This is due to the coupling of the electron to the background sea of photons and electron-positron pairs, as in the case of the hydrogenic electron. This prediction from quantum electrodynamics of the 'anomalous magnetic moment' of the electron was also in remarkable agreement with the experimental measurement of this effect.

In spite of these agreements between theory and experiment, it is still true that there has not been demonstrated any closed form for the expression of quantum mechanics, fully compatible with the rules of the theory of special relativity so as to include the radiation with the matter ('emitter' and 'absorber') both compatible with the rules of the quantum theory. Such a finite, consistent theory - *quantum electrodynamics* - has never been shown to exist.

4. Delayed-Action-At-A-Distance: All particles, No Fields

One very interesting approach to electrodynamics, proposed in the 1940s by John Wheeler and Richard Feynman, had the prospect of resolving the problem of *quantum electrodynamics*, after their theory would be quantized. With their theory, one dispenses with the photon participation altogether in the electrodynamic coupling between the emitter and absorber. It is replaced with the generalization of the 'retarded potential' for the expression of the signal between them.

The 'retarded potential' is conventionally used to explain the coupling of the physical 'emitter' (earlier) to that of the 'absorber' (later), where 'earlier' and 'later' refers to emitting the signal and absorbing the signal, respectively. The generalization of Wheeler and Feynman was to add the 'advanced potential' to the 'retarded potential', symmetrically.

The 'advanced potential', while being a bona fide solution of the Maxwell field equations, seems to imply a violation of causality because, here, the 'absorber' that we associate with 'effect', experiences the signal earlier and the 'emitter', associated with 'cause', experiences the signal later. That is to say, this seems to be a sequence of events in which an 'effect' precedes a 'cause'! There are no free photons in this theory. There is only the coupled interaction by way of the electromagnetic potential, propagating between the 'emitter' and the 'absorber'.

The retarded potential for the emitter follows as a solution of Maxwell's equations, in the following form of d'Alembert's equation:

$$\Box A_\mu^{(e)} = 4\pi j_\mu^{(e)} \tag{1}$$

where units have been chosen with $c = 1$ and where $\Box = \partial_o^2 - \nabla^2$ is the 'wave operator'. The solutions of this equation are a sum of the homogeneous part, representing the photon contribution, and the particular part, for the non-zero source. In the Wheeler-Feynman theory, the homogeneous part of the solution is rejected as incompatible with the starting premises of the theory. The particular solution of eq. (1) has two forms — the 'retarded' and the 'advanced' parts. The retarded solution has the form:

$$A_\mu^{(e)}(x) = \int j_\mu^{(e)} G_-(x' - x) d^4x' \tag{2}$$

In this solution, G_- is the 'retarded' Green's function for eq. (1):

$$G_-(x' - x) = |\vec{r}' - \vec{r}|^{-1} \delta[t' - t) - |\vec{r}' - \vec{r}|] \tag{3}$$

where δ is the Dirac delta function and $x \equiv (\vec{r}, t)$ is a four-dimensional spacetime point. The use of this function in the expression of the interaction implies that it propagates from $\vec{r} \rightarrow \vec{r}'$ in a positive time $(t' - t)$, at the speed of light, i.e. $|\vec{r}' - \vec{r}| = (t' - t) > 0$.

The 'advanced' solution of eq. (1) is equally valid; it has the same form as eq. (2), except that the retarded Green's function is replaced with the advanced Green's function:

$$G_+(x' - x) = |\vec{r}' - \vec{r}|^{-1} \delta[(t' - t) + |\vec{r}' - \vec{r}|] \tag{4}$$

The latter Green's function in the electromagnetic potential (2) implies that the interaction propagates from $\vec{r} \rightarrow \vec{r}'$ in the time $-(t' - t)$. That is to say, $|\vec{r}' - \vec{r}| = -(t' - t) < 0$.

Because of the linearity of the field equation (1), the average of the two Green's functions, (3) and (4), is also an admissible solution for the Green's function in the solution (2). Thus, Wheeler and Feynman take for the Green's function for this problem:

$$G(x' - x) = \tfrac{1}{2}(G_- + G_+) \tag{5}$$

That is, the solution of the field equation (1) is taken to be:

$$A_\mu^{(e)}(x) = \int j_\mu^{(e)}(x') G(x' - x) d^4x' \tag{6}$$

The interaction Lagrangian density for the 'emitter-absorber' is then given by the scalar coupling of the absorber four-density current, $j^{\mu(a)}$, to the electromagnetic potential (6) for the emitter, in the 4-space x:

$$\mathcal{L}_{int} = j^{\mu(a)}(x) A_\mu^{(e)}(x) = j^{\mu(a)}(x) \int j_\mu^{(e)}(x') \tfrac{1}{2}[G_-(x' - x) + G_+(x' - x)] d^4x' \tag{7}$$

This Lagrangian density then leaves no trace of 'photons' to describe the interaction between the emitter and the absorber. The total Lagrangian for n bodies then entails 4n degrees of freedom, $\{x_1, x_2, \ldots x_n\}$.

In a private conversation between this author and Richard Feynman, in the late 1950s, he expressed the hope that, upon quantization of their 'action-at-a-distance' theory, some modification might be found that would resolve the divergencies of quantum electrodynamics. One difficulty at the classical stage of their theory of 'emitter-absorber' is that, to utilize the advanced potential contribution, it is necessary to extend the model from a single atomic absorber to an infinite universe of absorbers. This conclusion, in turn, led to their need to entail statistics in the initially postulated purely electrodynamic problem. Feynman saw this requirement as a fundamental flaw of the theory.

Before going on to this author's resolution of the problem of electrodynamics, *also without photons*, a word should be said about the use of the 'advanced potential', vis à vis the symmetry requirements of the theory of relativity. It is my belief that the superposition of the advanced and retarded solutions symmetrically, as Wheeler and Feynman have done, is indeed *required* by the covariance restriction of relativity theory. My reasoning is as follows:

The retarded potential is associated with the emitter, from its own frame of reference - emitting a signal that propagates the interaction with the absorber at a finite time later, a distance away from it. But the principle of relativity requires that the emitter-absorber interaction must be expressed in a frame-independent manner. One must then be allowed to transform from the spacetime frame of the emitter to that of the absorber, without changing their law of interaction. In this case, what was formerly called by the name 'absorber' would now be called 'emitter', and vice versa. The electrodynamic interaction then propagates from what was formerly called 'absorber' to what was formerly called 'emitter', at the speed of light. From the frame of reference of the (formerly called) emitter, the latter is the action of the advanced potential on it. This argumentation then implies that in a fully covariant theory, it is *necessary* to represent the emitter-absorber interaction symmetrically, in terms of both the retarded and the advanced potentials. Thus, from any frame of reference, the interaction of the emitter and absorber must be represented in terms of the simultaneous propagation of the direct coupling between the emitter and absorber, in both directions - such as two trains simultaneously traveling between New York and Buffalo, each at the same speed in opposite directions.

The latter view fits in with the change of the concept of 'time', from the classical view to that of relativity theory. With the classical view, 'time' parameterizes the change from 'cause' to 'effect'. If indeed the physical causes and effects are absolute manifestations of a system of matter (i.e. not relative to any observer) then the time measure must be absolute as well. But this is not how we understand 'time' in the theory of relativity. Instead, with this view, 'time' is a relative measure, that is a function of the reference frame in which it is used to express a law of nature, covariantly. Thus, the cause and effect are *relative* according to the theory of relativity, as discussed above in regard to their connection to the advanced and retarded potentials - as a consequence of the *subjectivity* of the measure of time. In this case, there is no need for the free photon to represent the emitter-absorber interaction. Wheeler and Feynman used this idea in representing a system of charged particles, without any photons.

5. All Matter Fields: No Discrete Particles or Photons

The 'delayed-action-at-a-distance' theory that I have formulated, based on general relativity theory, does not suffer from the difficulties that Wheeler and Feynman encountered in their 'photon-less' approach of 'delayed-action-at-a-distance'. This author's theory is finite from the outset (as is the theory of Wheeler and Feynman, before their extension to a quantized version). In contrast with the quantum theory of electrodynamics, this author's theory is deterministic, nonlocal, nonlinear and a continuum field theory of matter. It is similar to the view of Wheeler and Feynman only in that the photon concept becomes superfluous.

One of Einstein's suggestions for any attempted formulation of a unified field theory that would maximize its predictions was to pay close attention not only to the geometrical logic of spacetime in its most general form, but also to express the algebraic logic of spacetime in its most general form. I have found that in this most general algebraic expression, the 'photon' concept automatically becomes superfluous. This conclusion follows for the following reason.

The algebraic logic that Einstein refers to is in reference to the underlying symmetry group in terms of its *irreducible representations*. The irreducible representations of the 'Einstein group' (the symmetry group that underlies general relativity theory) or those of the 'Poincaré group' (the symmetry group that underlies special relativity theory) are two-dimensional matrix sets, obeying the algebra of quaternions, whose basis functions are two-component spinor variables. The implication, then, is that the vector representation of the Maxwell field equations for electromagnetism, in special or general relativity, should be factorizable to a pair of two-component spinor field equations. This new, spin-½ representation for the law of electromagnetism should then incorporate all of the physical predictions of the *reducible* spin-one (vector) form of the theory, but it should make additional predictions that are outside of the scope of the vector representation of the theory.

The reason that the Maxwell, spin-one form of the law of electromagnetism factorizes to a spinor form is precisely the same reason that the scalar, spin-zero Klein Gordon equation factorizes to the spinor, Dirac equation. It is because, in both cases, the reflection symmetry elements are removed from the underlying symmetry group. This removal yields the most general form of the group representations, because the groups of relativity are with respect to relative motions alone - these are a set of *continuous transformations* that defines the covariance of laws in relativity theory. [Dirac's further elimination of the reflection-odd terms in the factorized version of the Klein-Gordon equation, the spin-½ coupled equations for quantum mechanics in special relativity, was accomplished by going to a bispinor form - a single, four-component spinor function, constructed from the initial two two-component spinor functions. The latter generalization, to eliminate the reflection-odd terms, is not required by the theory of relativity, since the latter's algebraic logic is in terms of *continuous groups*.]

The next question that arises is: If the basic variable for electromagnetism is a spin-$\frac{1}{2}$ field, so that the photon is eliminated from the description, since it is a spin-one particle, how does one explain anew all of the physical consequences previously explained in terms of photons, in modern physics? My answer is as follows:

The approach I take to the laws of physics, *in any domain*, is that they follow, fundamentally, from the premises of the theory of general relativity. With this view, and before one even proceeds to a curved spacetime geometry, there are implications that must hold even in the special relativity limit of the theory. Firstly, the theory is based on the continuous field concept to represent the physics of a closed system, from the outset. It follows from this feature of the philosophy of general relativity theory that the constituent matter fields for the closed system are the solutions of coupled, nonlocal, nonlinear field equations. That is to say, for an 'n-body system', one has n-coupled nonlinear spinor matter field equations wherein all of the spinor solutions, $\{\psi^{(1)}(x), \psi^{(2)}(x), .. \psi^{(n)}(x)\}$, are mapped onto a single four-dimensional spacetime x - there are no separate trajectories of things. Thus the theory is 'nonlocal'.

The correspondence principle, however, requires that in the limit of uncoupling, the field solutions $\psi^{(i)}$ become independent of each other, so that in this limit they are each mapped in their own spacetimes. This limit would correspond with the formulation of the 'action-at-a-distance' theory of Wheeler and Feynman. The latter limit of uncoupling in this theory corresponds precisely with the formal, Hilbert space structure of quantum mechanics. But the general theory of the matter fields in this view does not at all correspond with the structure of quantum mechanics because it is nonlocal, nonlinear and deterministic, in a curved spacetime. The theory of matter that gives back the formal structure of quantum mechanics, in a linear limit, has been found to be a generally covariant field theory of inertia. The electrodynamic interaction in this theory entails, *in part*, the same forms as shown in eqs. (1) - (7), with the current density forms replaced with bilinear forms in the Dirac four-component spinors for the matter fields, $j_\mu(x) = e\bar{\psi}(x)\gamma_\mu\psi(x)$ (or the corresponding terms in the two-component spinor form). There are additional parts to the interaction Lagrangian here that are odd under reflections in space and time, as well as contributions from the Lagrangian whose variation yields the spinor form of the electromagnetic equations.

Most experimental observations that are supposed to entail photons in their interpretation, also entail matter fields, breaking down to the coupled 'emitter' and 'absorber' components. As with the Wheeler-Feynman model in the classical stage, the electrodynamic interaction of the emitter and absorber does not require the participation of 'photons'. Instead, the potential is generalized from the retarded form to the symmetric sum of the retarded and the advanced potentials - though, in this approach in general relativity, this is in terms of nonlocal, nonlinear matter fields of a *closed system*.

Nevertheless, there are two physical situations that are conventionally interpreted in terms of the presence of photons without the participation of matter fields. One of these is the set of data associated with 'pair annihilation'. The other are the facts that explain the spectrum of blackbody radiation.

One feature of this theory that was found to explain both of these phenomena without the need of photons is the following: There was an exact solution found for the ground level of the bound electron and positron (or proton and antiproton) - a solution of their coupled, nonlinear spinor field equations. With the use of the formalism that emerges from Noether's theorem, this solution was found to correspond with null-energy, null momentum and null angular momentum, in all Lorentz frames (in the special relativity limit of the theory). That is to say, this 'truly' ground state of the pair has energy that is $2mc^2$ below the state where they would be (almost) free of each other

With the conventional quantum mechanical model, the ground state of the electron-positron pair ('positronium') is only a few electron-volts below the state where they would be free. In this author's nonlinear field theory the discovered ground state of the electron pair is, instead, $2mc^2$ below the free particle state. This would be around 1 Mev for the electron pair, and around 2Gev for the proton-antiproton pair.

It should be emphasized that the 'ground state' of the electron or the proton pair does not correspond with physical annihilation of matter. That is, these matter fields still exist - they would, for example, weigh 2mg near the Earth's surface. Further, the pairs would still be gravitationally sensitive to other matter. Indeed, a sea of such pairs in their true ground states could play the role of the 'dark matter' that is evoked in contemporary astrophysics, for example to explain the rotations of the galaxies.

Finally, the separate components of a given pair, in this ground state, must still couple, electromagnetically, to other charged matter, causing the pairs to become excited to other of their bound states above their ground state. Indeed, the transfer of around 1 Mev to an electron-positron pair in its ground state would cause it to ionize, revealing the separate electron and positron components, say in a cloud chamber. The latter event is seen experimentally, and interpreted as 'pair creation'. But there is no real creation of matter from a vacuum in this theory. Rather, when the pair is in its ground state it does not transfer energy and momentum to its surroundings - thus it is then 'invisible'. When it is ionized, the separate matter components are able to transfer energy and momentum to their surroundings, then making them visible.

What was found from this solution for the ground state of the pair is that it is in the form of two polarized currents, correlated with a 90° phase difference. The particle and antiparticle's currents are polarized in a plane that is perpendicular to the direction of propagation of their interaction with the charged matter of measuring apparatuses, on either side of their location, each current transferring energy equal to mc^2 to the respective apparatuses. This is indeed what is found experimentally and interpreted in terms of the two photons that are supposedly created when a pair annihilates. Yet, there are no photons in this interpretation of these data, in terms of the basis of general relativity theory.

The foregoing predictions then fully explain the experimental facts that are normally associated with 'pair annihilation', without the need of bona fide photons or any explanation in terms of a real annihilation (and creation)of matter and anti-matter fields.

With the foregoing results about pair annihilation from an exact solution of the matter field equations for their coupling, it follows that any region of space should be populated with some large density of such pairs in their ground states. The further analysis of a cavity full of pairs, in thermodynamic equilibrium with the walls, yields the Planck distribution function for the observed frequency spectrum, as a function of the equilibrium temperature. The use of Maxwell-Boltzmann statistics for such a gas of pairs was used - that is classical statistics, in which each of the pairs of the gas is 'tagged'. The result follows from the statistics of the interactions between these pairs in the cavity, in their 'true ground states', and the instruments that couple to them through a small window of the cavity, yielding the intensities as a function of the wavelengths of the contained gas of pairs. Thus, there is no need here to describe blackbody radiation in terms of the 'photon' model.

The assumption that all visible matter is embedded in a dense gas of pairs, in their newly discovered 'ground state', led to a number of other empirical facts from modern physics: 1) It correctly predicted the lifetimes of the excited states of atoms - the rate of decay depending on the coupling of the excited atoms to the sea of pairs in which it is embedded, 2) the inertial masses of observed elementary particle fields. The masses of the electron, muon and their respective antimatter counterparts have been determined quantitatively, thus far. In this case, the source of the inertial mass is the curvature of spacetime in its vicinity, which, in turn, is due to the density of pairs in which the observed matter is observed. This is an explicit manifestation of the Mach principle, wherein the inertial mass of any matter is a measure of its coupling to its total surroundings. It was found that the primary source that gives rise to the mass of an elementary particle are the electron and proton pairs that are in its immediate vicinity (a contribution that is much more significant than the distant stars - though the latter do contribute, infinitesimally!)

In the experiments that are supposed to entail photons as well as matter fields, in modern physics, the photon interactions of the conventional interpretation are replaced with direct interactions of matter fields, in the manner discussed earlier, using the symmetric sum of the advanced and the retarded potentials, thereby yielding the same results as those that evoke the photon model. Examples are the Compton effect, the photoelectric effect and bremsstrahlung.

Thus it is contended that this analysis, in terms of a deterministic, nonlocal, nonlinear spinor field theory of matter, that is fully compatible with the symmetry requirements of the theory of relativity, allows one to dispense with the need of the 'photon' - that the 'photon' becomes a superfluous concept. This is analogous to Einstein's discovery, in the early days of his theory of relativity, that the ether became a superfluous concept.

A bonus that comes with this paradigm shift - from the underlying concepts of the quantum theory to those of the theory of relativity with the elimination of photons, is that the infinities of relativistic quantum electrodynamics no longer appear and there is then no need for the renormalization scheme of calculation of the physical properties of elementary matter. This is indeed a distinct advantage since it reveals that the new theory of matter has bona fide finite solutions from the outset, thus that this theory is logically and mathematically consistent. Einstein's theory of relativity, as a theory of matter, is then a consistent theory in all domains, from the cosmological domain to that of elementary matter.

The 'photon-less' action-at-a-distance theory of Wheeler and Feynman is developed in: Wheeler, J.A. and Feynman, R.P. (1945) Interaction with the Absorber Mechanism of Radiation, *Reviews of Modern Physics* 17, 157 - 181; Wheeler, J.A. and Feynman, R.P. (1949) Classical Electrodynamics in Terms of Direct Interparticle Interaction, *Reviews of Modern Physics* 21, 425 - 433.

This author's research program that entails the 'photon-less' electrodynamics, also in terms of action-at-a-distance, but in the context of general relativity, is developed in: Sachs, M. (1986) *Quantum Mechanics from General Relativity - An Approximation for a Theory of Inertia*, Reidel Publishing Company, Dordrecht, Chapter 7.

The experimental confirmation of the dynamics of the two 'photons' that are correlated with pair annihilation was reported in: Wu, C.S. and Shaknov, I. (1950) The Angular Correlation of Scattered Annihilation Radiation, *Physical Review* 77, 136.

FLUCTUATION-DISSIPATION QUANTUM ELECTRODYNAMICS

G.F.EFREMOV, A.Y.CHEKHOV, L.G.MOUROKH, M.A.NOVIKOV
Radiophysics Department,
N.Lobachevsky State University,
Gagarin ave. 23,
603600 N.Novgorod, Russia

1. The central problem of the modern quantum theory of light is the investigation of electromagnetic processes in the different nonlinear media. The interaction with quantized electromagnetic field (photon thermal bath) is at once the fundamental mechanism of fluctuations and relaxation in the quantum systems and, thereby, this interaction defines the most considerable property of nonlinear media.

In this work we present the main thesises of fluctuation-dissipation quantum electrodynamics that makes possible the unified treatment of fluctuation and dissipative processes in the relativistic quantum systems. We shall obtain the master equations for variables of the electron-positron field interacting with heat electromagnetic one.

We use the approach based on the theory of open quantum systems [1,2] for our analysis. The basic idea of this theory is the separation of the system under study into two subsystem, one is dynamical with small number degrees of freedom and the other is macroscopic coupled with first and called the thermal bath. We consider quanta of excitation of fermion fields as a dynamical subsystem, interaction of which with photon thermal bath was turned on adiabatically in the infinitely distant time. In the initial moment dynamical subsystem can be in the arbitrary nonequilibrium state and the unperturbed photon thermal bath is in the thermodynamical equilibrium state with some temperature T.

Our main assumption implies that the unperturbed photon thermal bath potentials $A_j^0(\vec{r},t)$ and $A_0^0(\vec{r},t)$ have Gaussian statistics. It is true for free quantized electromagnetic field in the thermodynamical equilibrium state. The interaction of this field with electron-positron vacuum leads to specific nonlinear effects and the statistics of potentials becames nonGausian according nonlinear fluctuation-dissipation relations [3,4]. However, the chances of these nonlinear effects is negligible small and we shall consider photons as "good quasiparticle", i.e. the uncertainty of energy is much less than the full photon energy.

This assumption allows to circumvent the restrictions on the constant of interaction, distinguish times and frequences, to take into account the delay of interaction and the memory effects, and to investigate the interplay of relaxation and fluctuation processes.

2. We can write the full Hamiltonian of our system as

$$H = \int d^3\vec{r}\Psi^+(\vec{r},t)H_o\Psi(\vec{r},t) + F + H_{int}, \tag{1}$$

where

$$H_0 = c\vec{\alpha}(\vec{p} - \frac{e}{c}\vec{A}^{ex}(\vec{r},t)) + eA_0^{ex} + \beta mc^2$$

is Hamiltonian of the single Dirac electron, $\vec{A}^{ex}(\vec{r},t)$ and $A_0^{ex}(\vec{r},t)$ are the external potentials, Ψ^+ and Ψ are the creation and the annihilation operators of electron-positron field, α_j and β are the Dirac matricies, F is Hamiltonian of the photon thermal bath and H_{int} describes the interaction of dynamical subsystem with the thermal bath. We choose the Coulomb gauge for the thermal bath potentials:

$$div\vec{A}(\vec{r},t) = 0$$

. Canonically cojugate variables of dynamical subsystem may be found by taking derivatives of Hamiltonian with respect to $A_j^{ex}(\vec{r},t)$ and $A_0^{ex}(\vec{r},t)$

$$-\frac{\delta H}{\delta A_j(\vec{r},t)} = \frac{1}{c}J_j(\vec{r},t) \tag{2}$$

$$-\frac{\delta H}{\delta A_0(\vec{r},t)} = -\rho(\vec{r},t), \tag{3}$$

where

$$J_j(\vec{r},t) = ec\Psi^+\alpha_j\Psi \tag{4}$$

$$\rho(\vec{r},t) = e\Psi^+\Psi \tag{5}$$

Then we can write H_{int} as

$$H_{int} = -\frac{1}{c}\int d^3\vec{r}J_l(\vec{r},t)A_l(\vec{r},t) + \int d^3\vec{r}\rho(\vec{r},t)A_0(\vec{r},t) \tag{6}$$

It is shown in paper [5] that the time evolution of the thermal bath potentials has a following form if the unperturbed potentials $A_j^0(\vec{r},t)$ and $A_0^0(\vec{r},t)$ have Gaussian statistics:

$$A_j(\vec{r},t) = A_j^0(\vec{r},t) + \frac{1}{c}\int d^3\vec{r}_1\int dt_1 D_{jl}(\vec{r},t;\vec{r}_1,t_1)J_l(\vec{r}_1,t_1) \tag{7}$$

$$A_0(\vec{r},t) = A_0^0(\vec{r},t) + \int d^3\vec{r}_1 \int dt_1 D_{00}(\vec{r},t;\vec{r}_1,t_1)\rho(\vec{r}_1,t_1), \qquad (8)$$

where

$$D_{jl}(\vec{r},t;\vec{r}_1,t_1) = \langle \frac{i}{\hbar}[A_j^0(\vec{r},t), A_l^0(\vec{r}_1,t_1)]_-\rangle \eta(t-t_1) \qquad (9)$$

and

$$D_{00}(\vec{r},t;\vec{r}_1,t_1) = \langle \frac{i}{\hbar}[A_0^0(\vec{r},t), A_0^0(\vec{r}_1,t_1)]_-\rangle \eta(t-t_1) \qquad (10)$$

are the retarded Green functions of photon. Here $\eta(t-t_1)$ is the step function taking into account the causality principle automatically, $[...,...]_-$ is the commutator of operators, and $\langle ... \rangle$ means the average on the initial state.

In the case of Coulomb gauge $D_{j0}(\vec{r},t;\vec{r}_1,t_1) = 0$ and the interactions with vector and scalar potentials may be considered separately. In this work our concern will be only with vector potential because the scalar one does not lead to observable effects.

3. Let us write the equation for $\Psi(\vec{r},t)$ obtained from (1)

$$\dot{\Psi}(\vec{r},t) = \frac{1}{i\hbar}H_0\Psi(\vec{r},t) + e\frac{i}{\hbar}A_j(\vec{r},t)\alpha_j\Psi(\vec{r},t). \qquad (11)$$

After the substitution of the expression (7) this equation has a form

$$\dot{\Psi}(\vec{r},t) = \frac{1}{i\hbar}H_0\Psi(\vec{r},t) + i\frac{e}{\hbar}A_j^0(\vec{r},t)\alpha_j\Psi(\vec{r},t) +$$
$$i\frac{e}{\hbar c}\int dt_1 \int d^3\vec{r}_1 D_{jl}(\vec{r},t;\vec{r}_1,t_1)J_l(\vec{r}_1,t_1)\alpha_j\Psi(\vec{r},t) \qquad (12)$$

To eliminate the free thermal bath variables $A_j^0(\vec{r},t)$ we can use the quantum Furutsu-Novikov theorem [5]:

$$i\frac{e}{\hbar}\langle A_j^0(\vec{r},t)\alpha_j\Psi(\vec{r},t)\rangle = i\frac{e}{\hbar}\int dt_1 \int d^3\vec{r}_1$$
$$M_{jl}(\vec{r},t;\vec{r}_1,t_1)\langle \frac{\delta \alpha_j \Psi(\vec{r},t)}{\delta A_l^0(\vec{r}_1,t_1)}\rangle, \qquad (13)$$

where

$$M_{jl}(\vec{r},t;\vec{r}_1,t_1) = \langle A_j^0(\vec{r},t), A_l^0(\vec{r}_1,t_1)\rangle \qquad (14)$$

is the correlation function of the thermal bath. In the thermodynamically equilibrium state this function and the retarded Green function (9) are interrelated by means of fluctuation-dissipation theorem. It is suitable to write the given theorem for the Fourier-transforms of these functions:

$$M_{jl}(\vec{r},\omega) = 2\hbar(1 - exp\{-\frac{\hbar\omega}{T}\})^{-1}Im D_{jl}(\vec{r},\omega) \qquad (15)$$

The functional derivation in Eq.(13) is determined as following [3]:

$$\frac{\delta \alpha_j \Psi(\vec{r},t)}{\delta A_l^0(\vec{r}_1,t_1)} = \frac{i}{\hbar}\frac{1}{c}[\alpha_j \Psi(\vec{r},t), J_l(\vec{r}_1,t_1)]_-\eta(t-t_1) \tag{16}$$

In addition we define the fluctuation source

$$\xi^J(\vec{r},t) = e\frac{i}{\hbar}A_j^0(\vec{r},t)\alpha_j\Psi(\vec{r},t) - e\frac{i}{\hbar c}\int dt_1 \int d^3\vec{r}_1$$
$$M_{jl}(\vec{r},t;\vec{r}_1,t_1)[\alpha_j\Psi(\vec{r},t), J_l(\vec{r}_1,t_1)]_-\eta(t-t_1) \tag{17}$$

As a result we obtain the master stochastic equations for the dynamic subsystem variables:

$$\dot{\Psi}(\vec{r},t) = \frac{1}{i\hbar}H_0\Psi(\vec{r},t) + i\frac{e}{\hbar c}\int dt_1 \int d^3\vec{r}_1$$
$$(M_{jl}(\vec{r},t;\vec{r}_1,t_1)\frac{i}{\hbar}[\alpha_j\Psi(\vec{r},t), J_l(\vec{r}_1,t_1)]_-\eta(t-t_1) +$$
$$D_{jl}(\vec{r},t;\vec{r}_1,t_1)J_l(\vec{r}_1,t_1)\alpha_j\Psi(\vec{r},t)) + \xi^J(\vec{r},t) \tag{18}$$

We do not use any restriction on the constant of interaction deriving this equation. Therefore, we can investigate the effects of quantum electrodynamics in the all order of the perturbation theory. Eq.(18) is nonMarkovian and takes into account the delay of the interaction and the memory of the thermal bath. Besides, the explicit form of the fluctuation source (17) makes possible to calculate its any order corelation functions. By this means we can find any statistical characteristics of electron-positron field and, therefore, can study fluctuations in the quantum electrodynamics. Because of this, the proposed approach has some advantages over traditional ones and gives a better insight into the nature of electromagnetic interaction.

4. In this section we point the way for the possible application of this approach. The obtained equations can be simplified by means of some additional assumptions. It is well-known that the anticommutator

$$\Psi(\vec{r},t)\Psi^+(\vec{r}_1,t_1) + \Psi^+(\vec{r}_1,t_1)\Psi(\vec{r},t)$$

is the c-number for the free electron-positron field. Moreover, the average of $\Psi^+(\vec{r},t)\Psi(\vec{r}_1,t_1)\Psi(\vec{r}_2,t_2)$ on the single-particle states and on the vacuum one is zero. By vacuum is meant a physical vacuum when all states with negative energy are occupied. The statistics of Fermi-operators will be considered Gaussian if these conditions are fulfilled in the case of electron-positron excitations interacting with photon thermal bath. The reason of it that this exitations are "good quasiparticles" too. We can obtain the stochastic equation linearized on the operators Ψ on the basis of these conditions:

$$\dot{\Psi}(\vec{r},t) = \frac{1}{i\hbar}H_0\Psi(\vec{r},t) + i\frac{e^2}{\hbar}\int dt_1 \int d^3\vec{r}_1 \alpha_l \Psi(\vec{r}_1,t_1)$$

$$(\frac{i}{\hbar}M_{jl}(\vec{r},t;\vec{r}_1,t_1) + D_{jl}(\vec{r},t;\vec{r}_1,t_1))G_j(\vec{r},t;\vec{r}_1,t_1) + \xi^J(\vec{r},t), \qquad (19)$$

where

$$G_j(\vec{r},t;\vec{r}_1,t_1) = i(\alpha_j\Psi(\vec{r},t)\Psi^+(\vec{r}_1,t_1) + \Psi^+(\vec{r}_1,t_1)\alpha_j\Psi(\vec{r},t))\eta(t-t_1) \quad (20)$$

is a c-number.

Under given conditions we can find easy the correlation functions of fluctuation source. Knowing these functions we can obtain self-consistent integral equation for the functions $G_j(\vec{r},t;\vec{r}_1,t_1)$ from Eq.(19) both in the thermodynamical equilibrium state and in the strong nonequilibrim one. The solution of this equation will govern the Lamb shift, radiation damping and other.

The problems of renormalization can be studied beyong the framework of the perturbation theory by means of obtained equations. It should be noted that all main properties are determined by the photon Green function. One can find the expression of this function taking into account fluctuations of electron-positron vacuum in [G.F.Efremov, L.G.Mourokh, and M.A.Novikov "Statistical theory of photon in the nonlinear media and in vacuum".]

References

Schwinger, J.,(1961) Brownian motion of the quantum oscillator, *J. Math. Phys.*, **Vol. no. 2**, p. 407

Senitzky, I.R. (1960) Dissipation in quantum mechanics. The harmonic oscilator, *Phys.Rev.*, **Vol. no. 119**, pp. 670–679

Efremov, G.F. (1968) Fluctuation-dissipation theorem for nolinear media, *Zh. Eksp. Teor. Fiz.*, **Vol. no. 55**, pp. 2322–2333

Bochkov, G.N., Kuzovlev Yu.E.(1981) Nonlinear fluctuation-disspation relations and stochastic models in nonequilibrium thermodynamcs, *Physica A*, **Vol. no. 106**, pp. 443–520

Efremov, G.F. and Smirnov, A.Yu.(1981) Contribution to the microscopic theory of the fluctuatios of a quantum system interacting with a Gaussian thermostat, *Zh. Eksp. Teor. Fiz.*, **Vol. no. 80**, pp. 1071–1086

STATISTICAL THEORY OF PHOTON IN THE NONLINEAR MEDIA AND IN VACUUM

G.F.EFREMOV, L.G.MOUROKH AND M.A.NOVIKOV
Radiophysics Department, N.Lobachevsky State University,
Gagarin ave. 23, 603600 N.Novgorod, Russia

1. In this work the statistical theory of electromagnetic processes in the nonlinear relativistic media and in vacuum is proposed. This theory is based on the rigorous microscopic derivation of nonlinear stochastic Maxwell equations.

We separate the system under study into two subsystem. The long-wave part of quantized electromagnetic field with small number degree of freedom will be considered as dynamical subsystem and the nonlinear medium will play the role of the dissipative environment (thermostat).

We start from the relativistic theory for the unified description of the nonlinear media and the quantized fields. The solution of our problem is based on approach proposed in [1]. The expression

$$A_0(\vec{r}, t) = 0$$

is chosen as a gauge, where A_0 is a scalar potential. Then the energy of interaction of the quantized field and the medium can be written as

$$\hat{H}_{int} = -\frac{1}{c} \int d^3\vec{r} \hat{J}_l(\vec{r}, t) \hat{A}_l(\vec{r}, t), \qquad (1)$$

where $J_l(\vec{r}, t)$ is a component of the current density operator of the dissipative subsystem.

In this case the rigorous microscopic equations for dynamical subsystem variables are the Maxwell equations:

$$rot\, rot\, \hat{\vec{A}}(\vec{r}, t) + \frac{1}{c^2} \frac{\partial^2}{\partial t^2} \hat{\vec{A}}(\vec{r}, t) = \frac{4\pi}{c} \hat{\vec{J}}(\vec{r}, t) \qquad (2)$$

The Maxwell equations, obtained from microscopic approach, mainly imply that the phenomenological material equations which relate mean current density $\langle \vec{J} \rangle$ and mean vector-potential of field $\langle \vec{A} \rangle$ are substituted by operator relation

$$\hat{\vec{J}} = \hat{\vec{J}}^0 + \frac{1}{c}\hat{\varphi} : \hat{\vec{A}} + \frac{1}{c^2}\hat{\varphi}^{(2)} : \hat{\vec{A}}\hat{\vec{A}} + \frac{1}{c^3}\hat{\varphi}^{(3)} : \hat{\vec{A}}\hat{\vec{A}}\hat{\vec{A}}, \qquad (3)$$

where $\hat{\vec{J}}^0(\vec{r},t)$ is the unperturbed operator of current density of the medium without interaction with long-wave field. Linear and nolinear random responses (susceptibilities) $\hat{\varphi}, \hat{\varphi}^{(2)}, \ldots$ are described by rigorous microscopic relations:

$$\hat{\varphi} = \varphi_{\alpha\beta}(\vec{r},t;\vec{r}_1,t_1) = \frac{i}{\hbar}[J^0_\alpha(\vec{r},J^0_\beta(\vec{r}_1,t_1)]_-\eta(t-t_1)$$

$$\hat{\varphi}^{(2)} = \varphi^{(2)}_{\alpha\beta\gamma}(\vec{r},t;\vec{r}_1,t_1;\vec{r}_2,t_2) = \qquad (4)$$

$$= (\frac{i}{\hbar})^2[[J^0_\alpha(\vec{r},t),J^0_\beta(\vec{r}_1,t_1)]_-,J^0_\gamma(\vec{r}_2,t_2)]_-\eta(t-t_1)\eta(t_1-t_2)$$

Here $\eta(\tau) = 1$ if $\tau > 0$ and $\eta(\tau) = 0$ if $\tau < 0$. The following notation are included in (3) to put it into more instructive form

$$\hat{\varphi} : \hat{\vec{A}} = \int dt_1 d^3\vec{r}_1 \hat{\varphi}_{\alpha\beta}(\vec{r},t;\vec{r}_1,t_1)A_\beta(\vec{r}_1,t_1) \qquad (5)$$

The fluctuations of medium responses (sucseptibilities) are first of all responsible for excess nonequilibrium field fluctuations. Also, The parametric medium effects on the field changes its dynamics and, consequentaly, causes extra radiation energy dissipation in the system.

We shall restrict our consideration to the quadratic and cubic terms in Eq.(3). According nonlinear fluctuation-dissipation theorems [2,3] small nonlinearity of medium leads to weak nonGaussian (or quasiGaussian) statistics of current density $J^0(\vec{r},t)$. In this case it is possible to separate in Eq.(3) the dynamical terms containing the random variables of field only and to define the fluctuation sources [1].

$$\hat{\vec{J}} = \frac{1}{c}\varphi : \hat{\vec{A}} + \frac{1}{c^2}\varphi^{(2)} : \hat{\vec{A}}\hat{\vec{A}} + \frac{1}{c^3}\varphi^{(3)} : \hat{\vec{A}}\hat{\vec{A}}\hat{\vec{A}} +$$

$$+ \frac{1}{c}\psi^{(1)} : \frac{\partial \hat{\vec{A}}}{\partial J^{ex}} + \frac{1}{c^2}\psi^{(2)} : \hat{\vec{A}}\frac{\partial \hat{\vec{A}}}{\partial J^{ex}} + \frac{1}{c}\psi^{(3)} : \frac{\partial^2 \hat{\vec{A}}}{\partial J^{ex}\partial J^{ex}} + \hat{\xi}^J, \qquad (6)$$

where φ and ψ are the determinate mean values. For example,

$$\varphi = \langle \hat{\varphi} \rangle = \langle \varphi_{\alpha\beta}(\vec{r},t;\vec{r}_1,t_1) \rangle =$$

$$= \langle \frac{i}{\hbar}[J^0_\alpha(\vec{r},t),J^0_\beta(\vec{r}_1,t_1)]_-\rangle\eta(t-t_1) \qquad (7)$$

We give the explicit expression for functions ψ below. One can find the rigorous microscopic expression of fluctuation source ξ^J and the method to calculate its correlation functions in [1].

After substitution Eq.(6) to Eq.(2) we have nonlinear stochastic Maxwell equations making possible to investigate the wide range of electrodynamical effects in the nonlinear media. Generalized material equations (6) contain the terms with $\partial \vec{\tilde{A}}/\partial J^{ex}$ derivatives which describe a qualitative new dynamics type resulted from parametric effect of random responses to the field amplitude.

It should be noted that the rigorous microscopic derivation of Maxwell equations for nonlinear media was absent in spite of considerable development of quantum optics.

2. In this section we use the generalized material equations (6) for calculation of photon Green function. In the nonlinear dissipative medium photon is a quasiparticle excited by the external source $J^{ex}(\vec{r},t)$. In particular, it is true for physical vacuum. By the definition photon Green function is the linear response of the field on the external current.

$$\langle A_j(\vec{r},t)\rangle = \frac{1}{c}\int dt_1 d^3\vec{r}_1 D_{jl}(\vec{r},t;\vec{r}_1,t_1) J_l^{ex}(\vec{r}_1,t_1)$$

$$D_{jl}(\vec{r},t;\vec{r}_1,t_1) = \langle \frac{i}{\hbar}[A_j(\vec{r},t), A_l(\vec{r}_1,t_1)]_-\rangle \eta(t-t_1) \qquad (8)$$

In the case of weak field we have the linearized Maxwell equations averaging Eqs.(2),(6):

$$rotrot\langle\vec{\tilde{A}}(\vec{r},t)\rangle + \frac{1}{c^2}\frac{\partial^2}{\partial t^2}\langle\vec{\tilde{A}}(\vec{r},t)\rangle = \frac{4\pi}{c^2}\varphi^r:\vec{\tilde{A}} + \frac{4\pi}{c}\vec{\tilde{J}}^{ex}(\vec{r},t), \qquad (9)$$

where

$$\varphi^r = \varphi + \delta\varphi = \varphi + \psi^{(2)}D \qquad (10)$$

is a renormalized linear response. The explicit expression for $\delta\varphi$ has a form

$$\delta\varphi = \delta\varphi_{jl}(\vec{r},t;\vec{r}_1,t_1) = \int dt_2 d^3\vec{r}_2 \int dt_3 d^3\vec{r}_3 D_{mn}(\vec{r}_3,t_3;\vec{r}_2,t_2)$$

$$(\langle\frac{1}{2}[\tilde{\varphi}_{jm}(\vec{r},t;\vec{r}_3,t_3),\tilde{\varphi}_{nl}(\vec{r}_2,t_2;\vec{r}_1,t_1)]_+\rangle +$$

$$+\langle[\tilde{\varphi}^{(2)}_{jlm}(\vec{r},t;\vec{r}_1,t_1;\vec{r}_3,t_3), J_n^0(\vec{r}_2,t_2)]_+\rangle), \qquad (11)$$

where $\tilde{\varphi} = \hat{\varphi} - \langle\hat{\varphi}\rangle$ is a random response with zero mean value. We rewrite Eq.(9) using the Fourier transformation

$$\frac{1}{4\pi}((k^2-\frac{\omega^2}{c^2})\delta_{jl}-k_j k_l)\langle A_l(\vec{k},\omega)\rangle = \frac{1}{c^2}\chi^r_{jl}(\vec{k},\omega)\langle A_l(\vec{k},\omega)\rangle + \frac{1}{c}J_j^{ex}(\vec{k},\omega) \quad (12)$$

We can find the inverse Green function from this equation:

$$D_{jl}^{-1} = \frac{1}{4\pi}((k^2 - \frac{\omega^2}{c^2})\delta_{jl} - k_j k_l) - \frac{1}{c^2}\chi_{jl}^{\tau}(\vec{k},\omega),\qquad(13)$$

where

$$\chi_{jl}^{\tau}(\vec{k},\omega) = \chi_{jl}(\vec{k},\omega) + \delta\chi_{jl}(\vec{k},\omega) = \int d\tau e^{i\omega\tau}\int d^3\vec{r} e^{-i\vec{k}\vec{r}}\varphi_{jl}^{\tau}(\vec{r},t;\vec{r}_1,t_1)\quad(14)$$

Fluctuation corrections to photon Green function change the photon dispersion dependence $\omega(\vec{k})$ and give rise to additional fluctuational mechanism of photon damping, in particular. These effects can be calculated by means of obtained equations, that is the subject of specific consideration. The obtained results are true for strong nonequilibrium states, for example, for quantum amplifiers. Of special interest is the investigation of fundamental properties of photon in vacuum. It is well-known that electromagnetic vacuum fluctuations contribute to radiation damping and define the Lamb shift. Fluctuations of electron-positron vacuum determine the photon dispersion law according Eq.(11). Besides, renormilized photon Green function (13) governs the fundamental characteristics of electron-positron excitations. For more deteiles, see [G.F.Efremov, A.Yu.Chechov, L.G.Mourokh, and M.A.Novikov "Fluctuation-dissipation quantum electrodynamics"]

References

Efremov, G.F., Kazakov V.A. (1979) Contribution to derivation of nonlinear equation with fluctuation parameters for open quantum subsystem, *Izv.Vuz.Radiofiz.*, **Vol. no. 22**, p. 1236

Efremov, G.F. (1968) Fluctuation-dissipation theorem for nolinear media, *Zh. Eksp. Teor. Fiz.*, **Vol. no. 55**, p. 2322

Efremov, G.F. (1972) Contribution to theory of heat fluctuations in the nonequilibrium media, *Izv.Vuz.Radiofiz.*, **Vol. no. 15**, p. 1207

EINSTEIN-DE BROGLIE-PROCA THEORY OF LIGHT AND SIMULTANEOUS EXISTENCE OF TRANSVERSE AND LONGITUDINAL PHOTONS

SISIR ROY AND MALABIKA ROY
Physics and Applied Mathematics Unit
Indian Statistical Institute
Calcutta 700 035
INDIA
e-mail : sisir @isical.ernet.in

Abstract. In Einstein-de Broglie-Proca Theory (EBP), non-zero rest mass of the photon in the Maxwell vacuum with $\sigma \neq 0$ has been shown to be consistent with gauge invariance of the first and second kind if we introduce the fourth component of the current as $J^0 \sim B_0$. Here B_0 is the magnetic flux density associated with photon. The non-commutating B-field implies the existence of uncertainty relation like $\delta \hat{B}^{(1)} . \delta \hat{B}^{(2)} \geq \frac{1}{2} \mid B_0 \hat{B}^{(3)} \mid$. The simultaneous measurability of $B^{(1)}, B^{(2)}$ and $B^{(3)}$ are discussed within the theory of unsharp measurement in Quantum mechanics.

Key words : EBP Theory, Non-zero photon mass, $B^{(3)}$ field, unsharp measurement.

1. INTRODUCTION

The studies of the force law between electric charges dates back to the middle of the 1700's.With the advancement of quantum theory the inverse square law has been viewed as a consequence of the masslessness of the gauge particle, the photon. Recently, $Vigier^{(1)}$ has reviewed a substantial amount of evidence which leads to the conclusion that the photon has non-zero rest mass. These data include, for example,the direction dependent anisotropy of the frequency of light in cosmology and frequently observed anomalous red shifts. If the photon has non-zero mass m_γ, the Maxwell equations are replaced by Proca equation

$$\Box A_\mu = -\xi^2 A_\mu$$

where,

$$\xi = \frac{m_\gamma c}{h}$$

This is usually known as the Einstein-de Broglie-Proca equation $(EBP)^{(2)}$. The potential four vector A_μ of the de Broglie-Proca field is manifestly covariant and has four physically meaningful components, one time like and three space-like, of which two are transverse and one is longitudinal.The above equation is an expression of the Einstein-de Broglie theory of light and implies that gauge transformations of the first and second kind can no longer be interpreted as implying photon of zero rest mass. A small photon mass may also be generated in the context of string theory$^{(3)}$. $Barut^{(4)}$ constructed exact localized oscillating finite energy solutions of the massless wave equation which move like massive relativistic particles.Essentially this construction is simply a single massive "quantum particle" from light. Another attempt has been made by $Vigier^{(5)}$ so as to extend the Einstein-de Broglie theory of light considering Maxwell equations in vacuum with a non-zero conductivity coefficient($\sigma \neq 0$).

The present author along with his collaborators $^{(6)}$ started a systematic investigation to look into the different aspects of this framework in which non-zero conductivity coefficient gives rise to a dissipative term in the field equation.If we consider the propagation of the photon through this type of vacuum,the photon acquires a mass on the cosmological scale which can be related to the conductivity coefficient. The importance of this approach lies in the fact that the cosmological models can be tested in the laboratory experiment by measuring the conductivity coefficient.In this framework it also appears that the photon mass m_γ is not consistent with gauge invariance of the first and second kind.

In a series of recent papers $Evans^{(7)}$ and his collaborators studied the longitudinal solution of Maxwell field and the problem of gauge invariance.In the case of EBP theory, they have indicated that finite m_γ is consistent with gauge invariance of the first and second kind if and only if

$$A_\mu A_\mu \to 0$$

for $m_\gamma \neq 0$ which implies that $\phi = c|\vec{A}|$ where, $A_\mu = (\vec{A}, \frac{i\phi}{c})$. Here ϕ is the scalar potential and $\vec{A}$ the vector potential of the de Broglie-Proca field. The above condition has been shown to be consistent with the Lorentz condition but is inconsistent with a massless gauge i.e. the traditional coulomb gauge. But the dispersion relation (in its covariant form) which we have deduced [6] in EBP theory with $\sigma \neq 0$ is not consistent with the condition $A_\mu A_\mu \to 0$, for $m_\gamma \neq 0$. However, this has been shown to be consistent with coulomb gauge. In this paper, we shall show that the EBP theory with $\sigma \neq 0$ will be consistent with the gauge principle if we introduce fourth component of current as non-zero i.e., $J^\mu = (\sigma \vec{E}, J^0)$ in vacuum instead of $J^\mu = (\sigma \vec{E}, 0)$. Here J^0 is related to E_0 through the $\vec{B}^{(3)}$ field introduced by $Evans^{[6]}$. It comes directly from the original Maxwell equations without introduction of scalar and vector potentials.

In section II of this paper, we shall briefly recapitulate the EBP theory with non-zero σ for our convenience. Then the gauge invariance and $A_\mu A_\mu \to 0$ condition will be studied for this extended theory in section III. In section IV the uncertainty relation among $\vec{B}^{(1)}, \vec{B}^{(2)}$ and $\vec{B}^{(3)}$ fields and the possibilities of simultaneous measurement are discussed. It gives rise to an interesting situation where the transverse as well as the longitudinal photons can be observed. Finally the possible implications for $J^0 \sim E_0 \sim B_0$ will be discussed in section V.

2. MAXWELL EQUATIONS IN VACUO WITH $\sigma \neq 0$

If we endow the vacuum with non-zero conductivity i.e., $\sigma \neq 0$, the Maxwell equations should be rewritten in the form

$$\begin{aligned} div\vec{E} &= 0 \\ curl\vec{H} &= \sigma\vec{E} + \epsilon_0\chi_e \frac{\partial \vec{E}}{\partial t} \\ div\vec{H} &= 0 \\ curl\vec{E} &= -\mu_0\chi_m \frac{\partial \vec{H}}{\partial t} \end{aligned} \quad (1)$$

where,
ϵ_0 denotes vacuum dielectric constant.
μ_0 denotes permeability constant.
χ_e is the relative dielectric constant.
χ_m is the relative permeability constant.

Again,
$$\nabla \times \nabla \times \vec{E} = -\nabla^2 \vec{E}$$

which together with Maxwell equations give

$$\nabla^2 \vec{E} = \frac{\epsilon_0 \chi_e \chi_m \mu_0}{c^2} \frac{\partial^2 \vec{E}}{\partial t^2} + \sigma \mu_0 \chi_m \frac{\partial \vec{E}}{\partial t} \quad (2)$$

The second term in the right hand side of equation (2) indicates that there will be dissipation of energy during the propagation of the photon. After simple calculations, the phase velocity v_p and group velocity v_g can be written as

$$v_p = n\left(1 - \frac{\sigma^2}{8(\epsilon_0 \chi_e)^2 n^4 \omega^2}\right) \quad (3)$$

$$v_g = n\sqrt{1 + \frac{\sigma^2}{4(\epsilon_0 \chi_e)^2 n^4 \omega^2}} \quad (4)$$

where, the dispersion law gives $k^2 = \omega^2 + \frac{\sigma^2}{4}$

Now,
(i) If $\sigma = 0$ and $n = 1$ then, $v_p = V_g$ and,
(ii) If $\sigma \neq 0$ $\frac{\sigma}{\omega} \to 0$, and
$n = 1$, then $V_g = c$

It is now clear from equation (3) and (4) that for $n < 1$ and $\sigma \neq 0$, no superluminal transmission is possible in this type of vacuum. This has profound significance in describing the behaviour of quantum particles especially in resolving the paradox, like EPR one. So taking v_g as the velocity of the photon in the de Broglie relation,

$$E = h\nu = \frac{m_\gamma c^2}{\sqrt{(1 - \frac{v^2}{c^2})}} \quad (5)$$

we get the mass of the photon as

$$m_\gamma^2 = h^2 \omega^2 (1 - n^2) - \frac{\sigma^2 h^2}{4 n^2 \epsilon_0^2 \chi_e^2} \quad (6)$$

which for Maxwell vacuum i.e., for $n = 1, \sigma = 0$, m_γ vanishes. This is consistent with classical electrodynamics. Several authors[8] have calculated the effective photon mass in terms of the Hubble constant (H) as

$$m_\gamma \simeq \frac{hH}{2} \sim 10^{(-65)} gm. \quad (7)$$

It is to be noticed that our mass formula reduces to

$$m_\gamma \sim \frac{1}{2} \frac{\hbar\sigma}{n\epsilon_0\chi_e}$$

for small ω and $\sigma/\omega \to 0$ Comparing this relation with (7), we get the following relation between σ and H :

$$\sigma \sim H$$

So, by measuring the conductivity coefficient in the laboratory we can estimate the value of Hubble constant. It gives rise to a new test of cosmological theories at laboratory scale.

The relation (7) has been studied in [9] to explain the redshift mechanism at cosmological scale. *Shtyrkov*[9] started a systematic investigation on cosmological redshift and the tired photon hypothesis. He considered a wave equation similar to (2) but with a fricton term in vacuum like damped oscillator.

Using plane wave solutions we get the following dispersion relation in a covariant form,

$$(|\vec{k}|^2 - n^2k_0^2)A^\mu(k) = \chi_\mu\mu_0[g^{\mu\nu} + \frac{(n^2-1)\eta^\mu\eta^\nu}{n^2}]J_\nu(k) \qquad (8)$$

where,

$$J_\mu = (\sigma\vec{E}, 0)$$

$$A^\mu = (\vec{A}, \frac{i\phi}{c})$$

$$k_\mu = (\vec{k}, k_0)$$

and η^μ is the unit time like vector where as $\eta = (\vec{0}, 1)$ for the medium at rest, it is evident from (11) that

$$\vec{A} \neq 0 \text{ but } \phi = 0 \text{ for } \sigma \neq 0$$

which is nothing but the usual coulomb gauge. But the condition $A_\mu A_\mu \to 0$ is not consistent with the usual coulomb gauge. So it seems that the gauge principle has to be reinterpreted for $m_\gamma \neq 0$ with $\sigma \neq 0$ in the vacuum.

3. EXISTENCE OF $B^{(3)}$ FIELD AND GAUGE INVARIANCE WITH $\sigma \neq 0$

Longitudinal solutions of Maxwell equations in vacuo have not been considered as physically meaningful in the great majority of standard texts on electrodynamics. Recently, Evans and his collaborators have systematically studied the theory of phase-independent longitudinal electric and magnetic fields, which are solutions to the free space-time Maxwell equations and thus obey the differential form of Gauss theorem in free space time. The Proca equation which is

$$\Box A_\mu = -\xi^2 A_\mu \quad \text{where} \quad \xi = \frac{m_\gamma c}{\hbar}, \qquad (9)$$

m_γ being the mass of the photon, can be expressed as

$$\Box A_\mu = J_\mu^{eff} = (\vec{J}^{eff}, J^{eff}) = (-\xi^2 c^2 \epsilon_0 \vec{A}, -\xi^2 \epsilon_0 \phi) \qquad (10)$$

where $\vec{J}^{eff}$ and J^{eff} denote the effective three current and effective fourth component respectively. These equations imply that the photon mass is equivalent to the presence of an effective charge and current in free space. In our framework, up till now, we have taken $\vec{J} = \sigma \vec{E}$ and $J^{eff} = 0$. Let us now assume that $\vec{J} = \sigma \vec{E}$ and $J^0 \neq 0$ in which case J^0 is associated with B_0 i.e., magnetic flux density associated with photon.

Here, no electrostatic field can be generated out of $\vec{B}^{(3)}$, $\vec{E}$ can be thought of as to be produced by magnetic induction. In the conventional framework of Maxwell's equation $\vec{J} = \sigma(\vec{E} + \vec{E}')$ where $\vec{E}'$ consists of all non-electrostatic field and $\vec{E}$ is the field derivable from a potential. In our framework, $\vec{E} = 0$ and $\vec{J} = \sigma \vec{E}'$. It indicates that the current distribution and field distribution are entirely defined by the non conservative field and by the conductivity of the medium. The presence of this non-conservative field as produced by $\vec{B}^{(3)}$ is responsible for the loss of energy of the photon when it propagates through this type of vacuum, then it seems that the traditional interpretation of gauge transformation must be revised fundamentally within the context of extended EBP theory. Evans worked out this problem in a rigorous way and he started with a Lagrangian with the form

$$L = -\frac{F_{\mu\nu} F_{\mu\nu}}{4} + \frac{m_\gamma^2 A_\mu A_\mu}{4} \qquad (11)$$

for $m_\gamma = 0$ and $F_{\mu\nu} = \frac{\partial A_\nu}{\partial x_\mu} - \frac{\partial A_\mu}{\partial x_\nu}$.
For $m_\gamma \neq 0$,

$$A_\mu A_\mu = 0, A_\mu \neq 0 \qquad (12)$$

is the only alternative possibility. Conventionally, it is asserted that the invariance of L under gauge transformation of 2nd kind means that $m_\gamma = 0$. So EBP is not consistent with gauge invariance. But for $J^0 \sim B_0 \neq 0$ and $A_\mu A_\mu \to 0$ we have $n^2 \to \frac{1}{3}$. It is to be mentioned that in the radiation gauge, A_μ is not completely covariant, because A_0 is zero in the special Lorentz frame. This unsatisfactory feature has been discussed in recent literature [11]. Roy and $Evans^{(12)}$ have recently suggested a novel gauge theory which reconciles non-zero photon mass with type two gauge invariance i.e.,

$$A_0 \rightarrow FAPP \ |\vec{A}| \tag{13}$$

In this notation, FAPP denotes "for all practical purposes", so that

$$A_\mu A_\mu \rightarrow FAPP \ 0 \tag{14}$$

This conditiion (23) is a limiting form of, and an excellent approximation to, the condition introduced by $Dirac^{(13)}$ as

$$A_\mu A_\mu = constant \tag{15}$$

4. $\delta B^{(1)}.\delta B^{(2)} \geq \frac{1}{2}|B_0 \vec{B}^{(3)}|$ AND UNSHARP MEASUREMENT

It has been shown [7] that the quantized field operators $B^{(1)}$, $B^{(2)}$ and $B^{(3)}$, corresponding to classical $\vec{B}^{(1)}$, $\vec{B}^{(2)}$ and $\vec{B}^{(3)}$ fields, form angular momentum commutators in free space,

$$[\hat{B}^{(1)}e^{-i\phi}, \hat{B}^{(2)}e^{i\phi}] = -iB^{(0)}\hat{B}^{(3)} \tag{16}$$

$$[\hat{B}^{(2)}e^{i\phi}, \hat{B}^{(3)}] = -iB^{(0)}\hat{B}^{(2)}e^{i\phi} \tag{17}$$

$$[\hat{B}^{(3)}, \hat{B}^{(1)}e^{-i\phi}] = -iB^{(0}\hat{B}^{(1)}e^{-i\phi} \tag{18}$$

The fields can now be thought of in terms of creation and annihilation operators as usual.

In the usual Hilbert space formulation of Quantum Mechanics, the Heisenberg uncertainty principle shows that

$$\delta\hat{B}^{(1)}\delta\hat{B}^{(2)} \geq \frac{1}{2}|B^{(0)}\hat{B}^{(3)}| \tag{19}$$

where $\delta\hat{B}^{(1)}$ and $\delta\hat{B}^{(2)}$ are root mean square deviations. The right hand side is a rigorous lower bound on the product $\delta\hat{B}^{(1)}\delta\hat{B}^{(2)}$, a lower bound which is therefore defined by $\delta\hat{B}^{(3)}$. If $\hat{B}^{(3)}$ were zero, $\hat{B}^{(1)}$ and $\hat{B}^{(2)}$ would commute, then $\delta\hat{B}^{(1)} = \hat{0}$ and $\delta\hat{B}^{(2)} = \hat{0}$ simultaneously. The experimental observation of light squeezing[7] shows that this is consistent with data, therefore $\hat{B}^{(3)} \neq \hat{0}$. In the usual Hilbert space formalism, the relation (19) indicates that $\hat{B}^{(1)}$, $\hat{B}^{(2)}$ and $\hat{B}^{(3)}$ can not be measured simultaneously with high precision. The recent developments of unsharp observables[13] in the extended framework of Hilbert space formalism raise new interest to look

into the problem of simultaneous measurements of three spin observables. In generalized Quantum mechanics, certain positive operators, the effects (self adjoint with the spectrum within the interval[0,1]) are included to build positive operator valued (POV) measure, as generalized observable or unsharp observable. A normalized positive operator valued measure, which maps value sets $(I\!R, B(I\!R))$ into the set of bounded self adjoint positive operators $[L_s^+(H)]$, satisfies the followings :
(i) $a(E) \geq 0$ for all $E \epsilon B(I\!R), a(\phi) = 0, a(I\!R) = 1$
(ii) $a(U_i E_i) = \sum_i a(E_i)$ for the countable collection of pair wise disjoint sets $E_i \epsilon B(I\!R)$, the series converging in the weak operator topology.

One striking feature with this unsharp observable is the coexistence of formerly incompatible quantities. There is now the theoretical possibility of joint measurement for pairs of non-commuting observables. $Busch$[14,15] has shown that two non-commutative spin observables can be co-existent if they are replaced by their unsharp counter part. He presented a realistic model which shows that simultaneous measurement of two spin $\frac{1}{2}$ observables can be achieved. This model also represents joint measurement of three spin observables. We have generalized this approach [16] for a joint triple measurement of three unsharp observables whose corresponding vectors lie along Z-axis, in X-Y plane and in space depending on the parameter of the apparatus employed.
So, within this framework, $\hat{B}^{(1)}, \hat{B}^{(2)}$ and $\hat{B}^{(3)}$ can be measured simultaneously. It clearly establishes the fact that the transverse and longitudinal component of light can be measured simultaneously but not with high precision.

5. POSSIBLE IMPLICATIONS

More and more evidences are available now for finite photon rest mass, upon which is based the theory of Einstein and de Broglie. This theory has been extended by Vigier and independently by the present author to vacuum with $\sigma \neq 0$. It has several astrophysical consequences which have been well summerized by Vigier. But in the present paper, the condition $(A_\mu A_\mu \to 0)$ for the gauge invariance has been shown to be consistent with this extended EBP theory by introducing non-zero fourth component of J_μ. This fourth component is related to B_0 i.e. magnetic flux density assoiciated with one photon. It gives rise some new consequence. For example:
(i) The displacement current in the Maxwell vacuum can be generated by $\sigma \vec{E}'$. Again as $\sigma \vec{E}'$ is solely due to magnetostatic induction, this field can be identified with $\vec{B}^{(3)}$ field. This is a non-conservative force which immediately supports the hypothesis that the photon looses energy when it propagates through the Maxwell vacuum within EBP framework, considering

$\sigma \neq 0$. It is worthmentioning that Bartlett et.al.[15] already measured Maxwell displacement current inside a capacitor in free space. More refined experiment is needed to detect this kind of displacement current so as to confirm the existence of a magnetostatic field in vacuum.

(ii) The identification of magnetostatic field with Evans $B^{(3)}$ field raises interesting questions regarding the simultaneous measurability of $\hat{B}^{(1)}$, $\hat{B}^{(2)}$ and $\hat{B}^{(3)}$. This question of simultaneous measurability is the key problem in EBP theory of light. The theory of unsharp measurement in Quantum mechanics provides us a framework of joint measurement of $B^{(1)}$, $B^{(2)}$ and $B^{(3)}$ fields. It means the transverse components corresponding to $B^{(1)}$, $B^{(2)}$ and the longitudinal component corresponding to $B^{(3)}$ should be measureable simultaneously. This will be discussed elaborately in subsequent publications.

References

1. Vigier,J.P.(1993) Present Experimental Satus of the Einstein-de Broglie theory of light (Quantum Control and Measurement Ed.By H.Ezawa et.al.,Elsevier).

2. Einstein,A.(1916) Werk Deutsch. Phys.Ges **18**, 318;(1917) Phys.Zeit **18**,121.
   de Broglie,L.(1936) La Mechanique Ondulatoire Photon(Gauthier Villar, Paris).

3. Vachaspati,T. and Vilenkin,A.(1991) Large Scale Structure from Wiggly Cosmic Strings,Phys.Rev.Lett.**67**,1057.

4. Barut,A.(1989) $E = \hbar\omega$, ICTP preprint, IC/89/85.
5. Vigier,J.P.(1990) Evidence for Nonzero Mass Photon associated with a Vacuum Induced Dissipative Red-Shift Mechanism,IEEE Transaction on Plasma Science **18** Feb.

6. Kar,G. Sinha,M. and Roy,S.(1993) Maxwell Equations, Nonzero Photon Mass and Conformal Metric Fluctuation: Int.J.Theor.Phys. **32**,1052.

7. Evans,M.W.(1994) Modern Nonlinear Optics, Part 2, Ed.by Myron Evans and Kielich,S.(Advances in Chemical Physics,Series LXXXV,; Vigier,J.P. and Evans,M.W.(1994) Enigmatic Photon,**1**, (Kluwer,Dordrecht).

8. Fuli,L.(1981) An Estimate of Photon Rest Mass,Il Nuovo Cimento, **31**,289.
   Nieto,M.M.(1992) "Past, Present and possible future limitation on the photon rest mass" - LA-UR-92-4244.

9. Shtyrkov,E.I.(1992) Galilean Electrodynamics,**3**,N4,66.

10. Freedman,W.L. et.al.(1994) Distance to the Virgo Cluster Galaxy M100 From Hubble Space telescope observation of Cepheids, Nature (October 27).

11. Evans,M.W. and Vigier,J.P.(1994) The Enigmatic Photon,**1** and **2**,(Kluwer Academic Publishers.Dordrecht).

12. Roy,S. and Evans,M.W.(1995) Maxwell equations , Non-zero conductivity coefficient and existence of B(3) field, Found. Phys.(in the Press).

13. Dirac P.A.M.(1951) Is there an Aether?: Nature,**168**,906.; (1952) Is there an Aether ?, Nature,**169**,702.;(1951) The HamiltonianForm of Field Dynamics,Cand.J.Math.,**3**,1.

14. Busch,P.(1986) Unsharp Reality and joint measurements for Spin Observables,Phys.Rev **D,33**,2253.

15. Busch,P.(1987) Some realizable joint measurement of complementary Observables,Found. Phys.**17**,905.

16. Kar G. and Roy S.(1995) Unsharp spin-$\frac{1}{2}$ observables and CHSH inequalities, Phys.Lett.**A 100**,12.

17. Bartlett, D.F. and Corle, T.R.(1985) Measuring Maxwell Displacement Inside a capacitor, Phys.Rev.Lett.**55**,59.

ORIGIN, OBSERVATION AND CONSEQUENCES OF THE $B^{(3)}$ FIELD

M.W. EVANS[1]
Department of Physics and Astronomy
York University, 4700 Keele Street
Toronto, Ontario M3J 1P3 Canada

Dedicated to Jean-Pierre Vigier

A summary overview is given of the origin, methods of observation, and consequences of the $B^{(3)}$ field of free space electromagnetism, with suggestions for further work.

1. Introduction

The $B^{(3)}$ field of vacuum electromagnetism introduces a new paradigm of field theory, summarized in the cyclically symmetric equations linking it [1—8] to the usual transverse magnetic plane wave components $B^{(1)} = B^{(2)*}$. The $B^{(3)}$ field was first (and obliquely) inferred in January, 1992 at Cornell University from a careful re-examination of known magneto-optic phenomena [9,10] which had previously been interpreted in orthodoxy through the conjugate product $E^{(1)} \times E^{(2)}$ of electric plane wave components $E^{(1)} = E^{(2)*}$. In the intervening three and a half years its understanding has developed substantially into monographs and papers [1—8] covering several fundamental aspects of field theory. This paper is a summary overview of the origin, observation, and consequences of the fundamental $B^{(3)}$ field of electromagnetism in the vacuum.

2. Origin

The $B^{(3)}$ field originates [1—8] in an experimental observable of magneto-optics known to specialists as the conjugate product [11]. This observable is responsible, for

---

[1] Address for correspondence: *50 Rhyddwen Road, Craig Cefn Parc, Swansea SA6 5RA, Wales, Great Britain*

example, for the phenomenon of magnetization by light first confirmed experimentally by van der Ziel et al. [9]. If 3-D space is represented by unit vectors [1,2] in the complex basis

$$e^{(1)} \times e^{(2)} = ie^{(3)*}, \text{ et cyclicum,} \tag{1}$$

then $B^{(3)}$ is defined in vacuo by [1,2]

$$B^{(1)} \times B^{(2)} = iB^{(0)}B^{(3)*}, \text{ et cyclicum.} \tag{2}$$

On the left hand side the observable conjugate product appears in orthodox form [11], while on the right hand side it is expressed through the second order observable $iB^{(0)}B^{(3)*}$. Here $B^{(0)}$ is the scalar magnitude of the magnetic plane wave components $B^{(1)} = B^{(2)*}$, which, as usual, are transverse to the direction of light propagation. Equations (2) show that the standard vacuum Maxwell equations are self-inconsistent, because $B^{(3)}$ is evidently *longitudinal* in vacuo and is not a transverse magnetic flux density. Equations (2) are relations between infinitesimal generators of the rotation group O(3), with the startling consequences that both the Wigner little group [12] and the sector group symmetry [13] of vacuum electromagnetism become O(3). This implies that the photon, if particulate, is also massive. The E cyclics,

$$E^{(1)} \times E^{(2)} = -E^{(0)}(iE^{(3)})^*, \text{ et cyclicum,} \tag{3}$$

can be generated from the B cyclics (2) by fundamental tensorial duality [1,2] in vacuo. In contrast, to Eqs. (2), Eqs. (3) contain the unphysical $iE^{(3)}$ field, formally a purely imaginary quantity [1,2]. The conjugate product thus generates the real and physical $B^{(3)}$, but this is *not* accompanied by a real and physical $E^{(3)}$. The physical equations (2) conserve the known discrete symmetries [14].

Since $iB^{(0)}B^{(3)*}$ is a second order *observable*, then $B^{(3)}$, a physical and real magnetic flux density, acts also at first order under the right observational conditions [3], discussed in the following section. The $B^{(3)}$ field propagates in vacuo at the speed of light, F.A.P.P., and it is incorrect to think of it as a static magnetic field. The B cyclics, Eq. (2), are non-linear and non-Abelian [2], so that the self-generating source of $B^{(3)}$ is the cross product $B^{(1)} \times B^{(2)}$. Similarly, the source of one axis $e^{(3)}$, in 3-D space is the cross product of the other two, $e^{(1)}$ and $e^{(2)}$, emphasizing that the B cyclics originate in spacetime itself, be this flat, as in special relativity, or curved as in general relativity. It would have been quite natural to accept this if it had been realized prior to Maxwell's great work in the nineteenth century. (Phenomena of magnetization by light were of course unknown to Maxwell.) About 130 years later, we have become used to thinking of electromagnetism in vacuo as a 2-D phenomenon, so a sudden shift into 3-D paradigm appears startlingly new. If we accept $\boldsymbol{B}^{(3)}$, and base electromagnetism as a phenomenon on Eqs. (2), then the task of unification with gravitation becomes much easier, because both electroweak and gravitational fields become properties of curved space-time. Several fundamental inferences of this nature have been developed and are summarized in the Table [3].

A first attempt has been made by Evans and Vigier [2] to write the Maxwell equations more consistently in O(3), non-Abelian form, using methods [15] borrowed directly from general relativity. This work shows that in vacuo

$$\boldsymbol{B}^{(3)*} = -i\frac{e}{\hbar}\boldsymbol{A}^{(1)} \times \boldsymbol{A}^{(2)}, \qquad (4)$$

where $\boldsymbol{A}^{(1)} = \boldsymbol{A}^{(2)*}$ is the vector potential plane wave, e is the elementary charge, and $\hbar$ is Dirac's constant. Equation (4), in turn, reveals that the interaction of $\boldsymbol{B}^{(3)}$ with matter (e.g. a fermion) is dictated by the usual transverse vector potentials $\boldsymbol{A}^{(1)} = \boldsymbol{A}^{(2)*}$ through the minimal prescription [1—8]. Dirac's equation of motion should therefore be used to describe this interaction self-consistently [3]. If this is done correctly, the conditions of observation of $\boldsymbol{B}^{(3)}$ become properly defined, as described in the next section. The same thing can be done classically [1] with the *relativistic* Hamilton-Jacobi equation of motion. The relativistic factor reveals that the magnetization profile generated by the interaction of $\boldsymbol{B}^{(3)}$ with one fermion is in general a mixture of terms to first and second order in $\boldsymbol{B}^{(0)}$. In the *weak field limit* (visible frequency radiation of any practically accessible intensity), the profile is second order, and proportional to beam power density, I (W m^{-2}). In the *strong field limit* (intense, radio frequency pulses) the profile is expected to be proportional to $\sqrt{I}$. These considerations are developed as follows.

3. Observation

Equations (2) show that $i\boldsymbol{B}^{(0)}\boldsymbol{B}^{(3)*}$ is observable whenever $\boldsymbol{B}^{(1)} \times \boldsymbol{B}^{(2)}$ is observable [9,10], i.e., whenever circularly polarized electromagnetic radiation interacts with matter. If the latter can be represented for simplicity by an ensemble of N non-interacting electrons in a volume V, the relativistic Hamilton-Jacobi equation shows [1—8] that in general,

$$B^{(3)}_{\text{in sample}} = \frac{N}{V} \cdot \frac{\mu_0 e^3 c^2}{2m\omega^2} \left(\frac{B^{(0)}}{(m^2\omega^2 + e^2 B^{(0)2})^{1/2}} \right) B^{(3)}_{\text{free space}}, \qquad (5)$$

in which $B^{(3)}$ in free space is related to the beam power density I by

$$B^{(3)}_{\text{free space}} = \left(\frac{\mu_0}{c} I \right)^{1/2} e^{(3)} = \left(\frac{I}{\epsilon_0 c^3} \right)^{1/2} e^{(3)}. \qquad (6)$$

In these equations ϵ_0 and μ_0 are respectively the free space permittivity and permeability, c the speed of light in vacuo, e/m the charge to mass ratio of the fermion, and ω is the beam angular frequency. In the weak field (visible frequency) limit, $m\omega \gg eB^{(0)}$, and Eq. (5) reduces to

$$B^{(3)}_{\text{in sample}} \rightarrow \frac{N}{V} \left(\frac{\mu_0 e^3 c^2 B^{(0)}}{2m^2\omega^3} \right) B^{(3)}_{\text{free space}}. \qquad (7)$$

In the opposite strong field (radio frequency) limit,

$$B^{(3)}_{\text{in sample}} \rightarrow \frac{N}{V} \left(\frac{\mu_0 e^2 c^2}{2m\omega^2} \right) B^{(3)}_{\text{free space}}. \qquad (8)$$

The observable quantity is $B^{(3)}$ in the sample, which is different from $B^{(3)}$ in free space. Typically, as shown by van der Ziel et al. [9], the sample $B^{(3)}$ in the weak field limit is about 10^{-9} T (10^{-5} gauss), and to observe it requires a skillful inverse Faraday effect experiment [9,10]. On the basis of arguments in Sec. 2, these magneto-optic effects become fundamentally important. The observation has not been attempted to date of $B^{(3)}$ acting at first order (to order $I^{1/2}$), but is the straightforward result of the relativistic factor in Eq. (5), obtainable both from the Hamilton-Jacobi equation and the classical limit of the Dirac equation [3]. Such an experiment, carried out with circularly polarized radio-frequency pulses of high intensity, is of great interest. Many other variations are possible, using well-developed contemporary techniques in non-linear optics and magneto-optics.

TABLE 1. Theory of Electrodynamics

	Fundamental Concept	Standard Theory	New Theory
1	Structure of Theory	linear, Abelian	non-linear, Non-Abelian
2	Maxwell's Equations in Vacuo	$\frac{\partial F_{\mu\nu}}{\partial x_\nu} = 0$, $B^{(3)} = 0$	$B^{(1)} \times B^{(2)} = iB^{(0)}B^{(3)*}$, et cyclicum
3	Electromagnetic Sector Symmetry	$U(1) = O(2)$	$O(3)$
4	Poincaré Group Symmetry	$\hat{J}^{(3)}$ generator missing, unphysical little group	magnetic fields are rotation generators
5	Wigner Little Group	$E(2)$, unphysical planar, Euclidean	$O(3)$, physical space rotation
6	Source of $B^{(3)}$ at Observer Point R and Instant t	not considered	circling e at time $t - \frac{R}{c}$ earlier
7	Propagation of $B^{(3)}$	not considered	through the Liénard-Wiechert potentials $A^{(1)} = A^{(2)*}$
8	Gauge Group Definition of $B^{(3)}$	not considered	$B^{(3)*} = -i\frac{e}{\hbar}A^{(1)} \times A^{(2)}$
9	Free Space Four Tensor	$F_{\mu\nu}$, Abelian in space ((1), (2), (3))	$G_{\mu\nu}$, Non-Abelian in space ((1), (2), (3))
10	Photon Helicities	-1 and 1	-1, 0, 1
11	Translational Poynting Theorem	$\nabla \cdot N = -\frac{\partial U}{\partial t}$	same
12	Rotational Poynting Theorem	not considered	$\nabla \cdot J^{(3)} = -\frac{\partial U^{(3)}}{\partial t}$
13	Magnetic Fields	$B^{(1)} = B^{(2)*}$	$B^{(1)} = B^{(2)*}$, $B^{(3)}$
14	Electric Fields	$E^{(1)} = E^{(2)*}$	$E^{(1)} = E^{(2)*}$, $iE^{(3)}$

	Fundamental Concept	Standard Theory	New Theory
15	Vector Potentials	$A^{(1)} = A^{(2)*}$	$A^{(1)} = A^{(2)*}$, $iA^{(3)}$, A_0
16	Planck-Einstein Relation	$En = \hbar\omega$	$En = \hbar\omega = h_1\lambda$, $h_1 = ec\|B^{(3)}\|$
17	de Broglie Relation	$p = \hbar\kappa$	$p = \hbar\kappa = \dfrac{h_1\lambda}{c}$
18	Quantum of Energy	$\hbar\omega = \dfrac{1}{\mu_0}\int (B^{(1)} \cdot B^{(1)*} + B^{(2)} \cdot B^{(2)*})dV$	$\hbar\omega = \dfrac{1}{\mu_0}\int (B^{(1)} \cdot B^{(1)*} + B^{(2)} \cdot B^{(2)*} + B^{(3)} \cdot B^{(3)*})dV_1$
19	Quantum of Angular Momentum	$\hbar$	$\hbar = h_1 \dfrac{\lambda}{\omega}$
20	Quantum of Torque	$\hbar\omega$	$\hbar\omega = h_1\lambda$
21	Momentum Equivalence Condition	not considered	cyclic relations imply in vacuo $eA^{(0)} = \hbar\kappa$, the quantum of linear momentum
22	Mass of Photon	identically zero	$m = \dfrac{e\lambda_0\|B^{(3)}\|}{c}$
23	Gauge Invariant Lagrangian Mass Term	not considered	$En = \dfrac{V}{\mu_0}\left(\dfrac{Mc^2}{\hbar\omega}\right)^2 B^{(3)} \cdot B^{(3)*}$ in $O(3)$ gauge group
24	Gauge Conditions on Four Potential	1) transverse 2) Coulomb	1) not allowed 2) scalar, non-zero 3) $A_\mu A_\mu \to 0$
25	Field Quantization	canonical, beset with difficulties because A_μ is not covariant	direct quantization of cyclic field relations

	Fundamental Concept	Standard Theory	New Theory
26	de Broglie Theorem	not considered	$\hbar\omega_0 = h_1\lambda_0 = mc^2$
27	Interaction with Fermion	via $A^{(1)} = A^{(2)*}$ in minimal prescription, $A_0 = \phi = 0$	same as Standard Theory but finite scalar potential $A_0 = \phi \neq 0$
28	Magneto-optics	I dependence from $A_0 = 0$	$I^{1/2}$ dependence observable under the right conditions, $A_0 \neq 0$
29	Q.E.D.	no mass term	finite mass term
30	Observation of $B^{(3)}$ Field	not considered	routinely observable through $iB^{(0)}B^{(3)*}$ in magneto-optics

Notes N := Poynting vector; U = energy density; $J^{(3)}$ = radiation angular momentum; $U^{(3)} = \frac{1}{\mu_0}B^{(3)} \cdot B^{(3)*}$; λ := wavelength; λ_0 = rest wavelength; V = volume of radiation; M = mass of radiation; $\phi = A_0$ = scalar potential.

4. Consequences

Since $iB^{(0)}B^{(3)*}$ is a routine observable it is possible to conclude with currently available data that the two dimensional picture of vacuum electromagnetism is incomplete. There are several major consequences of this deduction if accepted (Table 1). For example, the gauge group symmetry of the electromagnetic sector of unified field theory becomes O(3), not U(1), requiring appropriate theoretical development. The Wigner little group [12] becomes O(3), the physical space rotation group, which replaces the unphysical E(2) of orthodoxy [15]. In special relativity, a massless particle can have two degrees of polarization only, so the existence in the vacuum of $B^{(3)}$ means that the photon if particulate is massive. Experimental efforts to measure limits on the photon mass are therefore shown to be well-founded in the B cyclic equations (2). The latter indicate the need to develop field equations of electromagnetism in an O(3) gauge symmetry [2], more completely, a Lorentz group gauge symmetry. When this is done, a self-consistent expression for $B^{(3)}$ is obtained (Eq. (4)) in terms of transverse potentials $A^{(1)}$ and $A^{(2)}$. Radiation theory must be developed to explain the mode of propagation of $B^{(3)}$ through the vacuum, and some inroads to this question have been made in the literature [3]. There are several interesting experimental consequences of a $B^{(3)}$ field, among these are the optical Aharonov-Bohm effect [1—8] and the inference that $B^{(3)}$ is the relict magnetic field in relativistic cosmology [3]. Satellite observations of the anisotropy in the cosmic background radiation would lead to a determination of $B^{(3)}$, and inter alia, show that $B^{(3)}$ is the seed field responsible for fundamental cosmic magnetic effects. If $B^{(3)}$ is accepted as indicating a massive photon, new impetus is given to interpretations of quantum field theory which rely on finite photon mass, for example theories developed by the Vigier school [16]. Finally in a list of consequences mentioned at random, field theory in the vacuum reduces to the theory of angular momentum commutators in quantum mechanics, allowing a straightforward route to quantization of Eqs. (2). In general, magnetic fields become proportional to rotation generators, electric fields to boost generators, expressed either in pseudo-Euclidean or Riemannian geometry. This realization leads to new ways of unifying quantized electroweak theory and quantized gravitation.

5. Acknowledgements

York University, Toronto, and the Indian Statistical Institute, Calcutta are warmly acknowledged for visiting professorships. The University of North Carolina at Charlotte is acknowledged for project support. Professors Sisir Roy and Stanley Jeffers are thanked for much interesting discussion. Last but by no means least, Professor A van der Merwe is thanked for his invitation to produce the volumes of *The Enigmatic Photon* in his prestigious series of monographs.

6. References

1. M. W. Evans and J.-P. Vigier, *The Enigmatic Photon, Vol. 1: The Field $B^{(3)}$* (Kluwer Academic, Dordrecht, 1994).
2. M. W. Evans and J.-P. Vigier, *The Enigmatic Photon, Vol. 2: Non-Abelian Electrodynamics* (Kluwer Academic, Dordrecht, 1995).
3. M. W. Evans, J.-P. Vigier, S. Roy, and S. Jeffers, *The Enigmatic Photon, Vol. 3: $B^{(3)}$ Theory and Practice* (Kluwer Academic, Dordrecht), in preparation.
4. M. W. Evans, *Physica B* 182, 227, 237 (1992); 183, 103 (1993); 190, 310 (1993); *Physica A* 214, 605 (1995).
5. M. W. Evans, *Found. Phys. Lett.* 7, 67, 209, 379, 467, 577 (1994); 8 (1995);*Found. Phys.* 24, 1519, 1671 (1994); 25, 175, 383 (1995); S. Roy and M. W. Evans, *Found. Phys.*, in press; ibid., submitted.
6. M. W. Evans and A. A. Hasanein, *The Photomagneton in Quantum Field Theory* (World Scientific, Singapore, 1994); M. W. Evans, *The Photon's Magnetic Field* (World Scientific, Singapore, 1992); M. W. Evans and S. Kielich, eds., *Modern Nonlinear Optics*, Vol. 85(2) of *Advances in Chemical Physics*, I. Prigogine and S. A. Rice, eds., (Wiley Interscience, New York, 1993).

7. M. W. Evans, S. Roy, and S. Jeffers, *Lett. Nuovo Cim.*, in press.
8. M. W. Evans, *Mod. Phys. Lett.* 7B, 1247 (1993); *J. Mol. Liq.* 55, 127 (1993).
9. J. P. van der Ziel, P. S. Pershan, and L. D. Malmstrom, *Phys. Rev. Lett.* 15, 190 (1965); P. S. Pershan, J. P. van der Ziel, and L. D. Malmstrom, *Phys. Rev.* 143, 574 (1966).
10. W. Happer, *Rev. Mod. Phys.* 44, 169 (1972); T. W. Barrett, H. Wohltjen, and A. Snow, *Nature* 301, 694 (1983).
11. G. H. Wagnière, *Linear and Nonlinear Optical Properties of Molecules* (VCH, Basel, 1993).
12. E. P. Wigner, *Ann. Math.* 40, 149 (1939).
13. J. C. Huang, J. Phys., G, Nucl. Phys. 13, 273 (1987).
14. L. H. Ryder, *Elementary Particles and Symmetries* (Gordon and Breach, London, 1986).
15. L. H. Ryder, *Quantum Field Theory.* (Cambridge University Press, Cambridge, 1987).
16. J.-P. Vigier, *Found. Phys.* 21, 125 (1991); Vigier Honorarium, ibid., 25, (1995).

OPTO-MAGNETIC EFFECTS IN NONABSORBING MEDIA: PROBLEMS OF MEASUREMENT AND INTERPRETATION

S. JEFFERS[a], M. NOVIKOV[b], and G. HATHAWAY[c]
[a]*Department of Physics and Astronomy, York University, Toronto, Canada.*
[b]*Institute for Physics of Microstructure, Russian Academy of Sciences, Nizhniy Novgorod, Russia.*
[c]*Hathaway Consulting Services, Toronto, Canada.*

1. Introduction

The phenomenon of magnetization by circularly polarized light in non-absorbing media was first studied in the classic work of Van der Ziel, et al, 1965. The physical mechanism which gives rise to such magnetization is essentially diffferent from the magnetization produced in media with optical absorption. The latter is produced by polarized electrons in the medium by the transfer of the angular momentum of the electromagnetic radiation to electrons via spin-orbit interaction. In the case of a non-absorbing medium this mechanism is absent. In this paper, we discuss some physical mechanisms which can produce this type of magnetization. One is the production of magnetisation through the nonlinear interaction of the light with the medium. Another possibility is the claim that electromagnetic radiation itself has a longitudinal, constant magnetic field which does the magnetising (Evans M.W., 1994).

Today many effects of magnetization by polarized light are recognised and now we can speak about a new field of the nonlinear optics which we will refer to as optomagnetics. Here we present a review of various methods of measuring small magnetic fields for the study of the optomagnetic effects. We are also concerned with particular experiments to investigate the predicted properties of the Evans-Vigier field (Evans, M.W., 1994, 1995).

2. Ultra-sensitive Methods of Measurement of Magnetic Fields

Some particular experimental problems arise in the study of optomagnetic effects. The main problem is the small value of these effects in particular for diamagnetic and paramagnetic media. Buckingham and Parlett (1994) have claimed that the nuclei in a fluid irradiated by a circularly polarized light beam experiences a static magnetic field proportional to $E \times E$. An order of magnitude estimate of the strength of this effective field for a circularly polarized laser beam of 10 W/cm^2 is around 10^{-11} Gauss giving a minute NMR shift of order 10^{-8} Hz. The improvement of the sensitivity of measurement of ultra-small magnetic field is thus very important. At present there are many methods available for the measurement of magnetic fields (Lenz J.E., 1990). We shall discuss those which are particularly relevant for optomagnetics.

Currently NMR (Nuclear Magnetic Resonance) quantum magnetometers are the most precise instrument for the measurement of magnetic fields (precision to within 0.002%) and their sensitivity is of the order of 10^{-8}T. They apply only to the study of optomagnetic effects in substances which have suitable atoms for NMR. If the results of W.S.Warren, et al will be confirmed this method of study of optomagnetic effects will have significant applications. The second quantum method of measuring of magnetic fields is the method of optical pumping (Cohen-Tanoudji et al,1969). Today the sentisitivity achieved by this technique is 10^{-14}T/(Hz)½. For the study of opto-magnetic effects, however, they may have limited value as they apply only to the gas phase.

In our opinion among quantum magnetometers the SQUID (Superconducting Quantum Interference Device) is the most promising device for optomagnetic applications. The strong impact of biomagnetism (Romani G.L. et al 1982.) has led to the highly sensitive, reliable and relatively inexpensive SQUIDs fabricated by integrated technology (Barrone A., 1984). The maximum sensitivity of such devices is set by by the intrinsic thermal noise and is about 2.10^{-14}T/(Hz)½ at the temperature of liquid helium. Today such devices have sensitivity near one. Apparently for the measuring of optomagnetic effects, it is reasonable to use the flux superconducting differential transformer. These can work without a specially constructed shielded room. Moreover for the SQUID method, it is very convenient to apply some modulation e.g. by using elliptically polarized light. The sensitivity can be increased by using lower temperatures (10^{-4}K is possible to day) and multiple SQUIDs (up to a hundred) (Wellstood F.C. et al 1987.).

Induction magnetometers (IM) are based on the phenomenon of electro-magnetic induction. Various methods can be used to change the magnetic flux. To study optomagnetic effects, it is convenient to achieve this by modulating either the polarization or the intensity of the light and using lock-in amplifier synchronous detection.

The sensitivity of IM is determined the thermal noise of the resistance of the coil. The minimum voltage in this case is (4kTR df)(½),where k is the Boltzmann constant, T the temperature of the coil, R the resistance of the coil and 1/df is the time of detection. For optimal geometry of coil and frequency modulation, the minimum magnetic field detectable by IM is about $10^{-12}T/(Hz)^{1/2}$ at room temperature. Here there is the problem of preamplifier noise. This problem can be solved, for example, by using the versatile superconducting femtovolt amplifier (Lukens J.E. et al, 1971.). In order to diminish both factors (R and T) we suggest using a superconducting resonance circuit. Today there are experimental data that show a reduction of ohmic losses by more then 10^5 (Falfeau P., et al.1994.). Here the use of resonance reduces the influence of preamplifier noses. Using the optimal coil and the modulating frequency for liquid helium temperatures one can achieve a sensitivity of less than $10^{-16}T/(Hz)^{1/2}$.

V.B.Braginsky, 1970 has shown that the sensitivity of measurement can be essentially improved by theapplication of mechanical oscillator with small damping. Such mechanical oscillator, for example, is the torque or mechanical pendulum. At present it may have a such oscillator with the quality Q more than 10^8. Today we do not know about application of such oscillators for the measuring of magnetic fields. Earlier torque balance was applied for the measure of magnetic anisotropy with great sensitivity (F.B.Humphrey, and A.B.Johnson, 1963).

The torque resonance magnetometer (TRM) is a device where the sample forms one of parts of a mechanical oscillator. The sample is placed in the high external (for example 10^2T) so the measurement magnetic moment of the sample is perpendicular to the external magnetic field. If light with (for example the right-left) circular modulation of polarization stimulates a magnetic moment in the sample then it will experience an alternating mechanical torque. This torque is equal to the vector product of magnetic moment and the external magnetic field. This method has a sensitivity to magnetic field of less than $10^{-16}T/(Hz)^{1/2}$.

As one can see from the above such method of increasing sensitivity with increasing of Q-factor also concern to electrical superconductor magnetometer(SCM).

As for the SCM it is very useful to apply the modulation principleof the measurement in TRM. In that case for the resonance modulation we can overcome the problem of the necessary sensitivity for measure of the torque oscillator amplitude, for example, by optical interferometric method. In principle sensitivity of TRM may be more than the SQUID magnetometer.

From indirect methods of measuring of OME the method of the Faraday effect has the great interest. Today this method has a big interest in the fibre-optic magnetometers (G.W.Day, and A.H.Rose, 1988). The induced magnetic field in OME nay be investigated through detection of the polarization-plane rotation of a weak probe laser beam. This method was applied successfully to the optical registration of the EPR spectrum with a great sensitivity (E.B.Alekcandrov, and V.S.Zapasski, 1977). This method is using a great sensitivity of today laser measuring of the rotation of plane polarization laser light(about 10^{-3} sec.).

This method was also successful to study of the inverse Faraday effect in semiconductors in the reflection configuration (Yu.P.Svirko, N.I.Zheludev, 1994). Using a femtosecond laser with a great intensity in this work authors have very useful method of the surface study. This method has one negative affect for investigation of OME. The effective value of polarization-plane rotation incorporate additional mechanisms rotation through the third-order nonlinearities as a four-wave mixing phenomenon. This effect has a great value for the resonance condition when the probe or pump frequencies and also their sum
or difference frequencies areccoincident with the energy levels of substance. For the first time that effect was observed in paper (P.E.Liao,and G.C.Bjorklund, 1976). Therefore this effect is necessary to take into account in the time of the investigation of OME by the Faraday magnetometer(G.L.J.A.Rikken, 1995).

3. Discussion

For the nonabsorbing media there are an unambiguous interrelation between the magnetoptical and optomagnetic effect. In that case the constants of the corresponding effect are equal. This may prove on base of the severe microscopic theory (G.F.Efremov, M.A.Novikov, 1994). The first example of such relation to be experimentally observed it is interrelation between the Faraday effect(EF) and an optically-induced magnetisation by the circularly polarized light in the media without absorption- the Inverse Faraday effect (IFE). In this respect the circularly polarized light is similar to the static magnetic field. Their time and spatial symmetries are identical. Moreover their action on the media without a light absorption is also identical. For the media with absorption this identity is broken. First of all in that case the light absorption breaks the thermodynamical equilibrium but the static magnetic field does not. Another magnetooptical effect, linear to magnetic field, the Magnetochiral birefringence(MB) or the nonreciprocal magnetooptical linear birefringence. The first such experiment in crystals without the center of symmetry was made in the paper (M.A.Novikov et al, 1977). This effect (MB) corresponds to another optomagnetic effect. It is Inverse Magnetochiral Birefringence (IMB). This effect gives a magnetisation, when a laser beam travels in a medium without the center of symmetry. In contrast to IEF the magnetisation from IMB may be in linear polarization light. Sign of this magnetisation changes with the change of direction of the light propagation. The theory of this effect for chiral liquids made in paper (G.Wagnier, 1989). In chiral liquids the magnetization from IMB does not depend on the light polarisation, and its direction is parallel to the direction of the light beam propagation.For this magnetization has the opposite sign for enantiomers. In crystals situation is more complicated (M.A.Novikov, to be published). The IMB should lead to a magnetization 10^{-2}-10^{-3} times smaller than the IFE. Today the IMB is not yet detectable.

Recently there have been claims that electromagnetic radiation is not completely described classically by Maxwellian electrodynamics (Evans M.W., 1992, 1993, Evans and Vigier, 1994, Ahluwalia, D.V, 1995), but its classical description has to be completed by the assertion of a longitudinal magnetic field which accompanies the usual transverse, field components. A phase free longitudinal component would not violate Maxwell's equations. The longitudinal field component designated by $B^{(3)}$ has, it is claimed, all the properties of a propagating magnetic field of constant amplitude. In the interaction between circularly polarized electromagnetic radiation and an electron, the z component of the angular momentum acquired by the electron is given according to Evans by:-

$$J_z = \frac{e^2 c^2}{\omega^2}\left[\frac{B^{(o)}}{(m_o^2 \omega^2 + e^2 B^{(o)2})^{1/2}}\right] B^{(o)}$$

where ω is the angular frequency of the radiation, e the electronic charge, c the velocity of light, m_o is the mass of an electron, and $B^{(o)}$ is the amplitude of the longitudinal magnetic field component. The magnetization due to $B^{(3)}$ for N non-interacting electrons in a plasma will be proportional to both N and J.

For
$$\omega \leq \frac{e\, B^{(o)}}{m_o}$$

then J and thus M become proportional to $B^{(o)}$. Consequently M will scale as the square root of I, the beam intensity.

and for the opposite limit

$$\omega \geq \frac{e\, B^{(o)}}{m_o}$$

J and M scale as $B^{(o)2}$ and thus as I, the beam intensity.

Thus depending on the intensity and angular frequency of the incident radiation, we may detect either a square root dependancy or linear dependancy on the beam intensity.

It has been estimated that for visible radiation of intensity 10^4 Wm^{-2}, the strength of this field would be of order 10^{-5} T. The physical existence of this field has been challenged by several authors on the grounds that (I) its existence would violate charge conjugation symmetry (Barron L.D., 1993) (ii) it is really not a static field as claimed (Grimes D.M. 1993) and (iii) it is not really fundamental (Lakhtakia A, 1993).

In this paper, we are concerned with methods to detect and measure these assumed magnetic field or effective magnetic field components. The measurement technique may either be direct as in the use of SQUID magnetometers or indirect in so far as the field may be detected through effects such as the Inverse Faraday Effect.

The Inverse Faraday Effect has been studied in plasmas (Deschamps et al, 1970, Talin et al,1975) and in the solid state (van der Ziel et al, 1965, Frey J et al, 1991) . The experiment of Deschamps et al 1970 is particularly relevant since, as has been pointed out (Evans and Vigier,1994), if extended and repeated it offers the prospect of detecting the characteristic signature of the assumed longitudinal magnetic field component of electromagnetic radiation which is a square root dependance on the incident intensity as discussed above.

In the original experiment of Deschamps et al 1970 high power pulses (few Mw in 12 microsecs) of microwave radiation (3 Ghz not 30 GHZ as reported in Evans and Vigier, 1994) ionised a low pressure He gas. The electrons so produced are driven into circular orbits by the electric field of the circularly polarised radiation and as a consequence produce an axial magnetic field. The field is detected through induction in a pick-up coil surrounding the ionised gas. The field strength was estimated from the induced signal detected. The dependance of the induced field on the intensity of the incident radiation was investigated by using two polarisers; changing the angle between them produced an intensity variation as cos (2θ). Over the range of intensity investigated, the relationship between incident strength and field was found to be linear. We show in Fig 1 the magnetization produced by the magnetic field as a function of the intensity of the incident circularly polarised microwave radiation. This plot pertains to the conditions of the Deschamps experiment i.e. frequency of 3 Ghz and input power of the order of 300 MW/m^2.

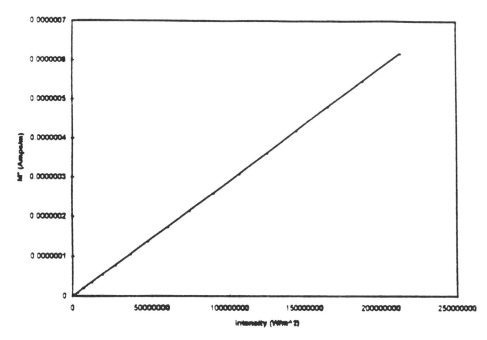

Figure 1. Magnetization produced in a plasma by irradiation with 3 Ghz radiation as a function of incident intensity for maximum intensity = 2.5×10^8 W/m^2.

Under these conditions the relationship between the magnetization and input power is linear as was observed in the actual experiment. We seek to establish the conditions under which the predicted square root dependance on input power could be observed. The square root dependancy will start to dominate if the strength of the B field increases and/or the frequency of the input power decreases for a fixed input power. Figure 2 shows the onset of the non-linear relationship for several thousand MW/m^2.

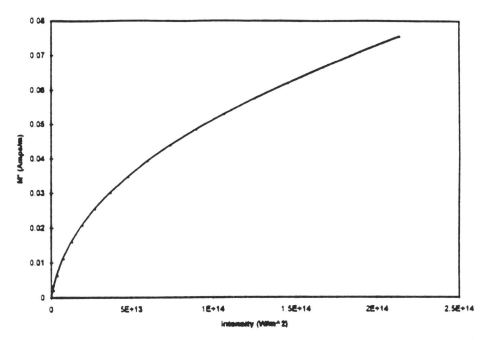

Figure 2. Magnetization produced in a plasma by irradiation with 3 Ghz radiation as a function of intensity for a maximum intensity = 2.5×10^{14} W/m^2.

These enormous powers are not readily available at microwave frequencies. It would be possible to increase the intensity in a Deschamps type of experiment as suggested by Evans and Vigier, 1994 but such methods as focussing etc could only boost the power by a relatively small factor. Alternatively one could reduce the frequency of the incident radiation. Figure 3 shows the same relationship computed for a frequency of 0.3 MHZ.

Here we suggest a number of possible improvements to an experiment of this type which, if implemented, could provide evidence for the predicted square root dependance. Care should be exercised in eliminating any effects which could produce an artifactual non-linear dependance. For example, Chian (1981) has drawn attention to the fact that in experiments of this type, the ions would be accelerated in the opposite direction and consequently produce an opposing field.

This effect is particularly pronounced at high field strengths when the ionised particles move at relativistic speeds. At relativistic speeds the two opposing fields will indeed cancel. In Fig 4 we reproduce Chian's calculations for values of $\mu=$ ratio of the mass of the electron to that of the ion produced= 0 and 1837 and 7348, corresponding to only electrons, ionised hydrogen and ionised helium.

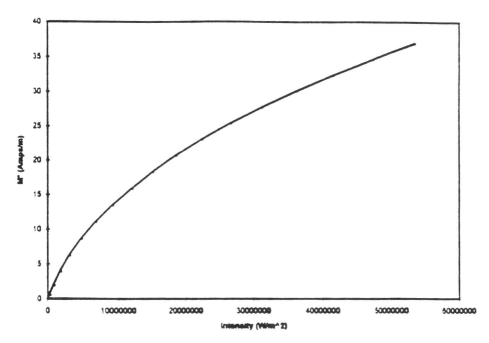

Figure 3. Magnetization produced in a plasma by irradiation with 0.3 MHZ as a function of incident intensity for a maximum intensity of 6×10^7 W/m^2

We have calculated that at incident field amplitudes E of order 2.89×10^5 V/m, that the induced magnetic field produced by the electrons in a helium plasma would be decreased by 10% due to the action of the ions. This corresponds to an intensity of 2.5×10^8 W/m^2 i.e. of the order of the range of intensity corresponding to Fig 1. The effect of an opposing field due to ions would be completely eliminated by using an electron beam in such an experiment rather than relying on ionisation. A further disadvantage of ionisation is that the degree of ionisation will not be independent of the strength of the ionising field. If an electron beam is used these effects are eliminated and further more, for a given incident field and induction coil, the induced voltage will be increased since many more electrons will be contributing to it. For example, if an electron gun were used delivering say 1 amp/cm^2 and the beam cross sectional area were 1 cm^2, then the total number of electrons exposed in a given pulse would be of the order of 10^{13} some two orders of magnitude higher than in the Deschamps experiment. The induced voltage could be further scaled up by using more turns on the induction coil. This experiment employs pulsed microwave circularly polarised radiation interacting with electrons to produce a detectable induced voltage in suitably arranged pickup coils.

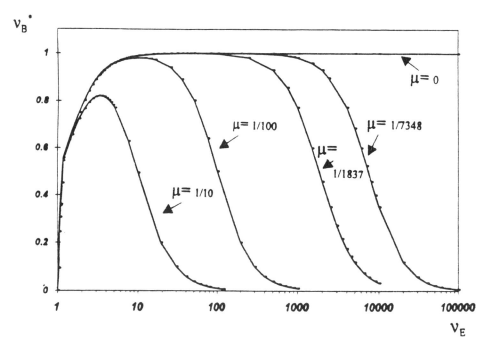

Figure 4. v_b versus v_e for different values of μ, the ratio of the mass of an electron to that of the ion producing it where $v_b = cB/m_0\omega c$ and $v_e = eE/m_0\omega c$. B is the net induced magnetic field and E is the electric field amplitude of the incident radiation.

An alternative scheme would be to use cw circulalry polarised microwave (or RF) radiation and use a pulsed electron beam. Such an arrangement would be possible, for example, by using the type of apparatus commonly used in undergraduate physics experiments to determine the ratio e/m. Here Helmholtz coils surround an glass bulb containing an electron gun and a low pressure gas. The electrons are accelerated perpendicular to the field produced by the Helmholtz coils and under the right conditions move in circular orbts of a few cms radius. Such an apparatus could be used to detect and measure B(3) by exposing the apparatus to cw circulalry polarized radiation, pulsing the electron gun and using the Helmhltz coils as pick-up coils to measure induced voltages.

The experiment of Rikken (1994) was specifically undertaken to provide evidence of the $B^{(3)}$ field. The experiment consists of using a linear polarised probe laser beam synchronised with a circularly polarised laser pulse , both irradiating a sample of benzene. The rotation of the plane of polarisation of the linearly polarised beam produced by an axial magnetic field is detected. The apparatus was calibrated using a static field produced by a solenoid surrounding the sample. The calibration indicated that an axial field as small as 4×10^{-5} T would have been detected by this method. No effect was observed. Evans (1995) has pointed to a number of difficulties with this experiment as far as deciding whether the $B^{(3)}$ actually exists or not (i) benzene is paramagnetic and thus there are no free electrons to produce the $B^{(3)}$ field (ii) the intensity levels and the angular frequency of the irradiation field were too low to produce an observable effect.

Conclusions

A survey has been given of a variety of techniques for investigating the magnetizing effects of circularly polarized electromagnetic radiation. Particular focus has been given to the experimental prospects for both detecting the predicted, longitudinal component of the magnetic field of electromagnetic radiation field and investigating its strength as a function of the intensity of the electromagnetic radiation.

Acknowledgements

The authors acknowledge many interesting discussions with Professors Evans and Sisir Roy.

References

Alekcandrov, E.B. and Zapasski, V.S. (1977) *Phy. Solid State* (Russian), **15**, 3083.

Ahluwalia, D.V., see paper in this proceedings.

Barron, L.D. (1993) *Physica B* **190**, 307-309.

Barrone, A. and Paterno, (1984) *Physics and the Application of the Josephson Effect*, J.Wiley.

Braginsky, V.B. (1970) *Physical Experiments with Probe Bodies*, (in Russian), Nauka.

Buckingham, A.D. and Parlett, L.C. (1994) *Science* **264**, 1748.

Chian, A.C.-L., (1981) *Phys. Fluids*, **24 (2)**, 369.

Cohen-Tanoudji, J., Du-Pont-Roc, Harouche, S., and Llac, F. (1969) *Phys.Rev.Lett.* **15**, 758.

Day, G.W and Rose, A.H. (1988) *Proc SPIE* **985**, 138.

Deschamps, J., Fitaire, M., and Lagoutte, M. (1970) *Phys.Rev.Lett.* **25,19**, 1330.

Efremov, G.F. and Novikov, M.A. (1994) *Laser Physics* **4**, 11.

Evans, M.W. (1992) *Physica B* **82**, 227.

Evans, M.W. (1993) *Physica B* **183**, 103.

Evans, M.W. and Vigier, J.-P. (1994) *The Enigmatic Photon*, Kluwer Academic Publishers.

Evans, M.W., Roy, S., and Jeffers, S. (1995) *Nuovo Cimento*, in press.

Falfeau, P. et al (1994) *Rev.Sci.Instr.* **65**, 1916.

Geng, Q. and Goto, E. (1993) *J.Appl.Phys.* **74 (10)**, 15.

Grimes, D.M. (1993) *Physica B* **191**, 367-371.

Humphrey, F.B. and Johnson, A.B. (1963) *Rev.Sci.Inst.* **34**, 388.

Lakhtakia, A. (1993) *Physica B* **191**, 362-366.

Lenz, J.E. (1990) *Proc. IEEE*, **78**, 973.

Liao, P.E. and Bjorklund, G.C. (1976) *Phys.Rev.Lett.*, **20**, 584.

Lukens, J.E., Warburtio, R.J., and Webb, W.W. (1971) *J.Appl.Phys.*,**42**, 22.

Novikov et al (1977) *JETPh Lett* (USSR) **25 (9)**, 405.

Novikov, M.A., to be published.

Rikken, G.L.J.A. (1995) *Optics Letts*, **20 (8)**, 846.

Romani, G.L., Williamson, S.J., and Lautia, L. (1982) *Rev.Sci.Instr.* **53**, 1815.

Svirko, Yu.P. and Zheludev, N.I. (1994) *J.Opt.Soc.Am.B*, **11**, 1388.

Talin, B., Kaftandjian, V.P., and Klein, L., (1975) *Phys.Rev.A* **11**, 2.

Van der Ziel, J., Pershan, P., Malstrom, L., (1965) *Phys.Rev.Lett.* **15**, 190.

Wagnier, G. (1989) *Phys.Rev.A*, **40**, 2437.

Wellstood, F.C., Urbina, Ca, and Clake, J. (1987) *Appl.Phys.Lett.* **50**, 772.

GENERALIZED EQUATIONS OF ELECTRODYNAMICS OF CONTINUOUS MEDIA

V.M.DUBOVIK AND B.SAHA
Laboratory of Theoretical Physics
Joint Institute for Nuclear Research, Dubna
141980 Dubna, Moscow region, Russia.
e-mail: dubovik@theor.jinrc.dubna.su
e-mail: saha@theor.jinrc.dubna.su

AND

M.A.MARTSENUYK
Department of Theoretical Physics
Perm' State University,
15 Bukirev Str., 614600 Perm' Russia.
e-mail: mrcn@pcht.perm.su

Development of quantum engineering put forward new theoretical problems. Behavior of a single mesoscopic cell (device) we may usually describe by equations of quantum mechanics. However if experimentators gather hundreds of thousands of similar cells there arises some artificial medium that one already needs to describe by means of electromagnetic equations. In the present work it is demonstrated that the inherent primacy of vector potential in quantum systems leads to a generalization of the equations of electromagnetism by introducing in them toroid polarizations. We mention some of their applications.

Key words: Toroid multipole moments, Toroidomagnetics, Electromagnetotoroidics, Toroidomagnetostatics, Magnetoelectronics

1. Introduction

This report is devoted to **Prof. Jean-Pierre Vigier** who has made very valuable contributions in the development of Quantum Mechanics and Electrodynamics. It is well-known that these two disciplines overlap within the scope of atomic physics. Electrodynamics of continuous media concerns much more intricate problems in both field and matter aspects. We show that usual equations of electromagnetic media are incomplete even in their fundamental representation and correct this oversight. It should be emphasized that we are not dealing with innovations based on additional (even plausible) hypotheses but with an inevitable modification imperatively following merely from the facts of our three-dimensional life.

Our report consists of two parts. The first one can be considered as some formal deductions. Its starting point is the demonstration of existence of the third family of multipole moments – the *toroid* one in multipole expansion of electromagnetic current (for their first strict introduction see ref. [1]). It was just the cause that made it necessary to modify the equations of electrodynamics of continuous media. Simply speaking, in addition to electrical polarization **P** and magnetization **M** one has to introduce toroid polarization **T** in the (vector!) equations. The impact of this operation is not trivial at all. The matter is that the toroid moments are the multipole sources of free-field potentials [2, 3], that are responsible in particular for effects like the Aharonov–Bohm one. In fact, we know that only quantum particles can serve as a detector of this potential. Thus, there arises a series of principle questions. For example, questions relating to the transition from quantum mechanical description of electromagnetic phenomena to the description with the help of classical equations. This concerns the profound physical problems that will not be discussed here (see, e.g. [4]). Now we will give a high-light of the history of the discovery of toroid moments and how they were associated with experiments to study the Aharonov–Bohm effect. Recently A. Tonomura and others [5] have observed the interference of electrons on a shielded ferromagnetic ring of mesoscopic size. Distribution of vector-potential created by the source mentioned needed computation. The most detailed calculation of this distribution was done by G. Afanas'ev [6]. Shortly after that it was noticed that the toroid dipole moment plays the role of a point-like source of this kind of distribution [2].

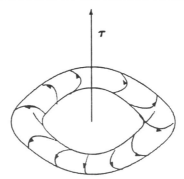

Figure 1. Poloidal current on the torus determines the toroid dipole moment. The simplest model of this is an ordinary toroidal coil with even number of windings.

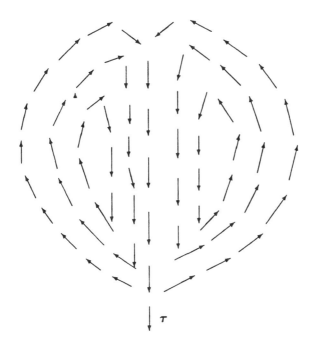

Figure 2. Poloidal lines of currents in a simply connected volume, the sphere (θ = const., where θ is the polar angle of spherical coordinate system.) It is a model of the Hill's vortex.

Let us explain what this moment is? From the geometrical point of view its model is the poloidal current on a torus [Fig. 1]. Macroscopically, its model is created by the usual toroid coil with an even number of windings. There is a hydrodynamical analog of this construction – Hill's vortex [Fig. 2].

It is easy to demonstrate how in the system of three particles one can emphasize all three dipole moments [7]. Suppose that a steady system consists of the sun (S), the earth (E) and the moon (M). Suppose the earth and the moon are oppositely charged and the sun is neutral [Fig. 3]. Then, in each given instant it may be convenient to describe the subsystem E-M by an electric dipole moment **d**. If the intrinsic angular velocity Ω_d in the E-M subsystem is high in comparison with its external rotation around S, $\Omega_m \ll \Omega_d$, we may observe the magnetic dipole **m** [Fig. 4]. If Ω_m increases, we have to take into account the toroid properties of the system, i.e. the toroid dipole τ.

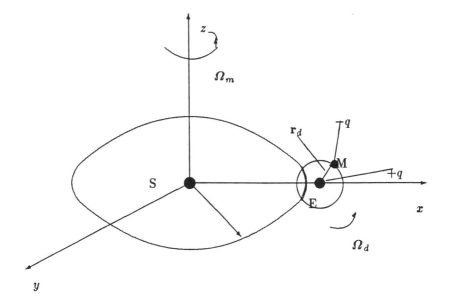

Figure 3. Microscopically we see two separate charges and at each instant the electric dipole moment $\mathbf{d} = q\,\mathbf{r}_q$.

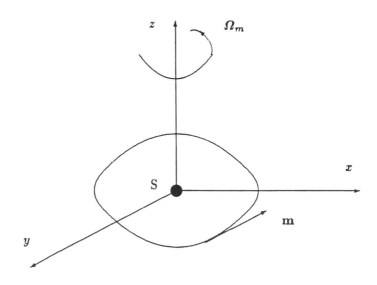

Figure 4. If we "see" the magnetic dipole (avaeage on the motion of the moon) $|\mathbf{m}| = \pi\,I_d\,r_d^2$, where $I_d = -q\Omega_d/2\pi$.

Remark: As far as in the atoms and nuclei their magnetic fluxes are mainly confined inside these many–body systems, they can possess great toroid moments. Moreover, it is not difficult to show that $[L_z, \hat{\tau}_z] = 0$ and, for example, in the external nonuniform and/or alternating fields we may observe the effect of atomic spectral line splitting additional to the Stark's and Zeeman's ones [1].

How does the dipole moment τ arise in electromagnetic current distribution? From the formal point of view 4-currents possess 4 scalar components and each of them can be expanded in multipole series. Implying one constrain, the current conservation condition, we should obtain three families of moments and in each of them will have its proper dipole. (Their definitions are given in Table – 1).

Table 1. The models and the definitions of three kinds of current dipole moments: a bit of the linear current, a ring with the circular current, the toroidal coil.

$\dot{\mathbf{d}} = \int \mathbf{j}\, dV$	$\mathbf{m} = \frac{1}{2}\int [\mathbf{r}\times\mathbf{j}]dV$	$\tau = \frac{1}{10}\int[\mathbf{r}(\mathbf{r}\cdot\mathbf{j}) - 2r^2\mathbf{j}]dV$
$\dot{\mathbf{d}} = I\mathbf{l}$	$\mathbf{m} = \frac{1}{2}I\mathbf{S}$	$\tau \approx \Phi\mathbf{S}, |\tau| = IV_{\text{torus}}$

A renowned soviet physicist Ya. B. Zeldovich [8] was the first to notice that the toroid coil was impossible to identify with any multipole moments, starting with quite different considerations. He assumed that as the third dipole one could take the classical analog of P-odd form-factor of spin 1/2 particle, named the anapole. However, further it was proved that this kind of assumption was not completely correct. For example, the anapole cannot radiate at all while the toroid coil can. The matter is that the anapole is some composition of electric dipole and actual toroid dipole giving destructive interference of their radiation.

Creation of a complete theory of multipole expansion appeared to be a very intricate problem. The first correct and vast article [1] on this topic was published only in 1974 and ref. [9] can be considered as the last one. In the theory of continuous media, there appear possibilities of introducing more families of multipole moments [9], e.g. for electric–dipole medium τ_e [10]. Four dipole moments $\mathbf{d}, \mathbf{m}, \tau$ and τ^e manifest all possible combinations of properties at inversions of space and time and form a complete vector basis of order parameters for describing crystalline substances [11]. By this time, possibilities and demands of physicists and technologists have been also increased. For example, now-a-days system of thousands of mesoscopic ferrite rings is being produced, studied and applied. However, the appropriate experimental and theoretical results in consideration don't coincide yet [12, 13].

What kind of equations can describe the properties and response of such magnetic medium? Magnetic field in this medium is confined inside the rings unlike outside of it there is only the distribution of free–field vector-potential. The first order equations for **E** and **B**, which are called the Maxwellian, are not obviously sufficient for this purpose. We offer a new two potential formulation (see also [14]).

2. Static dipole moment of toroid coil and a free-field potential created by it

We begin with a static problem. Let us first find the distribution of the vector potential **A** produced by a "point–like" poloidal current I. The toroid dipole moment of the toroidal coil is $\tau = I V$, where V is the volume of the coil (torus). In the (quasi)static case, the basic equation (with the gauge condition div**A** $= 0$, valid outside the source) has the form

$$\text{curl curl}\, \mathbf{A} = \text{curl curl}\, \boldsymbol{\tau}\, \delta(\mathbf{r}). \tag{1}$$

Its solution is a convolution of two distributions, the Green function and the δ- function, and is to be determined on a suitable test vector function. Thus we may get [2]

$$\begin{aligned}\mathbf{A} &= \text{curl curl}\, \boldsymbol{\tau} r^{-1} = \boldsymbol{\tau}\Delta(1/r) + \boldsymbol{\tau}\cdot\nabla\nabla(1/r) = \\ &= \frac{3\mathbf{r}\mathbf{r}\cdot\boldsymbol{\tau} - r^2\boldsymbol{\tau}}{r^5} + \boldsymbol{\tau}\delta(\mathbf{r}).\end{aligned} \tag{2}$$

We see that the toroid dipole moment τ produces the potential distribution **A**, just like **d** produces the electric field **E** and **m** the magnetic induction **B**. Therefore, for media

$$\mathbf{D} = \mathbf{E} + \mathbf{P} \quad \text{and} \quad \mathbf{B} = \mathbf{H} + \mathbf{M}. \tag{3}$$

What about **A** and τ?

3. Generalized equations of electromagnetism

Let us note that as early as 1977, V. Dubovik with his collaborators showed that the crystal media in general can hardly be described without introducing polar and axial toroid polarizations [10]. Even then, the question of generalizing the fundamental equations of electrodynamics of continuous media came to the light. First, they were presented at the seminar of LTPh, JINR in January 1991 and published in 1994 [14]. In this paper, we will present the inductive and deductive foundations of these equations and some of their consequences.

Let us write the magnetostatic equations in their two equivalent forms (see for example [15])

$$\text{curl}\,\mathbf{H} = \mathbf{j}, \tag{4a}$$

$$\text{div}\,\mathbf{H} = -\text{div}\,\mathbf{M}, \quad \mathbf{M} = \mathbf{M}(\mathbf{H}), \tag{4b}$$

$$\text{curl}\,\mathbf{H} + \text{curl}\,\mathbf{M} = \text{curl}\,\mathbf{B} = \mathbf{j}^{\text{free}} + \text{curl}\,\mathbf{M}, \tag{4c}$$

$$\text{div}\,\mathbf{B} = 0, \quad \mathbf{M} = \mathbf{M}(\mathbf{B}). \tag{4d}$$

It is easy to see that without introducing the vector-potential we cannot describe an arbitrary magnetic medium, e.g., consisting of closed chains of magnet, i.e., dipoles (e.g., in the form of a ring [Table 2]).

Table 2. Basic properties of the dipole moments under the spatial and temporal inversions and their interactions with external fields.

	$P(\mathbf{r} \to -\mathbf{r})$	$T(t \to -t)$	
d	−	+	$\mathbf{d}\cdot\mathbf{E}$
m	+	−	$\mathbf{m}\cdot\mathbf{B}$
τ^e and τ^m	−	−	$\tau\cdot\dot{\mathbf{D}}$ or $\tau\,\text{curl}\,\mathbf{B}$
τ^e	−	−	$\tau^e\cdot\dot{\mathbf{B}}$ or $\tau^e\,\text{curl}\,\mathbf{E}$

Really, in this case the macroscopic pattern of such a medium has $\mathbf{M} \equiv 0$ and in the absence of free currents we obtain $\mathbf{B} \equiv \mathbf{H}$. Then, the magnetostatic equations are trivialized:

$$\begin{aligned}\text{curl}\,\mathbf{B} &= 0, & \text{div}\,\mathbf{B} &= 0,\\ \text{curl}\,\mathbf{H} &= 0, & \text{div}\,\mathbf{H} &= 0,\end{aligned} \tag{4}$$

from where it seems that we should conclude, according to the Helmholtz theorem, that $\mathbf{B} \equiv 0$ and $\mathbf{H} \equiv 0$ all over the space. But it is not correct. The fact here is that $\mathbf{M} \equiv 0$ is taken on average, but each physical volume, occupied by a closed chain, becomes topologically non-trivial one!

So we rewrite the magnetostatic equation (4c) through **a** (an analog of **H**) *at the same time* adding the contribution of toroid polarization to the left- as well as right-hand sides, like in the transition from (4a) to (4c):

$$\text{curl}\,\text{curl}\,\mathbf{a} + \text{curl}\,\text{curl}\,\mathbf{T} = \mathbf{j}^{\text{free}} + \text{curl}\,\mathbf{M} + \text{curl}\,\text{curl}\,\mathbf{T}, \tag{5}$$

and at all points of space we introduce

$$\alpha := \mathbf{a} + \mathbf{T}. \tag{6}$$

In the absence of free charges, we may put for α (an analog of $\mathbf{B}$) that $\operatorname{div} \alpha = 0$. Consequently, the fundamental equations of toroidomagnetostatics can be written as

$$\operatorname{curl} \operatorname{curl} \alpha = \mathbf{j}^{\mathrm{free}} + \operatorname{curl} \mathbf{M} + \operatorname{curl} \operatorname{curl} \mathbf{T}, \tag{7}$$

$$\operatorname{div} \alpha = 0, \tag{8}$$

with the relation (6), that holds all over the space. The latter system is the analog of the equations for $\mathbf{B}$. The inverse reduction of this system may be considered as an analog of equations (4a) and (4b) for $\mathbf{H}$, naturally appears as (4a) and (4b)

$$\operatorname{curl} \operatorname{curl} \mathbf{a} = \mathbf{j}^{\mathrm{free}} + \operatorname{curl} \mathbf{M}, \tag{9}$$

$$\operatorname{div} \mathbf{a} = -\operatorname{div} \mathbf{T}. \tag{10}$$

Employing the Helmholtz theorem, we obtain the solution to the latest system in the form:

$$\mathbf{a}(\mathbf{r}) = \int \frac{\mathbf{j}^{\mathrm{free}}(\mathbf{r}')}{|\mathbf{r}-\mathbf{r}'|} d^3 r' + + \int \frac{\operatorname{curl} \mathbf{M}(\mathbf{r}')}{|\mathbf{r}-\mathbf{r}'|} d^3 r' + \nabla \int \frac{\operatorname{div} \mathbf{T}(\mathbf{r}')}{|\mathbf{r}-\mathbf{r}'|} d^3 r'. \tag{11}$$

The first two terms give the contribution, which is usually denoted as magnetic vector potential $\mathbf{A}(\mathbf{r})$; in our notation, the *total* vector-potential is $\alpha = \mathbf{a} + \mathbf{T}$. Obviously, the definition of the magnetic field is also changed due to toroid polarization (which can be nonhomogeneous for the given concrete medium, at least in the form of the surface effect [11]) as follows

$$\beta := \operatorname{curl} \alpha = \operatorname{curl} \int \frac{\mathbf{j}^{\mathrm{free}}(\mathbf{r}')}{|\mathbf{r}-\mathbf{r}'|} d^3 r' + \operatorname{curl} \int \frac{\operatorname{curl} \mathbf{M}(\mathbf{r}')}{|\mathbf{r}-\mathbf{r}'|} d^3 r' + \operatorname{curl} \mathbf{T}(\mathbf{r}). \tag{12}$$

As one can see, the first two terms gives the usual definition of magnetic field (according to the old terminology, the magnetic induction) in a medium, where as the magnetic field in the much of crystals, (e.g., without inversion center of a cell [3]) or on the surface of a crystal, studied by the experimentallists (e.g. by means of magneto–optical devices), will be also contributed by the third term.

Remark, naturally, the toroidomagnetostatic equation turns into the wave equation if one adds to $\operatorname{curl} \operatorname{curl} \alpha$ the term $\ddot{\alpha}$

$$\Box \alpha = \operatorname{curl} \mathbf{M} + \operatorname{curl} \operatorname{curl} \mathbf{G}. \tag{13}$$

Immediate generalization of equations of electromagnetism may be schematically made as follows. If in a given medium there are no free charges and currents, it can be described by the usual transverse equation:

$$\operatorname{curl} \mathbf{D} + \dot{\mathbf{B}} = \operatorname{curl} \mathbf{P}, \quad \operatorname{div} \mathbf{D} = 0, \tag{14}$$

$$\operatorname{curl} \mathbf{B} - \dot{\mathbf{D}} = \operatorname{curl} \mathbf{M}, \quad \operatorname{div} \mathbf{B} = 0. \tag{15}$$

We may now transit to the 2–potential formulation through $\boldsymbol{\alpha}^m$ and $\boldsymbol{\alpha}^e$ and introduce electric and magnetic toroid polarizations $\mathbf{T}^e$ and $\mathbf{T}^m$ through substitution [14]:

$$\mathbf{D} \Longrightarrow -\dot{\boldsymbol{\alpha}}^m + \operatorname{curl} \boldsymbol{\alpha}^e, \quad \operatorname{curl} \mathbf{P} \Longrightarrow \operatorname{curl} \mathbf{P} + \operatorname{curl} \operatorname{curl} \mathbf{T}^e, \tag{16}$$

$$\mathbf{B} \Longrightarrow \dot{\boldsymbol{\alpha}}^e + \operatorname{curl} \boldsymbol{\alpha}^m, \quad \operatorname{curl} \mathbf{M} \Longrightarrow \operatorname{curl} \mathbf{M} + \operatorname{curl} \operatorname{curl} \mathbf{T}^m. \tag{17}$$

Then, we obtain

$$\ddot{\boldsymbol{\alpha}}^e + \operatorname{curl} \operatorname{curl} \boldsymbol{\alpha}^e = \operatorname{curl} \mathbf{P} + \operatorname{curl} \operatorname{curl} \mathbf{T}^e, \tag{18}$$

$$\ddot{\boldsymbol{\alpha}}^m + \operatorname{curl} \operatorname{curl} \boldsymbol{\alpha}^m = \operatorname{curl} \mathbf{M} + \operatorname{curl} \operatorname{curl} \mathbf{T}^m. \tag{19}$$

If we choose a gauge condition $\operatorname{div} \boldsymbol{\alpha}^{e,m} = 0$, we may again obtain the form (13).

4. Conclusion

"... *it is impossible to introduce electrodynamics of "matter in general"*
- from the book by Russian academician **E. A. Turov (1983)**

It should be noticed that we do not consider contributions of high multipole moments in the Maxwell equations. Here, we develop only the macroscopic description of electromagnetotoroidic dipole media. There arose a large field of activity to model the material equations of concrete media. We did not consider here the problem of alignment of microscopic toroid moments by crystalline fields. Ideal static toroid moments do not interacts with each other at all. However, toroidization can appear due to dynamical effects (see e.g. [16]). Among the latest machinery we point out the articles [17, 18] that directly precede applications of toroid moments in the area of high technologies.

References

1. Dubovik, V.M. and Cheshkov, A.A. (1974) Multipole Decomposition in Classical and Quantum Field Theory and Radiation, *Sov. J. Part. Nucl.*, **5**, 3, 318-337.
2. Dubovik, V.M. (1989) On vector-potential distributions outside the toroidal solenoids, *JINR Rapid Communications*, 3[36]-89, 39-41.

3. Dubovik, V.M. and Tugushev, V.V.(1990) Toroid moments in electrodynamics and solid-state physics, *Phys. Rep.*, **187**, 4, 145-202.
4. Resta, R. (1994) Macroscopic polarization in crystalline dielectrics: the geometric phase approach, *Revew of Mod. Phys.*, **66**, 3, 899-915.
5. Peshkin, M. and Tonomura, A. (1989) The Aharonov–Bohm effect, *Springer Ferlag*, Berlin.
6. Afanas'ev, G. N. (1987) Closed analytical expressions for some useful sums and integrals involving Legendre functions, *J. Comp. Phys.*, **69**, 196-208;
   Afanas'ev, G. N. (1988) The scattering of charged particles on the toroidal solenoid, *J. Phys. A*, **21**, 2055-2110.
7. Dubovik, V.M. and Shabanov, S.V. (1993) The gauge invariance, radiation and toroid order parameters in electromagnetic theory, in Lakhtakia, A. (ed.), *Essays on the Formal Aspects of Electromagnetic Theory*, World Scientific, Singapore, pp. 399-474.
8. Zel'dovich, Ya.B. (1958) Electromagnetic interaction under parity-non-conservation, *Sov. Phys. JETP*, **6**, 1184.
9. Dubovik, V.M. and Kurbatov, A.M. (1994) Multipole interactions of dipole and spin systems with external fields, in Barut, A.O., Feranchuk, I.D., Shnir, Ya.M. and Tomil'chik, L.M. (eds), *Quantum Systems: New Trends and Methods*, World Scientific, Singapore, pp. 117-124.
10. Dubovik, V.M., Tosunyan, L.A. and Tugushev, V.V. (1986) Axial toroid moments in electrodinamics and solid state physics, *Sov. Phys. JETP*, **63**, 2, 344-351.
11. Dubovik, V.M., Krotov, S.S. and Tugushev, V.V. (1987) Toroid current structures in ferro- and antiferromagnets, *Krystallographia* **32**, 3, 540-549.
12. Altland, A. and others (1992) Persistant currents in an ensemble of isolated mesoscopic rings, *Ann. Phys.*, **219**, 148-186.
13. Kamenev, A. and Gefen, Y. (1995) (Almost) everything you always wanted to know about the conductance of mesoscopic systems, *Intern. J. of Mod. Phys. B*, **9**, 7, 751-802.
14. Dubovik, V.M. and Magar, E.N. (1994) Inversion Formulas for the Decompositions of Vector Fields and Theory of Continuous Media, *J. Mosc. Phys. Soc.*, **3**, 1-9.
15. Vlasov, A. A.(1955) *Macroscopic electrodynamics*, Moscow.
16. Dubovik, V.M., Martsenuyk, M.A. and Martsenuyk, N.M. (1995) Reversal of magnetizationof aggregates of magnetic particle by a vorticity field and use of toroidness for recording information, *J. of Mag. and Mag. Mat.*, **145**, 211-230.
17. Dubovik, V.M., Martsenuyk, M.A. and Martsenuyk, N.M. (1993) Toroid polarization of aggregated magnetic suspensions and composites and its use for information storage, *Phys. Part. Nucl.*, **4**, 453-484.
18. Dubovik, V.M., Lunegov, I.V. and Martsenuyk, M.A. (1995) Toroid response in nuclear magnetic resonance, *Phys. Part. Nucl.*, **26**, 1, 72-100.

MAXWELLIAN ANALYSIS OF REFLECTION AND REFRACTION

R. D. PROSSER[a], S. JEFFERS and J. DESROCHES[b]
[a] *16 Elaine Court, 123 Haverstock Hill, London NW3 4RT*
[b] *Department of Physics and Astronomy, York University, North York, Ontario, Canada*

The solution to Maxwell's Equations for a limited Gaussian beam incident on a partially reflecting boundary is used to study the spatial distribution of the amplitude, phase and Poynting vector for the relected and refracted beams. This analysis demonstrates that the energy of the reflected light comes from one side only of the incident beam and that all the transmitted energy comes from the other side.

1. Introduction

Everyone has observed that when light falls on glass a part is transmitted and a part reflected. But what is it that determines which part is reflected and which transmitted?

Newton I, (1704) was probably the first person to consider this problem in detail. Towards the end of Part 2 of his *Opticks*, he writes that "the disposition of any ray to be reflected I will call its 'Fits of easy reflexion' and those of its disposition to be transmitted its 'Fits of easy transmission'". The theory of Fits was not regarded as satisfactory, however, even in the eighteenth century. Brougham (1796) saw in it "the smoke of unintelligible theory" while Knox (1815), claimed that it was "generally admitted to be inadequate". Newton himself seems also to have had some reservations, for after developing his theory of Fits he writes, "But whether this hypothesis be true or false I do not here consider. I content myself with the bare Discovery that the Rays of Light are by some cause or other alternately disposed to be reflected and refracted for many vicissitudes".

During the 300 years that have elapsed since Newton first addressed the reflection problem much progress has been made. Classical electromagnetic theory through the well-known Fresnel equations relates reflection and transmission coefficients to the refractive index. The quantum theory of light assigns a probability of reflection that is related to the classically derived reflection coefficient, but the question as to which photons are reflected remains unanswered and indeed according to the orthodox interpretation cannot, in principle, be answered.

In this context, it is interesting that a slight development of the classical theory gives a significant insight into this question. The development consists in finding the solution to Maxwell's equations for a spatially limited beam and constructing the Poynting vectors that delineate the direction of energy flow. It thus appears that all of the reflected energy comes from one side only of an obliquely incident beam.

This conclusion should be testable with fairly simple apparatus and if confirmed would have significant implications for the interpretation of a number of the important historic experiments in classical physics.

2. Poynting Vectors for a Limited Light Beam

We have previously studied diffraction and interference effects using classical electromagnetic theory which yields the spatial distributions of the amplitudes, phase and Poynting vectors of the diffracted radiation, (Jeffers S, Prosser R.D, Hunter G, Sloan J , 1994). For reflection and refraction, the exact solution to Maxwell's equations for an infinite beam incident on an infinite glass interface is well known and given in many textbooks. The solution for a limited Gaussian beam is given by Prosser (1985).

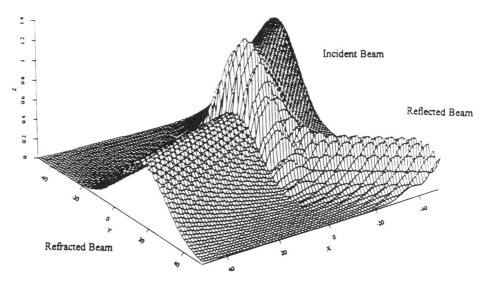

Figure 1. The spatial distribution of electric field amplitude for an incident Gaussian beam incident at an angle of 40 degrees.

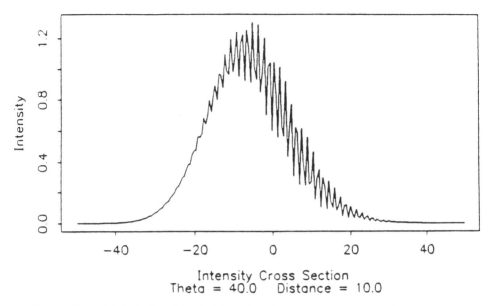

Figure 2. The spatial distribution of electric field amplitude in a plane parallel to the interface and spaced 10 wavelengths back on the side of incidence. This shows evidence for interference effects between the incident and reflected beams.

The solution gives the components of the electric vector E and the magnetic vector H at every point in the incident, reflected and transmitted light beams. This enables the Poynting Vector, defined as ExH, to be calculated at any point. The direction of the resulting Poynting Vectors is shown in Figure 3.

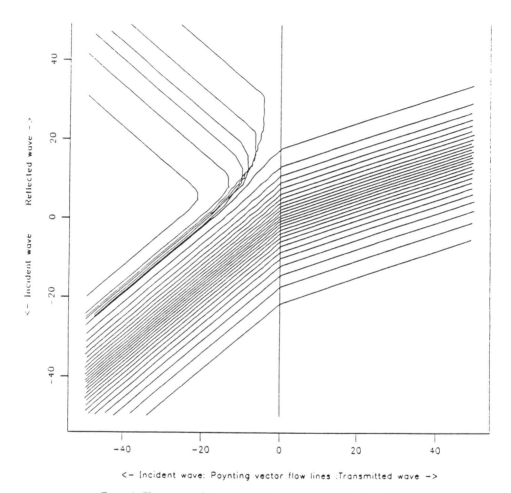

Figure 3. Electromagnetic energy flow lines defined by the Poynting vectors.

The calculations are for an angle of incidence of 40 degrees, relative index of refraction = 2, and width of the Gaussian incident beam = 10 standard deviations. The plots show how the energy of the incident beam divides into two sets of flow lines, one corresponding to the transmitted beam and the other to the reflected beam. All of the flow lines that form the reflected beam start on the top side of the incident beam. Since the Poynting Vector represents the direction of energy flow, we conclude that all of the energy in the reflected beam comes from the top side of the incident beam. The reflected flow lines fan out in such a way that the reflected beam has the same width as the incident beam, however, thus creating the impression that the reflected energy could have come from any part of the incident beam. We name this probably false impression the "usual understanding", since although it is an assumption which is never stated explicitly, it is frequently implicit in discussions of experiments involving reflection from glass or of the closely related phenomena involving semi-transparent mirrors and beam splitters.

3. Experimental possibilities

The experimental investigation of the solution should be possible using microwave frequencies of about 3 cm in conjunction with a glass sheet about 3 m in width. A detector consisting of a dipole antenna connected to a diode and a voltmeter should enable the electric field amplitude to be plotted as in Figure 2. Replacing the dipole with a directional loop of Yagi antenna should enable the direction of energy flow to be plotted as in Figure 1. There will be reflections from both sides of the glass, but provided the glass is not more than about 2 mm thick the two reflections will be almost in phase and the pattern of the result should not be significantly affected. Bearing in mind that in this classical domain Maxwell's equations are known to be valid, any discrepancy between the experimental and theoretical results would be extremely surprising.

4. Photon trajectories ?

In the quantum description, the energy in electromagnetic radiation is carried by photons. Since the Poynting vector delineates the direction of energy flow, it is natural to assume that photons, if localised, follow the Poynting vector trajectories. For many years it has been unfashionable to speak in terms of particle trajectories on account of the widespread adherence to the Copenhagen interpretation of quantum theory which allows only observable initial and final states to be described. The inability of this interpretation to deal satisfactorily with the problem of measurement, however, led to new hidden variable interpretations by Bohm (1952) and Vigier (1987) in which particle trajectories are included. In the light of the experiment of Mizobuchi and Ohtake, (1992) based on the suggestion by Ghose, Home and Agarwal, (1992), the status of the Prnciple of Complementarity is challenged on the basis of empirical evidence. The interpretation of this experiment (Ghose and Home, 1992) which allows for particles to have *simultaneously and objectively* wave properties is consistent with our suggestions.

However, it must be noted that there are now a significant number of optical effects including photon anti-bunching, squeeezed states, sub-Poissonian photon statistics and the violation of various classical statistical inequalities which do not admit of any classical description (see, for example, Perina, Hradil and Jurco, (1994)).

Einstein (1909) suggested that Maxwell's equations might yield pointlike singular solutions as well as waves. A stable electromagnetic packet was first discovered by Thomson in 1936; the same solution was discovered independently again by Ricard (1978). This packet has zero intensity on its axis rising to a maximum at a radius of about 1 wavelength and a length that depends upon the coherence properties of the source. It is circularly polarized and propagates without change in shape. It is possible, therefore, that this stable wavepacket could represent the photon.

The photon wavepacket can be represented as a Fourier superposition of infinite waves of appropriate temporal and spatial frequencies. These will sum to zero except in the region where the photon wavepacket has non-zero electric and magnetic fields. We assume that all the Fourier components have to satisfy the boundary conditions at the glass surface and at the collimation apparatus that limits the beam. These restrictions cause the centre of the wavepacket to follow one of the family of trajectories shown in Figure 1. The effect is similar to that found in the case of an infinite beam incident upon a glass surface, where boundary conditions necessitate the creation of reflected and transmitted beams, but the details lie beyond the scope of this paper. An analysis of this type has been given for the case of double slit interference, however. (Jeffers S., Prosser R.D., Berseth W., Hunter G., Sloan J., 1994)

5. Discussion

This paper presents a testable theory of reflection which has significant implications. In this theory, reflection is a deterministic process; for an inclined incident beam all the reflected light comes from one side of the beam and the transmitted light from the other. This conclusion contrasts markedly with the usual understanding which assumes that reflection is controlled by some unknown statistical process and that both the transmitted and reflected light could have come from any part of the incident beam.

A similar analysis can be applied to beam splitters which are widely used in physical experiments and instruments, (e.g. the Michelson interferometer), and a similar result obtained, namely that half the incident beam is reflected and the other half transmitted. Beam splitters are used in some fundamental experiments designed to illuminate physical processes. One example is the delayed choice experiment of Wheeler (1980) which leads to the paradoxical result that it is possible to modify the past. This experiment involves two beam splitters and is analysed according to the usual understanding of reflection. Analysis via the present theory would lead to a different conclusion.

Another interesting aspect of the new theory concerns the reflection of very low intensity beams such that only one photon is present in the apparatus, on average, at any moment. How is the photon constrained to follow one of the designated trajectories? The answer that we have given is in terms of the requirement that the unmanifest Fourier components should satisfy the electromagnetic boundary conditions at the glass surface and at the collimating apparatus that forms the Gaussian beam. This suggests that undivided wholeness previously associated with quantum behaviour is present also in the classical phenomenon of reflection.

6. References

Bohm, D. (1952) *Phys.Rev*. **85**, 166, 180.

Brougham, H. (1796) Experiments on the inflection, reflection and colours of light, *Phil.Trans*. **86**, 1.

Einstein, A. (1909) *Physikalische Zeitschrift* **10**, 817-825.

Ghose, P., Home, D. and Agarwal, G.S., (1991) *Phys. Lett. A* **153**, 403, and (1992) *Phys. Lett.. A* **168**, 95.

Ghose, P. and Home, D., (1992) *Founds Phys*, **22**, 12, 1992.

Jeffers, S., Prosser, R.D., Hunter, G. and Sloan, J. (1994) Classical Electromagnetic Theory of Diffraction and Interference: Edge, single slit and double slit solutions, in *Waves and Particles in Light and Matter*, A. van der Merwe and A. Garrucio, eds, Plenum Press, New York.

Jeffers, S., Prosser, R.D., Berseth, W., Hunter, G. and Sloan, J. (1995) Maxwellian Analysis of the Pulsed Microwave Double Slit Experiment, *Proceedings of the Second Conference on Ultra-Wide Band Short Pulse Electromagnetics*, L.Carin and L.B. Felson, eds, Plenum Press, New York.

Knox. J. (1815) On some phenomena of colours exhibited by thin plates, *Phil.Trans*. **105**, 162.

Mizobuchi, Y. and Ohtake, Y. (1992) *Phys Lett A* **168**, 1-5.

Newton, I. (1704) *Opticks, Part 2*, Dover, 281.

Perina, J., Hradil, Z. and Jurco, B. (1994) *Quantum Optics and Fundamentals of Physics*, Kluwer Academic Publishers, Dordrecht.

Prosser, R.D. (1985) in *Open Questions in Quantum Physics*, G. Tarozze, A. van der Merve, eds, Reidel, 121-122.

Ricard, J. (1984) *Annales de la Fondation Louis de Broglie* **9**, 101-124.

Thomson, J.J. (1936) *Nature* **137**, 232.

Vigier, J.-P. (1987) in *Quantum Uncertainties - Recent and Future Experiments and Interpretations*, W.M. Honig, D.W. Kraft and E. Panarella, eds, Plenum, New York.

Wheeler, J. A. (1980) Delayed Choice Experiments and the Bohr-Einstein Dialog, presented at the joint meeting of the American Philosophical Society and the Royal Society, London, June, 1980.

SOLUTIONS OF MAXWELL EQUATIONS FOR A HOLLOW CURVED WAVE CONDUCTOR

V.BASHKOV AND A.TCHERNOMOROV
Kazan State University, Kazan, Russia

In the present paper the idea is proposed to solve Maxwell equations for a curved hollow wave conductor by means of effective Riemannian space, in which the lines of motion of photons are isotropic geodesies for a 4-dimensional space-time. The algorythm of constructing such a metric and curvature tensor components are written down explicitly. The result is in accordance with experiment.

1. Parallel translation of polarization vector in geometrical optics of non-homogenous media

The discovery of law of parallel translation for vectors $\vec{e} = \frac{\vec{E}}{|E|}$ and $\vec{h} = \frac{\vec{E}}{|E|}$ of electric and magnific fields in media with slowly changing refraction factor is being rizen to the paper of S.M.Rytov [1] in 1938. It was shown that for the light beam of nonflat curve form there is a rotation of vectors $\vec{e}$ and $\vec{h}$ in respect to natural 3-facet, created by tangent $\vec{\tau}$, normal $\vec{n}$ and binormal $\vec{b}$ vectors to the curved beam. Setting :

$$\begin{cases} \vec{e} = \vec{n} \cos\varphi + \vec{b} \sin\varphi, \\ \vec{h} = -\vec{n} \sin\varphi + \vec{b} \cos\varphi, \end{cases} \quad (1)$$

we find

$$\frac{d\varphi}{ds} = \frac{1}{T}, \quad (2)$$

where S is the arc length, T is the tension radius. Formula (2) is known as Rytov's law.

In 1941 the paper of V.V.Vladimirsky [2] has appeared in which on the basis of "Rytov Law" there was predicted a global (topological) effect: the angle of rotation of polarized light beam plane, trajectory of which in optically non-homogenous media presents nonflat curve, equals integral of Gauss curvature over a region bounded by contour C, which is obtained by the top of the vector $\vec{\tau}$ on sphere of unit radius. The angle θ equals space angle π concluded inside of the conus drawn by the vector $\vec{\tau}$. This angle is coincided with topological Berry phase [3] for spiral photons, experimentally founded in recent works of Tomita and Chao [4] in spirally curved optical wave conducters. In the case of a flat curve $\theta = 0$ and after a beam gets initial direction the field vectors take their previous position.

In the first series of experiments the spiral was uniformly winded on cylinder. In the second series of experiments providing a fixed length P and radius r there were used non-uniform spirals of different shapes, modeled on computer. The space angle in this case is calculated by formula:

$$\Omega(C) = \int_0^{2\pi} \int_0^{\theta(\varphi)} \sin\theta d\theta d\varphi = \int_0^{2\pi} (1 - \cos\theta) d\varphi \qquad (3)$$

Berry [3] was searching for solution of classical wave equations for a curved wave conductor by means of local coordinate system in approximation of idea of locally connected modes and weakly directing wave conductor [4]. It was produced not taking into account radiation effects, connection with reflected modes and elastic-optical effects. It led in fact to the approximation of geometricai optics and could have been analized in easier way, taking into account that light trajectory represents a broken line, corresponding to multiple reflections from walls [5],[6].

In the present paper for interpritation of obzerved effect we find the exact solutions of Maxwell equations for hollow curved wave conductor in certain curved Riemannian space-time different from flat Minkowsky space-time of special theory of relativity. The choice of 4-dimensional Riemannian space is such that trajectories of light beams to be isotropic geodesics in this space, so the projection of this geodesics in 3-dimensional space were the lines corresponding to topology of the wave conductor.

So consider a wave conductor the axis line of which is described by equation:

$$\vec{r}_0 = \vec{i}\, a\cos t + \vec{j}\, a\sin t + \vec{k}\, bt \qquad (4)$$

in Cartesian coordinate system. It's easy to see that curvature of this curve is:

$$K = \frac{1}{R} = \frac{a}{a^2 + b^2}$$

and a torsion: $\chi = \frac{1}{T} = \frac{b}{a^2+b^2}$.

Any point of wave conductor can be described by radius-vector $\vec{r} = \vec{r}_0 + \vec{\rho}$, where

$$\vec{\rho} = \eta \vec{n} + \zeta \vec{b} \qquad (5)$$

MAXWELL EQUATIONS FOR A HOLLOW CURVED WAVE CONDUCTOR 161

$\vec{n}$ — normal vector to curve (5), $\vec{b}$ — binormal vector.

The boundary of wave-conducter is described by equation $\rho = \rho_0$. Let us introduce new variable $\xi \equiv s = \sqrt{a^2 + b^2}\, t$ which is a length of axis line arch. Then formula for any point inside the wave conductor is :

$$\begin{cases} x = (a - \eta)\cos\frac{\xi}{\sqrt{a^2+b^2}} - \zeta\frac{b}{\sqrt{a^2+b^2}}\sin\frac{\xi}{\sqrt{a^2+b^2}}, \\ y = (a - \eta)\sin\frac{\xi}{\sqrt{a^2+b^2}} + \zeta\frac{b}{\sqrt{a^2+b^2}}\cos\frac{\xi}{\sqrt{a^2+b^2}}, \\ z = \frac{b\xi - a\zeta}{\sqrt{a^2+b^2}}. \end{cases} \quad (6)$$

Let's go from coordinates (x, y, z) to locally curved coordinates (ξ, η, ζ).

The metric tensor of 3-dimensional space in this coordinate system being:

$$g_{\alpha\beta} = \begin{pmatrix} 1 + \frac{K}{a}\eta^2 + \chi^2\zeta^2 - 2K\eta & \chi\zeta & -\chi\zeta \\ \chi\zeta & 1 & 0 \\ -\chi\eta & 0 & 1 \end{pmatrix} \quad (7)$$

We suppose further that optical beams — lines of motion of photons — are curves, which tangent vectors are orthogonal to planes $\xi = \xi_0$. It's dictated by our desire to get a non-dispersed beam at the exit of wave conductor. Congruation of these lines is described by equations

$$\begin{cases} \xi = \xi \\ \eta = r\cos(\chi\xi + \varphi) \\ \zeta = r\sin(\chi\xi + \varphi) \end{cases} \quad (8)$$

where r and φ - some parameters, characterizing the certain lines of congruation. Going to coordinates (ξ, r, φ) we get for congruation lines corresdonding to optical beams :

$$\begin{cases} x = (a - r\cos(\chi\xi + \varphi))\cos\frac{\xi}{\sqrt{a^2+b^2}} - r\sin(\chi\xi + \varphi)\frac{b}{\sqrt{a^2+b^2}}\sin\frac{\xi}{\sqrt{a^2+b^2}}, \\ y = (a - r\cos(\chi\xi + \varphi))\sin\frac{\xi}{\sqrt{a^2+b^2}} + r\sin(\chi\xi + \varphi)\frac{b}{\sqrt{a^2+b^2}}\cos\frac{\xi}{\sqrt{a^2+b^2}}, \\ z = \frac{b\xi - ar\sin(\chi\xi+\varphi)}{\sqrt{a^2+b^2}}. \end{cases}$$

(9)

Transition to 4-dimensional space-time with coordinates (ξ, r, φ, t) is accomplished by replacing $\overset{0}{g}_{44} = 1$ in Cartesian coordinate system in Minkowsky space on $g_{44} = -g_{11}$, so that in local coordinate system the metric tensor of 4-dimensional space-time will be:

$$g_{ij} = \begin{pmatrix} (1 - rK\cos\phi)^2 + \chi^2 r^2 & 0 & -\chi r^2 & 0 \\ 0 & 1 & 0 & 0 \\ -\chi r^2 & 0 & r^2 & 0 \\ 0 & 0 & 0 & -(1 - rK\cos\phi) \end{pmatrix} \quad (10)$$

where $\phi \equiv \chi\xi + \varphi$.

It's easy to show that congruation of isotropic geodesics of this space-time after projecting gives the lines of optical beams – trajectories of photons in 3-dimensional space, defined by formula (9). Non-vanishing components of curvature tensor of this Riemannian space-time are:

$$\begin{cases} R_{1414} = K^2(1 - rK\cos phi) + \chi^2 K^2 r^2 - \chi^2 K^2 \cos\phi \\ R_{1424} = \chi K \sin\phi \\ R_{1434} = -\chi K r(Kr - \cos\phi) \end{cases} \quad (11)$$

Ricci tensor components:

$$\begin{cases} R_{11} = K^2 - \chi^2 \frac{K^2(\cos\phi - Kr)}{(1-rK\cos\phi)^2} \\ R_{12} = \frac{\chi K \sin\phi}{(1-rK\cos\phi)^2} \\ R_{13} = \chi K r \frac{(\cos\phi)-Kr}{(1-rK\cos\phi)^2} \\ R_{44} = -K^2 - \chi^2 \frac{K^2(\cos\phi - Kr)}{(1-rK\cos\phi)} \end{cases} \quad (12)$$

Scalar curvatur R is:

$$R = \frac{2}{(1-rK\cos\phi)^2}\left[K^2 + \chi^2 \frac{K^2(\cos\phi - Kr)}{(1-rK\cos\phi)^2}\right] \quad (13)$$

And Einstein tensor components:

$$\begin{cases} G_{11} = -2\chi^2 \frac{K^2(\cos\phi-Kr)}{(1-rK\cos\phi)^2} - \frac{\chi^2 r^2}{(1-rK\cos\phi)^2}\left[K^2 + \chi^2 \frac{K^2(\cos\phi-Kr)}{(1-rK\cos\phi)^2}\right] \\ G_{12} = \chi K \frac{\sin\phi}{(1-rK\cos\phi)^2} \\ G_{13} = \chi \frac{K^2(\cos\phi-Kr)}{(1-rK\cos\phi)^2} + \frac{\chi r^2}{(1-rK\cos\phi)^2}\left[K^2 + \chi^2 \frac{K^2(\cos\phi-Kr)}{(1-rK\cos\phi)^2}\right] \\ G_{22} = -\frac{1}{(1-rK\cos\phi)^2}\left[K^2 + \chi^2 \frac{K^2(\cos\phi-Kr)}{(1-rK\cos\phi)^2}\right] \\ G_{33} = -\frac{1}{(1-rK\cos\phi)^2}\left[K^2 + \chi^2 \frac{K^2(\cos\phi-Kr)}{(1-rK\cos\phi)^2}\right] \end{cases} \quad (14)$$

2. Solution of Maxwell equations

In general covariant formulation the Maxwell equations are:

$$\begin{cases} F_{[ij,k]} = 0 \\ F^{ij}_{,j} = 0 \end{cases} \quad (15)$$

where comma means covariant devivative, and F_{ij} – components of electromagnetic field tensor.

Writing down the system (15) in coordinates (ξ, r, φ, t) in Riemannian space (10) it can be shown that it has a partial solution corresponding to co-axial line with singularities on axis line $r = 0$ of the type:

$$\begin{cases} F_{12} = F_{24} = \frac{C_1}{r} e^{ik(\xi-t)}, \\ F_{13} = F_{34} = C_2 e^{ik(\xi-t)}, \\ F_{14} = F_{23} = 0. \end{cases} \quad (16)$$

For components of energy-momentum tensor of electromagnetic field we find:

$$\begin{cases} T_{11} = \frac{C_1^2 + C_2^2}{r^2}, \\ T_{14} = -\frac{C_1^2 + C_2^2}{r^2}, \\ T_{44} = \frac{C_1^2 + C_2^2}{r^2}. \end{cases} \quad (17)$$

3. Conclusion

Returning to 3-dimensional designations $\vec{E}$ and $\vec{H}$ by means of the known formula it can be shown from (16), that along photons lines (8) we found a vector of linear polarization rotates on the angle $\chi\xi$, what is well-coordinated with experimental data and caused by effect of topological phase of Vladimirski-Berry.

References

1. Rytov S.M., Docl. Academ.Nauc USSR, v.28,N 4 - 5, 1938, p.263.
2. Vladimirski V.V., Docl. Academ. Nauc USSR, v.31,N 3, 1941,p.222.
3. Berry M.V. Nature, 1987, v.326, N 6110, p.277.
4. Tomito A.,Chaio R.V., Phis. Rev. Lett., 1986, v. 57, N8, p.937.
5. Bialyniski-Birula I., Bialynski - Birula S., Phis.Rev.D., 1987, v.35, N8, p.2383.
6. Vinitzki S.I., Derbov V.L., Dubrovin V.M., Marcovski B.L., Stepanovski Yu.P., Procedings of the work conference on elaboration and construction of radiator and detector of gravitational waves, Dubna, 1989, p.74.

MAXWELL'S EQUATIONS DIRECTLY FROM THE DYNAMICS OF POINT PARTICLES

G.N. ORD

Department of Applied Mathematics
University of Western Ontario
London, Ontario N6A 5B7
CANADA

Abstract.

The Kac Stochastic Model of the Telegraph Equations may be modified to produce Maxwell's equations in 1+1 Dimensions. This shows that Maxwells's equations, while usually regarded as field equations governing electromagnetic waves, may also be interpreted as descriptions of ensembles of particles. Thus Maxwell's equations may be derived in the same way as recent derivations of the Schroedinger and Dirac equations. All three equations have the same features that the 'wave' aspect of solutions to the equations are produced by ensembles of classical particles, which themselves have no 'wave' attributes. The model then provides a many-particle simulation of the electromagnetic field in 1+1 dimensions.

Maxwell's Equations provide a well tested description of classical Electromagnetic fields. The underlying picture accompanying these equations is that of waves, and the idea of photons does not fit easily into this picture. Having said this, some recent work, [1, 2, 3, 4, 5] has shown that the Schroedinger and Dirac equations may be derived within classical physics as descriptions of correlations over ensembles of particles. These derivations show that it is possible to **simulate** single-particle quantum mechanics using ensembles of classical particles. In this note we show that, at least in 1+1 dimensions, Maxwell's Equations also describe the dynamical behaviour of ensembles of point particles constrained to move on light cones.

The stochastic model that produces Maxwell's Equations is a simple modification of a model due to Mark Kac[6], which he showed provided a basis for the Telegraph equations.

We start with a space-time lattice in 1+1 dimensions with lattice spacing Δx and Δt. Assume all particles move either to the left or right with speed c and there is a source of particles specified by $a(x,t)$.

If $F^{\pm}(x,t)$ represents particle density on the lattice then the Kac model gives

$$\begin{aligned} F^+(x, t+\Delta t) &= F^+(x - \Delta x, t) + a(x,t)\Delta t \\ F^-(x, t+\Delta t) &= F^-(x + \Delta x, t) + a(x,t)\Delta t. \end{aligned} \quad (1)$$

Where $a(x,t)$ is a source term for particles. Equations (1) just express the fact that each particle propagates at constant speed in a unique direction, and that particles are emitted in opposite directions with equal frequency. To lowest order in Δx and Δt these equations give

$$\begin{aligned} \frac{\partial F^+}{\partial t}\Delta t &= -\frac{\partial F^+}{\partial x}\Delta x + a(x,t)\Delta t \\ \frac{\partial F^-}{\partial t}\Delta t &= \frac{\partial F^-}{\partial x}\Delta x + a(x,t)\Delta t. \end{aligned} \quad (2)$$

Writing

$$\begin{aligned} A(x,t) &= \frac{1}{2}(F^+(x,t) + F^-(x,t)) \\ \phi(x,t) &= \frac{1}{2}(F^+(x,t) - F^-(x,t)) \end{aligned} \quad (3)$$

then in terms of A and ϕ, (2) implies

$$\frac{\partial A}{\partial t} = -c\frac{\partial \phi}{\partial x} + a(x,t) \quad (4)$$

$$\frac{\partial \phi}{\partial t} = -c\frac{\partial A}{\partial x}. \quad (5)$$

where $\frac{\Delta x}{\Delta t} = c$.

To decouple equations (4) and (5) differentiate the first with respect to t and the second with respect to x to give

$$\frac{\partial^2 A}{\partial t^2} = c^2 \frac{\partial^2 A}{\partial x^2} + \frac{\partial a}{\partial t}. \tag{6}$$

Similarly, eliminating A gives

$$\frac{\partial^2 \phi}{\partial t^2} = c^2 \frac{\partial^2 \phi}{\partial x^2} - c\frac{\partial a}{\partial t}. \tag{7}$$

Equations (5), (6) and (7) are equivalent to Maxwell's field equations in $1+1$ dimensions[7]. Here equation (5) is the Lorentz condition:

$$\frac{\partial A}{\partial x} + \frac{1}{c}\frac{\partial \phi}{\partial t} = 0 \tag{8}$$

. Writing

$$\frac{1}{c}\frac{\partial a}{\partial t} = 4\pi J \tag{9}$$

(6) becomes the wave equation for the 'vector' potential A

$$\frac{\partial^2 A}{\partial x^2} - \frac{1}{c^2}\frac{\partial^2 A}{\partial t^2} = -\frac{4\pi J}{c} \tag{10}$$

and similarly writing

$$\frac{1}{c}\frac{\partial a}{\partial x} = -4\pi\rho \tag{11}$$

(7) becomes the wave equation for the scalar potential ϕ

$$\frac{\partial^2 \phi}{\partial x^2} - \frac{1}{c^2}\frac{\partial^2 \phi}{\partial t^2} = -4\pi\rho. \tag{12}$$

The two definitions (9) and (11) imply that

$$\frac{\partial J}{\partial x} + \frac{\partial \rho}{\partial t} = 0 \tag{13}$$

which is the continuity equation.

Notice here that although we have a form of Maxwell's Field equations in $1+1$ dimensions, we obtained them only through the use of counting arguments. Equations (1) from which (9), (10) and (12) follow themselves arise from counting particles on a lattice [6].

Generalizing (9), (10) and (12) to $3+1$ dimensions is easy to accomplish formally. Writing A and J as three component vectors and replacing $\frac{\partial}{\partial x}$

and $\frac{\partial^2}{\partial x^2}$ by ∇ and ∇^2 respectively does the trick. However justifying this in terms of the stochastic model requires more work and will be presented elsewhere.

Since we generally think of Maxwell's field equations as a system of equations describing waves, we can ask at this point 'Where do the wave aspects come from in this description of particles?' The answer is that the quantities A and ϕ above are ensemble averages and they must derive their wave aspects from the ensemble. Each separate particle propagates simply as $x(t) = x_0 \pm ct$. The concept of 'wavelength' and 'frequency' come about from the Fourier decomposition of the initial conditions of the particle densities.

Unlike 'real photons' our particles have no analog of frequency, wavelength or interference properties by themselves, and only ensemble averages can be usefully described by the 'field' equations. This is very similar to the recent results on simulations of Quantum Mechanics[4, 5]. There the wave functions emerged as descriptions of expectations of correlations over ensembles of classical particles. Interference effects arose as a result of ensemble averages, and not (as in Quantum Mechanics) as a result of properties of the 'particles' themselves.

References

1. Ord, G.N. (1992) A Classical Analog of Quantum Phase, Int. J. Theo. Phys., Vol. **31**, pp. 1177-1195
2. McKeon, D.G.C. and Ord, G.N. (1992) Classical Time reversal and the Dirac Equation, Phys. Rev. Lett, Vol. **69**, pp. 3-4
3. Ord, G.N. (1993) Quantum Interference from Charge Conservation, Phys. Lett A , Vol. **173**, pp. 343-346
4. Ord, G.N. Schroedinger Dirac and Einstein without Quantum Mechanics, Annals of Physics, Under review.
5. Ord, G.N. Quantum Mechanics and Intrinsic Correlations in Brownian Motion, Phys. Rev. Lett., Under Review
6. Kac, Marc (1974) A Stochastic Model Related to the Telegrapher's Equation Rocky Mountain Journal of Mathematics, Vol. **4**,
7. Jackson, J.D. (1974) CLASSICAL ELECTRODYNAMICS. John Wiley & sons, 2'nd edition.

Acknowledgements

This work was financially supported by N.S.E.R.C. of Canada and the Channon Foundation.

OBTAINING THE SCHRÖDINGER AND DIRAC EQUATIONS FROM THE EINSTEIN/KAC MODEL OF BROWNIAN MOTION BY PROJECTION

G.N. ORD
Department of Applied Mathematics
University of Western Ontario
London, Ontario N6A 5B7
CANADA

Abstract.
It is well known that the Einstein and Kac stochastic models of Brownian motion provide microscopic models for the Diffusion and Telegraph equations respectively. Furthermore these classical partial differential equations may be transformed to the Schrödinger and Dirac Equations by analytically continuing them to imaginary time. However the formal nature of the analytic continuation removes any chance of using the Brownian models as microscopic models of quantum mechanics.

Some recent results have however shown that random walk models may be used to derive the quantum equations *without* invoking a formal analytic continuation. The quantum equations arise as *projections* and thus describe real measurable aspects of the space-time geometry of particle trajectories. This shows that there are in fact microscopic models which give rise to the quantum equations, although close inspection of the derivations show that these models provide *many-particle simulations* of quantum mechanics and do not qualify as microscopic models of quantum mechanics (equations plus interpretation) itself.

There is one intriguing exception to this. By allowing Brownian motion in time as well as space the Dirac equation may be derived as the equation governing a net charge. This constitutes a derivation of the Dirac equation which may possibly provide aspects of an interpretation of quantum mechanics, as well as the equation itself.

1. Introduction

Einstein's theory of the motion of Brownian particles has provided a well accepted microscopic model of diffusion for many years. Until recently the relationship between this model and quantum mechanics has been completely formal. Brownian motion provides a 'particle picture' basis for the diffusion equation, but this in turn is related to the Schrödinger equation through a formal analytic continuation, so the relationship between Brownian particles and quantum mechanics has been correspondingly vague.

Some recent work [1-5] has, however, changed this picture in a rather fundamental way. In section 3 of this article we show that a random walk model of Brownian motion produces the Diffusion equation as a description of particle densities, while at the same time correlations in the space-time geometry of these same Brownian particles obey the Schrödinger equation. The structural similarity between the classical and quantum equations arise because the latter is obtained from the microscopic model of the former by *projection*.

This result is of interest because it puts the mathematics of non-relativistic quantum mechanics in a completely classical context where the wave function is observable and is interpreted directly in terms of ensembles of particles. This is in sharp contrast to the situation in quantum mechanics where the interpretations of Schrödinger's equation are indirect to say the least.

We shall also note in passing that the same projection which extracts the Schrödinger propagator from the random walk model of diffusion proposed by Einstein, also extracts the Dirac propagator from the random walk model of Marc Kac.

2. The Diffusion Equation

The diffusion equation is:

$$\frac{\partial u}{\partial t} = D\frac{\partial^2 u}{\partial x^2} \tag{1}$$

where u is a concentration of particles and D is a positive real constant. The propagator of the Diffusion equation is the density of diffusing particles at point (x,t) given a point source of particles at the origin at time zero. The propagator on an unrestricted domain is:

$$u(x,t) = \frac{1}{\sqrt{4\pi Dt}}e^{-x^2/4Dt}. \tag{2}$$

which is the well known Gaussian distribution.

To derive the Diffusion equation from a random walk model consider a particle moving on a discrete lattice in space and time. The lattice spacings are respectively Δx and Δt. At each time step the particle hops either to the left or right one lattice spacing so that if $p(x,t)$ is the probability that the particle is at x at time t then

$$p(x, t + \Delta t) = \frac{1}{2} p(x - \Delta x, t) + \frac{1}{2} p(x + \Delta x, t) \qquad (3)$$

Expanding this in a Taylor series about the point (x,t) gives

$$p(x,t) + \frac{\partial p}{\partial t} \Delta t + \cdots = p(x,t) + \frac{\partial^2 p}{\partial x^2} \frac{(\Delta x)^2}{2} + \cdots \qquad (4)$$

and equating lowest order terms we have

$$\frac{\partial p}{\partial t} = \frac{\partial^2 p}{\partial x^2} \frac{(\Delta x)^2}{2 \Delta t}. \qquad (5)$$

Provided we refine the lattice in such a way that

$$\lim_{\Delta t \to 0} \frac{(\Delta x)^2}{2 \Delta t} = D \qquad (6)$$

the result is the Diffusion equation. The refinement criterion (6) is called diffusive scaling and is an observed property of real Brownian particles. It is also a mathematical characteristic of the random walks of the model.

3. The Schrödinger Equation

The Schrödinger equation:

$$\frac{\partial \psi}{\partial t} = i\hbar \frac{\partial^2 \psi}{\partial x^2} \qquad (7)$$

differs from the diffusion equation only in that the diffusion coefficient is imaginary. However this imaginary diffusion constant has until recently meant that there is no physical microscopic model of diffusing (or any other) particles which implies the Schrödinger equation as part of its description.

Until recently, all derivations of the Schrödinger equation from the dynamics of point particles have had to 'cheat' by invoking a formal analytic continuation at some point (eg. $t \to it$ or $D \to i\hbar$). What we shall now show is that the simple microscopic model of Brownian motion that we discussed above in fact *implies* the Schrödinger equation as part of its description. All we have to do to the model is to retain more information about the particle trajectories. To this end consider a trajectory in spacetime (Figure 1 a). We are considering exactly the same Brownian motion model that we considered previously, only this time we are going to save a

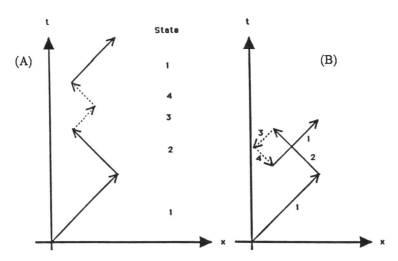

Figure 1. A random walk trajectory on a lattice. The Parity changes every two collisions.

bit more information as we do the calculation. Instead of just calculating the probability that a particle moves from the origin at time 0 to, say, x at time t, we are also going to be interested in which direction the particle is traveling (i.e. $+x$ or $-x$) as well as in something called parity.

As before we work in $1+1$ dimensions on a lattice with spacings Δx and Δt in space and time respectively. At each lattice site a Brownian particle makes a binary choice of whether to go left or right at the next step.

To keep track of the correlations in the trajectories we will assign states to the particles as they move between lattice sites. States one and three will correspond to particles moving to the right while states two and four will correspond to left moving particles. States one and three will be distinguished by the *parity* of the path. That is, states one and three will be separated by an odd number of transitions from right moving to left moving (Figure 1 a). Similarly states two and four will be separated by an odd number of left-to-right transitions. Thus, a particle starting in state one will enter state two at the first collision, three at the second collision, four at the third collision and back to one at the fourth collision. We can then write the difference equation for this as follows:

$$\begin{aligned}
p_1(x,t+\Delta t) &= \frac{\alpha}{2}p_1(x-\Delta x,t) + \frac{\alpha}{2}p_4(x+\Delta x,t) \\
p_2(x,t+\Delta t) &= \frac{\alpha}{2}p_2(x+\Delta x,t) + \frac{\alpha}{2}p_1(x-\Delta x,t) \\
p_3(x,t+\Delta t) &= \frac{\alpha}{2}p_3(x-\Delta x,t) + \frac{\alpha}{2}p_2(x+\Delta x,t) \\
p_4(x,t+\Delta t) &= \frac{\alpha}{2}p_4(x+\Delta x,t) + \frac{\alpha}{2}p_3(x-\Delta x,t)
\end{aligned} \quad (8)$$

Here $p_\mu(x,t)$ is proportional to the probability that a walker on a lattice with (x,t) spacings Δx and Δt respectively arrives in state $\mu (\mu = 1,2,3,4)$ at (x,t). Equation (8) is just a master equation for the ensemble of random walks. Notice that if we merge states one with three and two with 4, the resulting system is just the simple binary random walk with 2 direction states. The extra states considered in equation (8) do not affect the dynamics of the random walk, they just serve to help count paths according to direction and parity.

For normalization corresponding to interpretation of the p's as normalized probabilities, α would be 1. However, we leave α as an unspecified positive constant for the time being.

Equations (1) may be written

$$\vec{p}(x, t + \Delta t) = \mathcal{E}\vec{p}(x,t) \qquad (9)$$

with $\vec{p} = (p_1, p_2, p_3, p_4)^T$ and

$$\mathcal{E} = \frac{\alpha}{2} \begin{pmatrix} E^{-1} & 0 & 0 & E \\ E^{-1} & E & 0 & 0 \\ 0 & E & E^{-1} & 0 \\ 0 & 0 & E^{-1} & E \end{pmatrix} \qquad (10)$$

where E is a shift operator such that $Ep(x,t) = p(x + \Delta x, t)$.

Now let

$$\phi_1 = \frac{p_1 - p_3}{2}, \quad \phi_2 = \frac{p_2 - p_4}{2}, \quad z_1 = \frac{p_1 + p_3}{2} \quad \text{and} \quad z_2 = \frac{p_2 + p_4}{2}. \qquad (11)$$

The reason for choosing these new variables is that the ϕ's correspond to the expected difference in the number of even and odd parity paths to a given point. In effect they calculate a correlation inherent in paths generated by the binomial distribution. The z's calculate the expected sum of the number of paths to a given point distinguishing only direction of arrival and not parity.

In terms of these new variables equation (9) becomes

$$\begin{pmatrix} \phi_1(x, t+\Delta t) \\ \phi_2(x, t+\Delta t) \\ z_1(x, t+\Delta t) \\ z_2(x, t+\Delta t) \end{pmatrix} = \frac{\alpha}{2} \begin{pmatrix} E^{-1} & -E & 0 & 0 \\ E^{-1} & E & 0 & 0 \\ 0 & 0 & E^{-1} & E \\ 0 & 0 & E^{-1} & E \end{pmatrix} \begin{pmatrix} \phi_1(x,t) \\ \phi_2(x,t) \\ z_1(x,t) \\ z_2(x,t) \end{pmatrix}. \qquad (12)$$

Here it is very significant that the change of variables (11) makes the shift matrix block diagonal. It means that the correlations described by the ϕ's exist completely in an eigenspace of the full system and are *independent of the eigenvalues driving the dominant behaviour (diffusion) of the random walk*. Notice that this eigenspace of correlations is found as a *projection* of the full system.

The two projections give us

$$\begin{pmatrix} z_1(x, t+\Delta t) \\ z_2(x, t+\Delta t) \end{pmatrix} = \frac{\alpha}{2} \begin{pmatrix} E^{-1} & E \\ E^{-1} & E \end{pmatrix} \begin{pmatrix} z_1(x,t) \\ z_2(x,t) \end{pmatrix}. \quad (13)$$

for the dominant eigenspace and

$$\begin{pmatrix} \phi_1(x, t+\Delta t) \\ \phi_2(x, t+\Delta t) \end{pmatrix} = \frac{\alpha}{2} \begin{pmatrix} E^{-1} & -E \\ E^{-1} & E \end{pmatrix} \begin{pmatrix} \phi_1(x,t) \\ \phi_2(x,t) \end{pmatrix} \quad (14)$$

for the correlations.

One can verify that the equations (13) do give rise to diffusion by taking the limit of (13) as

$$\left\{ \Delta x \to 0, \quad \Delta t \to 0, \quad \frac{(\Delta x)^2}{\Delta t} \to 2D \right\}. \quad (15)$$

After a simple calculation one finds that the normalized solutions approach

$$\begin{aligned} Z(x,t) &= \frac{1}{2} \begin{pmatrix} 1 \\ 1 \end{pmatrix} \frac{1}{2\pi} \int_{-\infty}^{+\infty} e^{ipx} e^{-p^2 Dt} dp \\ &= \frac{1}{2} \begin{pmatrix} 1 \\ 1 \end{pmatrix} \frac{1}{\sqrt{4\pi Dt}} e^{-x^2/4Dt}. \end{aligned} \quad (16)$$

This is the usual density for diffusion except for the presence of two identical states. The states are identical in the continuum limit because the Markov chain governing the random walk is in equilibrium after a single step.

We now turn to the difference equations corresponding to the eigenspace of correlations, namely equations (14). The same techniques can be used for this case and the result is that the Fourier transform of the normalized propagator is

$$\begin{aligned} \Phi(p,t) &= \left(e^{ip^2 Dt} \frac{1}{2} \begin{pmatrix} 1 & i \\ -i & 1 \end{pmatrix} + e^{-ip^2 Dt} \frac{1}{2} \begin{pmatrix} 1 & -i \\ i & 1 \end{pmatrix} \right) \Phi(p,0) \\ &= \begin{pmatrix} \cos(p^2 Dt) & -\sin(p^2 Dt) \\ \sin(p^2 Dt) & \cos(p^2 Dt) \end{pmatrix} \Phi(p,0). \end{aligned} \quad (17)$$

If we write

$$\begin{aligned} \psi_+(p,t) &= \frac{i}{2} \phi_1(p,t) + \frac{1}{2} \phi_2(p,t) \\ \psi_-(p,t) &= \frac{-i}{2} \phi_1(p,t) + \frac{1}{2} \phi_2(p,t) \end{aligned} \quad (18)$$

(17) becomes

$$\begin{pmatrix} \psi_+(p,t) \\ \psi_-(p,t) \end{pmatrix} = \begin{pmatrix} e^{-ip^2 Dt} & 0 \\ 0 & e^{ip^2 Dt} \end{pmatrix} \begin{pmatrix} \psi_+(p,0) \\ \psi_-(p,0) \end{pmatrix}. \quad (19)$$

Taking $\Psi(p,0) = \frac{1}{\sqrt{2}} \begin{pmatrix} 1 \\ 1 \end{pmatrix}$ and transforming back to position space we have

$$\Psi(x,t) = \begin{pmatrix} \frac{e^{ix^2/4Dt}}{\sqrt{4\pi i Dt}} & 0 \\ 0 & \frac{e^{-ix^2/4Dt}}{\sqrt{-4\pi i Dt}} \end{pmatrix} \frac{1}{\sqrt{2}} \begin{pmatrix} 1 \\ 1 \end{pmatrix}. \quad (20)$$

By inspecting equation (20) we can see that each component of Ψ satisfies a Schrödinger equation, although the two component aspect is suggestive of a two component spinor.

It is important to notice that there is no formal analytic continuation involved in the above result. The i used in the linear combination (18) was for cosmetic purposes only. The oscillatory form of the propagator is apparent in (17) from the presence of the trigonometric functions. These in turn arise completely from the dynamical behaviour of the correlations on the lattice.

Testing whether the above is correct is as simple as counting trajectories on a lattice. In Figure (2) a little program has been used to count paths on the lattice and subtract contributions of even and odd parity. Note that on the lattice the propagator has a 'light cone' corresponding to the fixed hopping speed on the lattice. Near the light cone the approximation to the propagator is noisy because there are few paths to contribute to the approximation. For comparison, in the same figure the exact Feynman propagator evaluated on the lattice is plotted. There the noisy values near and outside the light cone are a result of the fact that the wavelength of the propagator is shorter than the lattice spacing in those regions. As the lattice is refined the light cone flattens and the area of smooth agreement between the calculations increases, approaching the entire half plane.

4. The Dirac Equation

One aspect of the derivation of the diffusion equation from Brownian motion which is unrealistic is the requirement that:

$$\lim_{\Delta t \to 0} \frac{(\Delta x)^2}{2\Delta t} = D \quad (21)$$

This diffusive scaling is fine down to scales that are of the order of the mean free path of the diffusing particles, but on smaller scales real Brownian

Expected Parity Excess

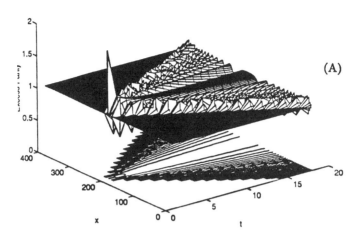

Real Part of Feynman Propagator

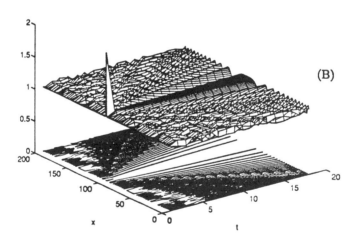

Figure 2. The Feynman Propagator by Path Counting and Evaluation

particles cease to have trajectories which obey this scaling. This simply means that in real fluids the diffusion equation is a bad approximation on very short time scales and very large distance scales.

Notice that the above limit suggests that the particle velocity on the scale of Δx is inversely proportional to scale:

$$\frac{\Delta x}{\Delta t} \sim \frac{D}{\Delta x}. \qquad (22)$$

This can only be true down to the scale of the mean free path of the particle, at which point the particle's speed would remain roughly of the order of the speed of sound in the fluid.

In the 1950's Mark Kac [6] showed that the Brownian motion model above, slightly changed to accommodate a mean free path and a finite propagation speed, produced the Telegraph equations.

$$\frac{\partial u_\pm}{\partial t} = \mp c \frac{\partial u_\pm}{\partial x} + a(u_\pm - u_\mp) \qquad (23)$$

These reduce to the Diffusion equation when the characteristic speed c goes to infinity in such a way that $2a/c^2 \to 1/D$. The solutions of the Telegraph equations basically amend the solutions of the diffusion equation by correcting them for the finite propagation speed in the system.

We can extend the Kac model in exactly the same way that we extended the previous random walk model. That is we leave the physical model alone and just keep track of the parity of particles. The result is that the calculation of the *same parity difference* as in the previous case gives rise to another projection. This time the projection is the Dirac equation rather than the Schrödinger equation, ie.

$$\frac{\partial u_\pm}{\partial t} = \mp c \frac{\partial u_\pm}{\partial x} + ia(u_\pm - u_\mp) \qquad (24)$$

It is important to note that this calculation shows that the Dirac equation is no longer just a mysterious creation by Dirac which happens to describe spin 1/2 particles. The Dirac equation itself is simply a view (projection) of a simple physical model (Brownian particles with finite speed cutoff) and for the first time may be studied by examining the object itself (Brownian motion) rather than the projection (Dirac equation).

Now I have been careful so far to draw a distinction between the equations of quantum mechanics and quantum mechanics itself (equations plus interpretation). At this point I hope it is clear that the equations themselves have an underlying microscopic model which puts them in the same category as the Diffusion and Telegraph equations. Notice that this means that most of the usual icons of quantum mechanics have simple, direct interpretations as projections of real physical features of the underlying model. For example, the wave function, the uncertainty principle, the de Broglie and Compton wavelengths all have simple interpretations as projections of features of the space-time geometry of Brownian paths.

Since all of these 'quantum mechanical objects' are observable features in the random walk models it makes sense to ask what distinguishes our models from quantum mechanics proper. The answer to this question can be understood by performing a gedanken double slit experiment with our diffusing particles.

Imagine first that we set up detectors at the screen to register a 1 for a particle of even parity, and a -1 for a particle of odd parity. Now imagine that we send a few billion random walk particles through the double slit apparatus and wait until the detectors have finished recording.

Given the Schrödinger (or Dirac) propagator of parity differences the eventual pattern taking into account the sign of the parity will approximate the characteristic interference pattern of the two slit experiment. This will be the case even though if we plot the particle density according to parity without doing the subtraction, we will simply get the classical result.

Now imagine that we turn down the intensity of particles so that only one particle goes through the apparatus at a time. If we tell the detectors to record parity differences then each particle contributes a plus or minus one to a detector. Adding these signals over a long period of time with a large number of particles will indeed show interference, but the individual detection event (a plus or minus one) is not like real quantum particles which form the interference pattern through an accumulation of scalar (plus one only) events.

Our scheme fails to reproduce real quantum mechanics because *it requires many of our Brownian particles to simulate a single particle wavefunction*. Our model lacks the important property of wave-particle duality. It is a many particle simulation of quantum mechanics.

5. The Dirac Equation and Brownian motion in Time

It is worth noticing that the above calculations had two peculiar features.

The first peculiarity was the fact that parity, which almost magically produced the quantum equations by projection, was itself unmotivated. We were left wondering why parity worked as it did.

Secondly we found that the projection produced by parity reproduced the quantum equations but represented only a many-particle simulation of quantum mechanics.

Now both these features may be changed by changing the geometry of the trajectories. Consider the trajectory in Figure 1 a. Note that the parity rule does not seem to reflect any obvious feature of the geometry of the trajectory. However consider what happens when we change the geometry of the trajectory so that the particle can only turn left on the space-time diagram. With this geometry odd parity sections of the trajectory correspond to the particle moving backwards in time. Notice that with this geometry the parity rule makes perfect sense. The new geometry does not conserve particle number but does conserve our parity excess provided all particles in our ensemble progress from the past to the future. With the new spiral geometry we call parity charge, since the 'corners' where the particle reverses it's direction in time may be interpreted as pair creation and annihilation as proposed by Feynman and Stuckelberg in the 1940's.

Really for us, the association with electromagnetic charge is unnecessary. Since all we are doing is counting paths, our 'charge' rule simply allows us to count the number of continuous trajectories, without being confused by loops.

The question at this point is what happens if you count all these spiral paths and project onto the eigenspace of conserved charge. The calculation has been done [1] and the result is that net charge projects out the Dirac equation!

Notice that this time the projection of a *single continuous space-time trajectory* results in a whole ensemble of our previous Brownian paths complete with the parity rule intact. Thus we cannot immediately rule this model out as a microscopic model of quantum mechanics (as opposed to just the equations of quantum mechanics). Here, one continuous trajectory represents a whole ensemble of temporally monotonic trajectories, and it may well be that a single trajectory covers enough of the Feynman ensemble to be correctly represented by a wave function. If this is the case then the model may give us some concrete insights into the theory of measurement.

6. Summary

To summarize, we have discussed two levels of microscopic models.

The first level provided microscopic models for the *equations* of quantum mechanics by constructing many-particle simulations of quantum mechanics itself. These calculations transplant the quantum *equations* directly into the classical world of the statistics of random walks. This will hopefully bring a new set of tools, techniques and scientists to the problem of solving the equations.

The second level extended the geometry of particle paths to include spiral trajectories in space-time. Here the projection was mediated by a net charge and the resulting equation was the Dirac equation. The calculations with space-time loops also provide a simulation of the equations of quantum mechanics, however here it is not yet known whether some of the measurement postulates of quantum mechanics are also provided by the model. That this may prove to be the case is actually *likely* since the many-particle aspect of the spiral trajectories provides a direct mechanism for wave-particle duality. In any case the spiral model can in principle be tested by counting trajectories on a lattice directly on a computer and it will be interesting to see whether there emerges a viable theory of measurement from the consideration of such trajectories.

References

1. Ord, G.N. (1992) A Classical Analog of Quantum Phase, Int. J. Theo. Phys., **Vol. 31**, pp. 1177-1195
2. McKeon, D.G.C. and Ord, G.N. (1992) Classical Time reversal and the Dirac Equation, Phys. Rev. Lett, **Vol. 69**, pp. 3-4
3. Ord, G.N. (1993) Quantum Interference from Charge Conservation, Phys. Lett A, **Vol. 173**, pp. 343-346
4. Ord, G.N. Schrödinger Dirac and Einstein without Quantum Mechanics, Annals of Physics, Under review.
5. Ord, G.N. Quantum Mechanics and Intrinsic Correlations in Brownian Motion, Phys. Rev. Lett., Under Review
6. Kac, Marc (1974) A Stochastic Model Related to the Telegrapher's Equation Rocky Mountain Journal of Mathematics, **Vol. 4**.

7. Acknowledgements

The author gratefully acknowledges financial support from N.S.E.R.C. of Canada, and the Channon Foundation.

DERIVATION OF THE SCHRÖDINGER EQUATION

THOMAS B. ANDREWS
3828 Atlantic Avenue
Brooklyn, NY 11224

Abstract.

The derivation of the time-independent Schrödinger equation is based on a new approach to basic physics. The concept of fundamental particles with associated wave properties as the basic elements of physics is abandoned. Instead, it is hypothesized that the universe is a pure wave system and the constructive interference peaks of the wave system are the elementary particles. Based on this hypothesis, an elementary particle moving in the rest system may be represented in the moving system by a stationary constructive interference peak. Then, the Schrödinger equation is obtained by a Lorentz transformation of the stationary constructive interference peak from the moving system to the rest system.

Two new results are obtained. First, two identical and independent Schrödinger equations are derived instead of one as expected. Applying both independent Schrödinger equations to an atom, an additional independent set of states is obtained. Therefore, the number of states now agrees with the experimentally determined number of states in atoms without consideration of the two spin states of the electron. Second, the electron in the ground state of the hydrogen atom is found to oscillate radially over a very small distance around the average position of the electron.

1. Introduction

Schrödinger [1] in 1926 derived the Schrödinger equation by solving the variational problem associated with the Hamiltonian partial differential equation. This variational problem can be regarded as Fermat's principle for wave propagation. Thus, from the very first, the Schrödinger equation was intimately connected with wave processes. Schrödinger, in fact, said he was tempted to associate the solution of the Schrödinger equation with a vibrational process within the atom.

Schrödinger in the same paper solved the Schrödinger equation in spherical coordinates and obtained the same energy levels as Bohr for the hydrogen atom. As is well known, the Schrödinger equation has proved to be enormously successful. It is now considered [2] that Schrödinger's paper was "undoubtably one of the most influential contributions ever made in the history of science."

However, Schrödinger's derivation can not be considered rigorous or basic. From Schrödinger's derivation, it is not clear to physicists even now why the Schrödinger equation works. Also, the interpretation of experiments designed to check quantum mechanics leads to severe theoretical problems as, for example, in the EPR photon polarization experiment [3]. Taken together, these observations strongly indicate current physical theory does not correctly model the physical world at a basic level. Consequently, a new approach to basic physics and quantum mechanics appears required.

As a first step, the concept of fundamental particles with associated wave properties is abandoned. In its place, a new approach is adopted which hypothesizes that the physical universe is a pure wave system or, equivalently, a vibrating system. Then, it is proposed that the constructive interference peaks of the wave system are the elementary particles. Thus, the particles are parts of the wave system with inherent wave properties.

It will be shown that the lowest energy state or eigenvalue of the wave system occurs when there is complete constructive interference between the wave modes of the wave system. As a result, the constructive interference peaks are stable entities since any reduction in the constructive interference would require a large global energy input.

Although an existing medium is normally required for wave motion, no medium is assumed to exist independently of the wave system. Instead, the medium is assumed generated by the interactions between the wave modes of the wave system. This avoids a requirement for a fundamental material existing prior to the wave system.

The wave system as a basic hypothesis has a theoretical advantage over almost any other conceivable hypothesis since a wave system is Lorentz invariant and conserves energy. These are the most important invariances in physics since they underlie modern physical theory and experimentally hold exactly. The derivation of the Schrödinger equation will be based on the Lorentz invariance of the wave system.

The paper has two main parts. In the first part, the wave system theory is developed and a general force equation is obtained which is essential to understanding the Schrödinger equation. In the second, the de Broglie wavelength and the two independent Schrödinger equations are derived.

2. Wave System Theory

The wave system hypothesis was suggested by the following observations:

1. The elementary particles exhibit interference effects which may only be explained reasonably by wave theory.

2. Since elementary particles identical to existing particles are produced in high energy collisions, the environment must dynamically maintain particle identity. A possible explanation of this phenomenon is that the particles and the environment are parts of a wave system. Then, the precise frequencies of the particles would depend on the eigenvalues of the wave system.

3. Hamilton's principle states that the variation of the time integral of the Lagrange function $L = \int (pq - H) dt$ is zero. J. Olsen [4] shows Hamilton's principle is identical to the conservation of phase. This implies elementary particles vibrate and stay in-phase with an underlying wave system.

A strong proof for the existence of a wave system has been given by H. Giorgi [5] using a symmetry argument. Giorgi has proved waves must exist in a linear, infinite system where the laws of physics are both translation and time symmetric. The proof is simple: Consider a system with the above symmetries. The system is governed by the representations of both the space and time translation groups. The solution of each representation is a complex exponential, given by $\exp(ikx)$ and $\exp(i\omega t)$ for space and time, respectively. Combining these solutions results in left and right progressive waves of the form $\exp i(\omega t \pm kx)$. When many of these waves are added together, a standing wave system is established.

C. Vassallo [6] has written a paper closely related to the above symmetry argument titled "Translational Invariance of Waveguides and Normal Modes." He also finds that fields invariant with space and time translations must vary as $\exp i(\omega t \pm kx)$. Furthermore, he shows that any bounded wave field which has sources can be represented by a system of wave modes which do not have sources. This result is non-intuitive and very important. It suggests that a system of bounded modes can bootstrap itself since there is no requirement for the existence of initial sources to produce the field.

2.1. STABILITY OF THE WAVE SYSTEM

Since the electron and the proton are experimentally stable particles, it is necessary to show that the constructive interference results in a globably stable wave system and, consequently, stable particles. Consider a single wave mode which interacts with the other wave modes parametrically. For simplicity, only one-dimensional wave motion equivalent to waves on a string will be considered. Assume the classical wave equation with variable parameters applies. given by

$$\frac{\partial}{\partial x}\left[\tau(x)\frac{\partial Y}{\partial x}\right] = \sigma(x)\frac{\partial^2 Y}{\partial t^2} \qquad (1)$$

where $\tau(x)$ and $\sigma(x)$ are the tension and mass density. The tension and mass density are variable and depend indirectly on their position in the wave system.

The stability of the wave system is based on two principles:

1. The wave system frequency is reduced when the mass density and tension are increased at the wave mode peaks.

2. The wave system frequency increases as the number of particles which mutually interact in the universe increase.

To apply the first principle. assume the mass density and tension are proportional to the local energy density of the wave system. This is an essential assumption of the theory. Then. the frequency of a wave mode will decrease as the constructive interference between the wave modes increases and produces higher energy concentrations at the peaks. Since the energy density at a peak is proportional to the square of the number of wave modes (10^{45}) constructively interfering. the decrease in the frequency of the wave mode is very large. If the natural frequency of the wave system without constructive interference is f_o, complete constructive interference reduces the frequency to

$$f_m = f_o/N. \qquad (2)$$

Details on the derivation of equation 2 are given in [7].

Since the number of wave modes increases as the distance between particles increases. the frequency could go to zero as N increases without limit. However. by the second principle. there is a lower bound on the system frequency determined by the number of mutually interacting particles in the system. This frequency is given by the eigenvalue equation [8]

$$f_p = \frac{1}{2\pi}\sqrt{n}\sqrt{k/m} \qquad (3)$$

where n is the effective number of particles interacting and k/m is the average interaction constant between any two particles. This vibrating system is special since it only has two discrete frequencies. one degenerate equal to f_p with $n-1$ modes and the other equal to $1/2\pi\sqrt{k/m}$. the frequency of a single element.

The frequency of the wave system may result from a physical process similar to the following: As the number of wave modes constructively interfering increases. the distance between particles in the universe increases.

DERIVATION OF THE SCHRÖDINGER EQUATION

This decreases the wave energy reflected from the particles back into the local system since the particles are further apart. In turn, this increases the mean interaction radius of the universe, the volume of the universe and, therefore, the number of interacting particles. The frequency, f_p, then increases since the number of particles increases. A stable frequency of the wave system is attained when $f_m = f_p$.

Provided the universe is in a stable state, the lower bound on the frequency of the wave system may be estimated from equation 3. Let $1/(2\pi)\sqrt{k/m} = c/(2R)$ where R is the mean radius of the universe. $c/(2R)$ estimates the minimum frequency which can exist in the universe. Then, assuming $n = 10^{82}$ and $R = 4 \times 10^{26}$ m, the frequency of the wave system is approximately 4×10^{22} hertz.

2.2. GENERAL FORCE EQUATION

Forces are the result of changes in the parameters. To derive a general force equation, set $Y(x,t) = S(x)T(t)$ in equation 1 to separate the wave equation into space-dependent and time-dependent equations. The angular frequency, ω, is determined by the space-dependent equation

$$\frac{d}{dx}\left[\tau \frac{dS}{dx}\right] + \sigma \omega^2 S = 0. \tag{4}$$

For small variations in the values of the parameters, the perturbative solution [9] of equation 4 is

$$\omega^2 = \omega_o^2 \left[1 - \frac{2}{L\sigma_o}\int_0^L (\sigma - \sigma_o)\sin^2(kx)\,dx \right. \tag{5}$$
$$\left. - \frac{2}{kL\tau_o}\int_0^L \frac{\partial \tau}{\partial x}\cos(kx)\sin(kx)\,dx\right]$$

where ω_o is the stable angular frequency, ω is the perturbed angular frequency and L is the length of the system (in general, $L = R$, the mean radius of the universe). The wave system angular frequency is reduced when σ and τ are larger at the constructive interference peaks than their average values, σ_o and τ_o. Simplifying equation 5 by assuming the last two terms are equal, taking the square root and setting $\sin^2(kx)$ and $\sin(kx)\cos(kx) = 1/2$, ω is given by

$$\omega = \omega_o[1 - \frac{1}{L\sigma_o}\int_0^L (\sigma_\delta - \sigma_o)dx] \tag{6}$$

where the subscript δ refers to the value of the mass density at a constructive interference peak.

Assume there is only a single particle in the system and the particle or constructive interference peak has a linear diameter equal to the wavelength, λ. Then, $\sigma_\varepsilon = m/\lambda$ and $\sigma_o = m/L$ where m is the mass of the particle. Integrating equation 6,

$$\omega = \omega_o[2 - \lambda\sigma_\varepsilon/(L\sigma_o)]. \tag{7}$$

Without the constructive interference peak, $\omega = 2\omega_o$. With the constructive interference peak, ω is reduced to ω_o, the equilibrium angular frequency of the wave system.

Setting $\omega = E/\hbar$ and $\omega_o = mc^2/\hbar$ in equation 7, the equilibrium energy E of the particle varies as

$$E = mc^2[2 - \lambda\sigma_\varepsilon/(L\sigma_0)]. \tag{8}$$

Equation 8 shows that when the energy density of a particle decreases, the equilibrium energy of the particle increases. In general, the force required to move a mass particle when the energy density is changing is given by

$$F = \frac{dE}{dx} = -\frac{mc^2\lambda}{L\sigma_o}d\sigma_\varepsilon/dx = -c^2\lambda d\sigma_\varepsilon/dx = -\lambda dI_\varepsilon/dx \tag{9}$$

where $c^2 d\sigma_\varepsilon = dI_\varepsilon$. dI_ε/dx is the change in the energy density of the particle with position. The type of force depends upon the process that changes I_ε.

3. de Broglie Wavelength

Consider two systems moving with respect to each other. From the viewpoint of the rest system, the moving system has a velocity v in the positive x direction. In the moving system, let two progressive waves with the same angular frequency, ω, and wave number, k, propagate in opposite directions. These progressive waves produce a stationary standing wave as viewed in the moving system, given by

$$Y = A/2\cos(\omega t' - kx') + A/2\cos(\omega t' + kx') = A\cos\omega t' \cos kx' \tag{10}$$

where x' and t' are the coordinates in the moving system. In the rest system, the equation for the standing wave system [10] may be obtained by a Lorentz transformation of equation 10. This results in

$$Y(v) = A\cos\gamma\omega(t - vx/c^2)\cos\gamma k(x - vt) \tag{11}$$

where $\gamma = 1/\sqrt{1 - (v/c)^2}$.

In the rest system, the second factor of equation 11 shows that the standing wave is moving with a velocity v. Apart from the γ factor, this is the same as expected for a galilean transformation. However, the first factor shows that the phase varies linearly with the distance x. This is a purely relativistic effect. It describes an envelope which sinusoidally modifies the amplitude and moves in the x direction with a velocity v. The wavelength of this envelope is the de Broglie wavelength, given by

$$\lambda_b = 2\pi c^2/(\gamma \omega v) = h/(\gamma m v) \qquad (12)$$

where $\omega = 2\pi m c^2/h$.

Because the constructive interference intensity at a particle must be the maximum possible in order to minimize its frequency, the particle is generally located near the peak of the de Broglie wave envelope. A strong restoring force, discussed later in section 4.1, invisibly keeps the moving particle at the peak of the de Broglie envelope. Quantum mechanics depends on this effect.

4. Schrödinger Equation

The basic idea for the derivation of the Schrödinger equation is due to J.R. Platt [11]. Platt proposed a standing wave model for elementary particles in 1946. He believed that the simultaneous appearance of mass and quantum mechanical effects in the model suggested that the model might be a starting point for a unified field theory. However, he did not develop the idea that the particle was a constructive interference peak and, therefore, was unable to show that the model resulted in stable particles. Nevertheless, he developed the method used here for the derivation of the Schrödinger equation.

The method is similar to that used for deriving the de Broglie wavelength. The starting point, however, is at the more basic level of the classical wave equation. Assume a particle is moving with the velocity v in the positive x direction with respect to the rest system. In the moving system, the particle is a stationary constructive interference peak vibrating at the angular frequency $\omega = mc^2/\hbar$. The stationary constructive interference peak is represented mathematically by the time-dependent part of the classical wave equation. Then, assuming that the particle remains in-phase with the de Broglie envelope, the Schrödinger equation is obtained by a Lorentz transformation of the time-dependent equation from the moving system with coordinates x' and t' to the rest system with coordinates x and t. The method depends critically on the wave modes vibrating at the same frequency.

The Lorentz transformation equations from the moving system to the rest system are

$$x = \gamma(x' + vt') \text{ and } t = \gamma(t' + v/c^2 x'). \tag{13}$$

Setting $Y(x', t') = S(x')T(t')$, equation 1 separates into a time-dependent part and a space-dependent part. The time-dependent part is

$$\partial^2 T(t')/\partial t'^2 + \omega^2 T(t') = 0 \tag{14}$$

where $T(t')$ depends only on the time in the moving system.

Applying the Lorentz transformation equations to equation 14, the following operational equation is obtained

$$[(\gamma \partial/\partial t + \gamma v \, \partial/\partial x)^2 + \omega^2] T(t') = 0. \tag{15}$$

Now substitute $S(x)T(t)$ for $T(t')$ in the operational equation where $T(t) = \exp(i\gamma\omega t)$. Carrying out the algebra, the operational equation (after cancelling $\exp(i\gamma\omega t)$) results in

$$\gamma^2 v^2 d^2 S/dx^2 + 2i\gamma^3 \omega v \, dS/dx + (1 - \gamma^4)\omega^2 S = 0. \tag{16}$$

Dividing by $\gamma^2 v^2$, setting $\omega = mc^2/\hbar$ and rearranging the order of the terms, we have

$$d^2 S/dx^2 + \gamma^2 m^2 v^2/\hbar^2 S$$
$$+ 2i\gamma mc^2/(v\hbar) \, dS/dx - 2\gamma^2 m^2 c^2/\hbar^2 S = 0. \tag{17}$$

The last two terms of equation 17 will cancel if the particle stays in-phase with the de Broglie envelope. To show this is the case, substitute the de Broglie wave for a free particle, given by $S = A\exp(-i\gamma mvx/\hbar)$, in the last two terms. The sum is given by

$$2\gamma^2 m^2 c^2/\hbar^2 S - 2\gamma^2 m^2 c^2/\hbar^2 S = 0. \tag{18}$$

Therefore, the two terms cancel.

The equation represented by the remaining two terms is the relativistic form of the Schrödinger equation. To obtain the non-relativistic form of the Schrödinger equation, set $\gamma^2 = 1$ and the kinetic energy $mv^2/2$ equal to $(E-V)$. The equation then is identical to the time-independent Schrödinger equation

$$d^2 S/dx^2 + 2m/\hbar^2 (E - V)S = 0 \tag{19}$$

where E is the energy eigenvalue and V is the potential of the particle.

However, the derivation omits another solution of the classical wave equation. In the derivation of the Schrödinger equation, the particle was

DERIVATION OF THE SCHRÖDINGER EQUATION

assumed to move in the positive x direction relative to the rest system. If the particle is assumed to move in the negative x direction, another independent Schrödinger equation can be derived. This conclusion is consistent with the general solution of the classical wave equation which is the sum of two independent solutions, given by $F(x - vt)$ and $G(x + vt)$ where F and G are arbitrary functions.

For the second solution, the Lorentz transformation equations from the moving system to the rest system are

$$x = \gamma(x' - vt') \text{ and } t = \gamma(t' - v/c^2 x'). \tag{20}$$

The second derivation from this point on is the same as the first derivation except one of the two terms which cancelled in the first derivation now differs in sign. However, the two terms still cancel because the particle is moving in the opposite direction.

The second, independent Schrödinger equation is identical to the first and has the same eigenvalues and the same number of states. However, because the general solution of the wave equation is the sum of the two solutions, the number of states will be doubled.

For a free particle moving in the positive x direction, the solution of the Schrödinger equation is a de Broglie envelope moving with the particle. For a particle in a bound state, the solution is a stationary de Broglie envelope made up of de Broglie envelopes moving in opposite directions. The two de Broglie envelopes represent the two independent Schrödinger equations.

A quantitative discussion of the motion of the electron in the ground state of the hydrogen atom is given in the next section. Qualitatively, assume the electron is located at the peak of the stationary de Broglie envelope and is moving radially away from the nucleus with a velocity v. As the electron moves past the peak of the de Broglie envelope, the velocity of the electron will start to decrease due to a large restoring force (and the smaller electrostatic force). The restoring force results from the decrease in the intensity of the de Broglie envelope. Within a very short distance, the electron's velocity is reduced to zero and it reverses direction. Similarly, as the electron moves toward the nucleus and past the peak, the electron's velocity will start to decrease due to the decrease in intensity of the de Broglie envelope. Again, within a short distance, the velocity is reduced to zero and the electron reverses direction. The electron then oscillates radially around the peak.

4.1. SCHRÖDINGER EQUATION APPLIED TO HYDROGEN ATOM

It is possible now to understand the application of the Schrödinger equation to the hydrogen atom. The three-dimensional equivalent of equation 19 in spherical coordinates determines the eigenvalues, E, associated with the motion of a single electron. Consider the hydrogen atom in s states. The Schrödinger equation for these states is

$$d^2R/dr^2 + 2/r\, dR/dr + 2m/\hbar^2(E + e^2/r)R = 0 \tag{21}$$

where $V = -e^2/r$, the electrostatic potential.

Since two independent Schrödinger equations are derived, the number of states for an electron in an atom is twice the number predicted by a single Schrödinger equation. Currently, it is assumed that the two spin states of the electron double the number of states in an atom. However, this assumption cannot be correct since the vibrational pattern of the stationary de Broglie envelope is not dependent on the spin states of the electron.

To simplify further discussion of the hydrogen atom, the substitution, $R = U/r$, reduces equation 21 to

$$d^2U/dr^2 + 2m/\hbar^2(E + e^2/r)U = 0. \tag{22}$$

This is of the same form as the equation for the motion of an electron in one-dimension. For the ground state of the hydrogen atom, the wave function or eigenstate, U, is

$$U = br\exp(-r/a) \tag{23}$$

where a and b are constants. The amplitude U in equation 23 is the stationary de Broglie envelope that the moving electron "sees" in the ground state of the hydrogen atom.

In states given by equation 22, the amplitude, U, is a function of the variable wavelength de Broglie envelopes. In turn, the wavelength is determined by the calculated velocity of the electron in the potential field, $E - V$. Since the wavelength changes as a function of the radius and the angular momentum is zero, I conclude that the electron only moves radially. Furthermore, the electron is localized within the hydrogen atom since it is, at all times, a constructive interference peak very much smaller than the hydrogen atom. The idea that the electron is spread out in a "probability" cloud is inconsistent with the wave system theory.

The motion of the electron in the hydrogen atom will be determined primarily by the strong restoring force centered on the peak of the wavefunction. The force is due to the reduction in intensity and can be estimated by using the general force law developed in section 2.2. The intensity of the de Broglie wave envelope is given by

$$U^2 = b^2 r^2 \exp(-2r/a) \tag{24}$$

DERIVATION OF THE SCHRÖDINGER EQUATION

where a equals 5.29×10^{-9} cm. b^2 is the only unknown and may be estimated by setting the peak intensity which occurs at $r = a$ equal to the energy density of an electron, given by

$$I = mc^2/\lambda_e \tag{25}$$

where λ_e is the linear diameter of the electron. Then, substituting $r = a$ in equation 24 and setting $U^2 = I$ results in the following equation.

$$b^2 = mc^2/(\lambda_e a^2 \exp(-2)). \tag{26}$$

λ_e is unknown but it is not necessary to know its value since it drops out later in the force calculation.

The force, F, required to move a particle in the positive direction when the energy density is decreasing is given by equation 9. However, since the restoring force is the negative of F, the restoring force is

$$F_r = \lambda_e dI_\delta/dx. \tag{27}$$

Substituting $d(U^2)/dr$ for dI_δ/dx, the restoring force on the electron is given by

$$\begin{aligned} F_r &= \lambda_e \frac{d}{dr}[b^2 r^2 \exp(-2r/a)] \\ &= 2mc^2/[a^2 \exp(-2)](r - r^2/a)\exp(-2r/a). \end{aligned} \tag{28}$$

The restoring force is strong in comparison with the electrostatic force. For example, if $r - a = 10^{-11}$ cm, the restoring force is approximately -0.6 dyne. In comparison, the electrostatic force at $r = a$ is approximately -8×10^{-3} dyne. Given the velocity of the electron at the peak intensity, $v = \pm\sqrt{2(E + e^2/a)/m} = \pm 2.19 \times 10^8$ cm/sec, the velocity versus position of the electron can be calculated. It is found that the electron will oscillate radially around $r = a$ with an amplitude of approximately 3×10^{-11} cm.

4.2. INTERPRETATION OF SCHRÖDINGER EQUATION RESULTS

It is clear experimentally that the Schrödinger equation correctly determines the eigenvalues and the wavefunctions. But, the actual motion of the electron near the peak depends primarily on the restoring force. This is paradoxical since the eigenvalues and wavefunctions are calculated on the basis that the electron velocity is determined entirely by the term $(E - V)$. When the electron has a different velocity distribution than implied by the Schrödinger equation, why does the Schrödinger equation give the correct eigenvalues?

This paradox may be resolved as follows: The eigenvalue calculated from the Schrödinger equation applies to the entire wavefunction and, in particular, to the peak of the wavefunction where the restoring force is zero. Consequently, the eigenvalue is certainly correct at the peak. At other positions, the lower intensity of the wavefunction increases the energy at which the electron is in equilibrium. This reduces the kinetic energy but the electron's total energy, E (relative to mc^2), remains constant. Therefore, E is independent of the restoring force and is the correct eigenvalue, irrespective of the actual velocity of the electron.

5. Conclusion

The Schrödinger equation derivation is based on the concept of a particle as a constructive interference peak and special relativity. The derivation is significant in a practical sense since it reveals for the first time the physical processes of quantum mechanics. For example, it was not realized previously that quantum mechanics is primarily a relativistic phenomenon. In addition, the discovery of two independent Schrödinger equations from purely theoretical considerations, if true, predicts the correct number of states for electrons in atoms.

References

1. Schrödinger, E. (1926) Quantisierung als Eigenwertproblem (1st communication). *Annalen der Physik*, **79**, pp. 361–376. Reprinted in English in Ludwig, G. (1968), *Wave Mechanics*, Pergamon Press, London, pp. 94–105.
2. Jammer, M. (1966) *The Conceptual Development of Quantum Mechanics*, McGraw-Hill Book Company, 267.
3. Andrews, T.B. (1994) On the Wave System Theory of the EPR Experiment, *Waves and Particles in Light and Matter*, Plenum Press, New York, pp.463–471.
4. Olsen, J. (1983) The Phase, *Am. J. Phys.*, **51**, pp. 180–181.
5. Giorgi, H. (1990) An Overview of Symmetry Groups in Physics, Talk presented at Yale University, Harvard preprint number HUTP - 90/A065.
6. Vassallo, C. (1978) Translational Invariance of Wave Guides and Normal Modes, *Am. J. Phy*, **46**, pp. 1022–1025.
7. Andrews, T.B. (1988) A Wave System Theory of Quantum Mechanics, *Physica B*, **151**, pp. 351–354.
8. Chen, F.Y. (1970) Similarity Transformations and the Eigenvalue Problem of Certain Far-Coupled Systems, *Am. J. Phys.* **38**, pp. 1036–1039.
9. Slater, J.C. (1947) *Mechanics*, McGraw-Hill Book Co., New York, pp. 188–192.
10. R.V.L. Hartley, (1950) Matter, a Mode of Motion, *Bell System Tech. J.*, **29**, pp. 350–368.
11. Platt, J.R. (1946) On the Representation of Mass by Standing Waves, Private Communication. Also, an abstract, A Suggestion Concerning Fundamental Particles, *Physical Review*, **70**, 110.

AN APPRAISAL OF STOCHASTIC INTERPRETATIONS OF QUANTUM MECHANICS

MILLARD BAUBLITZ, JR.
Boston University
College of General Studies
871 Commonwealth Avenue
Boston, MA 02215, USA

Recently it has been shown that a class of stochastic theories is inconsistent with experimental data from low temperature electron field emission measurements. A different objection to many stochastic theories, which has been raised by Wallstrom, is that attempted derivations of the Schrödinger equation often are untenable because of an unjustified assumption. On the other hand, the stochastic theory of Bohm and Vigier is not subject to Wallstrom's criticism, and their stochastic theory uses the quantum potential to explain satisfactorily the electron field emission results. Additional comments about the Bohm-Vigier stochastic theory and some related theories are presented.

1. Introduction

This paper primarily discusses the following two developments. (1) Low temperature electron tunneling data are in excellent agreement with the predictions of conventional quantum theory, and the tunneling data eliminate some stochastic interpretations of quantum mechanics. (2) Attempts at deriving the Schrödinger equation from stochastic models often have used an assumption which has not been justified. Other serious criticisms of local realistic theories, including local stochastic interpretations of quantum mechanics, have been based on Bell's Theorem and the optical measurements of Aspect et al.[1-3], but these issues have been discussed in detail in earlier publications and will not be addressed in the present paper.

It is important to emphasize some fundamental differences between standard quantum mechanics and certain alternative theories. In standard quantum mechanics the pure quantum state of a system is the complete description of the system. In other words, the wave function ψ for a system, which may include electrons and nucleons, specifies all of the objective properties of the system. Very different understandings of quantum phenomena are presented in the following three classes of alternative theories: the deterministic theory of Bohm, the classical stochastic theories, and the stochastic theories which modify classical mechanics. According to all three classes of alternative theories each electron is a small particle with a continuous trajectory and a well defined position at each instant of time[4,5]. The deterministic theory of Bohm is not a stochastic theory and will not be discussed in detail in this paper, but a critique of Bohm's deterministic theory will be published in another volume[6]. According to the stochastic theories' description of a randomly moving electron, for example, $|\psi(\mathbf{r},t)|^2$ is the probability density for the position of the electron at time t, but the ψ function is not the "entire story." In some stochastic theories stochastic differential equations are assumed explicitly[7,8] which, at least in principle, may provide supplementary information about the position or trajectory of a particle. Some differences between the two classes of stochastic theories will be described in the following paragraphs.

The classical stochastic theories assume that classical mechanics, the Maxwell equations, and Lorentz force law are rigorously true even for subatomic distances, but each particle's random fluctuations in position and momentum are caused by a postulated fluctuating submicroscopic medium. The Schrödinger equation is viewed essentially as a diffusion equation within these theories. Nelson[5] is perhaps the best known researcher who has aspired to explain quantum phenomena using stochastic processes without otherwise modifying classical physics, but hundreds of other articles by physicists with classical stochastic ideas have been published[7,9-13].

A different type of stochastic theory was proposed by Bohm and Vigier[4] and later developed by Bohm and Hiley[8]. Bohm and Vigier[4] proposed a physical model with random or stochastic motions for subatomic particles, but their theory also included a "quantum potential." If $\rho(\mathbf{r},t)$ is the probability of finding a subatomic particle with mass m at position $\mathbf{r}$ at time t, then the particle is influenced by the quantum potential U, in addition to the normal electromagnetic and gravitational potentials of classical physics. It is also sometimes convenient to express the quantum potential in terms of the real function R with $R^2 = \rho$.

$$U(\mathbf{r},t) = -\frac{\hbar^2}{4m}\left[\frac{\nabla^2 \rho}{\rho} - \frac{1}{2}\left(\frac{\nabla \rho}{\rho}\right)^2\right] = -\frac{\hbar^2}{2m}\frac{\nabla^2 R}{R} \qquad (1.1)$$

The quantum potential significantly alters classical mechanics and sometimes yields long-range forces or interactions; thus the theory of Bohm and Vigier[4] is *not* a classical stochastic theory. It will be seen that the quantum potential is more important than the stochastic fluctuations for understanding electron tunneling in solids within the context of the Bohm-Vigier theory.

In the next section of this paper attempts at explaining electron tunneling will be discussed from three different viewpoints: conventional quantum mechanics, the stochastic theory of Bohm and Vigier, and the classical stochastic theories. Section 3 will review a shortcoming of many derivations of the Schrödinger equation from stochastic theories. In Section 4 properties of the Bohm-Vigier stochastic theory and some related theories will be discussed. The conclusions will be presented in Section 5.

2. Tunneling

Tunneling is not only one of the more striking properties of quantum mechanics, but the use of tunneling also has led to important technological developments[15].

For example, the field emission of electrons from a cold metal surface under the influence of a strong electric field involves electron tunneling through the vacuum, and some applications of scanning electron microscopes require electron sources based on this principle of field emission.

Plummer and Gadzuk[16,17] have been two of the most prominent researchers in electron field emission, and they have used the following technique. At the temperature $T = 0$ K the electrons in a metal are filled to the Fermi energy E_F, and the vacuum state is normally a few electron volts above the Fermi energy. If a metal tip with a small radius is used as a cathode or emitter, then a very strong electric field can be produced at the surface of the metal tip and the electric potential outside the metal will be deformed so that an approximately triangular potential results. Electrons that tunnel through this barrier are attracted to the anode and may pass through a tiny probe hole in a screen. A third electrode may be placed some distance from the anode, and a retarding potential electron energy analyzer can be operated [17]. The collected current can be measured as a function of the voltage between the emitter and analyzer, and if this collected current is differentiated with respect to the bias potential, then a "total energy distribution" of the field emitted electrons can be obtained[17].

Physicists have proposed at least three different theoretical models in their attempts to explain electron tunneling. The most widely accepted explanation is, of course, the conventional quantum mechanical viewpoint. In quantum theory an electron's probability amplitude is represented by the Schrödinger wave function $\psi(\mathbf{r},t)$. Since an electron has a wave aspect, an analogy with classical optics is useful. In general, when a light wave strikes a glass window pane, some of the light will be reflected and some transmitted. In a similar way when an electron wave is incident on a barrier potential, there is some probability that the wave will be reflected and some probability that the wave will be transmitted.

A different interpretation of electron field emission arises from the stochastic theory of Bohm and Vigier[4] and the deterministic theory of Bohm[8]. These theories assume that each electron is a small particle with a continuous trajectory, but an electron in a field emission experiment is subject to the quantum potential in addition to the usual electric potential. The quantum potential often is very complicated, but it may lower significantly the total effective potential in the immediate vicinity of the electron as the electron crosses over the barrier. This lowering of the potential barrier has been demonstrated most convincingly in the numerical calculations and graphs of Dewdney and Hiley [14,8]. Although Dewdney and Hiley do not discuss the possibility of stochastic fluctuations in Reference 14, their graphs show the total potential that acts on an electron in a barrier penetration experiment. Since the quantum potential is an important part of the total effective potential that an electron "sees" according to both the Bohm-Vigier stochastic theory and one version of Bohm's deterministic theory[6], the two theories have a similar mechanism for electron field emission.

A third mechanism for traversing a potential barrier has been suggested in the classical stochastic theories. A particle cannot tunnel through a potential barrier according to classical physics, but in the classical stochastic theories a particle may acquire sufficient kinetic energy from the fluctuating background medium and escape over the barrier[13]. It must be emphasized that this escape of particles over a barrier is a selection process which favors the particles with high kinetic energies.

The preceding discussion has been qualitative, but it is possible to make quantitative predictions for electron transmission across a potential barrier for the three classes of theories. Bruinsma and Bak[18] have derived a rather general quantum theory of inelastic tunneling and have obtained expressions for $< \Delta E >$, the mean change in energy per tunneling electron. Let E be the initial energy of an electron, E' its energy after tunneling, and $< \Delta E > = < E' > - < E >$. The idealized case of elastic tunneling is described by the equation $< \Delta E > = 0$, but in realistic field emission experiments electrons may suffer inelastic collisions with phonons or other collective modes in the metal just before tunneling through the vacuum region. The quantum theory of inelastic tunneling by Bruinsma and Bak[18, 19] shows that inequalities (2.1) and (2.2) are always true at T = 0 K and are generally valid for low but nonzero temperatures.

$$< E' > \leq < E > \tag{2.1}$$

$$< \Delta E > \leq 0 \tag{2.2}$$

The physical import of inequalities (2.1) and (2.2) is that tunneling electrons at low temperatures either lose energy because of inelastic collisions or they tunnel elastically.

Although the detailed nature of the stochastic force in the Bohm-Vigier theory has not been specified, the quantitative predictions for the mean energy per tunneling electron should be the same for the Bohm-Vigier theory and standard quantum mechanics for typical field emission experiments with large numbers of electrons and a potential barrier height $V_0 >> < E >$, despite the different physical understandings of tunneling in the two theories. Agreement between the theories should occur because Bohm and Vigier postulated that the $\psi(\mathbf{r}, t)$ function satisfies the Schrödinger equation, and the probability density for an ensemble of electrons with the same wave function is $|\psi|^2$ in the long time limit[4]. Thus (2.1) and (2.2) should be valid both for the stochastic theory of Bohm and Vigier and for orthodox quantum mechanics.

Inequalities which describe electrons traversing a barrier within the context of the classical stochastic theories will be presented now. Again let V_0 be the maximum value of the barrier, and choose coordinates[20] so that the potential barrier is located at positions $x \geq 0$. Consider an ensemble ϵ composed of all free electrons initially on the left side of the barrier, $x \leq 0$. The average energy of an electron in this ensemble will be denoted by $< E >_\epsilon$. It is also possible to consider the subensemble S of those electrons which have crossed over the barrier potential by a later time t_2. The average energy of an electron in this subensemble is represented by $< E' >_S$. Each electron in this subensemble must have attained an energy greater than V_0, at least for a brief time, in order to surmount the potential barrier[13,20]. Unless the electrons experienced a subsequent correlated decrease in energy immediately after crossing the barrier, then

$$<E'>_s \geq V_0. \qquad (2.3)$$

The difference in mean energies, according to the classical stochastic theories, can be represented by $<\Delta E>_{CS} = <E'>_s - <E>_\epsilon$, and from (2.3) we find

$$<\Delta E>_{CS} \geq V_0 - <E>_\epsilon. \qquad (2.4)$$

If $<V_0> >> <E>$, then the pairs of inequalities (2.1)-(2.2) and (2.3)-(2.4) yield very different predictions. Quantum mechanics permits electrons to tunnel elastically through a barrier with no change in energy, but according to the classical stochastic theories electrons may cross over a potential barrier only if they have sufficiently large kinetic energies caused by fluctuations of the background medium.

Electron field emission experiments directly measure E', the energy of an electron after tunneling, and it is worthwhile to concentrate on field emission from the (100) surface of tungsten since it has been studied in detail. Plummer and Gadzuk [16,21] found the following value for the mean energy of electrons which had tunneled from the W(100) surface through the vacuum at T = 78 K.

$$<E'> = E_F - 0.2 \pm 0.1 \; eV \qquad (2.5)$$

Since the tunneling electrons originate with energies near the Fermi level[17], this experimental value agrees with the quantum mechanical prediction of inequality (2.1). The data of Plummer and Gadzuk also showed that more than 99 percent of the field emitted electrons had final energies $E' \leq E_F$. The few electrons with relatively high energies after tunneling, $E_F < E' < E_F + 1.5$ eV, correspond to the "tail" of the energy distribution and may have experienced electron-electron interactions, either during the tunneling process or in the electron beam outside the metal[16].

Although the empirical results in (2.5) agree with quantum mechanics, they are in conspicuous disagreement with inequality (2.3), which has been derived for a family of classical stochastic theories. Expression (2.3) states that $<E'>_s \geq V_0$, and a plausible value[20,22,23] for V_0 must be about 0.9 times the value of the work function. Since the work function[23] for W(100) is approximately 4.6 eV, the relevant classical stochastic theories require the inequality $<E'>_s \geq E_F + 4$ eV, which obviously is violated by the experimental data in (2.5).

A proponent of classical stochastic theories now must argue either that the effective barrier potential V_0 in field emission measurements of W(100) is about 4 eV less than the work function or that each electron after it escapes over the barrier must experience a correlated decrease in energy. Some proponents of classical stochastic theories have suggested informally that electrons may escape over a potential barrier and then lose their excess energy shortly after crossing the barrier. Such a mechanism takes place in classical Brownian motion, but a Brownian particle experiences a viscous force caused by the surrounding fluid. If a similar universal frictional force is postulated to rescue the classical stochastic theories, then such an idea would contradict experimental data. More complicated viscous forces might be hypothesized that begin to operate only after the electron crosses the potential barrier, but it is difficult to imagine how such *ad hoc* notions could be part of a consistent, credible theory of physics.

Although this section has discussed electron tunneling and has shown the inadequacy of classical stochastic theories, it is interesting that Nelson[24] has found "severe difficulties" in such theories based on his consideration of neutron interferometry.

3. The Schrödinger Equation and Stochastic Theories

A different criticism of many stochastic theories has been published by Wallstrom[25]. In the introduction of this paper it was noted that the Schrödinger equation has been regarded essentially as a diffusion equation within the framework of the classical stochastic theories, and there have been attempts[5,26] to derive the Schrödinger equation from frictionless diffusion processes. Wallstrom[25] has argued convincingly that the equations derived from the classical stochastic theories are not equivalent to the Schrödinger equation, and it is worthwhile to review briefly Wallstrom's argument.

The Schrödinger equation for a single particle of mass m is

$$i\hbar \frac{\partial \psi(\mathbf{r},t)}{\partial t} = -\frac{\hbar^2}{2m}\nabla^2 \psi + V\psi. \tag{3.1}$$

$V(\mathbf{r})$ is the scalar potential, and it has been assumed that there is no vector potential. If the wave function is written in the form

$$\psi = R\, e^{iS/\hbar} \tag{3.2}$$

where R and S are real functions, then the wave equation implies

$$\frac{\partial S}{\partial t} + \frac{1}{2m}(\nabla S)^2 + V + U = 0 \tag{3.3}$$

$$\frac{\partial R^2}{\partial t} + \nabla \cdot (R^2 \nabla S/m) = 0. \tag{3.4}$$

where U has the form of the quantum potential in equation (1.1). *If* the momentum of the particle is the gradient of the S function,

$$m\mathbf{v} = \nabla S(\mathbf{r},t) \tag{3.5}$$

then differentiation of (3.3) yields equation (3.6), and the continuity equation is obtained from (3.4).

$$\frac{\partial \mathbf{v}}{\partial t} + (\mathbf{v}\cdot\nabla)\mathbf{v} = -\frac{1}{m}\nabla U - \frac{1}{m}\nabla V \tag{3.6}$$

$$\frac{\partial \rho}{\partial t} + \nabla\cdot(\rho\mathbf{v}) = 0 \tag{3.7}$$

In some stochastic theories equations (3.6) and (3.7) have been derived from postulated stochastic processes and the mathematical steps shown above have been followed in reverse order to obtain the Schrödinger equation. Wallstrom[25] has emphasized correctly that S must be many valued for wave functions with angular momentum because of the occurrence of the factor $e^{im'\phi}$ where m' is an integer and ϕ is the azimuthal angle. Since S is many valued, $\mathbf{v}$ is only locally a gradient, and equations (3.6) and (3.7) are not equivalent to the Schrödinger equation unless the following quantization condition is assumed

$$\int_L \mathbf{v} \cdot d\mathbf{l} = 2\pi j \qquad (3.8)$$

where j is an integer and the integration is around any closed loop L. I am not aware of any derivation of (3.8) within the context of the classical stochastic theories, and it seems improbable that a justification of (3.8) can be given in such theories.

Although many attempted derivations of the Schrödinger equation must be rejected because of Wallstrom's argument, the Bohm and Vigier paper of 1954 assumed the Schrödinger equation rather than attempting to derive it. Thus their work[4] is not subject to Wallstrom's criticism.

4. Stochastic Theory of Bohm and Vigier

The previous sections have emphasized difficulties of the classical stochastic theories, but in this section some properties of the Bohm and Vigier theory[4] will be considered. The theory includes the conventional postulate that $\psi(\mathbf{r},t)$ satisfies the Schrödinger equation, but the theory also includes the distinctive assumption that each particle k has random fluctuations $\xi_k(\mathbf{r},t)$ in velocity, in addition to the local mean velocity given by $\nabla S/m$. If the terms are additive, then equation (4.1) results for the instantaneous velocity of particle k.

$$\mathbf{v}_k(\mathbf{r},t) = \frac{1}{m}\nabla S(\mathbf{r},t) + \xi_k(\mathbf{r},t) \qquad (4.1)$$

(The notation adopted above is closer to that of equation (9.29) of Bohm and Hiley [8]. The variable ξ referred to the particle's position in the paper of Bohm and Vigier[4].) The term ξ_k, which represents the random fluctuations in motion, is especially important since it allowed Bohm and Vigier[4] to show that "an arbitrary probability density ultimately decays" to $\rho = |\psi|^2$.

Take the total time derivative of (4.1) and obtain

$$\frac{d\mathbf{v}_k}{dt} = (\frac{\partial}{\partial t} + \dot{r} \cdot \nabla)\frac{1}{m}\nabla S + \frac{d\xi_k}{dt} \qquad (4.2)$$

In this case $\dot{r} = \nabla S/m$. Substitute this result, $\nabla[(\nabla S)^2] = 2(\nabla S \cdot \nabla)\nabla S$, and equation (3.3) into (4.2) to obtain (4.3).

$$m\frac{d\mathbf{v}_k}{dt} = -\nabla(V + U) + \frac{d\xi_k}{dt} \qquad (4.3)$$

This is effectively equation (9.32) of Bohm and Hiley[8], and it is similar to the equation of motion for a particle in Bohm's deterministic theory except that (4.3) has the stochastic term $d\xi_k/dt$. The Schrödinger equation and equation (4.3), taken together, may be viewed as the fundamental equations of the stochastic theory of Bohm and Vigier.

It is instructive to consider a special case of equation (4.3), s-states which are real stationary eigenstates of the Schrödinger equation. For these states the Schrödinger equation reduces to

$$-\frac{\hbar^2}{2m}\nabla^2\sqrt{\rho} + V\sqrt{\rho} = E\sqrt{\rho} \qquad (4.4)$$

with energy eigenvalue E. Divide (4.4) by $R = \sqrt{\rho}$, and the expression $U + V = E$ is obtained. Substitute this into (4.3) to obtain the following equation of motion for a particle in an s-state.

$$m\frac{d\mathbf{v}_k}{dt} = \frac{d\xi_k}{dt} \qquad (4.5)$$

It is evident that the terms in (4.3) from the quantum potential and the classical potential V have cancelled in (4.5). Equation (4.5) implies that if a particle in an s-state is at an improbable position at one instant of time, then the force on the particle will not return it to a more probable position unless the stochastic term $d\xi_k/dt$ depends on the wave function ψ for the particle.

Ever since the pioneering work of Bohm and Vigier[4] it has been recognized that the ψ function has a dual role: the quantum potential depends on ψ and the probability density also is related to ψ by the Born equation $\rho = |\psi|^2$. The preceding paragraphs indicate that the rapidly fluctuating stochastic force of Bohm-Vigier theory also might depend on ψ. More precisely, it may be stated from equation (4.3) that either the random forces on a particle due to fluctuations of the subquantum medium in the Bohm-Vigier theory depend on ψ, at least for particles in s-states, or the statistical moments for a particle's motion obtained from (4.3) for stationary quantum states will not always equal the corresponding statistical moments obtained from the Schrödinger ψ function. The result that both the quantum potential *and* the stochastic forces depend on ψ seems very surprising on first observation, but it is similar in some respects to ideas in Santamato's theory of geometric quantum mechanics[27]. Santamato has suggested a feedback mechanism between the geometry of space and a particle's motion. According to this viewpoint the geometry of space is determined by the presence of matter and energy, and "in turn, geometry acts as a 'guidance field' for matter"[27]. In this way Santamato has attempted to provide a deeper justification for the quantum potential.

5. Conclusions

The data from low temperature electron field emission experiments are in excellent agreement with the predictions of standard quantum mechanics, and the same data also eliminate the most plausible stochastic theories which retain classical mechanics and assume that electrons escape over the potential barrier instead of tunneling. Wallstrom's criticism casts serious doubt on most attempts to derive the Schrödinger equation from classical stochastic theories.

It is helpful to place these conclusions within a broader theoretical context. Conventional quantum physics agrees with all known experimental results, but the measurement problem[28] remains as a disconcerting feature of quantum theory. One possible solution to the measurement problem is that the Schrödinger wave function for a pure quantum state does not provide a complete description of a system. In Section 1 of this paper three classes of alternative theories were mentioned that propose supplementary equations for the trajectories of subatomic particles: classical stochastic theories, the deterministic theory of Bohm, and the stochastic theories which modify classical mechanics. In Section 2 it was demonstrated that the most plausible classical stochastic theories contradict experimental tunneling results and are fatally flawed. The deterministic theory of Bohm includes counterintuitive ideas such as an electron in an s-state being at rest[8]. Therefore any future attempts to solve the measurement problem by arguing for the incompleteness of the wave function's description of physical reality should concentrate on stochastic theories which modify classical mechanics. The theory of Bohm and Vigier[4] with the quantum potential is the best known example of such a theory.

Although the Bohm-Vigier theory appears to be the most promising of the three classes of alternative theories that have been mentioned, much work remains to be done so that it can be evaluated more fully. The properties of the postulated stochastic forces need to be specified, and equation (4.3) of this paper must be studied in detail. The first part of this paper emphasized that tunneling is a powerful tool for deciding between quantum mechanics and certain alternative theories. It is necessary to calculate, based on equation (4.3), properties of tunneling particles in a variety of situations and to compare these calculations with conventional quantum mechanical predictions. Two cases which deserve special consideration are tunneling experiments that measure properties of single electrons and tunneling processes in nuclear decays. The stochastic forces in (4.3) also need to be studied to see if they yield a rapid approach to quantum equilibrium and yet permit the correlations of pairs of particles in entangled states to persist over long distances in Bell-type experiments.

Acknowledgments: The author gratefully acknowledges Michael Crommie and Timothy Wallstrom for helpful discussions and Abner Shimony for his critical reading of this manuscript and comments.

6. References

1. Aspect, A., Dalibard, J., and Roger, G.: Experimental tests of Bell's inequalities using time-varying analyzers, *Phys. Rev. Lett.* **47** (1982), 1804-1807.
2. Clauser, J.F., and Shimony, A.: Bell's theorem: experimental tests and implications, *Rep. Prog. Phys.* **41** (1978), 1881-1927.
3. Greenberger, D.M., Horne, M.A., Shimony, A., and Zeilinger, A.: Bell's theorem without inequalities, *Am. J. Phys.* **58** (1990), 1131-1143; Kar, G., and Roy, S.: Unsharp spin-1/2 observables and CHSH inequalities, *Phys. Lett. A* **199** (1995), 12-14.
4. Bohm, D. and Vigier, J.P.: Model of the causal interpretation of quantum theory in terms of a fluid with irregular fluctuations, *Phys. Rev.* **96** (1954), 208-216.
5. Nelson, E.: Derivation of the Schrödinger equation from Newtonian mechanics, *Phys. Rev.* **150**

(1966), 1079-1085; Nelson, E.: *Quantum Fluctuations*, Princeton University Press, Princeton, 1985.
6. Baublitz, M. and Shimony, A.: Some queries concerning Bohmian mechanics, in J.T. Cushing, A. Fine, and S. Goldstein (eds.), *Bohmian Mechanics and Quantum Theory* , Kluwer Academic Publishers, Dordrecht (in press).
7. de la Pena, L. and Jauregui, A.: Stochastic electrodynamics for the free particle, *J. Math. Phys.* **24** (1983), 2751-2761.
8. Bohm, D., and Hiley, B.J.: *The Undivided Universe* , Routledge, London, 1993. Chapters 5 and 9, respectively, describes barrier penetration and stochastic theories.
9. Kalitsin, N.S.: On the interaction of an electron with the fluctuating electromagnetic field of the vacuum, *Zh. Eksper. Teor. Fiz.* **25** (1953), 407-409.
10. Marshall, T.W.: Statistical Electrodynamics, *Proc. Camb. Phil. Soc.* **61** (1965), 537-546; Franca, H.M., Marshall, T.W., and Santos, E.: Spontaneous emission in confined space according to stochastic electrodynamics, *Phys. Rev. A* **45** (1992), 6436-6442; Boyer, T.H.: A brief survey of stochastic electrodynamics, in A.O. Barut (ed.), *Foundations of Radiation Theory and Quantum Electrodynamics*, Plenum, New York, 1980, pp. 49-63. Many related articles can be found in the bibliographies of References 5, 7, and 10.
11. de la Pena-Auerbach, L. and Cetto, A.M.: Derivation of quantum mechanics from stochastic electrodynamics, *J. Math. Phys.* **18** (1977), 1612-1622.
12. Puthoff, H.E.: Ground state of hydrogen as a zero-point-fluctuation-determined state, *Phys. Rev. D* **35** (1987), 3266-3269.
13. McClendon, M., and Rabitz, H.: Numerical simulations in stochastic mechanics, *Phys. Rev. A* **37** (1988), 3479-3492.
14. Dewdney, C., and Hiley, B.J.: A quantum potential description of one-dimensional time-dependent scattering from square barriers and square wells. *Found. Phys.* **12** (1982), 27-48.
15. Esaki, L.: Long journey into tunneling, *Rev. Mod. Phys.* **46** (1974), 237-244.
16. Plummer, E.W., and Gadzuk, J.W.: Surface states on tungsten, *Phys. Rev. Lett.* **25** (1970), 1493-1495.
17. Gadzuk, J.W., and Plummer, E.W.: Field emission energy distribution (FEED), *Rev. Mod. Phys.* **45** (1973), 487-545; Plummer, E.W., Gadzuk, J.W., and Penn, D.R.: Vacuum- tunneling spectroscopy, *Physics Today* **28** (April 1975), 63-71.
18. Bruinsma, R., and Bak, P.: Quantum tunneling, dissipation, and fluctuations, *Phys. Rev. Lett.* **56** (1986), 420-423. Bruinsma and Bak and Baublitz in Ref. 19 used the notation $< \Delta E > = < E > - < E' >$ so inequalities (2.1) and (2.2) in this paper appear reversed from the corresponding inequalities in Ref. 19.
19. Baublitz, M.: Electron tunneling spectroscopy: A test of quantum mechanics, *Phys. Rev. A* **47** (1993), 2423- 2426.
20. Baublitz, M.: Electron field-emission data, quantum mechanics, and the classical stochastic theories, *Phys. Rev. A* **51** (1995), 1677-1679.
21. Plummer and Gadzuk do not quote a value for the mean final energy $< E' >$ of electrons which had tunneled through the barrier, but I have determined this numerical value based on data in Fig. 1 of [16]. The estimate of the uncertainty is mine.
22. Penn, D.R., and Plummer, E.W.: Field emission as a probe of the surface density of states, *Phys. Rev. B* **9** (1974), 1216-1222.
23. Vorburger, T.V., Penn, D., and Plummer, E.W.: Field emission work functions, *Surf. Sci.* **48** (1975), 417-431.
24. Nelson, E: Field theory and the future of stochastic mechanics, in S. Albeverio, G. Casati, and D. Merlini (eds.), *Stochastic Processes in Classical and Quantum Systems*, Springer- Verlag, Berlin, 1986, pp. 438-469.
25. Wallstrom, T.C.: Inequivalence between the Schrödinger equation and the Madelung hydrodynamics equations, *Phys. Rev. A* **49** (1994), 1613-1617.

26. Baublitz, M.: Derivation of the Schrödinger equation from a stochastic theory, *Prog. Theor. Phys.* **80** (1988), 232-244.

27. Santamato, E.: Geometric derivation of the Schrödinger equation from classical mechanics in curved Weyl space, *Phys. Rev. D* **29** (1984), 216-222.

28. Fehrs, M.H., and Shimony, A.: Approximate measurement in quantum mechanics.I, *Phys. Rev. D* **9** (1974), 2317- 2320; Shimony, A.: Approximate measurement in quantum mechanics.II, *Phys. Rev. D* **9** (1974), 2321-2323.

COMPATIBLE STATISTICAL INTERPRETATION OF QUANTUM BEATS

MIRJANA BOŽIĆ
Institute of Physics
P.O. Box 57, Beograd, Yugoslavia

1. Introduction

Quantum beat phenomena is a very good example of a quantum interference. Quantum beats are observed, for example, in:

i) the fluorescence from atoms whose electrons are excited, impulsively, into a coherent superposition $\Psi(\vec{r}_e, 0)$ of excited states $\Phi_1(\vec{r}_e)$ and $\Phi_2(\vec{r}_e)$, having slightly different energies [1,2,3]. After the excitation, the atom starts emitting light by spontaneous emission and decays back to the ground state $\Phi_g(\vec{r}_e)$. The intensity of the emitted light, detected by a photo multiplier, oscillates in time with frequency proportional to the energy difference of the two coherently superposed levels [1].

ii) the double resonance interferometric neutron experiment, suggested by Vigier and coworkers [4,5] and performed by Badurek, Rauch and Tuppinger (BRT) [6]. The number of neutrons in the two detectors behind the interferometer show real time-beat effect. The observed beating period corresponds to an energy difference of the interfering beams.

According to the interpretation based on the quantum theory of measurement, quantum beat signal, in the case of fluorescence, is due to the impossibility to tell whether the photon has been emitted from the level Φ_1 of from the level Φ_2 [7,8]. In the case of neutrons, according to the same theory, interference is due to the lack of information on the path of each particle inside the interferometer [9,10].

Therefore, quantum theory of measurement attributes an objectively existing phenomena to observer's lack of knowledge.

We propose here an explanation of quantum beat phenomena which is **objective** and therefore independent of the observer knowledge. In this explanation, which is based on the concepts and ideas of de Broglie-Einstein wave mechanics [11], quantum beat phenomena is due to the coherence in the electronic (neutron) wave function, which is a real superposition of real waves. We use in our explanation the so called de Broglian probabilities (de Broglian probability densities) associated with the events which we assume to happen objectively inside the interferometer (the atom) [12-16]. Those probabilities are determined by assuming that the particle and wave properties are compatible (coexistent). De Broglian probabilities supplement the set of well known probabilities of the standard quantum mechanics.

2. Quantum beats in the interference experiment with a large number of particles

Quickly after the first exciting neutron interference experiments were performed [17], researchers around Jean Pierre Vigier realized [4,5] that the neutron interferometer (Fig. 1a) offers new possibilities to formulate old questions of quantum mechanics in a new way, and to get more clear and convincing answers to those questions.

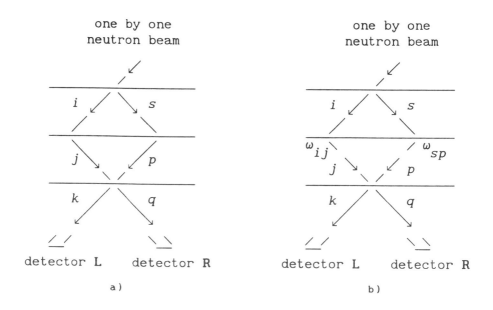

Fig. 1. a) The scheme of an empty neutron interferometer,
b) Double-resonance neutron interferometer

Being participants of the long standing debate between followers of de

Broglie-Einstein and Bohr-Heisenberg interpretation of quantum mechanics, Vigier and coworkers proposed a new version of the experiment with the neutron interferometer. In this experiment, two separate resonance coils are situated along two paths inside the interferometer. Those coils invert the spin state of each sub beam within the interferometer (Fig. 1b). The results of two variants of this experiment are reported [6]. In the first variant the frequencies of resonance coils are equal, in the second the frequencies ω_{ij} and ω_{sp} of two coils slightly deviate from each other, $\Delta\omega = |\omega_{ij} - \omega_{sp}|$. This latter experiment shows that the interference pattern is not stationary but oscillates in time with a beat period $T = 2\pi/\Delta\omega$ (Fig. 2).

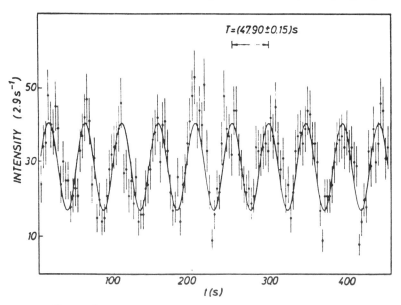

Fig. 2. Quantum beats in the double-resonance experiment [6].

Badurek, Rauch and Tuppinger conclude [6] that their results are in complete agreement with the formalism of the standard quantum mechanics, what means the Bohr-Heisenberg interpretation, according which wave and particle properties are not coexistent but complementary.

It is true that the BRT result, the sinusoidal oscillations in time of the number N of detected neutrons, when this number is large, is in agreement with the predictions of the standard quantum mechanics. But, we would like to point out that BRT experiment contains, in addition, the data which the standard quantum mechanics does not explain.

3. Small N interference experiments

In order to find the oscillatory behavior of the intensity, one has to observe neutrons in a long time interval, such that a large number of neutrons pass through the interferometer. If one would stop the experiment after 10 or 20 neutrons for example, it would not be possible to see these sinusoidal oscillations.

Absence of oscillations (fringes) when number of particles is small was clearly demonstrated in Young double-slit experiment with electrons [18]. The series of pictures for various (increasing) values of the number N of electrons show that the fringes emerge from "chaos" when N is very large.

The standard quantum mechanics describes only the regularities of the interference pattern when enough dots (events) are collected. It does not explain neither the appearance of single dots on the screen (single counts in the detector) nor the process of accumulation of the dots, counts. Barut used this fact in his argumentation about the incompleteness of Bohr-Heisenberg quantum mechanics: "Probabilistic quantum mechanics is incomplete because it cannot make statements about individual events" [19].

Barut's notion "individual events" is equivalent to the notion "particle properties" in de Broglie-Einstein description of micro-objects (quantons), were they are represented by waves and real particle aspect simultaneously. This description differs from Bohr's and Heisenberg's, were micro-objects correspond to probabilistic $\Psi(\vec{r}, t)$ waves or observed particles, never the two simultaneously [20].

With those facts in mind, we conclude that BRT conclusion that "the theoretical predictions about the outcome of the double-resonance experiment we have actually performed are identical for both interpretations" [6] is not quite appropriate. In deriving this conclusion Badurek, Rauch and Tuppinger have not used all their results.

It is true that the predictions of oscillations in intensity for large N are the same in both interpretations. But, what about experiments with small N? Quantum mechanics of Bohr and Heisenberg (standard quantum mechanics) does not explain the appearance of single dots on the screen in the Young experiment, neither it explains the patterns for small N, nor the time dependence of the intensity which BRT would plot if they would stop the experiment when the number of neutrons is small. The standard quantum mechanics even forbids us to ask for the theoretical explanation of the results of experiments for small values of N.

The followers of de Broglie-Einstein concepts and ideas take seriously the results of the experiments for small N and try to complete the standard quantum mechanics in such a way that it could explain and predict the results of interference experiments for arbitrary values of N. Common basic assumption of various approaches along this line of thought is that particle trajectories exist inside the interferometer [19-22]. The aim of certain approaches is to determine the form of the trajectories. The aim of the others is to build a theory, or interpretation in which this assumption would be incorporated.

The probabilistic scheme which we have proposed recently [12,13,15], contains also the assumption that particles travel inside the interferometer along certain trajectories. In this scheme we associated probabilities with those trajectories, called de Broglian probabilities.

4. Present, predicted and de Broglian probabilities associated with the events inside and at the exit from the neutron interferometer

In a neutron interferometer (Fig. 1a) a beam of monochromatic thermal neutron is split into two coherent beams which can be separated by several centimeters and subsequently recombined and brought to interference. Two detectors (**L** and **R**), situated behind the interferometer, measure the number of neutrons.

The wave function $\Psi(\mathbf{L}, t)$ in the detector **L** is equal to the sum of wave functions $\Phi_{ijk}(\mathbf{L}, t)$ and $\Phi_{spk}(\mathbf{L}, t)$, associated with the trajectories ijk and spk. Similarly, $\Psi(\mathbf{R}, t)$ is equal to the sum of wave functions $\Phi_{ijq}(\mathbf{R}, t)$ and $\Phi_{spq}(\mathbf{R}, t)$, associated with the trajectories ijq and spq.

$$\Psi(\mathbf{L}, t) = \Phi_{ijk}(\mathbf{L}, t) + \Phi_{spk}(\mathbf{L}, t)$$
$$\Psi(\mathbf{R}, t) = \Phi_{ijq}(\mathbf{R}, t) + \Phi_{spq}(\mathbf{R}, t) \tag{1}$$

The properties of an empty interferometer (no additional devices along the paths, Fig. 1a) are of special importance, and for that reason we shall denote by $\varphi_{ijk}(\mathbf{L}, t)$, $\varphi_{spk}(\mathbf{L}, t)$, $\varphi_{ijq}(\mathbf{R}, t)$ and $\varphi_{spq}(\mathbf{R}, t)$ wave functions associated with mentioned trajectories in an empty interferometer. In the case of an ideal symmetric interferometer those functions have the following properties:

$$\varphi_{ijk}(\mathbf{L}, t) = \varphi_{spk}(\mathbf{L}, t), \quad \varphi_{ijq}(\mathbf{R}, t) = -\varphi_{spq}(\mathbf{R}, t) \tag{2}$$
$$|\varphi_{ijk}(\mathbf{L}, t)| = |\varphi_{spk}(\mathbf{L}, t)| = |\varphi_{ijq}(\mathbf{R}, t)| = |\varphi_{spq}(\mathbf{R}, t)| \tag{3}$$

Consequently, wave functions $\Psi(\mathbf{L}, t)$ and $\Psi(\mathbf{R}, t)$ have very simple form

$$\Psi(\mathbf{L}, t) = 2 \cdot \varphi_{ijk}(\mathbf{L}, t), \quad \Psi(\mathbf{R}, t) = 0 \tag{4}$$

If it is necessary to take spin into account, one multiplies all wave functions associated with the trajectories by the spin eigenstate, usually denoted by $|+\rangle$ and $|-\rangle$.

Let us now consider an interferometer with various devices situated along the internal paths ij and sp: phase shifters characterized by χ_{ij} and χ_{sp}, static absorbers characterized by real transmission coefficients a_{ij} and a_{sp} ($0 < a_{ij} \leq 1, 0 < a_{sp} \leq 1$) and resonance coils ω_{ij} and ω_{sp} (Fig. 3).

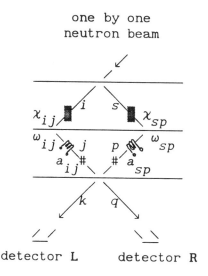

Fig. 3. Neutron interferometer with absorbers, phase shifters and resonance coils.

We shall assume that incident neutrons are polarized and that their spins are along the z-axis, in the positive direction. Neutron state in the detectors **L** and **R** are now described by the functions

$$\Psi(\mathbf{L}, t) =$$
$$= \varphi_{ijk}(\mathbf{L}, t) \cdot [e^{i\chi_{ij}}\sqrt{a_{ij}}e^{-i(E/\hbar - \omega_{ij})t} + e^{i\chi_{sp}}\sqrt{a_{sp}}e^{-i(E/\hbar - \omega_{sp})t}]|+\rangle$$
$$\Psi(\mathbf{R}, t) = \qquad (5)$$
$$= \varphi_{ijq}(\mathbf{R}, t) \cdot [e^{i\chi_{ij}}\sqrt{a_{ij}}e^{-i(E/\hbar - \omega_{ij})t} - e^{i\chi_{sp}}\sqrt{a_{sp}}e^{-i(E/\hbar - \omega_{sp})t}]|+\rangle$$

The number of detected neutrons is the detectors **R** and **L** is proportional to $|\Psi(\mathbf{L}, t)|^2$ and $|\Psi(\mathbf{R}, t)|^2$

$$I(\mathbf{L}, t) \sim |\Psi(\mathbf{L}, t)|^2$$
$$= |\varphi_{ijk}(\mathbf{L}, t)|^2 \{a_{ij} + a_{sp} + 2\sqrt{a_{ij}}\sqrt{a_{sp}}\cos[\chi_{ij} - \chi_{sp} + (\omega_{ij} - \omega_{sp})t]\}$$

$$I(\mathbf{R}, t) \sim |\Psi(\mathbf{R}, t)|^2$$
$$= |\varphi_{ijk}(\mathbf{L}, t)|^2 \{a_{ij} + a_{sp} - 2\sqrt{a_{ij}}\sqrt{a_{sp}} \cos[\chi_{ij} - \chi_{sp} + (\omega_{ij} - \omega_{sp})t]\} \quad (6)$$

where we have taken into account the property (3).

The sum of intensities is constant, as it should be

$$I(\mathbf{L}, t) + I(\mathbf{R}, t) = const \quad (7)$$

According to de Broglie's classification of probabilities, $|\Psi(\mathbf{L}, t)|^2$ and $|\Psi(\mathbf{R}, t)|^2$ are called **present probabilities**, since they describe events which objectively happen and may be determined without any preparation of the physical system. One simply put the detector at the point $\mathbf{L}$ behind the interferometer and counts the number of neutrons present at that point.

According to the same classification, coefficients a_{ij} and a_{sp} determine the so called **predicted probabilities** of the following events:

$\mathcal{F}_{ijk}(\mathbf{L}, t)$: particle arrives along the trajectory ijk to the detector $\mathbf{L}$, at the moment t, when the path sp is closed ($a_{sp} = 0$).

$\mathcal{F}_{ijq}(\mathbf{R}, t)$: particle arrives along the trajectory ijq to the detector $\mathbf{R}$, at the moment t, when the path sp is closed ($a_{sp} = 0$).

$\mathcal{F}_{spk}(\mathbf{L}, t)$: particle arrives along the trajectory spk to the detector $\mathbf{L}$, at the moment t, when the path ij is closed ($a_{ij} = 0$).

$\mathcal{F}_{spq}(\mathbf{R}, t)$: particle arrives along the trajectory spq to the detector $\mathbf{R}$, at the moment t, when the path ij is closed ($a_{ij} = 0$).

We shall denote by $\mathcal{P}_{ijk}(\mathbf{L}, t)$, $\mathcal{P}_{spk}(\mathbf{L}, t)$, $\mathcal{P}_{ijq}(\mathbf{R}, t)$ and $\mathcal{P}_{spq}(\mathbf{R}, t)$ the probabilities of events $\mathcal{F}_{ijk}(\mathbf{L}, t)$, $\mathcal{F}_{spk}(\mathbf{L}, t)$, $\mathcal{F}_{ijq}(\mathbf{R}, t)$ and $\mathcal{F}_{spq}(\mathbf{R}, t)$, respectively. It is easy to see that:

$$\begin{aligned}\mathcal{P}_{ijk}(\mathbf{L}, t) &= |\varphi_{ijk}(\mathbf{L}, t)|^2 a_{ij}, \quad \mathcal{P}_{spk}(\mathbf{L}, t) = |\varphi_{ijk}(\mathbf{L}, t)|^2 a_{sp} \\ \mathcal{P}_{ijq}(\mathbf{R}, t) &= |\varphi_{ijk}(\mathbf{L}, t)|^2 a_{ij}, \quad \mathcal{P}_{spq}(\mathbf{R}, t) = |\varphi_{ijk}(\mathbf{L}, t)|^2 a_{sp}\end{aligned} \quad (8)$$

It is important to point out that events $\mathcal{F}$ do not happen in the apparatus which we consider, but in two modified apparatuses, in one of these modified apparatuses the transmission coefficient a_{sp} is equal to zero, in the other, transmission coefficient a_{ij} is equal to zero. With this modification, the interferometer with two internal paths is reduced to the apparatus with one internal path.

De Broglian probabilities are the probabilities of elementary events which we assume to happen inside the considered interferometer (although we might be unable to detect them directly). Therefore, those probabilities should satisfy classical probability axioms and should help in explaining the statistics of observable events.

We assume that for the observable event $\mathcal{E}(\mathbf{L}, t)$ - neutron arrives to the detector $\mathbf{L}$ at the moment t, elementary events are:

$\mathcal{E}_{ijk}(\mathbf{L}, t)$: neutron arrives along the trajectory ijk to the detector $\mathbf{L}$, at the moment t

$\mathcal{E}_{spk}(\mathbf{L}, t)$: neutron arrives along the trajectory spk to the detector $\mathbf{L}$, at the moment t

We assume that for the observable event $\mathcal{E}(\mathbf{R}, t)$ - neutron arrives to the detector $\mathbf{R}$ at the moment t, the elementary events are:

$\mathcal{E}_{ijq}(\mathbf{R}, t)$: neutron arrives along the trajectory ijq to the detector $\mathbf{R}$, at the moment t

$\mathcal{E}_{spq}(\mathbf{R}, t)$: neutron arrives along the trajectory spq to the detector $\mathbf{R}$, at the moment t

From the above definitions it follows that the probabilities $P_{ijk}(\mathbf{L}, t)$, $P_{spk}(\mathbf{L}, t)$, $P_{ijq}(\mathbf{R}, t)$, $P_{spq}(\mathbf{R}, t)$ of elementary events $\mathcal{E}_{ijk}(\mathbf{L}, t)$, $\mathcal{E}_{spk}(\mathbf{L}, t)$, $\mathcal{E}_{ijq}(\mathbf{R}, t)$ and $\mathcal{E}_{spq}(\mathbf{R}, t)$, respectively, have to satisfy the following relations:

$$P(\mathbf{L}, t) = P_{ijk}(\mathbf{L}, t) + P_{spk}(\mathbf{L}, t)$$
$$P(\mathbf{R}, t) = P_{ijq}(\mathbf{R}, t) + P_{spq}(\mathbf{R}, t) \qquad (9)$$

It seems to us obvious that probabilities of elementary events should satisfy also the following relations:

$$\frac{P_{ijk}(\mathbf{L}, t)}{P_{spk}(\mathbf{L}, t)} = \frac{a_{ij}}{a_{sp}}, \quad \frac{P_{ijq}(\mathbf{R}, t)}{P_{spq}(\mathbf{R}, t)} = \frac{a_{ij}}{a_{sp}} \qquad (10)$$

From the above two sets of relations one derives the expression for de Broglian probabilities of elementary events:

$$P_{ijk}(\mathbf{L}, t) =$$
$$= |\varphi_{ijk}(\mathbf{L}, t)|^2 a_{ij} \left\{ 1 + \frac{2\sqrt{a_{ij}}\sqrt{a_{sp}}}{a_{ij} + a_{sp}} \cos[\chi_{ij} - \chi_{sp} + (\omega_{ij} - \omega_{sp})t] \right\}$$

$$P_{spk}(\mathbf{L}, t) =$$
$$= |\varphi_{ijk}(\mathbf{L}, t)|^2 a_{sp} \left\{ 1 + \frac{2\sqrt{a_{ij}}\sqrt{a_{sp}}}{a_{ij} + a_{sp}} \cos[\chi_{ij} - \chi_{sp} + (\omega_{ij} - \omega_{sp})t] \right\}$$

$$P_{ijq}(\mathbf{R}, t) = \qquad (11)$$
$$= |\varphi_{ijk}(\mathbf{L}, t)|^2 a_{ij} \left\{ 1 - \frac{2\sqrt{a_{ij}}\sqrt{a_{sp}}}{a_{ij} + a_{sp}} \cos[\chi_{ij} - \chi_{sp} + (\omega_{ij} - \omega_{sp})t] \right\}$$

$$P_{spq}(\mathbf{R}, t) =$$
$$= |\varphi_{ijk}(\mathbf{L}, t)|^2 a_{sp} \left\{ 1 - \frac{2\sqrt{a_{ij}}\sqrt{a_{sp}}}{a_{ij} + a_{sp}} \cos[\chi_{ij} - \chi_{sp} + (\omega_{ij} - \omega_{sp})t] \right\}$$

In the double resonance experiment of BRT there are no absorbers ($a_{ij} = a_{sp} = 1$). Consequently, de Broglian probabilities read:

$$P_{ijk}(\mathbf{L}, t) = P_{spk}(\mathbf{L}, t)$$
$$= |\varphi_{ijk}(\mathbf{L}, t)|^2 \{1 + \cos[\chi_{ij} - \chi_{xp} + (\omega_{ij} - \omega_{sp})t]\} = \frac{|\Psi(\mathbf{L}, t)|^2}{2}$$
$$P_{ijq}(\mathbf{R}, t) = P_{spq}(\mathbf{R}, t)$$
$$= |\varphi_{ijk}(\mathbf{L}, t)|^2 \{1 - \cos[\chi_{ij} - \chi_{xp} + (\omega_{ij} - \omega_{sp})t]\} = \frac{|\Psi(\mathbf{R}, t)|^2}{2} \quad (12)$$

5. Discussion and conclusion

It seems to us that we now have new quantities for the analysis of the validity of the statement, derived in the framework of B-H interpretation, that beats are the consequence of the lack of information on the path of each particle inside the interferometer.

For this sake let us compare de Broglian probabilities of two pairs of events, characterized by identical internal paths and different external paths. Such pairs of events are:

$$(\mathcal{E}_{ijk}(\mathbf{L}, t), \mathcal{E}_{ijq}(\mathbf{R}, t)) \quad \text{and} \quad (\mathcal{E}_{spk}(\mathbf{L}, t), \mathcal{E}_{spq}(\mathbf{R}, t)).$$

We see that the sum of probabilities of events in any one of those two pairs is time independent for any value of a_{ij} and a_{sp}.

$$\begin{aligned} P_{ijk}(\mathbf{L}, t) + P_{ijq}(\mathbf{R}, t) &= \text{time independent} \\ P_{spq}(\mathbf{R}, t) + P_{spk}(\mathbf{L}, t) &= \text{time independent} \end{aligned} \quad (13)$$

Those two relations support the view that the particle trajectories exist inside the interferometer because they are easily understood by taking this assumption into account. Namely, independently by which internal path the particle arrives to the third plate, it has to choose one of two possible external paths, and therefore the sum of two (time dependent) probabilities has to be time independent.

From (11) it follows that the ratio of probabilities of events in two pairs is independent of the internal path:

$$\frac{P_{ijk}(\mathbf{L}, t)}{P_{ijq}(\mathbf{R}, t)} = \frac{P_{spk}(\mathbf{L}, t)}{P_{spq}(\mathbf{R}, t)} = \frac{|\Psi(\mathbf{L}, t)|^2}{|\Psi(\mathbf{R}, t)|^2} \quad (14)$$

It follows from this fact that knowledge about the internal path of each particle is irrelevant for the behavior of the particle at the exit from the interferometer. Which one of the two paths, k or q, particle would prefer at time t, is determined by the ratio of $|\Psi(\mathbf{L}, t)|^2$ and $|\Psi(\mathbf{R}, t)|^2$. Therefore, beats have nothing to do with our knowledge about the internal path of each particle. Beats are the consequence of the existence of a wave which accompanies the particle and spread along both paths inside the interferometer. The component waves superpose at the third plate, and guide the particle towards one of the detectors at the exit from the interferometer [20].

In the case of one path interferometer (one of transmission coefficients, a_{ij} or a_{sp}, equals zero) there is no superposition of waves at the third plate and at the exit from the interferometer. As a consequence, $|\Psi(\mathbf{L}, t)|^2$ and $|\Psi(\mathbf{R}, t)|^2$ are time independent, so that time plays no role in the particle choice of the external path.

The analogous reasoning and procedure are applied to the explanation of beats in atomic fluorescence [23]. Basic quantities are de Broglian probability densities $P_i(\vec{r}_e, t)$, where the index i takes the values g (denotes the atomic ground state) and 1 and 2 (denote two excited states to which atoms are excited coherently, at $t = 0$). $P_i(\vec{r}_e, t)$ is the probability density that the electron energy is E_i and that electron is situated around $\vec{r}_e$ at the moment t, when the electron wave function $\Psi(\vec{r}_e, t)$ is a linear superposition of two excited states and of the ground state. Probability density $P_g(\vec{r}_e, t)$, which is directly related to the intensity of fluorescence, is determined by the coherent state $\Psi(\vec{r}_e, t)$ as a whole, i.e. by $|\Psi(\vec{r}_e, t)|^2$, which oscillates in time with frequency $\omega_b = |E_1 - E_2|/\hbar$. This frequency is identical to the frequency of beats in the fluorescence signal.

References

1. J.N. Dodd, W.K. Sandle, D. Zissermann, Study of resonance fluorescence in cadmium: Modulation effects and lifetime measure, *Proc. Phys. Soc.* London **92** (1967) 497.
2. S. Haroche, J.A. Paisner, and A.L. Schawlaw, Hyperfine quantum beats observed in Cs vapor under pulsed dye laser excitation, *Phys. Rev. Lett.* **30** (1973) 948.
3. I.A. Sellin, J.R. Mowal, R.S. Peterson, P.M. Griffin, R. Lanbert, and H.H. Haselton, Observation of coherent electron-density-distribution oscillations in collision-averaged foil excitation of the $n = 2$ hydrogen levels, *Phys. Rev. Lett.* **31** (1973) 1335.
4. C. Dewedney, P. Gueret, A. Kyprianidis, and J.P. Vigier, *Phys. Lett.* A **102** (1984) 291.
5. J.P. Vigier, *Pramana J. Phys.* **25** (1985) 397.

6. G. Badurek, H. Rauch, and D. Tupinger, Neutron interference double-resonance experiment, *Phys. Rev.* **A34** (1986) 2600.
7. W.W. Chow, M.O. Scully, J.O. Stoner, Jr, Quantum beat phenomena described by quantum electrodynamics and neoclassical theory, *Phys. Rev. A* **11** (1975) 1380.
8. P. Meystre, M. Sergent III, *Elements of Quantum Optics*, Springer-Verlag, Berlin-Heidelberg, 1990.
9. M.O. Scully, H. Walther, Quantum optical test of observation and complementarity in quantum mechanics, *Phys. Rev. A* **39** (1989) 5229.
10. M.O. Scully, B.G. Englert and H. Walther, Quantum optical test of complementarity, *Nature*, **351** (1991) 111.
11. L. de Broglie, *Etude critique des bases de l'interprétation actuelle de la mechanique ondulatoire*, Gauthier-Villars, Paris, 1963.
12. M. Božić and Z. Marić, Probabilities of de Broglie's trajectories in Mach-Zehnder and in neutron interferometers, *Phys. Lett. A* **158** (1991) 33.
13. M. Božić, Z. Marić and J.P. Vigier, De Broglian probabilities in the double-slit experiment, *Found. Phys.* **22** (1992) 1325.
14. M. Božić, Compatible statistical interpretation of interference in double-slit interferometer, in A. van der Merwe and A. Garuccio (eds) *Waves and Particle in Light and Matter*, Plenum, New York.
15. M. Božić and Z. Marić, Probability and Interference, in G. Lochak and P. Lochac (eds), *Courants, amers et ecueils en microphysique*, Fondation Louis de Broglie, Paris (1994) pp. 89.
16. M. Božić and D. Arsenović, Rabi oscillations described by de Broglian probabilities, in M. Barone and F. Selleri (eds), *Frontiers of Fundamental Physics*, Plenum, New York (1994) pp. 503.
17. U. Bonse and H. Rauch (eds), *Neutron Interferometry*, Clarendon Press, Oxford, 1979.
18. A. Tonomura, *Electron Holography*, Springer-Verlag, Berlin, 1993; A. Tonomura, J. Endo, T. Matsuda, T. Kawasaki, H. Ezawa, *Am. J. Phys.* **57** (1989) 118.
19. A.O. Barut, The deterministic wave mechanics. A bridge between classical mechanics and probabilistic quantum theory, in L. Acardi (ed), *Interpretation of quantum theory*, Acta Encyclopedia, 1993.
20. J.P. Vigier, New theoretical implications of neutron interferometric double resonance experiments, *Physica B* **151** (1988) 386.
21. N.C. Petroni, Conditioning in quantum mechanics, *Phys. Lett. A* **160** (1991) 107.
22. F. Selleri and G. Tarozzi, *Il Nuovo Cimento B* **43** (1978) 31.
23. M. Božić, Z. Marić and D. Arsenović, Quantum beats described by de Broglian probabilities, to be published in the *Proceedings of NATO Advanced Study Institute on electron theory and quantum electrodynamics*, Edirne, Turkey, 1994.

STOCHASTIC NON-MARKOV MODEL OF QUANTUM MECHANICAL BEHAVIOUR

A.T.GAVRILIN
Department of Radiophysics,
Nizhniy Novgorod State University,
Gagarin avenue, 23, Nizhniy Novgorod 603600, Russia.

In stochastic quantum mechanics the evolution of particle's position is modeled by a diffusion where the drift is defined as a determinate function of this position and time. The latter forms the contradiction with incompatibility of the position and the linear momentum considered as quantum observables. In this report the more general framework of stochastic quantization is proposed which is deprived of the mentioned drawback.

Let $(\Omega, \mathcal{F}, P)$ be a complete probability space and $\{\mathcal{F}_s, -\infty < s < \infty\}$, $s \leq t$, $\mathcal{F}_s \subseteq \mathcal{F}_t \subseteq \mathcal{F}$ nondecreasing family of σ-algebras, containing all sets of vanishing P measure [1,2] Assume that $\nu(t,\omega)$ is the random process with continuous sample trajectories such that its increments for disjoint intervals are statistically uncorrelated and $\nu(t,\omega) - \nu(s,\omega) \sim N\left(0, \frac{\hbar}{m}|t-s|\right)$, where $\hbar$ is Plank's constant, m is the mass of the particle.

For a fixed interval of observation $[0,T]$ the processes $\nu_+(t,\omega) = \nu(t,\omega) - \nu(0,\omega)$ and $\nu_-(t,\omega) = \nu(T,\omega) - \nu(T-t,\omega)$ are Wiener processes with reference to σ-algebras $\mathcal{F}_{[0,t]} = \{\mathcal{F}_s, 0 \leq s \leq t\}$ and $\mathcal{F}_{[t,T]} = \{\mathcal{F}_s, t \leq s \leq T\}$ respectively.

Postulate that a particle's coordinate admits the stochastic differentials

$$d\xi(t) = \frac{1}{m} p_+(t,\omega)\, dt + d\nu_+(t,\omega), \tag{1'}$$

$$d\xi(t) = \frac{1}{m} p_-(t,\omega)\, dt + d\nu_+(T-t,\omega), \tag{1''}$$

where random processes $p_+(t,\omega)$ and $p_-(t,\omega)$, in turn, admit the stochastic differentials

$$dp_+(t,\omega) = Q_+(t,\omega)\, dt + d\eta_+(t,\omega), \tag{2'}$$

$$dp_-(t,\omega) = Q_-(t,\omega)\, dt + d\eta_-(t,\omega), \tag{2''}$$

with the processes $Q_+(t,\omega)$, $Q_-(t,\omega)$ adapted to the flows $\mathcal{F}_{[0,t]}$, $\mathcal{F}_{[t,T]}$ respectively and satisfying the conditions

$$P\left\{\int_0^T |Q_\pm(t,\omega)|\,dt < \infty\right\} = 1, \quad \int_0^T E|Q_\pm(t,\omega)|^2\,dt < \infty,$$

and with the square-integrable martingales $\eta_\pm(t,\omega)$ orthogonal to $\nu_\pm(t,\omega)$.

Note that the processes $p_\pm(t,\omega)$ are not adapted to the families of σ-algebras $\mathcal{F}_t^\xi$ generated by $\xi(t)$. The latter implies nonmeasurability of the processes $Q_\pm(t,\omega)$ with respect to these σ-algebras. Therefore, neither the process $Q_+(t,\omega)$ nor $Q_-(t,\omega)$ can be identified with a classical force because even knowledge of all trajectory $\xi_0^T = \{\xi(t), 0 \le t \le T\}$ does not enable one to determine exactly $Q_\pm(t,\omega)$. For this reason let us express $Q_\pm(t,\omega)$ as sums

$$Q_\pm(t,\omega) = F(t,x_t) + G_\pm(t,\omega),$$

where $F(t,x)$ is the classical force field and $G_\pm(t,\omega)$ are additional "quantum" forces tending to zero when $\hbar \to 0$.

Under above conditions the relations (1) admit the so-called minimal Ito representation [2]

$$d\xi(t) = \frac{1}{m}\hat{p}_+(t,\xi_0^t)\,dt + d\tilde{\nu}_+(t), \qquad (3')$$

$$d\xi(t) = \frac{1}{m}\hat{p}_-(t,\xi_t^T)\,dt + d\tilde{\nu}_-(T-t), \qquad (3'')$$

and relations (2) give rise to the basic equations of optimal nonlinear filtering [3]

$$d\hat{p}_+(t) = \hat{Q}_+\left(t,\xi_0^t\right)dt + \frac{1}{\hbar}\left(\widehat{p_+^2}\left(t,\xi_0^t\right) - \left(\hat{p}_+\left(t,\xi_0^t\right)\right)^2\right)d\tilde{\nu}_+(t), \qquad (4')$$

$$hatdp_-(t) = \hat{Q}_-\left(t,\xi_t^T\right)dt + \frac{1}{\hbar}\left(\widehat{p_-^2}\left(t,\xi_t^T\right) - \left(\hat{p}_-\left(t,\xi_t^T\right)\right)^2\right)d\tilde{\nu}_-(T-t), \qquad (4'')$$

where

$$\widehat{p_+^k}\left(t,\xi_0^t\right) = E\left\{p_+^k(t,\omega)\big|\mathcal{F}_{\le t}^\xi\right\}, \quad \widehat{p_-^k}\left(t,\xi_t^T\right) = E\left\{p_-^k(t,\omega)\big|\mathcal{F}_{\ge t}^\xi\right\},$$

$$\hat{Q}_+\left(t,\xi_0^t\right) = E\left\{Q_+(t,\omega)\big|\mathcal{F}_{\le t}^\xi\right\}, \quad \hat{Q}_-\left(t,\xi_t^T\right) = E\left\{Q_-(t,\omega)\big|\mathcal{F}_{\ge t}^\xi\right\},$$

$$\mathcal{F}_{\le t}^\xi = \sigma\{\xi(s), 0 \le s \le t\}, \quad \mathcal{F}_{\ge t}^\xi = \sigma\{\xi(s), t \le s \le T\},$$

$\tilde{\nu}_+(t)$ and $\tilde{\nu}_-(t)$ are the Wiener processes with respect to σ-algebras $\mathcal{F}^\xi_{\leq t}$ and $\mathcal{F}^\xi_{\geq T-t}$ respectively, which weakly equivalent to $\nu_\pm(t,\omega)$, i.e. have the same diffusion coefficient $\frac{\hbar}{m}$.

The equations (3) and (4) with certain starting conditions define the probabilistic evolution of the particle's position. However, for an actual determination of this evolution it is necessary to have a comprehensive description of quantum forces $G_\pm(t,\omega)$ in order to enable one to express the functionals $\hat{Q}_\pm(t,\cdot)$, $\widehat{p^2_\pm}(t,\cdot)$ in an explicit form. Unfortunately, the absence at present of such the mathematical model compels to be restricted to incomplete description of the microobject's behaviour.

Reduce the process $\xi(t,\omega)$ in the next way. Let $x \in R$ and $t \in (0,T)$ and introduce the designations for conditional expectations

$$\overline{p}_+(t,x) = E\left\{\hat{p}_+(t,\xi_0^t)\Big|\xi(t)=x\right\}, \quad \overline{p}_-(t,x) = E\left\{\hat{p}_-(t,\xi_t^T)\Big|\xi(t)=x\right\},$$

$$\overline{Q}_+(t,x) = E\left\{\hat{Q}_+(t,\xi_0^t)\Big|\xi(t)=x\right\}, \quad \overline{Q}_-(t,x) = E\left\{\hat{Q}_-(t,\xi_t^T)\Big|\xi(t)=x\right\},$$

$$D_{p+}(t,x) = E\left\{\widehat{p^2_+} - (\hat{p}_+)^2\Big|\xi(t)=x\right\},$$

$$D_{p-}(t,x) = E\left\{\widehat{p^2_-} - (\hat{p}_-)^2\Big|\xi(t)=x\right\}.$$

Then allowing for that the increments $d\tilde{\nu}_+(t)$, $d\tilde{\nu}_-(t)$ are independent of $\mathcal{F}^\xi_{\leq t}$, $\mathcal{F}^\xi_{\geq t}$ respectively, we obtain from (3) and (4) the equations:

$$dx(t) = \frac{1}{m}\overline{p}_+(t,x_t)\,dt + d\tilde{\nu}_+(t), \tag{5'}$$

$$dx(t) = \frac{1}{m}\overline{p}_-(t,x_t)\,dt + d\tilde{\nu}_-(T-t), \tag{5''}$$

$$d\overline{p}_+(t,x_t) = \overline{Q}_+(t,x_t)\,dt + \frac{1}{\hbar}D_{p+}(t,x_t)\,d\tilde{\nu}_+(t), \tag{6'}$$

$$d\overline{p}_-(t,x_t) = \overline{Q}_-(t,x_t)\,dt + \frac{1}{\hbar}D_{p-}(t,x_t)\,d\tilde{\nu}_-(T-t). \tag{6''}$$

The solution of the system (5), (6) (if it exists) represents a certain random process $x(t,\omega)$. Call the ensemble of fictious particles, the trajectories of which coincide with sample paths of the process $x(t)$, a quantum system.

It can be shown by standard techniques that the probability density $\rho(t,x)$ of $x(t)$ satisfies the equations

$$\frac{\partial \rho}{\partial t} = -\frac{\partial}{\partial x}\left(\frac{\overline{p}_+}{m}\rho\right) + \frac{\hbar}{2m}\frac{\partial^2 \rho}{\partial x^2}, \tag{7'}$$

$$\frac{\partial \rho}{\partial t} = -\frac{\partial}{\partial x}\left(\frac{\overline{p}_-}{m}\rho\right) - \frac{\hbar}{2m}\frac{\partial^2 \rho}{\partial x^2} . \qquad (7'')$$

It must be emphasized here that in contrast to markoffian case the equations (7) are nonlinear because as we shall see later the averaging drifts $\overline{p}_\pm$ depend on $\rho(t,x)$.

Assuming time reversibility of the quantum system evolution we have the following transformation properties of the drift coefficients under time inversion:

$$\overline{p}_+(\cdot,x) \to -\overline{p}_-(\cdot,x), \qquad \overline{p}_-(\cdot,x) \to -\overline{p}_+(\cdot,x) . \qquad (8)$$

If was $\overline{p}_+(t,x) = \overline{p}_-(t,x)$ then (8) would coincide with the transformation property of the classical momentum. However, $\overline{p}_+(t,x)$ is not identical with $\overline{p}_-(t,x)$ but differs by the magnitude

$$2p_k(t,x) \equiv \overline{p}_+(t,x) - \overline{p}_-(t,x) = \hbar \frac{1}{\rho(t,x)} \frac{\partial \rho(t,x)}{\partial x} . \qquad (9)$$

Introduce also the symmetrized "momentum"

$$p_c(t,x) = \frac{1}{2}(\overline{p}_+ + \overline{p}_-) = \overline{p}_+ - \frac{\hbar}{2}\frac{1}{\rho}\frac{\partial \rho}{\partial x} = \overline{p}_- + \frac{\hbar}{2}\frac{1}{\rho}\frac{\partial \rho}{\partial x} .$$

Then the pair of equations (7) can be transformed to

$$\frac{\partial \rho}{\partial t} + \frac{\partial}{\partial x}\left(\frac{p_c}{m}\rho\right) = 0 , \qquad (10)$$

$$\frac{\partial p_k}{\partial t} = -\frac{\hbar}{2m}\frac{\partial^2 p_c}{\partial x^2} - \frac{1}{m}\frac{\partial}{\partial x}(p_c p_k) . \qquad (11)$$

In order to define the conditional expectations of quantum forces

$$G_\pm(t,x) \equiv E\left\{G_\pm(t,\omega)\Big|\xi(t)=x\right\}$$

write the "crossed" Ito formulas:

$$d\overline{p}_+(t,x_t) = \left(\frac{\partial \overline{p}_+}{\partial t} + \frac{\overline{p}_-}{m}\frac{\partial \overline{p}_+}{\partial x} - \frac{\hbar}{2m}\frac{\partial^2 \overline{p}_+}{\partial x^2}\right)dt + \frac{\partial \overline{p}_+}{\partial x}d\tilde{\nu}_- \equiv$$

$$\equiv Q_{+-}(t,x_t)\,dt + \frac{\partial \overline{p}_+}{\partial x}d\tilde{\nu}_- ,$$

$$d\overline{p}_-(t,x_t) = \left(\frac{\partial \overline{p}_-}{\partial t} + \frac{\overline{p}_+}{m}\frac{\partial \overline{p}_-}{\partial x} - \frac{\hbar}{2m}\frac{\partial^2 \overline{p}_-}{\partial x^2}\right)dt + \frac{\partial \overline{p}_-}{\partial x}d\tilde{\nu}_+ \equiv$$

$$\equiv Q_{-+}(t,x_t)\,dt + \frac{\partial \overline{p}_-}{\partial x}\,d\tilde{\nu}_+ \,.$$

Combining these relations with the equations (6) one has

$$G_+(t,x) = G_-(t,x) = \frac{1}{2}\left[\overline{Q}_+(t,x) - Q_{+-}(t,x) - Q_{-+}(t,x) + \overline{Q}_-(t,x)\right] =$$

$$= \frac{\hbar}{m}\frac{1}{\rho}\frac{\partial}{\partial x}\left(\rho\frac{\partial p_k}{\partial x}\right) = \frac{2p_k}{m}\frac{\partial p_k}{\partial x} + \frac{\hbar}{m}\frac{\partial^2 p_k}{\partial x^2}\,.$$

Now the equation (6') can be rewritten as

$$d\overline{p}_+(t,x_t) = \left(F(t,x_t) + \frac{2p_k}{m}\frac{\partial p_k}{\partial x} + \frac{\hbar}{m}\frac{\partial^2 p_k}{\partial x^2}\right)dt + \frac{1}{\hbar}D_{p+}(t,x_t)\,d\tilde{\nu}_+(t)\,.$$

Comparing this equation with Ito formula for $\overline{p}_+(t,x)$

$$d\overline{p}_+(t,x_t) = \left(\frac{\partial \overline{p}_+}{\partial t} + \frac{\overline{p}_+}{m}\frac{\partial \overline{p}_+}{\partial x} + \frac{\hbar}{2m}\frac{\partial^2 \overline{p}_+}{\partial x^2}\right)dt + \frac{\partial \overline{p}_+}{\partial x}d\tilde{\nu}_+(t)\,,$$

we obtain

$$\frac{\partial p_c}{\partial t} + \frac{p_c}{m}\frac{\partial p_c}{\partial x} = F(t,x) + \frac{p_k}{m}\frac{\partial p_k}{\partial x} + \frac{\hbar}{2m}\frac{\partial^2 p_k}{\partial x^2}\,. \qquad (12)$$

If one introduces the potential $V(t,x)$ ($F(t,x) = -\frac{\partial V(t,x)}{\partial x}$), then (12) becomes

$$\frac{\partial p_c}{\partial t} + \frac{p_c}{m}\frac{\partial p_c}{\partial x} = -\frac{\partial}{\partial x}\left[V(t,x) - \frac{\hbar^2}{2m}\frac{1}{\sqrt{\rho}}\frac{\partial^2 \sqrt{\rho}}{\partial x^2}\right],$$

or

$$\frac{\partial S}{\partial t} + \frac{1}{m}\left(\frac{\partial S}{\partial x}\right)^2 + W(t,x) = 0\,, \qquad (13)$$

where

$$S(t,x) = \int p_c(t,x)\,dx\,, \qquad W(t,x) = V(t,x) - \frac{\hbar^2}{2m}\frac{1}{\rho}\frac{\partial^2 \rho}{\partial x^2} + C(t)\,,$$

$C(t)$ is an arbitrary function of time.

As is known the pair of equations (10), (13) is equivalent to the Schrödinger equation for the complex function $\Psi(t,x) = \sqrt{\rho(t,x)}\exp\left[\frac{i}{\hbar}S(t,x)\right]$:

$$i\hbar\frac{\partial \Psi(t,x)}{\partial x} = -\frac{\hbar^2}{2m}\frac{\partial^2 \Psi(t,x)}{\partial x^2} + [V(t,x) + C(t)]\,\Psi(t,x)\,. \qquad (14)$$

Thus, $\Psi(t,x)$ represents the formally aggregated construction by means of which the system of equations (10), (13) is linearized. Call to mind that the derivation of the equation (14) has been realized by means of reduction of complete stochastic model. This reduction consists in refusal for any instant t from information about, say, the previous history of a particle but with retention of some mean its behaviour characteristics in a neighborhood of t. In order to decide whether these characteristics form a sufficient statistic [4] for estimating future behaviour of a particle one must to know a probabilistic structure of the process $Q_+(t,\omega)$. Under a lack of such knowledge, evidently, there is no reason for affirmative answer this question.

References

Fisk, D.L.: Quasi-martingales, *Trans. Amer. Math. Soc.*, **120**(1965), 369–389.

Liptser, R.S., Shiryaev, A.N.: *Statistics Of Random Processes. I. General Theory*, Springer, New-York, 1977.

Liptser, R.S., Shiryaev, A.N.: *Statistics Of Random Processes. II. Applications*, Springer, New-York, 1978.

Wald, A.: *Statistical decision functions*, Wiley, New-York, 1950.

ESSENTIALLY PURE PARTICLE FORMULATION OF QUANTUM MECHANICS

S. R. VATSYA
Centre for Research in Earth and Space Science
York University
4700 Keele Street
North York, Ontario, Canada   M3J 1P3

Abstract. An extension of the classical action principle is used to describe the motion of a particle. This extension assigns many, but not all, paths to a particle. Properties of the particle paths are shown to impart wave like behaviour to a particle in motion and to imply various other assumptions and conjectures attributed to the formalism of Quantum Mechanics. The Klein-Gordon and other similar equations are derived by incorporating these properties in the path-integral formalism.

1. Introduction

This paper describes a recent approach to mechanics which was developed independently of any direct considerations of the behaviour of particles in experimental settings including the double slit experiment. However, its implications lead one to consider the following experimental observations and somewhat unorthodox conclusions that might be drawn from them.

In the double slit experiment, photons, electrons and other physical entities that are normally considered particles, demonstrate their particle nature if observed individually. However, if many are allowed to pass through the slits, together or one after the other, then an interference-like pattern of intensity emerges on the screen [1]. Since a wave would produce such a pattern, it is assumed that each particle also has a wave character. Quantum Mechanics accepts this duality by attaching a probability wave with a particle in motion i.e. the wave determines the probability of finding a particle in a certain space-time region. This fusion of wave and particle nature creates most of the logical difficulties with Quantum Mechanics [1,2].

While a wave would produce the intensity distribution of the type observed in the double slit experiment, the converse is not necessarily true i.e. the observation of this pattern does not prove that it was produced by a wave. A closer scrutiny of the experimental observations suggests an alternative possibility.

Conclusions based on relevant observations identify the observed entity as a particle when emitted or absorbed. Experiments designed to reveal its wave nature during travel, observe each individual with particle like attributes. Therefore it may be possible to describe the experimental observations by associating a particle like trajectory with each of the entities. These observations on a number of particles suggest the possibility of the existence of a collection of paths out of which each particle takes one, probably randomly. This collection must be endowed with some characteristics which are responsible for the inclusion of more paths ending about the bright regions and exclusion of others. Therefore it appears more reasonable to build a theory of mechanics by characterizing the collection of particle paths rather than attempting to fuse mutually exclusive wave and particle behaviours. If this view is adopted, then the effect of an observation on its outcome must be the result of the disturbance suffered by the particle and hence, must be described in this manner. This philosophy has its origin in Fermat's principle of stationary time in light and Hamilton's principle of stationary action in classical mechanics. Both of these theories are geometrical in nature instead of mechanical.

The present approach to mechanics is based on an extension of Hamilton's action principle, obtained by a process of completion in the framework of the gauge transformations. The implications of this formulation are in accordance with the inferences that could be drawn from the experimental observations as discussed above. To be precise, the extension yields a collection of infinitely many, but not all, paths for a particle to follow which are endowed with some properties by virtue of the fact that they are the solutions of the extended principle. These properties are shown to describe the behaviour of particles in a double slit experiment and in the Aharonov-Bohm experiment without invoking the usual assumption of probability waves or the formalism of Quantum Mechanics. The results are shown to justify Feynman's path integral formulation and used in this framework to derive a generalized Schrödinger type equation. Properties of the particle paths are used to reduce the general equation into infinitely many four dimensional equations, one of them being the Klein-Gordon equation.

This formulation yields as results, the assumptions underlying the standard Quantum Mechanics and various other intuitive conjectures usually attributed to the formalism of Quantum Mechanics. However there are some differences between the consequences of the present formulation and the standard Quantum Mechanics which are indicated and the direction for further investigations are suggested.

2. The Gauge Mechanical Principle

The classical action principle characterizes the particle path(s) by requiring the action to be stationary i.e. $\delta S = S_{BA}(\rho') - S_{BA}(\rho) = 0$, up to first order, where $S_{BA}(\rho)$ is the action associated with the curve $\rho(AB) = x(\tau)$ in a manifold $\mathcal{M}$, or in $\mathcal{M}'$ obtained from $\mathcal{M}$ by including τ as an additional coordinate [3, Ch. 1.1]. In $\mathcal{M}'$, this takes the form:

$$\exp(\alpha S_{ABA}(\rho_c)) = 1 \tag{1}$$

up to the first order in $d\sigma$ where ρ_c traces the path ρ' from (A, τ_1) to (B, τ_2) and then inverted ρ from (B, τ_2) to (A, τ_1), and $d\sigma$ is the area enclosed by ρ_c. For a charged particle in an electro-magnetic field which may be described by the Lagrangian $L = L^P - \phi_\mu \dot{x}^\mu$, where $L^P = \frac{1}{2}m(\dot{x}_\mu \dot{x}^\mu + 1)$, and ϕ_μ are the electro-magnetic potentials, (1) is equivalent to $U^P{}_{ABA}(\rho_c) = U_{ABA}(\rho_c)$ up to the first order in $d\sigma$, with $U^P{}_{BA}(\rho) = \exp(\alpha S^P{}_{BA}(\rho))$ where $S^P{}_{BA}(\rho)$ is the free particle part of the action and

$$U_{BA}(\rho) = \mathrm{Exp}\left(\alpha \int_{\rho(AB)} \phi_\mu(x) dx^\mu\right)$$

is the gauge group element associated with $\rho(AB)$ with the Lie algebra element $\alpha \phi_\mu dx^\mu$ [4].

It is clear that the classical description of motion limits itself to a characterization of particle-path(s) in terms of the Lie algebra elements, equivalently the infinitesimal gauge group elements, associated with curves of the type ρ_c, which is accurate only up to the first order. Additional information that may be available in the global group elements is not utilized in the action principle. Therefore a description in terms of the group elements should be expected to be more complete. Since the group elements are defined for all curves, the restriction to closed curves ρ_c, also becomes redundant. To achieve appropriate generality consistent with the domain of definition of the gauge group elements, the action principle should be extended to

$$\kappa^{-1}(B)\exp(\alpha S_{BA}(\rho))\kappa(A) = 1 \tag{2}$$

where κ is as yet an undetermined function which cancels out for the closed curves. The characterization of particle paths by (2) has been termed the gauge mechanical principle [5]. Its solutions will be called the physical paths which a particle is allowed to follow. If $\rho(AB)$ is a physical path then A and B will be termed the equivalent or physical points.

It should be remarked that there is no logical deficiency or inconsistency in the action principle itself. The argument here is that the action principle provides an incomplete description of motion in gauge group theoretical terms. Prejudice in favour of the group elements in comparison with the Lie algebra elements, in favour of the global in comparison with the local, is a matter of metaphysical conviction.

In the above, we have provided arguments to justify the present extension of the classical action principle, not a derivation of the gauge mechanical principle. These arguments are to some extent irrelevant as far as the matter of the extension is concerned. The fact that (2) reduces to (1) with appropriate restrictions is sufficient to prove that (2) is an extension of the action principle. Furthermore, the gauge mechanical principle by itself may be made the basis of a formulation of mechanics whether it is an extension of the action principle or not. All that is required is that it provide an adequate description of the motion of particles. In the remainder of this section we clarify the principle further and present its alternative statements.

First we relate the gauge mechanical principle with Newton's second law of motion. Some such relation should be expected as the action principle is equivalent to Newton's law. We limit here to the motion of a charged particle in an electro-magnetic field which illustrates the relation without cluttering the concepts with unnecessary generalities.

Eq. (2) may be expressed as

$$U^P{}_{BA}(\rho) = \kappa(B) U_{BA}(\rho) \kappa^{-1}(A) \qquad (3)$$

For infinitesimal closed curves, the right side of (3) is equal to $(1 + \alpha F_{\mu\nu} d\sigma^{\mu\nu})$ where $F_{\mu\nu}$ are the components of the field tensor, and the left side reduces to $(1 - (p_\mu \dot{p}_\nu - \dot{p}_\mu p_\nu) d\sigma^{\mu\nu})$ where p_μ are the components of the canonical momentum. This equality is equivalent to the Lorentz equation, equivalently, Newton's second law [6].

The gauge mechanical principle may also be interpreted in terms of Weyl's original notion of gauging a rigid measuring rod. According to Weyl's proposal, a rod of length Φ_A at A, transported along $\rho(AB)$, has length $\Phi_{BA}(\rho) = U_{BA}(\rho)\Phi_A$ at B. Eq. (3) may be expressed as

$$\Phi^P_{BA}(\rho) = \kappa(B)\Phi_{BA}(\rho) \qquad (4)$$

where $\Phi^P_{BA}(\rho) = U^P_{BA}(\rho)\Phi^P_A$, and $\Phi^P_A = \kappa(A)\Phi_A$. Weyl's gauge transformations determine the effect of a field on the rigid measuring rod. One may take another rod of length Φ^P_A and transport it along a given curve $\rho(AB)$. Let Φ^P_{BA} be its length at B determined as above without any reference to the field. The gauge mechanical principle requires that Weyl's gauge and the present gauge must return essentially in the same relation as they began with at A for $\rho(AB)$ to be a particle path. It is not necessary to set $\Phi_A^P = \Phi_A$ as it would limit generality without adequate justification. It is sufficient that a precise map between Φ_A^P, Φ_{BA}^P and Φ_A, Φ_{BA} be available. This is consistent with (2) and (3), as the equality (2) for closed curves implies only the group equivalence (3) for general curves.

The function κ in the above appears as a requirement of the mathematical generality as there is no justification for imposing further restrictions. However, for a physical theory, κ must have a clearer physical significance which we discuss below.

The elements U and U^P appearing in (3) pertain to the interiors of the respective curves. As such there is no consideration of the initial physical state of a particle or of local interventions at B or elsewhere. Obviously the physical paths for two particles in different physical states should be expected to be different in the same field. Therefore it is legitimate to interpret κ as representing the physical state of the particle. Interaction with the detecting instrument is local in nature and has a direct impact on the physical state of the particle. Therefore such interactions are also included in κ by way of the physical state of the particle. A precise computation of κ is not necessary for a variety of experimental situations. For example, in the double slit experiment, the particles passing through two slits at A and A' are prepared by the same physical process and are identical in every other respect. Therefore, it is legitimate to conclude that particles at A and A' are in the same physical state even if it may not be precisely defined. Hence, we may set $\kappa(A) = \kappa(A')$. Similarly, two beams meeting at B interact with the same instrument. Therefore B is not only geometrically the same point for two paths $\rho(AB)$ and $\rho(A'B)$, it is also physically equivalent. Therefore κ has the same value for two beams at B. This will be found sufficient for the description of the behaviour of particles in the double slit experiment. The same comment applies to various other experimental situations.

Consider a free particle travelling from A to B along $\rho(AB)$ without any interactions including the intrusion of a detecting instrument. In this situation, the physical state of the particle must remain unchanged. Therefore we shall assume that for a free particle $\kappa(A) = \kappa(B)$ for all points A and B i.e. κ is constant. This extends Newton's first law. The effects of interactions on κ may also be computed. A detailed description of such computations is beyond the scope of this article but it is indicated below to an extent necessary for the clarity of the gauge mechanical principle.

Since for a free particle κ is constant, any change in its value must be a result of an interaction. In standard interventions, a precise value of the interaction is unknown e.g. a detecting instrument but the instantaneous change in the classical momentum may be computed or estimated with sufficient accuracy. This information is sufficient to compute the change ΔS in the action caused by the interaction. The change in the value of κ is then given by $\exp(\alpha \Delta S_{BA}(\rho))$.

The value of α still remains undetermined. For a free particle, the physical paths are defined by $\exp(\alpha S_{BA}(\rho)) = 1$. In general, the action $S_{BA}(\rho)$ is real and non-zero. Therefore α must be purely imaginary which may be set equal to i in appropriate units.

The representations of the gauge mechanical principle given by (2), (3) and (4) are essentially equivalent. Reference to the gauging of the measuring rod is inconsequential for the following developments. Reference to one of the representations, therefore, will include others as well.

Consider a source-detector system with source at A and detector at B. A curve $\rho(AB)$ will be called monotonic if the parameter value increases or decreases monotonically along the curve. By convention, τ will be assumed to increase from A to B. A particle starting at A and confined to $\rho(AB)$ is observable at B if and only if $\rho(AB)$ is physical. If θ is the intensity associated with $\rho(AB)$ at A then the intensity transmitted to B by this path must be equal to θ.

Consider a configuration of two curves $\rho(AB)$ and $\rho'(AB)$ with $\rho_c(ABA)$ being the union of $\rho'(AB)$ and $\rho(BA)$. According to the present prescription, if (4) is satisfied then this is a physical configuration. Since the evolution parameter increases from A to B along both of the curves, particle must travel from A to B along $\rho(AB)$ and $\rho'(AB)$. Therefore $\rho(AB)$ and $\rho'(AB)$ offer equally likely alternatives for the transmission of a particle from A to B, even if $\rho(AB)$ and $\rho'(AB)$ may not be physical. The case of the alternatives of the type $\rho(AB)$ and $\rho'(CB)$ is treated similarly. Such configurations of trajectories are referred to as the interfering alternatives. The intensity of particles transmitted to B by the equally likely alternatives must be equal to the sum of the intensities at A and C associated with the respective trajectories. Such a union of paths is indistinguishable from a pair of monotonic physical paths since $\kappa(B)$ may be adjusted such that $\rho'(CB)$ and $\rho(AB)$ are both physical which does not alter the relevant physical content.

3. Physical Paths

Consider a physical system described by a Lagrangian $L(\dot{x}, x)$ with ρ_s being the resulting classical path. For a free particle, $L = L^P$. The action $S_{B'A'}(\rho')$ along a trajectory $\rho'(A'B')$ in a small neighbourhood of $\rho(AB)$ is given by

ESSENTIALLY PURE PARTICLE FORMULATION 229

$$(S_{B'A'}(\rho') - S_{BA}(\rho)) = \int_{\rho(AB)} \delta x^\mu \left[\frac{\partial L}{\partial x^\mu} - \frac{d}{d\tau}\frac{\partial L}{\partial \dot{x}^\mu}\right] d\tau$$
$$+ \left[\frac{\partial L}{\partial \dot{x}^\mu}\delta' x^\mu - H\delta'\tau\right]_A^B + O(\delta^2) \quad (5)$$

where $\delta' x^\mu$, $\delta'\tau$ correspond to the variation of the end points A, B to A', B', and H is the Hamiltonian. The term $O(\delta^2)$ is the integral along $\rho(AB)$ of an argument, containing functions of second or higher order in (δx) and $(\delta \dot{x})$.

If $\rho = \rho_s$, then the first term on the right side of (5) is equal to zero. Hence $S_{B'A'}(\rho') = S_{BA}(\rho_s)$ for some values of $\delta' x = O(\delta^2)$. Therefore the trajectories in a δx neighbourhood of a physical classical path $\rho_s(AB)$ are also physical and their end points are confined to a (δ^2) neighbourhood of B. Thus the intensity transmitted by paths in a δx neighbourhood of a classical trajectory is concentrated in (δ^2) neighbourhoods of the end point. Let ρ be a path transmitting intensity outside (δ^2) neighbourhood of B. Since ρ is not a solution of the Euler-Lagrange equation, the first term in (5) dominates which is $O(\delta x)$. Repeating the above argument, we have that the intensity transmitted by trajectories in a δx neighbourhood of ρ is spread over a δx neighbourhood of points outside (δ^2) neighbourhood of B. Further, the magnitude of the first term in (5) increases as ρ is removed farther from the classical trajectory. Therefore the contribution to the intensity decreases accordingly. Thus the intensity should be expected to be higher near the points equivalent to A and to decrease away from them, creating a wave-like pattern over a uniform background. Similarly, it can be shown that on a macroscopic scale, the particles from A to B travel along narrow beams centered about the classical trajectories [5].

The interfering alternatives play a prominent role in the double slit experiment. In this setup, identical particles are allowed to pass through two slits at A and A', and collected on a distant screen at a point B. The following treatment is valid in the presence of a field. As explained above, the particle paths may be assumed concentrated about the classical trajectories from A to B and from A' to B. If one of the beams is blocked, then the intensity observed in a neighbourhood of B should behave as above. However, if the intensity is transmitted by both of the beams, then a multitude of the interfering alternatives is allowed. Existence of such paths and their influence on the intensity distribution is studied next.

In view of the physical equivalence of A and A' and that of the particles, one has that $\kappa(A) = \kappa(A')$, $\Phi_A^P = \Phi_{A'}^P$, and hence $\Phi_A = \Phi_{A'}$. However, because of an interaction with the detecting instrument at B, $\kappa(B)$ may not be equal to $\kappa(A)$. For the interfering alternatives, the value of $\kappa(B)$ is the same for both of the monotonic segments (Sec. 2). Substitutions in (4) show that these paths are the solutions of

$$exp\left[i\left(\int_{\rho(AB)} dS(x,\tau) - \int_{\rho'(A'B)} dS(x,\tau)\right)\right]\Phi_A = \Phi_A \quad (6)$$

For the classical trajectories $\rho = \rho_s$ and $\rho' = \rho'_s$, (6) is solved by

$$\left(S_{BA}(\rho_s) - S_{BA'}(\rho'_s)\right) = 2\pi j$$

where j is an arbitrary integer and the action in this case is Hamilton's principal function or the arc-length in $\mathcal{M}$. Classical paths are characterized by a constant velocity $\bar{u}$. This reduces the solution to $\Delta r = 2\pi j/m\bar{u}$, where Δr is the difference between the path-lengths of $\rho_s(AB)$ and $\rho'_s(A'B)$. Therefore $\rho_s(AB)$ and $\rho'_s(A'B)$ are interfering alternatives whenever $\Delta r = 2\pi j/m\bar{u}$.

Let $B(\varepsilon)$ be the point on the screen such that

$$\left(S_{B(\varepsilon)A}(\rho_s) - S_{B(\varepsilon)A'}(\rho'_s)\right) = 2\pi(j+\varepsilon) \quad (7)$$

for a fixed j and each $0 \leq \varepsilon \leq 1/2$. From (5), $S_{CA}(\rho) = S_{B(\varepsilon)A}(\rho_s)$, $S_{C'A'}(\rho') = S_{B(\varepsilon)A'}(\rho'_s)$, for ρ, ρ' in δx neighbourhoods of ρ_s, ρ'_s respectively, where C and C' vary over a (δ^2) neighbourhood of $B(\varepsilon)$ on the screen for a fixed ε. Therefore, by varying the paths over a (δx) width of the beam and over a (δ^2) neighbourhood of $B(\varepsilon)$ it is possible to satisfy

$$(S_{DA}(\rho) - S_{DA'}(\rho')) = 2\pi(j+\varepsilon)$$

for most of the paths. For $\varepsilon = 0$, this implies that there is a large concentration of interfering alternatives reaching about $B(0)$ and hence the intensity in a (δ^2) neighbourhood of $B(0)$ is almost equal to the intensity in δx neighbourhoods of $\rho_s(AB(0))$ and $\rho'_s(A'B(0))$. For $\varepsilon \neq 0$, the configuration of the paths $\rho_s(AB(\varepsilon))$ and $\rho'_s(A'B(\varepsilon))$ is obviously non-physical. From the above argument, a large number of paths in δx neighbourhoods of $\rho_s(AB(\varepsilon))$ and $\rho'_s(A'B(\varepsilon))$ are excluded from combining to form the interfering alternatives and hence unable to transmit the intensity in a (δ^2) neighbourhood of $B(\varepsilon)$. There are paths capable of transmitting intensity about $B(\varepsilon)$ for $\varepsilon \neq 0$, but it can be shown that the amount of intensity that is concentrated in a (δ^2) neighbourhood of $B(0)$ is spread over a $\delta x(\varepsilon)$ neighbourhood of $B(\varepsilon)$. Consequently, a rapid decrease in the intensity is expected as ε increases away from zero. Similar arguments show that $O(\delta x(\varepsilon))$ increases with ε, implying a decrease in the intensity as ε increases further [5].

As explained before, if one of the beams is blocked, the interference pattern is destroyed. Also, such a distribution should not be expected to result if the equivalence of A and A' is violated. This situation arises when an attempt is made to observe the particle anywhere along the trajectory. Interaction with the detecting instrument changes the classical momentum of the particle say by ΔP. It is straight forward to estimate the change ΔS in the action which is very large for the macroscopic trajectories. This enables one to estimate κ. This can be used to conclude that the intensity transmitted to B by the interfering alternatives is negligible. Hence, the two beams transmit intensity as the classical beams of particles. Similar arguments may be used to describe other details of the double slit experiment.

For additional properties of the physical paths, consider the Aharonov-Bohm effect [5,7]. In this experiment, the electrons travel along paths $\rho(ACB)$ and $\rho(ADB)$. The closed curve $\rho_c(ACBDA)$ encloses a magnetic field but the electrons are shielded from it. The interfering alternatives, which are the major contributors in this case also, are the solutions of:

$$\exp\left[i\oint(dS^P(x,\tau) - \phi_\mu dx^\mu)\right]\Phi_A = \Phi_A \qquad (8)$$

where the integration is along $\rho_c(ACBDA)$.

It is clear that ϕ_μ-dependent part in (8) is $U_{ABA}(\rho_c)$ which is equal to $\exp(iF(\phi))$ where $F(\phi)$ is the magnetic flux enclosed by ρ_c. As ρ_c is distorted, $F(\phi)$ remains unchanged as long as the distorted closed path encloses the flux, which covers all of the paths of significance here. Let ϕ change to $\hat{\phi}$ such that:

$$\begin{aligned}(F(\phi) - F(\hat{\phi})) &= \oint(\phi_\mu - \hat{\phi}_\mu)dx^\mu \\ &= 2\pi j \end{aligned} \qquad (9)$$

with an arbitrary integer j. Whenever (9) is satisfied, for each curve ρ_c, (8) and hence its solutions remain unchanged. Therefore the experimental observation with $\hat{\phi}_\mu$ must be the same as with ϕ_μ. Thus the interference pattern on the screen should repeat itself periodically as the potential is varied continuously. Existence of the interference pattern is deduced as in the case of the double slit experiment.

Let $\phi_\mu(\varepsilon)$ be a one parameter family of potentials with $0 \leq \varepsilon \leq 1$, such that $(F(\phi(1)) - F(\phi(0))) = 2\pi$, i.e., ε covers one period. The intensity patterns corresponding to $\phi_\mu(0)$ and $\phi_\mu(1)$ are indistinguishable. Let the solutions of (8) with ϕ_μ replaced by $\phi_\mu(\varepsilon)$ be $\{\rho(\varepsilon)\}$. Owing to the continuity of $F(\phi(\varepsilon))$ with respect to ε, $\{\rho(\varepsilon)\}$ should vary continuously, implying a continuous variation of the corresponding interference pattern. As ε approaches one, the distribution of the intensity must return to the same as for $\varepsilon = 0$. Thus, each interference fringe should be expected to shift as ε varies from zero to one, from its position to the original location of the next. This conclusion agrees with the experimental observation [7].

Above considerations show that the wave-like behaviour of a particle in motion is a result of the properties of the physical paths. However, there is a crucial difference as described below. Consider the double slit experiment. If the intensity pattern on the screen is a result of a wave motion, then there must be a point of zero intensity in between two bright regions. According to the present formulation, a point of minimum intensity exists but it can be seen that there must be some physical paths reaching every point on the screen, resulting in some intensity everywhere. If accurate enough determination of the intensity can be made, it may be possible to test whether the present theory or Quantum Mechanics provides a better description of motion. Nevertheless, major contribution to the intensity in the present formulation is the same as predicted by the wave motion. Thus one may use the results from the wave theory in building a theory of mechanics, at least approximately. While the above considerations justify use of the results from the theory of waves, it is only for convenience rather than a physical attribute of the particles.

4. Equation of Motion

Since it is impossible to assign a unique trajectory to a particle, as an alternative, one may describe its motion in terms of the intensity of the particles transmitted to a region in $\mathcal{M}$ or $\mathcal{M}'$ by the physical trajectories. While such a theory is possible, it will require quite intricate computations for which a machinery is not yet developed. An approximate theory may be developed by exploiting the wave-like behaviour of the particles deduced in Sec. 3. In addition to simplifying the manipulations, this relates the present formulation with Quantum Mechanics which is instructive in itself.

Present formulation associates a phase-factor equal to $\exp(iS_{BA}(\rho))$ with $\rho(AB)$. The phases associated with a multiplicity of paths are shown in Sec. 3 to interfere in a manner that imparts wave-like properties to the particles in motion. Having yielded its basic assumptions, the gauge mechanical principle finds a natural expression within the framework of Feynman's path integral formalism [1,5,8]. However, only the physical paths should be included in the computation of the total contribution. This together with the standard methods yields the following equation [5,6].

$$i\frac{\partial \psi}{\partial \tau} = -\frac{1}{2m}\Pi_\mu \Pi^\mu \psi \qquad (10)$$

with $\psi(x,0) = -\psi(x, 2\pi/m)$ where $\Pi_\mu = (i\partial/\partial x^\mu \cdot 1 + \phi_\mu)$. In view of this boundary condition, (10) decomposes into

$$\begin{aligned}\Pi_\mu \Pi^\mu \psi_n &= (2n+1)m^2 \psi_n \\ n &= 0, \pm 1, \pm 2, \cdots\end{aligned} \qquad (11)$$

For $n = 0$, (11) reduces to the Klein-Gordon equation.

If all trajectories are allowed to contribute, the resulting equation is still (10) but without the boundary condition i.e. the original generalized Schrödinger equation conjectured by Stückelberg [9]. Feynman [10] used this equation to deduce the Klein-Gordon equation by an arbitrary restriction on the solution. Present treatment relates (10) with the Klein-Gordon equation quite naturally providing additional support for the assumption (4).

The above procedure has also been used to develop an equation of motion in a Riemannian space where the resulting theory is conceptually clearer [11]. The next major step in constructing a complete theory of mechanics in the present framework would be to abandon the path-integral formalism and compute the intensity transmitted by the physical paths directly by solving the functional equations. Comparisons with other theories e.g. Bohmian mechanics is desirable. The studies of other phenomena e.g. tunneling and behaviour of the correlated particles, even with the level of accuracy of Sec. 3, should prove instructive. Also, the physical implications of the additional equations arising here should be investigated.

5. Concluding Remarks

In this article, the classical action principle is extended in the framework of the gauge transformations. Solutions of the resulting equation, termed the physical paths, impart wave-like properties to a particle in motion which is the basic assumption underlying Quantum Mechanics. The Aharonov-Bohm effect is described to a large extent solely by (4) which also yields the basic assumptions of the path-integral formalism resulting in a systematic derivation of the generalized Schrödinger equation. Thus the present formulation develops a coherent theory unifying various treatments attributed to the existing Quantum Mechanics without involving its usual assumptions.

Classical description of motion is quite accurate at the macroscopic scale. Quantum Mechanics modifies these results only slightly but conceptually it is fundamentally different. It also appeals to experimental observations for its underlying assumptions without offering conceptual clarity. The present formulation extends Classical Mechanics yielding these assumptions and various conjectures in a coherent framework. Thus the gauge mechanical principle offers a more satisfactory basis for the formulation of mechanics. In particular, it eliminates the need for a direct assumption of wave nature of a particle in motion which underlies the well known difficulties with Quantum Mechanics. It is pertinent to remark that while the present theory associates a somewhat objective meaning to a particle in motion, an element of randomness remains in the availability of the equally likely, infinitely many paths.

Quantum Mechanics results as an approximation to the present theory, presumably quite accurate. Deviations from Quantum Mechanics are pointed out, and directions for further investigations, and to construct a more accurate and complete theory, are indicated.

Acknowledgements

The author is thankful to Patrick A. O'Connor for helpful discussions, encouragement and substantial logistical support without which this presentation would not have been possible.

References

1. R.P. Feynman and A.R. Hibbs, Quantum Mechanics and Path Integrals, McGraw-Hill, New York, 1965, pp. 2-9.
2. J. Horgan, Quantum Philosophy, Scientific American, July 1992, pp. 94-104.
3. H. Rund, The Hamilton-Jacobi Theory in the Calculus of Variations, Van Nostrand, London, 1966.
4. H. Weyl, Ann. Phys. (Leipzig) 59 (1919) 101; H. Weyl, Space Time Matter (Translated by H.L. Brose) Dover, New York, 1951, Ch. IV, Sec. 35.
5. S.R. Vatsya, Can. J. Phys., 73 (1995) 85.
6. S.R. Vatsya, Can. J. Phys., 67, (1989) 634.
7. Y. Aharonov and D. Bohm, Phys. Rev. 115 (1959) 485. R.G. Chambers, Phys. Rev. Lett. 5 (1959) 3. A. Tonomura, et al Phys. Rev. Lett. 56 (1986) 792.
8. R.P. Feynman, Rev. Mod. Phys. 20 (1948) 367.
9. E.C.G. Stückelberg, Helv. Phys. Acta. 14 (1941) 322; Ibid. 15 (1942) 23.
10. R.P. Feynman, Phys. Rev. 80 (1950) 440.
11. S.R. Vatsya, Mechanics of a charged particle on the Kaluza-Klein background, Can. J. Phys. 73 (1995).

A FUNDAMENTAL FORCE AS THE DETERMINISTIC EXPLANATION OF QUANTUM MECHANICS

BILLIE JACK DALTON
Department of Physics, Astronomy and Engineering Science
St. Cloud State University
720 Fourth Avenue South
St. Cloud, Minnesota 56301-4498

ABSTRACT

In a recent monograph [1], the author has proposed a new fundamental force as the root explanation of the relationship between particle distributions and wave intensities as well as the relationship between the energy momentum of particles and the periods and wavelengths of waves. This new force involves a nonlinear relationship between the relativistic particle acceleration and the components of waves which are treated as fields. Extensive numerical calculations were given to illustrate and verify these claims for a variety of time-independent wave solutions. The work presented here extends this previous work to time-dependent wave packet reflection. Using the fundamental force, the calculations show that starting with randomly chosen initial particle positions and momentum, the particle distribution is re-shaped to match that of the traveling wave-packet. Upon partial reflection of the packet the particles redistribute themselves to match the new wave intensity involving interference of the incident and reflected waves. The fundamental force imposes a constraint on the metric tensors that generally cannot be satisfied by a Minkowski flat space. For a solution of this dilemma, I propose a Riemann field metric with metric pulses that propagate with the waves. Some implications of the latter are briefly discussed.

1. A Deterministic Challenge

I am constantly reminded that in all areas of modern physics the ensemble data can be accumulated from counting individual events. In fact, our laboratories abound with particle detectors which count rather localized individual events [2]. For me, to **not** account for localized individual particles and their transport mechanism is to irresponsibility ignore experimental reality.

In addition to individual particle counting, we have at least three observations that indicate both waves and particles are involved together on a fundamental level [3]. The first two are direct particle-wave connections. 1) Ensemble particle distributions exhibit intricate details that can be described by wave intensities calculated with a quadratic form indicated here by $\tilde{\Psi}\Psi$. 2) Measured values of the energy and momentum of particles are directly connected to the frequency and wavelength of the waves used to explain the observed particle distributions (e.g. the photoelectric effect). 3) Many observed discreet values for quantities such as energy and angular momentum can be accounted for as consequences of either boundary conditions on waves or symmetries characterizing waves.

The two above mentioned particle-wave connections appear to be somewhat global in that they hold for all types of particles. Because of this, it seems reasonable to expect them to be consequences of some common cause, such as my proposed fundamental force [1], rather than properties of a particular wave equation.

To construct a combined particle and field theory that can explain individual particle detection, as well as the observed wave-particle ensemble connection, is a <u>deterministic challenge</u> to physicists, and one that will likely last most of us a lifetime. For the rest who choose to ignore the challenge of this problem, the common *a priori* probability theories offer an easy out. For me however, it is difficult to ignore particle detectors counting individual particles.

2. Fundamental Force

The particle is modeled as a "very localized entity" with a point value of four-momentum and space-time position. I use n_μ to represent particle momentum components, and p_μ to represent wave <u>ray</u> momentum components. The definition of the latter will generally differ for different types of fields, and it is the particular definition of these quantities that puts in much of the character of the fields. The symbol q is used to represent the line parameter. For the infinitesimal displacements dx_μ, we have the relations $dx_\mu = n_\mu dq$. The usual dot notation is used to indicate a derivative with respect to q. From my monograph [1] we have the following fundamental force.

$$\dot{n}_\mu = \dot{p}_\mu - F(n_\mu - p_\mu) \qquad (1)$$

In these equations F is a positive quadratic expression of the field Ψ (such as $\tilde{\Psi}\Psi$ where $\tilde{\Psi}$ is an appropriate adjoint) and has units inverse to the line parameter q. The exact expression for F will in general differ for different fields. In the calculations in my monograph [1] and the ones presented here I used the following form for p_μ.

$$p_\mu = -i\frac{\hbar}{2p_{ref}}(\tilde{\Psi}\frac{\partial}{\partial x^\mu}\Psi - (\frac{\partial}{\partial x^\mu}\tilde{\Psi})\Psi)/(\tilde{\Psi}\Psi) + e\frac{A_\mu}{p_{ref}} \qquad (2)$$

The vector potentials $A_\mu = \tilde{\Psi}\bar{A}_\mu\Psi/(\tilde{\Psi}\Psi)$ are necessary if this equation is to be invariant under gauge transformations [4], or covariant under diagonal nonlinear realizations of the Lorentz group [5][6]. For non zero mass, the reference momentum constant p_{ref} is set to mc. The form in (2) as well as other forms for the Dirac spinor field have been used in earlier deterministic studies[7][8][9][10]. The appropriate formula to use for the p_μ and function F in (1) for other fields, in particular the electromagnetic field, as well as inclusion of certain properties of quantum field theory, is a subject of ongoing research by the author.

There is an important difference between this and earlier deterministic theory. In the earliest deterministic studies the particle momentum was directly identified with the p_μ of the fields [7][8]. Later, Bohm and Vigier [11] and others[12][13] studied a stochastic fluctuation model in which the p_μ represented the average components of the particle. In [1] I have shown with numerical calculations that many small stochastic momentum contributions along the particle's path actually tend to destroy the wave intensity patterns. This can be explained by the central limit theorem with many small stochastic fluctuations accumulating as equivalent initializations with a normal, rather than uniform distribution [1].

In the theory presented in [1] and here, the particle's momentum is a separate entity and connected to the p_μ through the fundamental force (1). The natural trend of the fundamental force is to change the particle's momentum towards the p_μ. This is a limiting process in which the two may never become exactly equal. This is important when considering the sensitivity of the nonlinear equations in (1). This side of the big bang, normal transitions and evolution of the fields produce, through the p_μ and function F, sufficient initial phase space range to generate, through the fundamental force, the observed distributions.

Following [1] the limit $n_\mu \to p_\mu$ will be called the **attractor limit** of the fundamental force equations. The second wave-particle connection mentioned above (e.g. the photoelectric effect) is a consequence of this particular attractor trend of the nonlinear equations in this fundamental force. As demonstrated in a wide variety of calculations in [1] and in the following pages, the first of the above mentioned wave-particle connections (i.e. the ensemble connection) is likewise a consequence of this fundamental force.

3. Test of the Fundamental Force

To demonstrate that this force can bring about the wave intensity patterns in particle distributions, we consider the following equations.

$$\dot{n}_i = \dot{p}_i - \tilde{\Psi}\Psi(n_i - p_i) \tag{3}$$

$$\dot{n}_0 = -n^i \dot{n}_i / n_0 \tag{4}$$

The units are absorbed in $\tilde{\Psi}\Psi$. Here, invariance of the Minkowski space quadratic form $n^\mu n_\mu$ is imposed at the cost of giving up the attractor limit on the energy component. This is likely not in general consistent with physical observation. However, these simpler equations are convenient for examining the force trends.

We will refer to the $\dot{p}_i$ term as the p-dot term in the following and in the figures. With only this term in the force and setting $n_i = p_i$ at some initial particle location, this theory reduces to the follow-the-wave theory of earlier deterministic work [7][8]. We shall demonstrate with detailed calculations that with only the p-dot term, the force cannot bring about, from random patterns, the desired wave intensity pattern in the particle distributions, nor in general, preserve it even if the pattern is initially put in. The extent to which the p-dot term alone can preserve the distribution depends upon the definitions used for p_μ and F.

3.1. TRAVELING GAUSSIAN WAVE-PACKET

In this case I use a wave described by the following formula.

$$\Psi = \Psi_0 \exp i(k^0 x_0 + k^1 x_1) \exp\left(-\frac{(k^0 x_0 + k^1 x_1)^2}{2\delta^2}\right) \qquad (5)$$

As in [1], the calculations here are done one particle at a time. The final particle position where the histogram count is made is determined at the value of the line parameter q where the wave-packet center (Z in the figures) reaches the same pre-selected location. The particle momentum units are expressed in terms of $k = 2\pi/\lambda$, and all distance units are given in terms of the wavelength λ. All of the particles are initialized at $x_0 = 0$, with x_1 and n_1 chosen uniformly randomly from the range $-1 \leq x_1 \leq 1$ and $0.2 \leq n_1 \leq 1.8$ respectively. To initially satisfy the Minkowski invariant condition I initially set $n_0 = n_1$ and choose n_1 randomly.

To see how the force can correct the particle's momentum to that of the wave, we can compare the root mean square (Rms.) deviation between the two versus the distance (Z) that the wave center has traveled. TABLE 1. shows the calculated Rms. values for seven distances. The Rms. deviations for the energies are identical to the ones shown in TABLE 1. From these results it is clear that the force has a strong trend to bring about the attractor limit. It is this feature of the force that can explain the second observed wave-particle connection discussed above. The photoelectric effect is no accident!

Figure 1. a) and *Figure 1. b)* show the particle distributions when the wave-packet center has moved to 0.2 and 5 wavelengths respectively. Even though many of these particles are not exactly near the attractor limit in *Figure 1. b)*, the ensemble distribution is well on its way to matching the wave intensity profile. This change in distribution is due completely to the second term in (3) since the p-dot term is exactly zero because of the simple phase form in (5).

TABLE 1. Root mean square deviation (Rms.) between the particle and wave momentum versus the distance (Z) the wave center has traveled. Z is in wavelength units. Rms. is in wave momentum units.

Distance	Rms. momentum spread
0.2	0.43
1.0	0.39
5.0	0.29
10.0	0.18
20.0	0.068
100.0	0.0040
300	0.0004

To further examine this force, consider a reflector placed at 350.5λ with 80% amplitude reflection. The particle histograms are taken at the value of the line parameter q where the wave-packet center is at 350λ, just 0.5λ before the reflector. *Figure 2.* gives the results for this case. For the results in *Figure 2. a)*, the full force is used until the particles reach 300λ, after which only the p-dot term is used. At 300λ the particles are very near the attractor limit and the distribution has the Gaussian shape of the wave intensity. From the shape of the distribution in *Figure 1. a)*, it is clear that, although the p-dot term brings about a great deal of detail, it does not produce a distribution that matches the wave intensity. This distribution directly reflects information in the phase for two interfering waves. The phase simply does not have all of the amplitude information although there is a lot of correlation. As the wave changes the p-dot term alone cannot preserve the wave intensity pattern in the particle distribution.

In *Figure 2. b)* I show the distribution obtained by using the full force all the way. The excellent agreement between the particle distribution and the wave intensity in this case clearly demonstrates that the fundamental force can bring about the high order observed in particle distributions.

3.2. DOUBLE SLIT SCALAR DIFFRACTION

For this case the slit width is 5λ and the slit separation, center to center, is 15λ, and the "detector screen" where the histogram count is made is placed at 120λ. This location is in the Fresnel region. As in [1] I use the Raleigh-Sommerfield formula with Neumann boundary conditions. For the electromagnetic field one should use a full vector approach. However, this scalar case will illustrate how the fundamental force can bring about the wave intensity pattern in the particle distribution. It also offers an opportunity to further examine the contribution of the p-dot term. The calculated results are presented in *Figure 3*. For these calculations, the

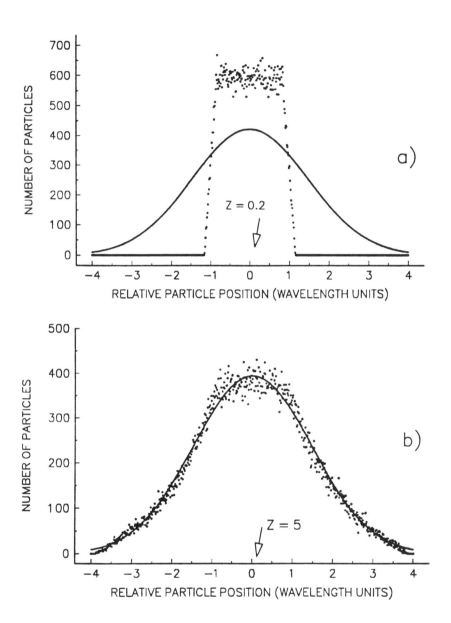

Figure 1. Evolution of particle distributions versus the wave intensity with Gaussian shape. The dot histogram is the particle distribution and the solid line is the wave intensity. *a)* Distribution at 0.2λ, *b)* Distribution at 5λ.

DETERMINISTIC EXPLANATION OF QUANTUM MECHANICS

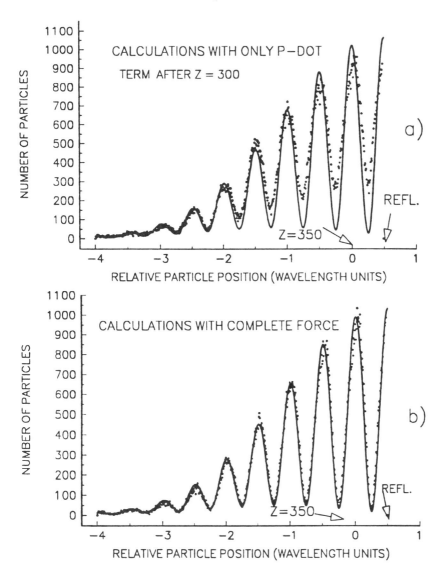

Figure 2. Comparing particle distributions with wave intensity patterns for barrier reflection of a Gaussian wave-packet. The amplitude reflection is 80%. The barrier is located 0.5λ beyond the wave-packet center located at 350λ. *a)* Distributions obtained using the p-dot term only after 300λ. *b)* Distributions obtained using the full force.

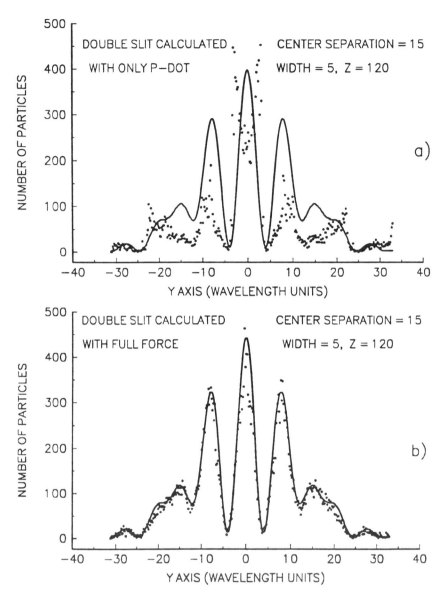

Figure 3. Comparing particle distributions with wave intensity patterns for double slit scalar diffraction. *a)* Distributions obtained using the p-dot term only, *b)* Distributions obtained using the full force.

particle positions were chosen uniformly randomly across the slit openings and at 0.2λ in front of the slits. At each point along the particle path the force terms were calculated by numerical integration over the slit openings. More details of the numerical techniques are given in [1]. The initial magnitude of the momentum of the particle is set to that of the wave. The initial direction of the momentum is perpendicular to the slit openings.

The calculated distribution using the p-dot term only is shown in *Figure 3. a)*. From this figure it is clear that, although the p-dot term does put in much information, it does not give the same profile as the wave intensity shown by the solid line in the *Figure 3*. The distribution obtained using the full force is shown in *Figure 3. b)*. This distribution is in much better agreement with the wave intensity profile. The slight deviations are due to initializing at 0.2λ in front of the slits in conjunction with the numerical approximations used. Initializing closer to the slits as well as using a finer grid on the numerical integration is extremely costly in computer run time. It is clear from this calculation that the double slit diffraction pattern can be reproduced deterministically with the fundamental force, one particle at a time.

3.3. STERN-GERLACH

Because the Stern-Gerlach experiments are so often used to justify the probability models of quantum mechanics I include a deterministic calculation to compare with the wave intensity predictions. The details of the solution is given in [1]. For the results in *Figure 4*. the particles were initialized at 3000λ before the gradient magnetic field and counted on the screen at 3000λ beyond this point at the magnetic field exit. The initial y positions and momentum directions were chosen uniformly randomly in the ranges $-5\lambda \leq y \leq 5\lambda$ and $-16.5^0 \leq \theta \leq 16.5^0$ Here, θ is measured from the Gaussian beam axis and y is perpendicular to it. $|C^+|^2 = 0.4$ and $|C^-|^2 = 0.6$. The Gaussian beam waist was 10λ. From this figure it is clear that the Stern-Gerlach observations can be deterministically explained with the fundamental force.

3.4. HYDROGEN

For an example involving three dimensions, I include in *Figure 5* the spherical shell electron distribution obtained from a deterministic calculation in comparison with the wave intensity profile. The p-dot term is zero for this case so that the distribution is brought about completely by the second term in the fundamental force. The initial momentum and positions were chosen uniformly randomly from the ranges $0.0 \leq |\vec{n}| \leq 3.0$ and $0.0 \leq |\vec{r}| \leq 50.0$. The initial momentum directions were chosen uniformly randomly from an isotropic range. The agreement between the deterministic distribution and the wave intensity pattern clearly indicates that electron distributions in Hydrogen can be explained deterministically, by the fundamental force, one particle at a time.

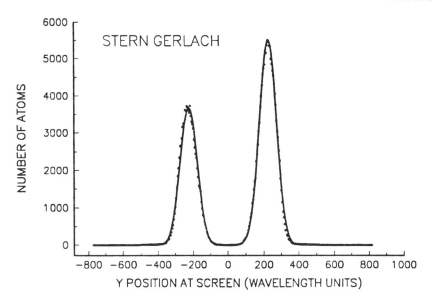

Figure 4 Calculated distribution on the detector screen for deterministic atom trajectories for an unsymmetrical Stern-Gerlach example. The solid line represents the wave intensity.

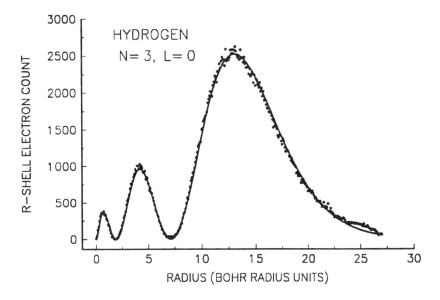

Figure 5. Radial distribution histogram for the Ψ_{300} state. Initial momentum and position chosen uniformly randomly from the ranges $0.0 \leq |\vec{n}| \leq 3.0$ and $0.0 \leq |\vec{r}| \leq 50.0$. The solid line is the wave intensity [14].

4. Geometric Implications

In the attractor limit the n_μ approach the p_μ. In this limit, invariance along the particle path of the Minkowski geometric form $n_\mu n^\mu$ implies invariance of the form $p_\mu p^\mu$. Because the p_μ are derived from waves, the latter form is rarely invariant. One exception is a perfect plane wave for which the p_μ are constant. One could impose the Minkowski constraint by not imposing the attractor limit for one of the p_μ as in (4) above. This is not consistent with physical observation which indicates that the particle energy is associated with the wave frequency and the momentum is associated with the wavelength. This wave-particle connection is strong throughout modern physics observations. One might even say that it is a foundation point of modern physics.

I propose that the solution to this dilemma is replacement of the Minkowski space with a Riemann space. For the latter, the invariance condition

$$\frac{d}{dq}(g^{\mu\nu} n_\mu n_\nu) = 0 \tag{6}$$

implies, in the attractor limit (AL), the following constraint.

$$\frac{d}{dq}(g^{\mu\nu} p_\mu p_\nu)_{AL} = 0 \tag{7}$$

In the attractor limit, the $g^{\mu\nu}$ become the field metric tensors. This is a constraint relation between the $g^{\mu\nu}$ and the wave ray momentum components p_μ. It is well known that general relativity has been extremely successful. In it the metric tensors depend upon the mass. However, the mass of particles is associated directly with one of the Casimir invariants of the Poincare' group. Explanation of this connection between the metric of general relativity and the waves of elementary particles has challenged physicists for a good part of this century. The possibility that the above constraint could lead to the latter connection as well as other implications is worth exploring. Let us write the metric tensors as a sum of two parts;

$$g^{\mu\nu} = g_0^{\mu\nu} + h^{\mu\nu} \tag{8}$$

where $g_0^{\mu\nu}$ is a locally flat metric ($\dot{g}_0^{\mu\nu} \cong 0$) and $h^{\mu\nu}$ is a metric pulse that travels with the wave that guides the particle via the fundamental force. With this, the constraint condition (7) can be rewritten as follows

$$(\dot{h}^{\mu\nu} p_\mu p_\nu + 2 g^{\mu\nu} \dot{p}_\mu p_\nu)_{AL} = 0 \tag{9}$$

The existence of metric pulses even though small for each particle, could have significant effects. Here, I briefly mention two. Since the pulses travel with the wave and particle, the metric in a region can "leak out" with the particles leaving the region, or "leak in" if particles enter a region. If light particles and neutrinos have metric pulses that travel with them, this could have some significant affect on the average metric, or changes in it. In a supernova explosion, for instance, the metric strength could possibly dissipate with the vast neutrino burst.

A second interesting possibility directly involves the invariance of the quantity $(g^{\mu\nu} p_\mu p_\nu)_{AL}$ in the attractor limit along the particle path. This constraint along the particle path could act as a local constraint on the fields themselves.

The fact that the fundamental force can bring about the wave intensity order in the particle distributions is a consequence of the high sensitivity of the nonlinear equations in conjunction with stability trends of the particular form of the equations. Normal field transitions provide sufficient initial phase space range to generate, through the fundamental force, the observed distributions in modern physics. From the wide variety of distributions calculated in [1] and this work, it should be clear that for any wave with a well defined intensity F and components p_μ, the above fundamental force can generate the two wave-particle connections, one particle at a time.

5. References

1. Dalton, B. J. (1994) *Deterministic Explanation of Quantum Mechanics Based on a New Trajectory-Wave Ordering Interaction*, North Star Press of St. Cloud Inc., P.O. Box 451, St. Cloud, Minnesota 56302 USA.
2. De Broglie, L (1964) *The Current Interpretation of Wave Mechanics: A Critical Study*, Elsevier, Amsterdam. In this book, professor De Broglie provides some perceptive insight into measurements of local events.
3. See Ch. 3. of [1].
4. Aitchison, I. J. R. and Hey, A. J. G. (1982) *Gauge Theories in Particle Physics*, Hilger, Bristol.
5. Dalton, B. J. (1982) *In. J. Theor. Phys.* **21**, 765.
6. Dalton, B. J. (1987) in *Group Theoretical Methods In Physics*, ed. R.Gilmore, World Scientific, New Jersey, 578.
7. Bohm, D. and Hiley, B. J. (1993) *The Undivided Universe*, Routhledge, London, References to earlier work can be found in this book and in [8].
8. Holland, P. R. (1993) *The Quantum Theory of Motion*, Cambridge University Press.
9. Bohm, D. (1953) *Prog. Theor. Phys.* **9**, 273.
10. Holland, P. R. (1992) *Foundations of Physics*, **22**, (10), 1287.
11. Bohm, D. and Vigier, J-P. (1954) *Phys. Rev.* **96**, 208.
12. Nelson, E. (1996) *Phys. Rev.* **150B**, 1079.
13. Bohm, D. and Hiley, B. J. (1989) *Physics Reports*, **172**, 93.
14. The University of Minnesota Super Computer Institute , Minneapolis MN USA, is acknowledged for providing computer time for this particular calculation.

A GEOMETRIC APPROACH TO THE QUANTUM MECHANICS OF DE BROGLIE AND VIGIER

W.R. WOOD
Faculty of Natural and Applied Sciences
Trinity Western University, 7600 Glover Road
Langley, British Columbia V2Y 1Y1, Canada

AND

G. PAPINI
Department of Physics, University of Regina
Regina, Saskatchewan S4S 0A2, Canada

Abstract. Following de Broglie and Vigier, a fully relativistic causal interpretation of quantum mechanics is given within the context of a geometric theory of gravitation and electromagnetism. While the geometric model shares the essential principles of the causal interpretation initiated by de Broglie and advanced by Vigier, the particle and wave components of the theory are derived from the Einstein equations rather than a nonlinear wave equation. This geometric approach leads to several new features, including a solution to the de Broglie variable mass problem.

1. Introduction

It is a pleasure to acknowledge the role that Professor Vigier has played in the development of the casual interpretation of quantum mechanics [1]. His demonstration of an explicit solitonic solution [2] has made de Broglie's conception of a double solution [3] a reality. As well, his extensive work [4] on issues relating to relativistic causal or stochastic models has been very helpful in our own efforts to formulate the principles of the causal interpretation within a geometrical framework.

In the geometric theory discussed here, a particle is represented by a thin shell or bubble solution to the Einstein equations rather than a solitonic solution to a nonlinear wave equation. The Gauss-Mainardi-Codazzi (GMC) formalism (a familiar tool in general relativity) is used to facilitate the analysis of the dynamics of the bubble. The junction conditions in the GMC formalism provide a tensorial description of the balance of energy and momentum across the thin shell. As a consequence, the geometric model provides a framework by which the influence of external fields, such as the wave field $\psi(x)$, on the motion of the particle can be rigorously analyzed.

In the classical theory of general relativity, the guidance mechanism is well-known: the geometry, which is determined by the distribution of matter, in turn, governs the motion of the matter itself. However, this classical guidance mechanism becomes insignificant when applied to particles at the microscopic scale where de Broglie's guidance principle is required to explain quantum effects such as the interference pattern in the two-slit experiment. It appears that a theory of gravitation whose domain of validity encompasses the microscopic scale is required if the desired guidance mechanism is to be given a geometric interpretation. A natural candidate is Weyl's conformally invariant theory of gravitation and electromagnetism [5]. Weyl generalized the Riemannian geometry of general relativity by supposing that a vector parallel transported around a closed circuit would also experience a change in length according to the formula $\delta\ell = \ell\kappa_\mu \delta x^\mu$. The vector field κ^μ, together with the metric tensor $g_{\mu\nu}$ that is defined modulo an equivalence class, comprise the fundamental fields of the new geometry. The choice of Weyl geometry is also strongly supported by Santamato's demonstration [6] of a (nonrelativistic) "quantum force" associated with κ_μ in his stochastic theory of "geometric quantum mechanics" [7]. In the model presented here, the geometry of Weyl is used to express the principles of the causal interpretation of quantum mechanics in a fully relativistic form.

Apart from providing a means to investigate the self-consistency of the dynamical aspects of the causal interpretation, the geometric model also offers several new interesting features. For example, by formulating the theory in the context of curved spacetime, new opportunities arise for considering the role that nonlocal interactions may play in a relativistic causal theory. As well, the geometric model provides a resolution to the problem of de Broglie's variable mass.

2. The GMC Formalism

In Weyl geometry, one introduces a gauge-covariant calculus [8] based on the gauge-covariant derivative $\bar{\Box}$ and a semimetric connection $\bar{\Gamma}^\alpha_{\mu\nu}$, where an overbar is used to distinguish objects from their Riemannian counterparts. In the GMC formalism, a timelike hypersurface Σ, which represents the history of the thin shell, divides spacetime into two four-dimensional regions (V^I and V^E), both of which have Σ as their boundary. The intrinsic metric on Σ is given by $h_{\mu\nu} = g_{\mu\nu} - n_\mu n_\nu$, where n^μ is a unit spacelike ($n_\mu n^\mu = 1$) vector field normal to Σ. The extrinsic curvature tensor in Weyl geometry is defined by $\bar{K}_{\mu\nu} = K_{\mu\nu} + h_{\mu\nu} n^\alpha \kappa_\alpha$. The development of the GMC formalism in Weyl geometry ultimately yields the equations [9]

$$n_\mu n^\nu G^\mu{}_\nu = -\frac{1}{2}(^3R + K_{\mu\nu}K^{\mu\nu} - K^2) - D_\mu \kappa^\mu + 2h_\mu{}^\nu \kappa^\mu \kappa_\nu + 2K n^\mu \kappa_\mu, \quad (1)$$

$$n_\mu h_\alpha{}^\nu G^\mu{}_\nu = D_\alpha K - D_\mu K^\mu{}_\alpha, \quad (2)$$

$$h^\alpha{}_\mu h_\beta{}^\nu G^\mu{}_\nu = {}^3G^\alpha{}_\beta + (K^\alpha{}_\beta - h^\alpha{}_\beta K)_{,n} - K K^\alpha{}_\beta + \frac{1}{2} h^\alpha{}_\beta (K_{\mu\nu}K^{\mu\nu} + K^2)$$
$$- 2(K^\alpha{}_\beta - h^\alpha{}_\beta K)n^\lambda \kappa_\lambda + 2h^\alpha{}_\beta h_\mu{}^\nu \kappa^\mu \kappa_\nu. \quad (3)$$

The intrinsic stress-energy tensor on Σ, which is defined by

$$S^\mu{}_\nu \equiv \lim_{\varepsilon \to 0} \int_{-\varepsilon}^{\varepsilon} T^\mu{}_\nu dn, \quad (4)$$

corresponds to the distributional part of $T_{\mu\nu}$. The junction conditions for the gravitational field are given by $h^\alpha{}_\mu h_\beta{}^\nu S^\mu{}_\nu = \gamma^\alpha{}_\beta - h^\alpha{}_\beta \gamma$ and $n_\mu S^\mu{}_\nu = 0$, where the jump in the extrinsic curvature is denoted by $\gamma^\mu{}_\nu \equiv [K^\mu{}_\nu]$, $\gamma \equiv \gamma^\mu{}_\mu$, and $g_{\mu\nu}$ and κ^μ are assumed to be continuous across Σ, but their normal derivatives discontinuous. It is also assumed that $\kappa_\mu = 0$ in the interior geometry V^I so that length integrability is established in the spacetime region occupied by the particle. Using (1) and (2), the jump in the equations $n_\mu G^\mu{}_\nu = n_\mu T^\mu{}_\nu$ yields the intrinsic tensor equations

$$D_\mu(h^\mu{}_\alpha h_\nu{}^\beta S^\alpha{}_\beta) + [n_\alpha h_\nu{}^\beta T^\alpha{}_\beta] = 0, \quad (5)$$

$$\{K^\mu{}_\nu\} S^\nu{}_\mu + [n_\mu n^\nu T^\mu{}_\nu] = 0, \quad (6)$$

where $\{K^\mu{}_\nu\}$ denotes the average of $K^\mu{}_\nu$ across Σ. Equations (5) and (6) describe the balance of stress-energy-momentum between neighboring external fields and the thin shell. It is this balance that governs the dynamical behavior of the thin shell. Indeed, the requirement that the fields in the exterior Weyl space join at Σ in accordance with the junction conditions places constraints on the motion of the bubble since Σ represents the *history* of the thin shell. Within the context of the causal interpretation of quantum mechanics, it is particularly significant that the interplay between the particle and wave aspects of the problem is an inherent feature of the present geometric formulation.

3. The Geometric Model

Once the assumption is made that elementary particles follow trajectories that are influenced in part by a wave field $\psi(x)$, it is only natural to consider the new field on equal footing with the gravitational and electromagnetic fields in a unified manner. In fact, the transfer of energy and momentum required in the guidance process suggests use of a tensorial formulation that would, hopefully, yield the Einstein-Maxwell theory in the classical limit. The fact that the hypothesized guidance mechanism is effective at the microscopic scale, while the corresponding mechanism in general relativity is significant only at large scales, suggests beginning with a conformally invariant theory to integrate quantum effects into a geometric theory. In this regard, it is of interest to note that under the local conformal transformation $g_{\mu\nu} \to \rho^2 g_{\mu\nu}$, the scalar curvature transforms as $R \to R + \frac{6}{\rho}\Box_\mu \Box^\mu \rho$, where the derivative term in ρ is the covariant generalization of the quantum potential in the causal interpretation of quantum mechanics. The conformally invariant geometry introduced by Weyl is particularly attractive because it also provides a geometric interpretation for the electromagnetic field.

For our purposes, the modified Weyl-Dirac theory [8]

$$I_c = \int \left\{ -\frac{1}{4} f_{\mu\nu} f^{\mu\nu} + |\beta|^2 \bar{R} + k|\bar{\Box}_\mu \beta \, \bar{\Box}^\mu \beta| + \lambda |\beta|^4 \right.$$
$$\left. + \rho \gamma^\mu (\bar{\Box}_\mu \rho - \varepsilon \rho \varphi_{,\mu}) \right\} \sqrt{-g} d^4 x, \qquad (7)$$

where $\varepsilon = \pm 1$, k and λ are real arbitrary constants and ρ, φ and κ_μ are real fields, is convenient because it gives the complex scalar field $\beta = \rho e^{i\varphi}$ a geometrical status as well as maintaining a theory that is linear in the scalar curvature. This latter point is essential when the particle is associated with a region of Riemannian space where the conformal symmetry of the exterior Weyl space is broken and the Gauss-Mainardi-Codazzi (GMC) formalism [9] is used to join the interior and exterior regions. The constraint, $\kappa_\mu = -(\ln \rho)_{,\mu} + \varepsilon \varphi_{,\mu}$, is introduced [8] to allow for quantization of flux and leads to a topologically nontrivial electrodynamics with $\varepsilon f_{\mu\nu} = \varphi_{,\nu\mu} - \varphi_{,\mu\nu}$.

3.1. THE FIELD EQUATIONS

The field equations that follow from the action (7), given here in terms of the Riemannian fields, are [8]

$$\Box_\nu f^{\mu\nu} = 4(k - 3\varepsilon)\rho^2 \varphi^{,\mu} \equiv j^\mu, \qquad (8)$$

$$G_{\mu\nu} = \frac{1}{2\rho^2} E_{\mu\nu} + I_{\mu\nu} + \frac{1}{2}\lambda g_{\mu\nu}\rho^2 + H_{\mu\nu} \equiv T_{\mu\nu}, \qquad (9)$$

$$\frac{1}{3}(\varepsilon k - 3)\varphi_{,\mu}\varphi^{,\mu} = -\frac{1}{6}(R + 2\lambda\rho^2) + \frac{1}{\rho}\Box_\mu\Box^\mu\rho \qquad (10)$$

and

$$(k - 3\varepsilon)\Box_\mu(\rho^2\varphi^{,\mu}) = 0, \qquad (11)$$

where $E_{\mu\nu}$ is the usual Maxwell tensor,

$$I_{\mu\nu} = \frac{2}{\rho}(\Box_\nu\Box_\mu\rho - g_{\mu\nu}\Box_\alpha\Box^\alpha\rho) - \frac{1}{\rho^2}(4\rho_{,\mu}\rho_{,\nu} - g_{\mu\nu}\rho_{,\alpha}\rho^{,\alpha}) \qquad (12)$$

and

$$H_{\mu\nu} = -2(\varepsilon k - 3)(\varphi_{,\mu}\varphi_{,\nu} - \frac{1}{2}g_{\mu\nu}\varphi_{,\alpha}\varphi^{,\alpha}). \qquad (13)$$

Taking the trace of (9) one recovers (10), while (11) follows from the conservation equation associated with (8). The theory also contains the wave equation [10]

$$(\Box^\lambda + i\kappa^\lambda)(\Box_\lambda + i\kappa_\lambda)\psi - \frac{\lambda}{3}|\psi|^2\psi - \frac{1}{6}R\psi = 0. \qquad (14)$$

In fact, if one writes $\psi = \rho e^{i\chi}$ and defines $\chi_{,\mu}$ according to $\alpha\varphi_{,\mu} \equiv \chi_{,\mu} + \kappa_\mu$ with $\alpha^2 \equiv (\varepsilon k - 3)/3$, then the imaginary part of (14) yields (11), while the real part coincides with (10) which can be expressed as

$$(\chi_{,\mu} + \kappa_\mu)(\chi^{,\mu} + \kappa^\mu) = -\frac{1}{6}(R + 2\lambda\rho^2) + \frac{1}{\rho}\Box_\mu\Box^\mu\rho. \qquad (15)$$

Since φ, χ, and κ_μ are real fields, α must be a real constant. In the causal interpretation of quantum mechanics, equation (15) is identified as the Hamilton-Jacobi equation for a system of momentum $\chi^{,\mu} + \kappa^\mu = Mu^\mu$, so that

$$Mu_\mu = \alpha\varphi_{,\mu}. \qquad (16)$$

From (15) one finds

$$M^2 = \frac{\lambda}{3}\rho^2 + \left(\frac{R}{6} - \frac{1}{\rho}\Box_\mu\Box^\mu\rho\right), \qquad (17)$$

which is the square of the de Broglie mass in the present model.

3.2. THE PARTICLE-WAVE SOLUTION

For $\alpha\varphi_{,\mu} = Mu_\mu$, the tensor $H_{\mu\nu}$ is seen to represent a perfect (irrotational) fluid with equal pressure and energy density: $H_{\mu\nu} = -6M^2(u_\mu u_\nu + \frac{1}{2}g_{\mu\nu})$. In the present geometric model, $H_{\mu\nu}$ is identified with the Madelung fluid in the causal interpretation. The particle is represented by a static, spherically symmetric thin shell solution to the Einstein equations when the Madelung fluid tensor $H_{\mu\nu}$ is neglected.

Application of the GMC formalism requires the determination of the interior and exterior line elements

$$ds^2_{I,E} = -e^{\nu_{I,E}}dt^2_{I,E} + e^{\mu_{I,E}}dr^2 + r^2(d\theta^2 + \sin^2\theta d\phi^2), \tag{18}$$

as well as the intrinsic stress-energy tensor $S_{\mu\nu}$ on the timelike hypersurface Σ. In the interior space it is assumed that $\kappa_\mu = 0$ and that the scalar field acquires a constant value $\rho = \rho_0$ which breaks the interior conformal invariance and fixes the scale of the particle. Under these conditions, the interior metric is given by [9]

$$e^{-\mu_I} = 1 + \frac{1}{6}\lambda\rho_0^2 r^2 = e^{\nu_I}, \tag{19}$$

so that the interior space is de Sitter ($\lambda < 0$), Minkowski ($\lambda = 0$) or anti-de Sitter ($\lambda > 0$). The exterior metric, expressed in terms of the arbitrary function $\rho(r)$, is given by [9]

$$e^{-\mu_E} = \left(1 + r\frac{\rho'}{\rho}\right)^{-2}\left[1 - \frac{2m}{\rho r} + \frac{q^2}{4\rho^2 r^2} + \frac{1}{6}\lambda\rho^2 r^2\right] \tag{20}$$

and

$$e^{\nu_E} = (\ell_0\rho)^{-2}\left[1 - \frac{2m}{\rho r} + \frac{q^2}{4\rho^2 r^2} + \frac{1}{6}\lambda\rho^2 r^2\right], \tag{21}$$

where m, q and ℓ_0 are integration constants and a prime denotes differentiation with respect to r. When it is assumed that $g_{\mu\nu}$ and κ^μ are continuous across Σ, but their normal derivatives discontinuous, the surface stress-energy tensor is found to take the form [9]

$$h^\alpha{}_\mu h^\nu{}_\beta S^\mu{}_\nu = -2\sigma h^\alpha{}_\beta, \tag{22}$$

where $\sigma \equiv [n^\mu(\ln\rho)_{,\mu}]$; that is, the surface stress-energy tensor is induced when the normal derivative of $\ln\rho$ across Σ is discontinuous. The intrinsic stress-energy tensor (22) is characteristic of a domain wall of surface energy 2σ, where $h_{\mu\nu}$ is the intrinsic metric on Σ. For $\sigma > 0$, the bubble is under a surface tension that opposes the Coulomb repulsion due to the surface charge. In this way, the particle finds its origin in the field ρ that (i) fixes the scale in V^I, (ii) ensures conformal invariance in V^E, and (iii) induces the surface tension needed for stability.

Taking $\varphi_{,\mu} = 0$ in V^I and $h_\mu{}^\nu \varphi_{,\nu}$ discontinuous across Σ allows the bubble to be embedded in the Madelung fluid in accordance with (5) and (6), while the surface stress-energy tensor (22) remains unchanged. For $n^\mu \varphi_{,\mu} = 0$, the normal component of the exterior fluid momentum at Σ takes the form $n_\mu H^\mu{}_\nu = -3M^2 n_\nu$. From this it follows that $n_\mu h_\alpha{}^\nu H^\mu{}_\nu = 0$ and $n_\mu n^\nu H^\mu{}_\nu = -3M^2$. As a consequence, the Madelung fluid tensor $H_{\mu\nu}$ does not contribute to (5), the time component of which, in the rest frame of the thin shell, governs the transfer of energy between the particle and its neighboring fields [12]. While the particle does not draw energy from $H_{\mu\nu}$ as one might expect [2], it can be shown that, if $[\rho_{,n}]$ varies on Σ, then $I_{\mu\nu}$ will transfer energy to the thin shell such that $\sigma \neq$ constant.

3.3. THE GUIDANCE CONDITION

Although the Madelung fluid doesn't serve as an energy source for the particle, it does influence the motion of the thin shell through (6) which represents Newton's second law [12]. In this manner, the bubble acquires a new dynamical nature as it is guided in its motion by the fluid in V^E. The realization of this guidance process can be seen by considering the dynamical behavior of a fluid element at a point P on the exterior surface of the bubble, where the four-velocity of the fluid element is denoted $u_f^\mu(P) = dz^\mu(P)/ds_E$. By construction, the metric tensor is continuous across Σ at P and consequently, $ds_E^2(P) = ds_\Sigma^2(P)$. Hence, at any point P on Σ, the four-velocity of the thin shell is given by

$$u_p^\mu(P) = \left.\frac{dz^\mu}{ds_\Sigma}\right|_P = \left.\frac{dz^\mu}{ds_E}\right|_P = u_f^\mu(P) = \frac{\alpha}{M}\varphi^{,\mu}(P) \qquad (23)$$

which is recognized as the guidance formula advanced by de Broglie. In the present approach, the validity of the guidance condition can be extended beyond holding only at a given point by noting that the motion of the fluid along its worldline from P to a subsequent point P', at which the fluid element and the bubble are still in contact, can be viewed as the result of a conformal transformation induced by the factor [13]

$$\xi^2 = 1 - M^2 \left(\frac{\bar{\Box} M}{ds}\right)^{-2} \left(\frac{\bar{\Box} u_\mu}{ds}\right)^2. \qquad (24)$$

The metric tensor at P' in V^E can therefore be obtained from its corresponding value at P by applying a conformal transformation with the same factor ξ^2 and, by continuity, the intrinsic metric at P' is also determined. The resulting identity, $h_{\mu\nu}(P') = \xi^2 h_{\mu\nu}(P)$, leads to the conclusion that the bubble and fluid must move in step. Consequently, the Hamilton-Jacobi equation (15) may be applied to the particle itself, as required in the causal interpretation of quantum mechanics.

3.4. NONLOCAL EFFECTS AND CURVED SPACETIME

A novel feature of the bubble model presented above is the manner in which the interior space V^I is made distinct from the exterior space V^E. This property not only makes it possible to break the conformal invariance in the interior space, whereby standards of length can be introduced into the theory while Weyl's geometric interpretation of the exterior electromagnetic field is preserved, but *the locality requirements of the exterior space need not be imposed in the interior space*. In particular, nonlocal influences that have been observed[1] in experiments employing correlated particles may simply be a consequence of the fact that, while the world tubes of the correlated particles diverge after the disintegration process, they actually share a common past geometry that affords nonlocal interactions. In this way, nonlocal effects could be explained without denying the objective reality of elementary particles or compromising the principles of relativity (in curved spacetime).

The suggestion that the separation of V^I and V^E plays an essential role in seeking to understand the intriguing nonlocal EPR-type correlations does not require any nonlocal effects to occur in the exterior Weyl space. This situation is clearly not in keeping with the idea that it is the quantum potential (that exists in V^E in the present model) that is responsible for nonlocal phenomena. Bohm *et al* [15] have argued that the invariance of the quantum potential under a scaling of $\psi(x)$ by an arbitrary *constant* plays a fundamental role in the nonlocal nature of the theory. However, this invariance property is reminiscent of the global phase invariance of pre-gauge field theories that also "contradicts the letter and spirit of relativity" [16], and as a consequence is replaced by local phase invariance. In the present geometric model, the generalized quantum potential in (17) is invariant under the conformal transformation

$$\tilde{g}_{\mu\nu} = \sigma^2 g_{\mu\nu}, \quad \tilde{\rho} = \sigma^{-1}\rho \tag{25}$$

for the arbitrary *function* $\sigma^2(x) > 0$. As mentioned earlier, by requiring invariance under the local scaling (25) one is naturally led to a conformally

---

[1] It is interesting to note, however, that Squires[14] has challenged the conclusion that the empirical evidence implies nonlocality by considering the time involved in the actual measuring process in an Aspect-like experiment.

invariant theory. For the Weyl-Dirac theory considered above, information regarding the particle's environment is propagated in the exterior spacetime via the tensor field $I_{\mu\nu}$ in a local manner.

3.5. A SOLUTION TO DE BROGLIE'S VARIABLE MASS PROBLEM

An outstanding issue in the de Broglie-Vigier causal interpretation of quantum mechanics [3] has been the problem associated with the reality of the variable mass M. Within the context of second-order wave equations, this problem manifests itself in the mathematical existence of negative probability densities and negative energy solutions. When the usual probabilistic interpretation is applied, these solutions cannot, in general, be given a physically meaningful interpretation. In contrast, when a particle follows a timelike causal trajectory, the situation changes radically. In this case, positive energy solutions are necessarily correlated with positive values of M and positive probability densities and the sign of the energy remains fixed along the trajectory [17]. However, general solutions of the Klein-Gordon equation do not ensure the reality of M.

This deficiency is overcome in the present geometric model [18] due to the existence of the timelike thin shell solution to Einstein's equations which can be embedded in the Madelung fluid according to the junction conditions discussed above. While spacelike and timelike directions are distinguished in any relativistic theory, a geometric theory permits these directions in spacetime to be related to the motion of matter via Einstein's field equations. That is, for a given foliation of spacetime, the GMC formalism requires the various timelike and spacelike components of $G_{\mu\nu}$ to be equated to the corresponding components of $T_{\mu\nu}$. In this way, a link is established between the properties of spacetime and matter that allows one to address the issue of whether or not a given four-vector that is associated with matter is timelike. It is due to the absence of this geometric structure that the possibility of spacelike four-momenta in de Broglie's guidance formula $P_\mu = Mu_\mu$ cannot be excluded in previous formulations of the causal interpretation derived from a scalar wave equation. By basing the theory on the field equations (8)-(11) (from which the wave equation (14) is then identified), one is not bound to demonstrate that all possible generic solutions to the wave equation must be physically meaningful as is the case when the causal interpretation is based solely on a wave equation. It is the field equations (8)-(11) that determine the set of physically acceptable solutions in the geometric approach. In the geometric model discussed above, the timelike nature of Mu_μ can be demonstrated as follows.

The constraint in (1) can be written in the gauge $\kappa'_\mu = \kappa_\mu + (\ln \rho)_{,\mu}$ so that $\kappa'_\mu = \varepsilon\varphi_{,\mu}$. This is permissible due to the gauge covariance of the theory and this particular gauge is viable since ρ must be greater than one in the model in order for the thin shell to be under a surface tension that balances the Coulomb repulsion [9]. For the static solution of Section 3b, $\kappa'_\mu \kappa^{\mu\prime} < 0$ and as a consequence $\varphi_{,\mu}$ is timelike. It then follows from (16) that M must be real since α is real. This result, obtained in the static case, also holds true in a frame comoving with the thin shell, and is therefore quite general. Indeed, due to the covariant nature of M under conformal tranformations that preserve the sign of the line element, M^2 must be positive in general.

The condition for $\varphi_{,\mu}$ to be timelike can also be expressed within the context of the theorem of Frobenius. In terms of differential forms, equation (16) is given by

$$u = \frac{\alpha}{M} \, d\varphi \equiv h \, d\varphi. \tag{26}$$

The condition for u^μ to be orthogonal to hypersurfaces of constant φ, and hence for $\varphi_{,\mu}$ to be timelike, is given by $du \wedge u = 0$. Recognizing that, due to the multivalued nature of φ, $d\varphi$ is not closed even though it is an exact 1-form, the condition for timelike $\varphi_{,\mu}$ becomes

$$d^2\varphi \wedge d\varphi = 0. \tag{27}$$

Equation (27) is satisfied in the static case considered above, where $d^2\varphi \sim dx^0 \wedge dx^1$ and $d\varphi \sim dx^0$.

For timelike $\varphi_{,\mu}$, it follows that the Maxwell current j^μ in (8) is also timelike without having to impose this as an auxiliary condition. In the present theory, the Maxwell current is proportional to the Klein-Gordon current [10] associated with the wave equation, $j^\mu_{KG} = \alpha \rho^2 \varphi^{,\mu}$, and is therefore also timelike and as such does not suffer from the difficulties normally associated with the current for a second order wave equation. It should be noted, however, that ψ in the present theory is a physical field and not immediately identifiable with a probabilistic wave function. In addition, the time component of the current can be made positive by choosing the positive sign of the radical in the definition of α. It then follows that positive (negative) energy particles will correspond to positive (negative) values of M and positive (negative) values of j^μ_{KG}. In this regard, it is interesting to observe that the equation of motion

$$\frac{\bar{\Box}}{ds}(M u_\mu) = -\varepsilon \alpha f_{\mu\nu} u^\nu - \bar{\Box}_\mu M \tag{28}$$

is invariant under charge conjugation and time reversal transformations as discussed by Dirac [19]. In addition, equation (28) is invariant under $M \to -M$ together with time reversal. This indicates that, in the present theory, negative energy particles may be interpreted as positive energy particles moving backward in time.

4. Summary

Although the geometric formulation of the causal theory presented above is in an early developmental stage, it nevertheless demonstrates that it is possible to inject the principles of the causal interpretation of quantum mechanics into a fully relativistic geometric theory in Weyl space. In the authors' opinion, this is an essential step towards obtaining a satisfactory causal theory of quantum phenomena. By formulating the problem within the context of a theory of gravitation, whereby the description of the transfer of energy-momentum becomes an inherent feature, it becomes possible to demonstrate that the guidance principle is dictated by the physics rather than the physicist. As well, the geometric model provides a basis upon which issues such as the reality of the de Broglie variable mass and nonlocal interactions can be addressed.

Acknowledgements

This work was supported in part by the Natural Sciences and Engineering Research Council of Canada. The authors are grateful for the assistance provided by S. Jeffers for the International Organizing Committee. One of the authors (G.P.) wishes to thank Dr. K. Denford, Dean of Science, University of Regina, for continued research support.

References

1. P. R. Holland, The Quantum Theory of Motion: An account of the de Broglie-Bohm causal interpretation of quantum mechanics (Cambridge University Press, Cambridge, 1993).
2. J. P. Vigier, Found. Phys. 21 (1991) 125.
3. L. de Broglie, Nonlinear Wave Mechanics (Elsevier, Amsterdam, 1960).
4. For a comprehensive list of references to Vigier's publications and work see Ann. Inst. Henri Poincaré 49 (1988) 261.
5. H. Weyl, Sitzung. d. Preuss. Akad. d. Wiss. (1918) 465.
6. E. Santamato, Phys. Rev. D29 (1984) 216; J. Math. Phys. 25 (1984) 2477.
7. E. Santamato, Phys. Rev. D32 (1985) 2615; Phys. Lett. A130 (1988) 199. For a detailed discussion of Santamato's approach see: C. Castro, Found. Phys. 22 (1992) 569.
8. D. Gregorash and G. Papini, Nuovo Cim. B63 (1981) 487; G. Papini, in: High-energy physics, eds. S. L. Mintz and A. Perlmutter (Plenum Press, New York, 1985) p. 179.
9. W. R. Wood and G. Papini, Phys. Rev. D45 (1992) 3617.
10. G. Papini, in: Proceedings of the fifth Grossmann meeting, eds. D. G. Blair and M. J. Buckingham (World Scientific, Singapore, 1989) p. 787.
11. W. R. Wood and G. Papini, Found. Phys. Lett. 6 (1993) 207.
12. C. Barrabès and W. Israel, Phys. Rev. D43 (1991) 1129.

13. G. Papini and W. R. Wood, in Proceedings of the Conference on Structure: from Physics to General Systems, M. Marinaro and G. Scarpetta, eds. (World Scientific, Singapore, 1992), p. 94.
14. E. J. Squires, Phys. Lett. A 178 (1993) 22.
15. D. Bohm, B. J. Hiley and P. N. Kaloyerou, Phys. Rep. 144 (1987) 321.
16. L. H. Ryder, Quantum Field Theory (Cambridge University Press, Cambridge, 1985).
17. A. Kyprianidis, Phys. Lett. A 111 (1985) 111; C. Dewdney, P. R. Holland, A. Kyprianidis and J. P. Vigier, included note in: W. Mückenheim, Phys. Rep. 133 (1986) 337.
18. G. Papini and W. R. Wood, Phys. Lett. A 202 (1995) 46.
19. P. A. M. Dirac, Proc. R. Soc. Lond. A 333 (1973) 403.

STABLE ORBITS AS LIMIT CYCLES

A mathematical model
suggested by Bohm's interpretation of quantum mechanics

S. BERGIA
Dept. of Physics, Univ. of Bologna and I.N.F.N., Sezione di Bologna
Via Irnerio, 46, 40126 Bologna

1 Introduction

Bohm's formulation is today accepted as a legitimate reformulation and reinterpretation of Quantum Mechanics (QM), and one that allows a description of many a typical quantum phenomenon in terms of particle trajectories in the space-time continuum, even though these trajectories must be thought as determined by the joint action of the classical and of a non-local quantum potential (QP). The description in terms of the QP has permitted a suggestive visualization of quantum phenomena such as double slit experiments (Philippidis, Dewdney and Hiley 1979; for a discussion of a wider class of phenomena, see Vigier 1987, Dewdney 1987). Figures depicting the latter have deservedly become a favourite piece in presentations of the subject. The QP formulation has also provided us with a new ontological view of apparent non-local interactions in terms of a new quality of wholeness. All this is beautifully illustrated in the recent book of Bohm and Hiley (to be quoted hereafter as BH 1993 [1]).

Given these premises, which all concern *stationary phenomena*, neophytes are eager to find in the book equivalent illustrations of *non-stationary processes*. Their desire is likely to be to some extent satisfied by the treatment of barrier penetration as a transition process [2]. They may however feel disappointed by the lack of similar descriptions of atomic transitions.

They would like, in particular, to see them visualized in terms of single particle trajectories, like in the previous cases. Does Bohm's version of QM allow such a visualization? This would mean overthrowing a prohibition dating from the times of Bohr's atom, in which a transition between stationary states is a "quantum jump", an expression conveying the idea that "there are no phenomena which correspond to any state in between" [3]. Bohm and Hiley suggest that a "more

---

[1] Until very recently I did not have the opportunity to get acquainted with P. H. Holland's book (Holland 1993); this is the reason why I will make only scanty references to it.
[2] BH 1993, Sec. 5.1, after Dewdney and Hiley 1982.
[3] BH 1993, Sec. 2.5.

complete description" could perhaps be achieved "at a deeper more complex level" in which the transition process could be "treated as continuos and analizable" [4].

As a matter of fact, transitions undergone by an electron in an atom *are* discussed in their book. As a first step, an Auger-like effect, in which energy is taken up by an additional particle that was originally also in a bound state of the atom, is analyzed [5]. The discussion, however, essentially addresses the question as to why only some members of the ensemble undergo transition. The subsequent treatment of more complex cases, in which a number of final cases is present, is aimed at showing the emergence of mutually exclusive channels into one of which the particle must enter [6]. But still no visualization in terms of single particle trajectories arises.

It seems in fact that, in order to implement "a more complete description", allowing a visualization in the sense outlined, an extension of Bohm's approach ought to be considered.

The crucial point is what F. Fer has called "the Hamiltonian origin of Schrödinger equation", which makes it incapable of describing an irreversible evolution [7]. And in fact, in ordinary QM, in order to describe atomic transitions, one must go through the following two essential steps: 1) Add to the Hamiltonian a term describing the interaction between charged matter and the electromagnetic field, and develop the standard time-dependent perturbation treatment of the problem; 2) Compute the transition probability either in terms of quantum electrodynamics or in the semiclassical approximation supplemented by Einstein's theory of stimulated and spontaneous emission [8]. Nothing of this sort is automatically built in in Bohm's scheme, nor could it be, until it limits itself to give an alternative version of the quantum mechanical formulation of the bound state problem. It therefore seems that, in order to fulfil the purpose, one should envisage the way of mirroring steps (1) and (2) in the latter.

An attempt at implementing this program has been carried out by Dalton [9]. An analogous attempt carried out in terms corresponding to the program just outlined has apparently failed to give satisfactory results [10].

Lam and Dewdney have tackled the problem in very general terms, i. e. by considering a two-level "atom" placed in a quantized cavity scalar field. The field, which is treated à la Bohm, that is as a pure field (a boson is a quantized excitation of a field mode, and its particle- like character is revealed in the interaction with quantized matter), plays the role of the external agent from and to which energy is transferred. The role of the two-level "atom" is taken by an electron confined

---

[4] *ibidem*.
[5] BH 1993, Sec. 5.3.
[6] BH 1993, Sec. 5.4.
[7] Fer 1977, p. 40.
[8] See, for instance, Sakurai 1976.
[9] B. J. Dalton, private communication; Dalton's dynamical scheme, however, shows, already at the Hamiltonian level, important differences with Bohm's approach (B. J. Dalton 1994, p. 15, 1995, p. 3).
[10] M. Bozić, private communication.

to an infinite potential well [11]. It is assumed that, before the atom is placed in the cavity, the quantum field is in a single-mode state (of coordinate q), and that, when the atom is placed within the cavity, the field-matter coupling is described by a potential of the form

$$V_{int}(q,x) = q \sin(k[x + x_0]), \qquad (1)$$

where k is the wave vector of the mode and x_0 the position of the origin of the infinite well within the cavity. A super-Schrödinger equation describes the evolution of the wave equation $\chi(q, x, t)$. Application of Bohm's scheme to this case leads to the equations

$$\frac{dq}{dt} = \frac{\partial S(q, x, t)}{\partial q} \qquad (2)$$

and

$$\frac{dx}{dt} = 2 \frac{\partial S(q, x, t)}{\partial x} \qquad (3)$$

for the rates of change of the system's coordinates. The paper's main conclusion appears to be that, in Bohm's version of QM, and of quantum field theory, the interaction of a quantized scalar field with matter can be described in terms of the continuous trajectory of a well-defined system point in configuration space, so that the transition of the field-matter system, caused by the field-matter interaction, does not occur via a discontinuous quantum jump but takes place causally and continuously [12]. Bohr's old prohibition appears therefore definitively overcome. Moreover, electron trajectories are computed in their framework, from Eqs.(2) and (3), revealing some interesting features (in particular the nonlocal dependence of the electron velocity upon the form of the scalar field over the whole cavity).

What else should one then look for? The possibility of other interesting features is in fact suggested by Bohm and Hiley. After their discussion of atomic transitions in presence of several final states, they observe that "there must be some limiting points or curves which divide the initial conditions of the particles which will enter one channel from those that will enter the other (as happened in the case of barrier penetration)". In general, they will be in the overall configuration space and may be "quite complex in form". They had on the other hand previously shown [13] that, for non-vanishing angular momenta, more specifically for p-states, stationary states are characterized by a set of circular motions in ordinary space. The intuitive picture one gets of atomic transition processes is thus in terms of continuos trajectories which, depending on the position in ordinary space in which the particle finds itself at some initial time, will eventually tend to one or another of a set of nested circles. This kind of situation is explicitly depicted by Bohm and Hiley [14]. The general behaviour thus described is similar, they point out, "to that obtained in the study of non-linear equations whose solution contain what are

---

[11] Lam and Dewdney 1994b; for the pure field case, see Lam and Dewdney 1994a.
[12] Lam and Dewdney 1994b, p. 59.
[13] BH 1993, Sec. 3.5.
[14] BH 1993, Sec. 5.4, pp. 91-92.

called stable limit cycles" (equations for particle motions in their interpretation are likewise clearly non linear [15]). The difference with that standard situation is that here one has a whole set of possible limit cycles, corresponding the the orbits associated with the quantum stationary states. It seems worth pointing out that another difference between the two situations stems from the fact that, as already mentioned, initial conditions in Bohm's case are in the overall configuration space rather than in ordinary space. The analogy is therefore somewhat imperfect.

Can one reach the conclusion that stable orbits are indeed formally obtained as limit cycles presented by solutions of particle motions as described in Bohm's version of QM [16]? This question seems worth of further investigation. Rather than carrying it on within a general framework such as that outlined by Lam and Dewdney, the choice has been made here to check whether a simplified version of the theory, formulated in terms of a skeleton mathematical model, would permit to grasp the essential aspects of the problem.

The main purpose of this paper is the presentation of such a model and the discussion of its limits.

The model, as will be seen, is based on: a) The replacement of (1) and (2) above by a classical hopefully equivalent mechanism; b) Very specific assumptions about the wave packets describing the atomic electron; these assumptions can be only partially justified as approximations to a correct treatment.

It is therefore clear that the model, although suggested by Bohm's formulation, cannot claim to be derived from it, and that it cannot therefore be taken too seriously. It should rather be considered as providing the ground for a preliminary investigation of the perspectives of the full theory, a pedagogical tool suited for this aim.

However, to the extent that it presents many shortcomings and achieves its purpose only in a very limited sense, it dos not make one fully confident about the possibility that the basic idea could be indeed be a feature of Bohm's formulation. I thought it worth presenting anyway because it permits to single out the features the full theory should take care of.

The plan of the paper is as follows: in Sec. 2, I briefly report on some antecedents to the basic idea, formulated in the framework of a modified version of classical mechanics and of stochastic electrodynamics, and comment on the features that distinghish these version from the one that is here pursued; in Sec. 3, I introduce the essential aspects of the mode and briefly discuss limits and perspectives.

---

[15] This is due to then fact that, "even though the equation for the wave function is linear the quantum potential can be a highly non-linear function of the coordinates of the particle." (BH 1993, pp. 65-78)

[16] adequately completed in order to account for the interaction with the electromagnetic field. It seems worth stressing that the very notion of limit cycle is foreign to a strictly Hamiltonian dynamics, since it is contrary to the idea of conservation of measure in phase space expressed by Liouville's theorem. This point has been particularly emphasized by Fer and Lochak (Fer 1977, pp. 52-53; Lochak 1978, p. 12).

2 Stable orbits as limit cycles: antecedents

The idea that stable orbits in atoms could be identified with limit cycles of *some* dynamics is not new. Several years ago, F.Fer expressed the idea that quantum stable orbits could be obtained from a non-Hamiltonian form of *classical* dynamics, in fact from a hereditary dynamics. In his book (Fer 1977), Fer has even formulated a heuristic explicit model in which "orbits" in phase space arise as limit cycles of a certain differential equation [17].

Fer's model is expressed in terms of the differential equation

$$\frac{d^2x}{dt^2} = f(E)\frac{dx}{dt} + \omega^2 x = 0, \tag{4}$$

where E is defined by

$$E = \frac{1}{2}\omega^2 x^2 + \frac{1}{2}\left(\frac{dx}{dt}\right)^2, \tag{5}$$

and $f(E)$ is a function, which need only be defined for $E \geq 0$, characterized by a finite, or infinite denumerable, set of zeros (the first one of first order, all the others of second order - the character of the first zero is linked with the stability of the lowest level), corresponding to the energy eigenvalues. Fer then shows that particles in initial states characterized by an energy eigenvalue remain, according to Eqs. (4), (5), on a stable orbits in the system's phase plane, whereas particles in initial states with energies lying in between two eigenvalues spiral toward the next inner orbit, which thus acts as a limit cycle [18].

Another framework where it has recently been put forward (de la Peña and Cetto 1993b) that stationary orbits of bound electrons could be obtained as limit cycles of the theory's dynamical equations is stochastic electrodynamics (SED) [19].

These attempts are carried out within an essentially *classical* framework. The idea here is to investigate whether a version of *quantum* mechanics that admits particle trajectories could entail similar conclusions, thus corroborating the general views expressed by Bohm and Hiley I referred to in the Introduction. I should however stress that the model I am going to present gives an image of the processes involved which is closer to that provided by Fer's model than to the one foreshadowed by Bohm and Hiley, inasmuch as it is limited to a one-dimensional case and the attempt at visualising limit cycles is pursued in the phase plane of the system rather than in ordinary or configuration space.

---

[17] See also Lochak 1978 for a discussion of this general attitude and for references to work in the field.

[18] Fer 1977, pp. 42-45.

[19] For a short historical account, see de la Peña and Cetto 1993a. I may recall that Cetto and de la Peña have recently proposed and started developing a new version of SED, after the first one failed in reproducing the hydrogen spectrum (Cetto and de la Peña 1991).

3 The model

In conformity with Bohm's formulation, I will initially consider, as describing the motion of a particle (an electron) bound by a potential V, the equation

$$m\frac{d\vec{v}}{dt} = -\nabla(V) - \nabla(Q), \qquad (6)$$

or, in the one-dimensional case

$$m\frac{d^2x}{dt^2} = -\frac{\partial V}{\partial x} - \frac{\partial Q}{\partial x}, \qquad (7)$$

where Q is the quantum potential. The general idea is that Eqs. (6,6') describe motion in a non-stationary state; stationary states may arise from solutions to (6,6') presenting a limit cycle. Just like in Fer's model, to deal with the simplest possible case, the "atom" shall be represented by a harmonic oscillator. Hence $\frac{\partial V}{\partial x} = kx$, $k > 0$; in the absence of damping and of a quantum potential, the classical oscillator would then move with the angular frequency $\omega_0 = \sqrt{k/m}$.

The first feature of the model is the attempt to replace, or mimick, the action of a time-dependent Hamiltonian in producing transition between levels by a damping term $\beta\dot{x}$, which is to be added to the classical potential V. However, one should also reproduce the fact that the perturbation acts for a finite time interval till the transition is completed. This feature can be forced into the model by requiring a dependence of β on the particle energy just as in Fer's heuristic model.

A particle injected, so to say, in the field of the "nucleus", with given initial position and velocity, in a state of global negative energy of the "atom", will be accompanied by its own wave packet. The packet will be essentially characterized by its mean energy. The second basic assumption of the model consists in choosing, as representative of particles with energies close to a sequence of energies $E_{\bar{n}}$, wave packets with mean energy $E_{\bar{n}}$. One such set of wave packets is provided, with a necessary specification, by the sequence

$$\Psi_{\bar{n}}(x,t) = (\frac{m\omega}{\pi\hbar})^{1/4} e^{-\frac{1}{2}\xi - i\alpha_{\bar{n}}\cos\omega t)^2 + \frac{\alpha_{\bar{n}}^2}{4} - i[\frac{1}{2}\omega t - \xi)\alpha_{\bar{n}}\sin\omega t]}. \qquad (8)$$

where m is the particle mass, $\xi = \sqrt{\frac{m\omega}{\hbar}}x$, and $\alpha_{\bar{n}}$ is chosen as

$$\alpha_{\bar{n}} = (2\bar{n} + 1)^{\frac{1}{2}}, \quad \bar{n} \ \ integer. \qquad (9)$$

The wave packets (7) are obtained assigning definite values to the coefficients c_n in a generic superposition of eigenstates of the harmonic oscillator with angular frequency ω, and by making use of the identity [20]

$$\sum_{n=0}^{\infty} \frac{H_n(\xi)}{n!} \eta^n = e^{\xi^2 - (\eta - \xi)^2}. \tag{10}$$

In each wave packet of the sequence, the probability distribution has a maximum whose position depends on $\bar{n}$. The corresponding energy is

$$E_{\bar{n}} = \hbar\omega(\bar{n} + \frac{1}{2}). \tag{11}$$

It must be recalled (this is the necessary specification alluded to above) that this result [(Eq. (10)] depends on the use of the Stirling approximation, $logn! = nlogn - n$, (in n, not in $\bar{n}!$), which is of course valid for $n \gg 1$.

The wave packets of the sequence are coherent states of the harmonic oscillator [21]. They do in fact form a denumerable subset of the continuum set of these states, corresponding to a real denumerable subset of their eigenvalues [22]. They have the well-known property that the center of the packet oscillates in time, between the points $\pm a = \sqrt{\frac{\hbar}{m\omega}} \alpha_{\bar{n}}$ [23] with the angular frequency (ω) of the classical oscillator [24].

According to the definition

$$Q = -\frac{\hbar}{2m} \frac{\nabla^2 R}{R}, \tag{12}$$

with the sequence of wave packets $\{\Psi_{\bar{n}}\}$, Eq. (7), is associated the sequence [25] of quantum potentials

$$Q_{\bar{n}} = \frac{\hbar\omega}{2}[1 - (\sqrt{\frac{m\omega}{\hbar}} x - \alpha_{\bar{n}} cos\omega t)^2]. \tag{13}$$

Note that the quantum potentials (12) are evaluated for the unperturbed, that is undamped, oscillator, driven by the potential V such that $\frac{\partial V}{\partial x} = kx$, except for

---

[20] See, for instance, Chahoud and Castelvetri 1970, p. 98, or, in slightly different notations, Holland 1993, p. 165 ff.; the distribution of initial particles according to a denumerable set of wave packets is less arbitrary than it appears at first sight: in some way one is thus selecting initial conditions close to those of stationary states.

[21] See, for instance, Klauder and Sudarshan 1968, p. 109 ff., Galindo and Pascual 1990, p. 148 ff.. The relevance of this point was first pointed out to me by F. Cannata.

[22] I wish to thank A. Terenzi for fixing this point for me.

[23] Holland 1993, p. 166.

[24] As has been stressed by Holland, the harmonic oscillator stationary bound states are an instance of quantum states that have no classical limit; the nonclassical features persist if one superposes a few neighbouring eigenfunctions in the high n limit; in order to achieve the classical oscillation of angular frequency ω one must "wipe out the nodes" by superposing eigenfunctions to form a packet such as those of the sequence of coherent states above (Holland 1993, pp. 249-251).

[25] See Holland 1993, p. 250, for the expression of the quantum potential associated with a wave packet whose center oscillates with angular frequency ω between the points $\pm a = \sqrt{\frac{\hbar}{m\omega}} \alpha_{\bar{n}}$.

the fact that the angular frequency has not been fixed to the value $\omega_0 = \sqrt{k/m}$. Damping has the side effect of shifting the frequency, so that it appears to make sense the choice of letting ω in Eqs. (7), (10) and (12) differ from ω_0 [26]. The effect of the perturbation, basically described by the damping term, is thus partially ascribed to the quantum potential.

Note the the quantum potentials (12), according to

$$-\frac{\partial Q_{\bar{n}}}{\partial x} = m\omega^2 x + \gamma_{\bar{n}} \cos\omega t, \quad \gamma_{\bar{n}} = \sqrt{m\hbar\omega^3}\alpha_{\bar{n}}, \tag{14}$$

are responsible for two quantum force terms, respectively linearly dependent on [independent of] x and independent of [dependent on] $\bar{n}$. Thus the non-linearity of the problem, foreseen by Bohm and Hiley, does non manifest itself in the model, replaced, to some extent, by the feedback nature of the QP action. The choice $\omega = \omega_0$ would make the first term in (13) exactly cancel the classic elastic force term $-kx$, a feature that seems undesirable [27]. Better said, the two terms should exactly cancel *only* in correspondence of stable orbits. This suggests that one should force into the model an energy dependence of the quantum potentials, similar to the one that seems necessary to assume for the damping coefficient β (with the implication $\omega = \omega_0 \iff \beta = 0$) [28]. An energy dependence of the quantum potentials is justifiable on the basis of the following argument: particles injected in the field of the nucleus with mean energies (10) will loose energy as they spiral toward the next inner orbit (if they indeed do so); on the other hand, they determine their own quantum potentials; an energy dependence of the latter seems therefore to be a necessary feature of the model.

Substituting the classical and quantum force terms in Eq. (6') gives a sequence of familiar-looking equations of motions, whose solutions, for given initial conditions $x(0) = x_0, \dot{x}(0) = \dot{x}_0$, are a combination of the homogeneous solution, an oscillation with the (small) angular frequency $\sqrt{\omega_0^2 - \omega^2}$, damped in time according to the factor $e^{-\frac{\beta}{2m}t}$, and of a particular solution, an undamped oscillation with angular frequency ω.

The overall picture we get is thus the following: particles with energy around one of the the values (10) are acted upon by the corresponding quantum potential of the sequence, which, together with the classical potential, determines a family of solutions $\{x_{\bar{n}}\}$, and of corresponding "trajectories" in the phase plane; the damping of the homogeneous solution is what basically introduces the effect looked for, that is a form of spiralling of each trajectory of the family in the phase plane (hopefully) toward the next inner stable orbit. Once the latter is reached (if it is

---

[26] Note, however, that what would seem the natural choice, that is $\omega_0^2 - \omega^2 = \beta^2/4m^2$ spoils what seems to be a desirable behavior of the solution of the model equations; only the much weaker assumption $\omega_0^2 > \omega^2$ is made in the following.

[27] "...in a non-stationary state, the balance between ∇V and ∇Q will not hold." (BH 1993, Sec. 3.5, p. 43) Difficulties arising in connection with this point have been first pointed out to me by M. Ferrero.

[28] The implication would hold with the natural choice $\omega_0^2 - \omega^2 = \beta^2/4m^2$. I have mentioned the reasons that make this prescription unpracticable.

reached at all), the particle oscillates driven by the $\bar{n}$-dependent driving term of the quantum potential, with angular frequency ω_0.

The doubts expressed through the expressions in brackets in the last paragraph are justified: the basic question is in fact: do all the particles in a family actually reach a state of definite energy, given *exactly* by expression (10)? [29]. Not quite so. One gets in fact for the final energy a rather complicated expression, which simplifies to

$$E_{\bar{n}} = \frac{\hbar \omega^5}{2(\omega_0^2 - 2\omega^2)^2}(2\bar{n} + 1), \qquad (15)$$

in the case of small damping ($\beta \to 0$), and reduces to expression (10) only in the limit as $\omega \to \omega_0$. One could argue that the two limits ought in fact to be taken, according to what has ben said on the necessary energy dependence of both he damping term and the quantum potentials, and that they should actually be taken at the end of the procedure, but this only adds to the conclusion that the model, beside being largely ad hoc and probably not even quite self-consistent, is also largely incomplete. As I said, it was conceived as a tool which might provide some insight as to the possibility stable orbits could indeed arise as limit cycles in Bohm's formulation, once completed in order to account for the interaction between matter and the electromagnetic field. Due to its many drawbacks, its very limited success, and, last but certainly not least, its only partial relationship with the full theory, it seems to leave the question wide open.

Work is in progress (S. B., F. Cannata, M. Ferrero, and A. Terenzi) in order to improve on the present limits in the formulation of the model, and, on more solid ground, to tackle the problem in terms of the full theory.

Acknowledgments

I wish to thank F. Cannata, M. Ferrero and A. Terenzi for helpful discussions and advice, M. Bozić, B. J. Dalton, and C. Dewdney for information about previous work in the field.

References

Klauder, J. R., and Sudarshan, E. C. G. (1968) *Fundamentals of Quantum Optics*, W. A. Benjamin, Inc., New York.
Chahoud, J., and Castelvetri, P. (1970) *Meccanica Quantistica*, C.L.U.B., Bologna.
Sakurai, J.J.J. (1976) *Advanced Quantum Mechanics*, Addison-Wesley, Città.
Lochak, G. (1978) L'irréversibilité en physique, *Microphysique*, 3-16.
Philippidis, C., Dewdney, C., and Hiley, B. J. (1979) Quantum Interference and the Quantum Potential, *Nuovo Cimento* **52B**, 15-28.

---

[29] Particles in the family are supposed to have energies *around* the average value (10); no distinction is drawn between particles with energies above and below this value; this is of course a very unsatisfactory feature of the model.

Dewdney, C., and Hiley, B. J. (1982) A Quantum Potential Description of One-Dimensional Time-Dependent Scattering From Square Barriers ans Square Walls, *Foundations of Physics* **12**, 27-48.

Vigier, J. P. (1987) Theoretical implications of time- dependent double resonance neutron interferometry, in W. M. Honig, D. W. Kraft, and E. Panarella (eds.), *Quantum Uncertainties, Recent and Future Experiments and Interpretations*, Plenum Press, New York and London.

Dewdney, C. (1987) Calculations in the causal interpretation of quantum mechanics, in W. M. Honig, D. W. Kraft, and E. Panarella (eds.), *Quantum Uncertainties, recent and Future Experiments and Interpretations*, Plenum Press, New York and London.

Galindo, A., and Pascual, P. (1990) *Quantum Mechanics I*, Springer-Verlag, Berlin.

Cetto, A. M., and de la Peña, L. (1991) Titolo, *Foundations of Physics Letters* **4**.

Bohm, D., and Hiley, B. J. (1993) *The undivided universe, An ontological interpretation of quantum theory*, Routledge, London and New York.

Holland, P.R. (1993) *The quantum theory of motion*, Cambridge University Press, Cambridge.

De la Peña, L., and Cetto, A. M. (1993a) Planck's law as a consequence of the zeropoint radiation field, Preprint IFUNAM FT93-023.

de la Peña, L., and Cetto, A. M. (1993b) Is quantum mechanics a limit cycle theory?, Preprint IFUNAM FT93-030, to appear in the Proceedings of the Oviedo Symposium on Fundamental Problems in Quantum Physics (Aug. - Sept. 1993), M. Ferrero, and A. van der Merwe (eds.), Kluwer Academic Publishers, Dordrecht, in press.

Lam M. M., Dewdney, C (1994a) The Bohm Approach to Cavity Quantum Scalar Field Dynamics. Part I: The Free Field, *Foundations of Physics*, 3-27.

Lam M. M., Dewdney, C (1994b) The Bohm Approach to Cavity Quantum Field Dynamics. Part II: The Interaction of the Field with Matter, *Foundations of Physics*, 29-60.

Dalton, B. J. (1994) *Deterministic Explanation of Quantum Mechanics Based on a New Trajectory-Wave Ordering Interaction*, North Star Press of St. Cluod, Inc. St Cloud.

Dalton, B. J. (1995) A fundamental force as the deterministic explanation of quantum mechanics, Preprint presented at this conference.

THE CORRESPONDENCE PRINCIPLE: PERIODIC ORBITS FROM QUANTUM MECHANICS

DANIEL PROVOST

Laurentian University,
Department of Physics and Astronomy,
Sudbury, Canada P3E 2C6

dprovost@neutrino.phys.laurentian.ca

1. Introduction

The discrete energy levels of a bound quantum Hamiltonian system $H(\hat{q},\hat{p})$ exhibit both short–range and long–range correlations [1]. These correlations can be explained with the help of the corresponding classical Hamiltonian system $H(q,p)$ when $\hbar$ is small[1]. The short–range correlations are generic in nature and depend on whether the underlying classical dynamics is regular or chaotic. For example, the nearest–neighbor spacing distribution (NNS) of the energy levels obeys GOE statistics if the classical phase space is chaotic and obeys Poisson statistics if it is regular[2]. A GOE behavior of the NNS statistic is commonly used as a definition of quantum chaos when only experimentally determined energy levels are available.

In contrast to the short–range correlations in the spectra, the long–range correlations are non-generic in nature and depend on the particular form of the Hamiltonian under study. Gutzwiller's trace formula[3] describes semi-classically these long–range correlations in terms of classical periodic orbits:

$$d_{osc}(E) = \frac{1}{\pi\hbar} \sum_{po} \mathcal{A} \cos(\frac{1}{\hbar}S - \frac{\pi}{2}\mu) \qquad (1)$$

---

[1] By short–range we mean of the order of the mean level spacing, whereas by long–range we mean of the order of several mean level spacings

where $d_{osc}(E)$ is the oscillating part of the density of states. The sum is over all the classical periodic orbits, S is the action of a particular orbit, μ its Maslov index, and the amplitude $\mathcal{A}$ depends on its stability.

To demonstrate these long–range correlations in a given spectra, one usually takes the Fourier transform of the oscillating part of the density of states $d_{osc}(E)$ with respect to E [4–7]. The success of this approach depends on how fast the period of the orbits change as a function of the energy – this can be seen from Eq. (1) by expanding the classical action S about some energy E^*:

$$S(E) = S(E^*) + \tau(E^*)(E - E^*) + \cdots \qquad (2)$$

so that the Fourier transform of $d_{osc}(E)$ over an energy interval about E^* should give a distinct peak at $\tau_j(E^*)$ if the period of the j-th orbit is approximately constant on that energy interval. The widths of these peaks about the periods of classical orbits are however inversely proportional to the energy interval under consideration and consequently the resolution of these peaks is poor for generic, non–scalable systems. For special systems for which the action S scales with energy, other Fourier transforms are usually more informative [8–13].

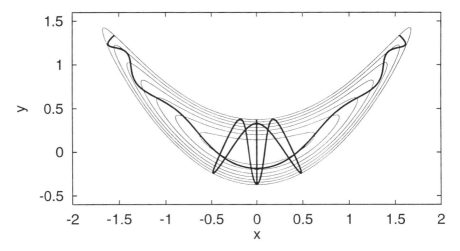

Figure 1. Equipotential energy contours of the Nelson Hamiltonian for $E = 0.02$ to 0.14 in steps of 0.02. The vertical orbit, a symmetric rotation, and a symmetric libration at $E = 0.14$ are also shown.

In this paper we demonstrate the influence of classical periodic orbits on the quantum spectra and on the quantum eigenfunctions of the following model Hamiltonian system:

$$H(\mathbf{q}, \mathbf{p}; \alpha) = \frac{p_x^2}{2} + \frac{p_y^2}{2} + \frac{\mu x^2}{2} + (y - \frac{\alpha x^2}{2})^2 \qquad (3)$$

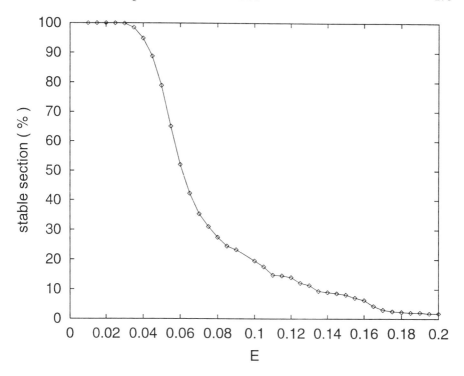

Figure 2. Percentage of the Poincaré surface of section $x = 0$ that is regular as a function of energy.

where $\mu = 0.1$ and α is the coupling strength.

2. The Nelson Hamiltonian

A simple scaling argument shows that a variation of the coupling strength can be interpreted as keeping the coupling strength fixed and varying $\hbar$:

$$E_n[\hbar; \alpha] = \frac{1}{\alpha^2} E_n[\hbar \alpha^2; 1] \qquad (4)$$

where $E_n[\hbar; \alpha]$ is the n–th eigenvalue of $H(\mathbf{q}, -i\hbar \partial_\mathbf{q}; \alpha)$. We henceforth fix the coupling constant at $\alpha = 1$ and vary $\hbar$.

The Hamiltonian given by Eq. (3) with $\alpha = 1$ has been studied extensively in recent years[5, 7, 18, 19, 20] and has been dubbed the Nelson Hamiltonian. Figure 1 shows a contour plot of the Nelson potential $V(x, y) = H - \frac{1}{2}p^2$ for energies from 0.02 to 0.14 in steps of 0.02. The vertical orbit, a symmetric rotation, and a symmetric libration are also shown.

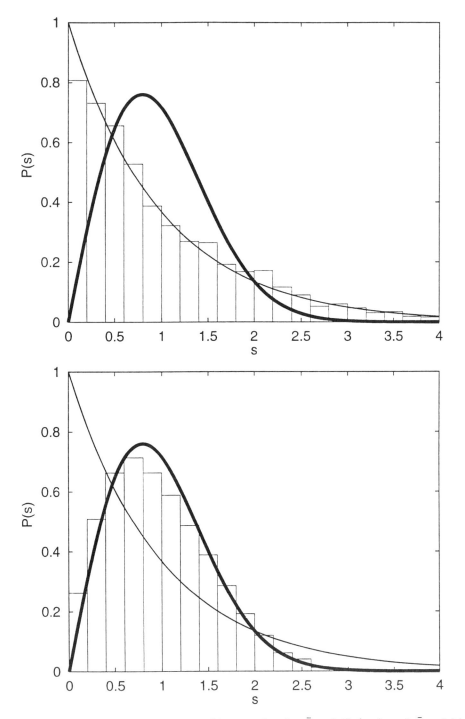

Figure 3. Distribution of nearest neighbor spacing for $\bar{E} = 0.02$ (top), and $\bar{E} = 0.14$ (bottom). A chaotic spectra should follow GOE statistics (thick line) and a regular spectra should follow Poisson statistics.

In Fig. 2 we plot the percentage of the Poincaré surface of section $x = 0$ that is regular as a function of energy[25]. We see that the phase space is mostly regular for energies up to about 0.04 and is mostly chaotic for energies above 0.13.

This transition from regular to chaotic manifests itself quantum mechanically in the NNS distribution. The system has the reflection symmetry $x \leftrightarrow -x$. We therefore restrict our quantum calculations to the even eigenvalues. For a given $\hbar$ we obtain the exact even quantal eigenvalues by using the basis $\phi_m^{(x)}(x)\phi_n^{(y)}(y - x^2/2)$, where $\phi_m^{(x)}$ and $\phi_n^{(y)}$ are harmonic oscillator wave functions appropriate for the bottom of the well[19]. In Fig. 3 we plot the NNS distribution for $\bar{E} = 0.02$ (top) and $\bar{E} = 0.14$ (bottom). For the top figure we considered all energy levels with energies between 0.019 and 0.021 with $\hbar$ values ranging from $1/\hbar = 500$ to $1/\hbar = 1000$ in steps of 2. The resulting histogram fits well the Poisson curve and this implies that the underlying classical phase space at this energy is mostly regular. For the bottom figure we considered all levels with energies between 0.139 and 0.141 with $\hbar$ values ranging from $1/\hbar = 100$ to $1/\hbar = 300$ in steps of 0.5. Here the GOE curve fits best, indicating that the classical motion at these energies is mostly chaotic.

3. Periodic orbits from quantum spectra

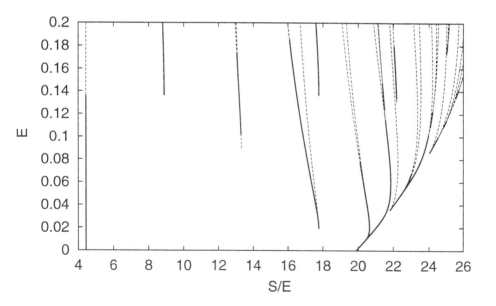

Figure 4. Energy–scaled action S/E as a function of energy E for periodic orbit families. A solid line indicates where a family is stable and a dashed line when it is unstable.

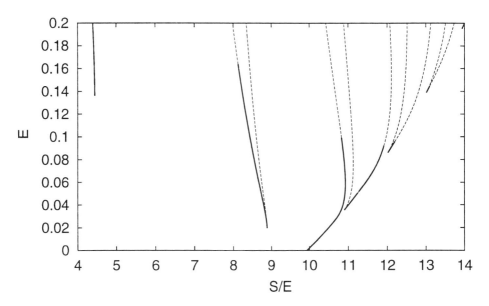

Figure 5. $S/2E$ vs E plot for symmetric librations.

We demonstrate the influence of classical periodic orbits on the quantum spectra by Fourier transforming $\hbar d_{osc}(E;\hbar)$ with respect to $1/\hbar$[25]. According to Eq. (1) this must give peaks at the classical actions of the periodic orbits. This has recently been done for the hydrogen atom in a strong magnetic field [14–17]. This system is special however since they found that only the periodic orbits that start and end at the nucleus gave peaks.

The energy–smoothed even density of states is given by

$$d^{(+)}(E;\hbar) = \sum_i f_\epsilon(E - E_i^{(+)}) \qquad (5)$$

where the smoothing function $f_\epsilon(E)$ is taken to be of the Lorentzian type:

$$f_\epsilon(E) = \frac{\epsilon}{\pi} \frac{1}{E^2 + \epsilon^2} \qquad (6)$$

and $\epsilon = \hbar/T^*$, with T^* fixed. The oscillating part of the even density of states is defined as

$$d_{osc}^{(+)}(E;\hbar) = d^{(+)}(E;\hbar) - d_{avg}^{(+)}(E;\hbar) \qquad (7)$$

We calculated the average even density of states $d_{avg}^{(+)}(E;\hbar)$ semiclassically[19, 21]. This has the advantage of telling us when the actual numerical diagonalizations fail. We obtained[19]

$$d_{avg}^{(+)}(E) = \frac{1}{2\sqrt{2\mu}} \frac{E}{\hbar^2} \left(1 + \frac{\sqrt{\mu}\hbar}{2E} - \frac{\hbar^2}{12\mu E} + \mathcal{O}(\hbar^3) \right) \tag{8}$$

The first term is the usual Thomas–Fermi density of states.

Gutzwiller's trace formula for the oscillating part of the even density of states contains three parts:

$$2\pi\hbar d_{osc}^{(+)}(E) = \frac{1}{\pi\hbar} \sum_{ppo} \sum_{n=1}^{\infty} \tilde{f}(n\tau) \frac{\tau}{\sqrt{|tr(\mathcal{M}^{(n)}) - 2|}} \cos[n(\frac{\bar{S}}{\hbar} - \frac{\pi}{2}\mu_m)] \bigg|_{\substack{periodic \\ orbits}}$$

$$+ \sum_{ppo} \sum_{n \text{ odd}} \tilde{f}(n\tau/2) \frac{\tau}{\sqrt{|tr(\tilde{\mathcal{M}}^{(n)}) - 2|}} \cos[\frac{n}{2}(\frac{\bar{S}}{\hbar} - \frac{\pi}{2}\mu_m)] \bigg|_{\substack{symmetric \\ librations}} \tag{9}$$

$$+ \sum_{n=1}^{\infty} \tilde{f}(n\tau) \frac{\tau}{\sqrt{|tr(\mathcal{M}^{(n)}) + 2|}} \cos[n(\frac{\bar{S}}{\hbar} - \frac{\pi}{2}\mu_m)] \bigg|_{\substack{vertical \\ orbit}}$$

where τ is the period of the primitive periodic orbit and $\tilde{f}(t) = e^{-t/T^*}$, the Fourier transform of $f_\epsilon(E)$. The first term is the usual sum over all primitive periodic orbits and their repetitions. The amplitude term is written out explicitly in terms of the stability matrix $\mathcal{M}$ of the periodic orbit. The second term is a sum over all the symmetric librations[22] except for the orbit that lies on the symmetry line $x = 0$. This last orbit is a boundary orbit and its contribution[23] is the third term in Eq. (9). The Fourier transform in $1/\hbar$ of the oscillating part of the even density of states must therefore show additional peaks at odd multiples of half the action of symmetric librations. Physically, these additional peaks can be understood if we look at the classical motion in the fundamental domain[23] where it is seen that half of a symmetric libration becomes a full periodic orbit.

The periodic orbits come in 1–parameter families. In Fig. 4 we plot the energy E as a function of the energy–scaled action S/E for periodic orbit families. A solid line indicates where a family is stable and a dashed line when it is unstable. In Fig. 5 we plot E as a function of half of the energy–scaled action of periodic orbit families that are also symmetric librations.

In Fig. 6 and 7 we show the results of the finite Fourier transform in $1/\hbar$ of $\hbar d_{osc}(E; \hbar)$ for different values of E. The maximum amplitude in each of the frames has been set to unity. The location of a periodic orbit's contribution at multiples of its energy–scaled action S/E is denoted by a vertical line of height 1 and the location of a symmetric libration's extra contribution at odd multiples of half of its energy–scaled action is denoted by a vertical line of height 0.8. In Fig. 6 we considered values of $\hbar$ ranging

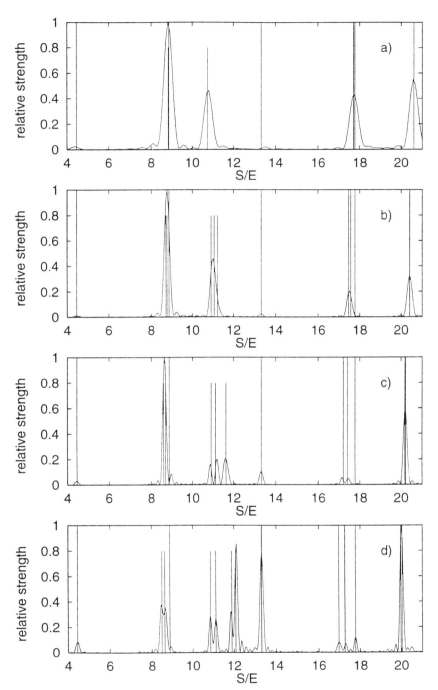

Figure 6. Fourier transform in $1/\hbar$ for a) $E = 0.03$, b) $E = 0.05$, c) $E = 0.07$, d) $E = 0.09$. The location of a periodic orbit's energy–scaled action S/E is denoted by a vertical line of height 1 and the location of a symmetric libration's extra contribution at $S/2E$ is denoted by a vertical line of height 0.8.

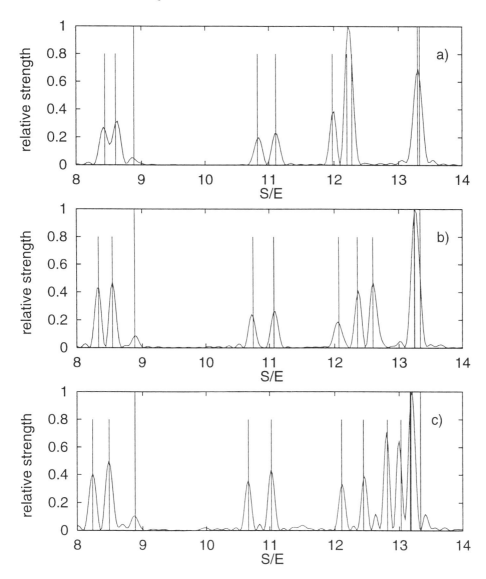

Figure 7. Fourier transform in $1/\hbar$ for a) $E = 0.10$, b) $E = 0.12$, c) $E = 0.14$.

from $1/\hbar = 100$ to $1/\hbar = 400$ in steps of 0.5. and we set $T^* = 40$. In Fig. 7 $\hbar$ ranged from $1/\hbar = 10$ to $1/\hbar = 240$ in steps of 0.5, and we set $T^* = 20$. Whether we were in the mostly regular, mixed, or mostly chaotic regime, the influence of the periodic orbits on the spectra is clearly seen. The signature of bifurcations of periodic orbits in the quantum spectra is also evident. The large peak at $\bar{S}/E \simeq 13.2$ is due to the symmetric rotation shown in Fig. 1.

4. Periodic orbits from quantum wavefunctions

A similar analysis can be done on the quantum wavefunctions themselves[26]. The energy–smoothed spectral probability density $\Delta(\mathbf{q}, E; \epsilon)$ is defined as

$$\Delta(\mathbf{q}, E; \epsilon) = \sum_k f_\epsilon(E - E_k) |\langle \mathbf{q}|\psi_k\rangle|^2 \tag{10}$$

$$= \Delta_{osc}(\mathbf{q}, E; \epsilon, \hbar) + \Delta_{avg}(\mathbf{q}, E; \epsilon, \hbar) \tag{11}$$

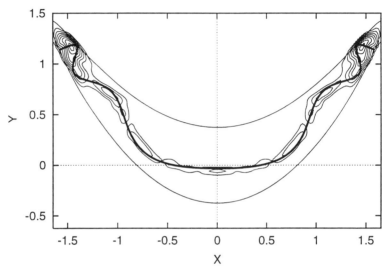

Figure 8. Fourier transform in $1/\hbar$ of the energy–smoothed spectral probability density for $E = 0.14$ and $S = 12.89$.

The average part of energy–smoothed spectral probability density is given by[19]:

$$\Delta_{avg}(\mathbf{q}, E; \epsilon, \hbar) = \frac{1}{2\pi\hbar^2} \frac{1}{2} \mathrm{erfc}\left(\frac{V - E}{\epsilon\sqrt{2}}\right) \tag{12}$$

where erfc is the complementary error function. The oscillating part of the $\Delta(\mathbf{q}, E; \epsilon)$ is given by Bogomolny's formula as a sum over periodic orbits[24, 19, 20]:

$$\Delta_{osc}(\mathbf{q}, E; \epsilon, \hbar) = \frac{1}{(2\pi\hbar)^{3/2}} \sum_{po} \mathcal{A} \tilde{f}(\tau) \cos\left(\frac{1}{\hbar}(S + W q_2^2) - \frac{\pi}{2}\mu - \frac{\pi}{4}\right) \tag{13}$$

So a Fourier transform in $1/\hbar$ of $(2\pi\hbar)^{3/2}\Delta(\mathbf{q}, E; \epsilon, \hbar)$ should give the periodic orbits in coordinate space. This is indeed the case as can be seen in Fig. 8 and 9. Details will be published shortly[26].

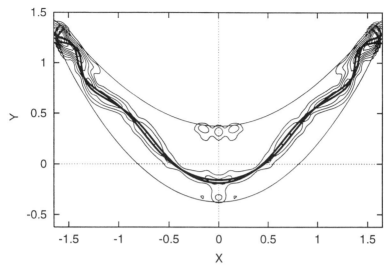

Figure 9. Fourier transform in $1/\hbar$ of the energy–smoothed spectral probability density for $E = 0.14$ and $S = 13.02$.

5. Conclusion

In conclusion, we showed that the spectra of a smooth Hamiltonian system contains information about the periodic orbits of the underlying classical system. This correspondence is shown to exist whether the classical phase space is mostly regular, mixed, or mostly chaotic. We also showed that the quantum spectra senses through the NNS statistics whether the classical dynamics is regular or chaotic. Finally we extracted the imprints of periodic orbits from the quantum wavefunctions.

References

1. By $\hbar$ small we mean the dimensionless quantity formed from a combination of physical parameters with $\hbar$ in the numerator. For an introduction to semiclassical mechanics, see for example A. M. Ozorio de Almeda, *Hamiltonian Systems: Chaos and Quantization*, Cambridge University Press (1988); M. C. Gutzwiller, *Chaos in Classical and Quantum Mechanics*, Springer–Verlag (1991).
2. M.L. Mehta, *Random Matrices and the Statistical Theory of Energy Levels* (New York: Academic Press, 1972); A. Bohigas, in *Chaos and Quantum Physics*, edited by M. J. Giannoni, A. Voros, and J. Zinn–Justin (North Holland, Amsterdam, 1991).
3. M. C. Gutzwiller, J. Math. Phys. **12**, 343 (1971); R. Balian and C. Bloch, Ann. Phys. (N.Y.) **69**, 76 (1972), M. C. Gutzwiller, in *Chaos and Quantum Physics*, edited by M. J. Giannoni, A. Voros, and J. Zinn–Justin (North Holland, Amsterdam, 1991).
4. C. P. Malta and A. M. Ozorio de Almeida, J. Phys. A **23**, 4137 (1990).
5. C. P. Malta, M. A. M. de Aguiar, and A. M. Ozorio de Almeida, Phys. Rev. A **47**, 1625 (1993).
6. M. Brack, R. K. Bhaduri, J. Law, M. V. N. Murthy, Phys. Rev. Lett. **70**, 568 (1993).

7. M. Baranger, M. R. Haggerty, B. Lauritzen, D. C. Meredith, and D. Provost, CHAOS **5**, no. 1 (1995).
8. B. Eckhardt, G. Hose, and E. Pollak, Phys. Rev. A **39** (1989) 3776.
9. T. Szeredi and D. A. Goodings, Phys. Rev. Lett. **69**, 1640 (1992).
10. M. Kuś, F. Haake, and D. Delande, Phys. Rev. Lett. **71**, 2167 (1993); M. Kuś, F. Haake, and B. Eckhardt, Z. Phys. B **92**, 221 (1993).
11. M. Kollmann, J. Stein, U. Stoffregen, H.-J. Stöckman, and B. Eckhardt, Phys. Rev. E **49**, R1 (1994).
12. R. Scharf and B. Sundaram, Phys. Rev. E **49**, R4767 (1994).
13. K. M. Atkins and G. S. Ezra, Phys. Rev. A **50**, 93 (1994).
14. D. Wintgen, Phys. Rev. Lett. **58**, 1589 (1987).
15. M. L. Du and J. B. Delos, Phys. Rev. A **38**, 1896 (1988); **38**, 1913 (1988).
16. A. Holle, J. Main, G. Wiebusch, H. Rottke, and K. H. Welge, Phys. Rev. Lett. **61**, 161 (1988); J. Main, G. Wiebusch, K. H. Weldge, Comments At. Mol. Phys. **25**, 233 (1991).
17. D. Delande, in *Chaos and Quantum Physics*, edited by M. J. Giannoni, A. Voros, and J. Zinn-Justin (North Holland, Amsterdam, 1991).
18. M. Baranger and K. T. R. Davies, Ann. Phys. (N.Y.) **177**, 330 (1987); M. A. M. de Aguiar, C. P. Malta, M. Baranger, and K. T. R. Davies, Annals of Physics **180**, 167 (1987); M. Baranger, K. T. R.Davies, and J. H. Mahoney, Annals of Physics **186**, 95 (1988).
19. D. Provost, Ph. D. Thesis, M.I.T. (1993).
20. D. Provost and M. Baranger, Phys. Rev. Lett. **71**, 662 (1993); D. Provost, in *Coherent States: Past, Present and Future*, edited by D. H. Feng, J. R. Klauder, and M. R. Strayer, (World Scientific, 1994), D. Provost, Phys. Rev. E **51**, 5396 (1995).
21. A. Bohigas, S. Tomsovic, and D. Ullmo, Phys. Rep. **223**, 43 (1993).
22. A libration is a periodic orbit that is time reversal invariant.
23. B. Lauritzen, Phys. Rev. A **43** 603 (1991); J. M. Robbins, Phys. Rev. A **40**, 2128 (1989).
24. E. B. Bogomolny, Physics D **31**, 169 (1988).
25. D. Provost, Phys. Rev. E **51**, R841 (1995).
26. D. Provost, *in preparation*.

NEUTRON INTERFEROMETRIC EXPERIMENTS ON QUANTUM MECHANICS

G. BADUREK
Institut für Kernphysik, Technische Universität Wien
Schüttelstraße 115, A-1020 Wien, Austria

1. Introduction

Since its invention in 1974 [1] neutron interferometry has been established as an almost ideal tool to test fundamental principles of quantum mechanics with massive particles on a macroscopic space-time scale. After a brief introduction into the basic principles of perfect crystal neutron interferometry some selected topics are reviewed which our group could contribute to this active and still growing field of research, ranging from experimental verification of the 4π period of spinor wave functions, demonstration of the quantum mechanical principle of linear superposition of states, macroscopic quantum beating of the neutron wave function (the neutron analogue of the AC Josephson effect), elucidation of the difference between probabilistic and deterministic absorption of matter waves to interferometric measurements of geometric phases. Additionally we report on the recently performed first successful verification of the topological nature of the scalar Aharonov-Bohm effect by means of neutron spin precession, which represents an alternative version of neutron interference [2].

2. Perfect Crystal Neutron Interferometry: Basic Principles

In the standard three-plate neutron interferometer, cut out monolithically from a perfect silicon crystal, dynamical diffraction at the first crystal plate splits the incoming beam into two coherent components propagating along different trajectorys (denoted by *I, II*) over a distance of several centimeters before they are recombined again. Their lateral separation (~2-4 cm) is large enough that the associated wave functions can be altered independently from each other by either, nuclear, electromagnetic or gravitational interaction. Thereby observable modifications of the interference pattern, i.e. the intensity distribution of the two beams leaving the last crystal plate of the interferometer, are induced. As usual in neutron interferometry the beam propagating in forward direction is denoted as "O-beam", whereas the diffracted one is called "H-beam" [1].

Because of symmetry reasons in an ideal empty interferometer the wave function components of the O-beam B are equal ($\psi_O^I = \psi_O^{II}$). For the H-beam a somewhat more complicated relation exists, but since both beams are complementary to each other due to particle number and angular momentum conservation, in what follows it is sufficient to condsider solely the forward beam.

A planparallel polycrystalline sample inserted between two of the crystal plates causes a phase shift $\chi = N b_c \lambda \Delta D$ between the interfering subbeams, where N is the particle density and b_c the coherent nuclear scattering length of the phase shifter atoms and ΔD is the geometric path difference of the two subbeams in the phase shifter plate. Upon rotation of the latter around a vertical axis the intensity of the forward beam oscillates according to

$$I_O \propto |\psi_O^I + \psi_O^{II}|^2 \propto 1 + \cos\chi \ . \tag{1}$$

In practice, a normalized fringe contrast $\Gamma \equiv (I_{max} - I_{min})/(I_{max} + I_{min})$ of the O-beam of up to about 80% can be achieved.

3. 4π-Periodicity of Spinor Wave Functions

In classical physics it is taken for granted that the rotation of an object by an integral multiple of 2π returns the object to its initial state. However, according to quantum theory a 2π rotation of a spin-½ particle leads to a sign reversal of its wave function, a result obviously inconceivable within the framework of classical physics. Until it could be verified explicitely for the first time by neutron interferometry [3,4], the validity of this 4π spinor periodicity has been tacitly assumed in quantum mechanics. There the neutron spin in one arm of the interferometer undergoes a well-defined number of Larmor rotations in the gap field of an electromagnet (Fig. 1). The action of a magnetic field on the spinor wave function is described by the unitary operator

$$U_R = \exp(-i\vec{\sigma}\vec{\alpha}/2) = \cos(\alpha/2) - i\vec{\sigma}\hat{\alpha}\sin(\alpha/2) \ , \tag{2}$$

where $\vec{\sigma}$ is the Pauli spin operator and $\vec{\alpha} = \alpha\hat{\alpha}$ is the rotation vector. The rotation angle α is the accumulated precession phase in the field $\vec{B}$ along the beam trajectory

$$\alpha = \gamma \int B dt = \frac{\gamma}{v} \int B ds \ , \tag{3}$$

where $\gamma = -1.833 \times 10^8$ s^{-1}T^{-1} is the gyromagnetic ratio of the neutron and v its velocity. The integrations extend over the interaction time and length, respectively.

If the rotation operator U_R acts, for instance, on the wave function ψ^{II} in path II one obtains $\psi^{II} = U_R\psi^{II} = U_R\psi^I$ and after superposition of the two coherent subbeams at the last crystal plate the intensity behind the interferometer therefore oscillates according to

$$I_O \propto |\psi_O^I + \psi_O^{II}| \propto 1 + \cos(\alpha/2) . \qquad (4)$$

This means that a 2π rotation of the neutron spin leads to an observable modification on the interference pattern and that a rotation of 4π is necessary, indeed, in order to return to the initial state.

It is worth to remember that this experiment can be performed with unpolarized neutrons although it is based on rotations in spin space. The reason is that the magnetic phase shift for the up and down spin states differ just in sign but not in magnitude, which is irrelevant for the cosine-dependence of the final intensity. However, if an additional scalar and hence spin-independent phase shift χ is introduced between the two interfering beams, this opposite signs of α lead to a polarization of the superposed beam. This interference-induced polarization has also been verified [5].

4. Spin State Superposition

Interferometry with polarized neutrons allows to demonstrate explicitely the quantum mechanical principle of linear superposition of spin states. Provided the two interfering beams can be prepared in different spin states, say "up" and "down" with respect to a quantization axis defined by an external magnetic field, quantum mechanics predicts that the combined state is not a mere mixture but a coherent superposition of "up" and "down". This means that the final state should be polarized perpendicular to the quantization axis and hence the emerging beams behind the interferometer should have properties which -from a classical point of view- neither of their constituents had. Choosing the z-axis as quantization axis and denoting the interfering states by $|I\rangle = |\uparrow_z\rangle$ and $|II\rangle = \exp(i\chi)|\downarrow_z\rangle$, the linear superposition can be written as

$$|O\rangle = \tfrac{1}{2}|\uparrow_z\rangle + \tfrac{1}{2}e^{i\chi}|\downarrow_z\rangle = \frac{e^{i\chi/2}}{\sqrt{2}}\left\{\cos\tfrac{\chi}{2}|\uparrow_x\rangle - i\sin\tfrac{\chi}{2}|\downarrow_x\rangle\right\} , \qquad (5)$$

which obviously shows that the polarization of the O-beam (and because of conservation of angular momentum also that of the H-beam) is confined to the xy-plane although the interfering states are polarized parallel and antiparallel, respectively, to the z-axis.

However, Eq. (5) only holds if the orthogonality of the two interfering states is achieved by means of d DC spinflipper which invert the spin state of one of

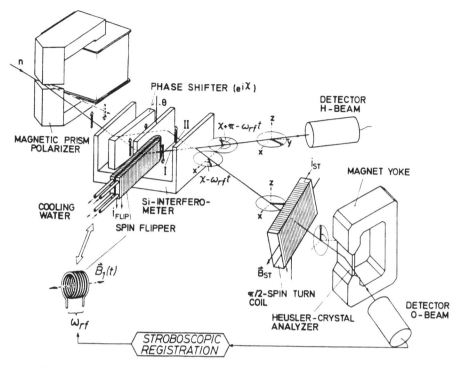

Figure 1. Experimental setup to demonstrate linear superposition of spin states [6].

the two subbeams via a well-defined Larmor precession around a static magnetic field, as represented by the rotation operator (2). Since in that case the interaction Hamiltonian $\mathcal{H} = -\vec{\mu}\vec{B}$ between the neutron magnetic momement and the magnetic field does not explicitely depend on time ($\partial \mathcal{H}/\partial t = 0$), the total energy of the neutrons is conserved during the spinflip process, although the latter is inevitably associated with a change of potential Zeeman energy by an amount $\Delta E_p = 2\mu B_0$ ($\vec{B}_0$...guide field required to avoid depolarization of the beam). This change of potential energy corresponds to a shift of the neutron wavelength $\Delta \lambda = \Delta E_p m \lambda^3/h^2 \approx 10^{-8}$ Å, where m is the neutron mass and h is Planck's constant. It is found that in spite of its extreme smallness this wavelength shift of one of the interfering beams is sufficient to cause a significant loss of coherence and consequently a reduction of the observable interference contrast. This disturbing effect can by considered as an obvious demonstration of the outstanding sensitivity of perfect crystal neutron interferometry.

If alternatively the static spin turn device is replaced by a resonance radiofrequency (rf) spinflipper, the interaction Hamiltonian is purely time-dependent and therefore commutes with the momentum operator $\vec{p} = -i\hbar \vec{\nabla}$. Thus the wavelength of the neutrons is conserved upon spin reversal but their total energy is either increased or decreased by $\Delta E = \pm \hbar \omega_r$ due to resonant absorption or emission of rf photons. Consequently the two interfering beams are not in the same energy state. The resulting nonstationary fringe pattern can be detected only stroboscopically. Similar to the static flipper case it finally turns out that the polarization of the beams behind the interferometer should have no component along the initial z-direction, but is instead expected to rotate in the xy-plane with the frequency $\omega = \omega_r = \Delta E/\hbar$ whithout being driven by an applied magnetic field.

The intrinsic angular width of perfect crystal Bragg reflections is only about 2 seconds of an arc. This feature allows to polarize the incident neutrons very elegantly by magnetic birefringence at prismatically shaped magnetic fields [7]. Symmetric passage through such a magnetic field prism leads to an angular separation δ of the two spin states, given by

$$\delta = \frac{2\mu B}{E} \tan(\frac{\Phi}{2}) , \qquad (6)$$

where E is the kinetic energy of the neutrons and Φ the prism apex angle. With thermal neutrons fields of about 1 T are sufficient to achieve an angular splitting twice as large than the reflection width of the crystal. In Fig. 1 the experimental setup is sketched which was used to demonstrate the principle of linear spin state superposition both with a static [8] and a Rabi-type [9] spinflipper. A magnetic crystal analyzer was used in combination with a static $\pi/2$-spin-turn coil to analyze the polarization of the O-beam both along and perpendicular to the initial spin direction. For the sake of brevity only the stroboscopically measured fringe patterns are plotted in Fig. 2. They clearly demonstrate that no interference oscillations occur upon varying the phase diffence between the interfering beams

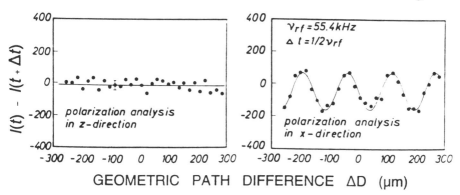

Figure 2. Stroboscopically measured interference patterns for Rabi-type spinflip in one arm of the interferometer. The final spin state obviously has no component along the initial z-direction [9].

if the z-component of the polarization is analyzed. As expected from quantum theory, the emerging beams behind the last crystal plate of the interferometer are indeed in a state perpendicular to that of their constituents.

This nonstationary superposition of spinor states has raised some discussions whether it is possible, at least in principle, to observe an interference pattern and nevertheless to know which path the neutron has taken by detecting its passage through the rf coil via the photon exchange. Of course, this would be completely contradictionary to the Bohr-Heisenberg ("Copenhagen") interpretation of quantum mechanics. A simple, but striking argument against such a possibility is based on the uncertainty relation $\Delta N^2 \Delta \varphi^2 \geq \frac{1}{4}$ ($N \geq 4$) between the number of photons N and the phase φ of the rf field (for a qualitative explanation it is not necessary to use the fully correct formulation of the number-phase uncertainty relation [10]). The stroboscopic detection scheme evidently requires the knowledge of the phase of the rf field, which is no problem as long as the field behaves like a classical external field. But according to Poisson statistics the photon number of such a classical field fluctuates with $\Delta N = N^{1/2} \gg 1$ and hence it is not possible to detect the emission or absorption of a single photon upon neutron passage. Obviously a reduction of the field amplitude diminishes the particle number uncertainty, but this does not help since then the phase fluctuations increase accordingly and the stroboscopic detection method fails. By the way, from the impossibilty to obtain simultaneously the path information and the interference pattern one can draw the conclusion that the energy transfer within the rf flipper can by no means be considered as measurement process.

5. Macroscopic Quantum Beating: Neutron Josephson Effect Analogue

Following a suggestion of Vigier and his group [11], who claimed that our experiments should give new insights into the particle-wave duality on the basis of a causal stochastic interpreation of quantum mechanics, an experiment was realized where in each of the two coherent subbeams of the interferometer a separate rf coil was mounted (Fig. 3) [12]. Thereby the most interesting situation arises if the two coils are operated at slightly different resonance frequencies $\omega_{r1} \neq \omega_{r2}$. If a frequency difference $\Delta \omega = \omega_{r1} - \omega_{r2} = \gamma \Delta B_0$ is established which is much smaller than the rf resonance width, the corresponding difference $\Delta B_0 = B_{01} - B_{02}$ between the necessary static field components of this two Rabi-type flippers is negligible compared to the abolute field size. This simplifies the design of the required magnet system and it becomess relatively easy to adjust both coils to a spinflip efficiency sufficiently close to 100 %.

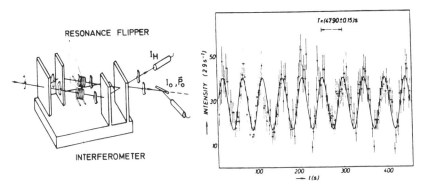

Figure 3. Quantum beat effect at extremely small frequency difference of the two rf flippers [12].

Then, after passage through the resonance flippers, the two interfering subbeams are in the same spin state but they differ in energy by an amount $\Delta E = \hbar \Delta \omega$ due to the exchange of photons of slightly different energy. As seen in Fig. 3, instead of a stationary interference pattern the final intensity therefore exhibits a typical beating behaviour, given by

$$I(t) \propto 1 + \cos(\Delta \omega t - \varphi_0) , \quad (7)$$

where φ_0 is an arbitrary constant initial phase shift. In this experiment the frequency splitting $\Delta \omega$ between the two two rf coils was as small as $2\pi \times 0.02$ s^{-1}, leading to a beating period $T = 2\pi/\Delta \omega = h/\Delta E$ of 47.90±0.15 s, that corresponds to an extremely small energy difference $\Delta E = 8.6 \times 10^{-17} \pm 2.7 \times 10^{-19}$ eV. One should bear in mind that this extremely high energy resolution, which is still waiting for application in fundamental, nuclear or solid state physics, is completely decoupled from the monochromaticity of the neutron beam, which in this experiment was 5.5×10^{-4} eV at a mean beam energy of 0.023 eV. Of course, the result can formally also be interpreted as being the effect of a time-dependent phase difference

$$\varphi(t) = \Delta \omega t = \frac{2\mu \Delta B_0}{\hbar} \cdot t \quad (8)$$

between the two rf fields, which follows from

$$\frac{\partial \varphi}{\partial t} \equiv \frac{\partial}{\partial t}(\varphi_1 - \varphi_2) = \omega_{r1} - \omega_{r2} = \frac{2\mu}{\hbar} \Delta B_0 . \quad (9)$$

But this inevitably leads to the above energy transfer interpretation, and vice versa.

Eqs. (8) and (9) indicate that an obvious formal similarity exists between this neutron quantum beat effect and the AC Josephson effect in superconducting tunnel junctions. There the phases of the Cooper-pair wave functions on both sides of the junction are related according to

$$\frac{\partial}{\partial t}(\Phi_1 - \Phi_2) = \frac{\partial \Phi}{\partial t} = \frac{2e}{\hbar} \Delta U , \quad (10)$$

which leads to the well-known oscillation behaviour of the tunnel current with a frequency $\omega = 2e\Delta U/\hbar$ that depends on the electric potential difference ΔU across the junction.

6. Probabilistic vs. Deterministic Particle Absorption

A particularly interesting situation arises if one of the two interering beams is attenuated by some factor a, say. This can be achieved by two principially different ways [13]. In the so-called "*probabilistic*" way a static absorber with a transmission probability a is introduced, which removes the neutrons either by a neutron-nuclear reaction or via some scattering process. The "*deterministic*" attenuation can be realized, for instance, by a rotating beam chopper which either fully absorbs or fully transmits the particles, depending whether or not they hit the absorbing section of the chopper wheel. The important conceptual difference between these two methods relies on the fact that in the case of probabilistic (or "stochastic") absorption it is not possible, not even in principle, to predict whether the neutron will be absorbed or not at any point in the absorbing region at any moment of time, whereas in the deterministic case it is known with certainty (at least in principle) what will happen with the neutron.

Intuitively one might argue that both absorption concepts lead to the same result. However, this is not true. To see this, let us consider the O-beam. A stochastic absorber in path I, say, reduces the corresponding probability amplitude by a factor $\sqrt{a}$, that is $\psi_O^I \rightarrow \sqrt{a}\psi_O^I$. The intensity of the superposed beam is then described by

$$I_O \propto |\psi_O^I + \psi_O^{II}|^2 \propto 1 + a + 2\sqrt{a}\cos\chi , \qquad (11)$$

which means that the observable interference oscillations are damped by a factor $\sqrt{a}$. With a deterministic beam attenuator, on the other hand, where the transmission is determined by the open-to-closed ratio of the beam chopper, $a = t_{open}/(t_{open} + t_{closed})$, one obtains a linear damping of the interference contrast

$$I_O \propto (1-a)|\psi_O^{II}|^2 + a|\psi_O^I + \psi_O^{II}|^2 \propto 1 + a + 2a\cos\chi . \qquad (12)$$

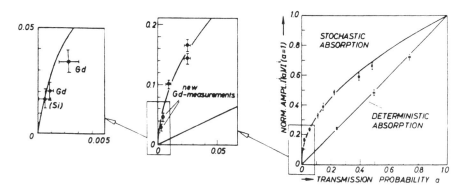

Figure 4. Normalized amplitude of the interference oscillations in the case of stochastic and deterministic attenuation of one of the interfering subbeams [13].

Fig. 4 shows the experimental verification of this predicted behaviour for three different attenuation regimes. In particular at very low transmission probability the difference between stochastic and deterministic absorption becomes dramatic. In the stochastic case an interference contrast still persists, whereas the fringe visibility for deterministic attenuation has long disappeared. By the way, with an absorbing lattice that diffracts neutrons off the beam and which can be rotated around the beam axis, a continous transition from one case to the other can be achieved [14]. In this context it is worth to note that in the quantum limit, where either the chopper period becomes shorter than the coherence time of the incident beam or the period of the absorbing lattice is shorter than the beam's coherence length, the attenuation factor a has to be defined more precisely. It turns out that it denotes the *nonremoval probability* of the neutrons from their initial phase space volume rather than it is simply the transmission probability of the absorber.

7. Scalar Aharonov-Bohm Effect

An electron passing through a Faraday cage, to which a pulse of electric potential is applied while the electron wavepacket is completely localized inside the cage, experiences no field and hence no force. In contrast to the classical intuition it should nevertheless acquire a phase shift proportional to its charge and the time integral of the scalar potential. Mainly from technological reasons this so-called scalar Aharonov-Bohm (AB) effect [15] has not yet been confirmed experimentally until now. The even more famous vector AB effect, which manifests itself in a topological phase shift of an electron wave packet that traverses a closed curve in a field-free region around a confined magnetic flux, has been verified by electron interference experiments (e.g. [16]). Both AB effects are illustrious examples of the non-local character of quantum theory and underscore that potentials are much more fundamental than forces.

The neutron analogue of the vector AB effect is the so-called Aharonov-Casher effect [17]. Here the neutron traverses a closed curve around an (infinitely long) charged line and thereby acquires a topological phase which is proportional to the product of its magnetic moment and the line charge density. This phase shift is expected to be typically only of the order of several mrad and in fact about two month of data acquisition were required to observe it in a neutron interferometric experiment [18]. The neutron analogue of the scalar AB effect can be realized by applying a uniform magnetic field pulse over a spatial region while the wavepacket propagates completely within this region. The accumulated scalar AB phase shift which has also been observed interferometrically [19], is

$$\Delta \Phi_{SAB} = \pm \frac{\mu}{\hbar} \int B(t) dt \ . \tag{13}$$

There the different signs refer to the two different neutron spin eigenstates. This phase shift is nondispersive,

$$\frac{\partial}{\partial k}\Delta\Phi_{SAB} = 0 \quad , \tag{14}$$

that means it is completely independent of the neutron wavenumber k. This phase nondispersivity is a common feature of all AB effects and their neutron counterparts. It is a direct consequence of their topological nature, which forbids any measurable positional shift or spreading of the wavepacket due to the absence of any force. Evidently monochromatic neutrons, like in perfect crystal interferometry, are inappropriate to prove the absence of classical forces via the phase nondispersivity. Indeed, none of the electron and neutron interferometric experiments performed so far were able to achieve this goal.

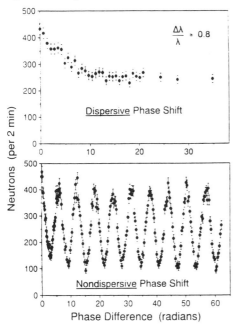

Figure 5. Spin precession of a polychromatic polarized neutron beam in a static magnetic field (*above*) and in a field pulsed according to the SAB effect requirements (*below*) [2].

Neutron spin precession, on the other hand, which can be considered as an implicit interference effect between the two spin states but without their spatial separation, allowed the first unambiguous verification of the nondispersive nature of the SAB phase shift [2]. There pulses of almost completely "white" polarized neutrons with a coherence length $\lambda^2/\Delta\lambda$ ($\Delta\lambda$ ...spectral width of the beam) of only about 3 Å are sent through a 56 cm long pulsed precession field region and subsequently analyzed with respect to their precession phase. If the field is turned on and off as long as the neutrons are completly localized inside the field region, no force acts upon the wavepackets and precession persists up to arbitrarily large phase angles (Fig. 5). In a static potential, however, acceleration and decelaration forces split the two spin state wavepackets in momentum space. Due to their small coherence length the associated displacement in real space causes a reduced overlapping and hence the interference effect, i.e. the precession of the polarization, vanishes rapidly with increasing field.

8. Geometric Phase Shift

During its evolution any quantal system, besides of the familiar dynamical phase, in general acquires also a component which is independent of the Hamiltonian and depends solely on the geometry of the evolution curve traced out in ray space. For a neutron spinor this so-called geometric phase turns out to be $-\Omega/2$, Ω denoting the solid angle spanned by the spin trajectory [20]. Noncyclic evolution curves can be treated as well by closing them by geodetic arcs on the spin sphere, which do not contribute any dynamical component to the total phase.

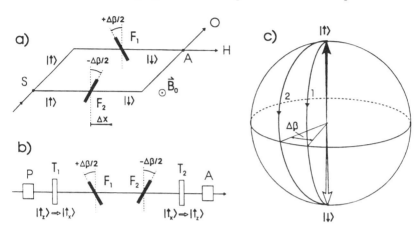

Figure 6. (a) Interferometer and (b) spin rotation setup to measure geometric phase shifts (F flipper, T $\pi/2$ -spin turn coil, S beam splitter, A analyzer, P polarizer). (c) The geometric phase shift $\Delta\beta = -\Omega/2$, where Ω is the solid angle the spin trajectories span on the unit sphere.

Very recently a polarized neutron interferometric experiment was performed in a Bombay-Missouri-Vienna collaboration [21] to identify the geometric and dynamic components of the total phase shift if two static flippers act upon the two coherent subbeam (Fig. 6a). A relative rotation $\pm\Delta\beta/2$ of the the flippers F_1 and F_2 around the (vertical) quantization axis produces a pure geometric phase $\Delta\beta$, whereas a relative translation Δx of the flippers along the respective beam path generates a pure dynamical phase shift $\Delta\phi = \gamma B_0 \Delta x / v$ due to the excess spin precession about the magnetic guide field $\vec{B}_0$. A reversal of the current of one flipper causes an observable phase jump of π, which is a direct consequence of the anticommutation relation between the orthogonal components of the Pauli spin operator. With exception of this last item, neutron spin polarimetry is likewise able to measure the purely geometric phase shift produced upon passage of a polarized neutron beam through two counter rotated π-flippers (Fig. 6b) [22].

9. Concluding Comments

The main intention of this paper is to show the versatility of neutron interferometry and to present a variety of applications, but by no means it claims to give a complete survey over the field of quantum interference with neutron waves. For instance, such important topics like gravity-induced interference and interferometry with very cold neutrons remained completely out of consideration here.

References

1. Rauch, H., Treimer, W., and Bonse, U.: Test of a single crystal neutron interferometer, *Phys. Lett.* **47A** (1974), 369-371.
2. Badurek, G. Weinfurter, H., Gähler, R., Kollmar, A., Wehinger, S., and Zeilinger, A.: Nondispersive phase of the Aharonov-Bohm effect, *Phys. Rev. Lett.* **71** (1993), 307-311.
3. Rauch, H., Zeilinger, A., Badurek, G., Wilfing, A., Bauspiess, W., and Bonse, U.: Verification of coherent spinor rotations of fermions, *Phys. Lett.* **54A** (1975), 425-427.
4. Werner, S.A., Colella, R., Overhauser, A.W., and Eagen, C.F., Observation of the phase shift of a neutron due to precession in a magnetic field, *Phys. Rev. Lett.* **35** (1975), 1053-1055.
5. Badurek, G., Rauch, H., Zeilinger, A., Bauspiess, W., and Bonse, U.: Phase-shift and spin-rotation phenomena in neutron interferometry, *Phys. Rev. D* **14** (1976), 1177-1181.
6. Badurek, G., Rauch, H., and Summhammer, J.: Polarized neutron interferometry: A survey, in G. Badurek, H. Rauch and A. Zeilinger (eds.), *Matter Wave Interferometry*, North-Holland, Amsterdam, *Physica B* **151** (1988), pp. 82-92.
7. Badurek, G., Rauch, H., Wilfing, A., Bonse, U., and Graeff, W.: Perfect-crystal neutron polarizer as an application of magnetic prism refraction, *J. Appl. Cryst.* **12** (1979), 186-191.
8. Summhammer, J., Badurek, G., Rauch, H., Kischko, U., and Zeilinger, A.: Direct observation of Fermion spin superposition by neutron interferometry, *Phys. Rev. A* **27** (1983), 2523-2532.
9. Badurek, G., Rauch, H., and Summhammer, J.: Time-dependent superposition of spinors, *Phys. Rev. Lett.* **51** (1983), 1015-1018.
10. Carruthers, P., and Nieto, M.M.: Phase and angle variables in quantum mechanics, *Rev. Mod. Phys.* **40** (1968), 411-440.
11. Dewedney, C., Gueret, P., Kyprianidis, A., and Vigier, J.P., Testing wave-particle dualism with time-dependent neutron interferometry, *Phys. Lett.* **102A** (1984), 291-294.
12. Badurek, G., Rauch, H., and Tuppinger, D.: Neutron interferometric double-resonance experiment, *Phys. Rev. A* **34** (1986), 2600-2608.
13. Rauch, H., Summhammer, J., Zawisky, M., and Jericha, E.: Low-contrast and low-counting-rate measurements in neutron interferometry, *Phys. Rev. A* **42** (1990), 3726-3732.
14. Rauch, H., and Summhammer, J.: Neutron-interferometer absorption experiments in the quantum limit, *Phys. Rev. A* **46** (1992), 7284-7287.
15. Aharonov, Y., and Bohm, D.: Significance of electromagnetic potentials in quantum theory, *Phys. Rev.* **115** (1959),485-110.
16. Tonomura, A., Osakabe, N., Matsuda, T., Kawasaki, T., Endo, J., Yano, S., and Yamada, H.: Evidence for Aharonov-Bohm effect with magnetic field completely shielded from electron wave, *Phys. Rev. Lett.* **56** (1986), 792-795.
17. Aharonov, Y., and Casher, A.: Topological quantum effects for neutral particles, *Phys. Rev. Lett.* **53** (1984), 319-321.
18. Cimmino, A., Opat, G.I., Klein, A.G., Kaiser, H., Werner, S.A., Arif., M., and Casher, A.: Observation of the topological Aharonov-Casher phase shift by neutron interferometry, *Phys. Rev. Lett.* **63** (1989), 380-383.
19. Allman, B.E., Cimminio, A., Klein, A.G., Opat, G.I., Kaiser, H., and Werner, S.A.: Scalar Aharonov-Bohm experiment with neutrons, *Phys. Rev. Lett.* **68** (1992), 2409-2412.
20. Aharonov, Y., and Anandan, J.: Phase change during a cyclic quantum evolution, *Phys. Rev. Lett.* **58** (1987), 1593-1596.
21. Wagh, A.G., Rakhecha, V.C., Allman, B., Kaiser, H., Hamacher, K., Jacobson, D.L., Werner, S.A., Summhammer, J., Badurek, G., and Weinfurter, H.: Experimental separation of geometric and dynamical phases in neutron interferometry, to be submitted to *Phys. Rev. Lett.*
22. Badurek, G., Weinfurter, H., Wagh, A.G., Riedler, P., and Summhammer, J.: Neutron polarimetric measurement of geometric phase, to be submitted to *Phys. Lett.*

ENIGMATIC NEUTRONS

Some Trends in the Development of Quantum Mechanics

V.K. IGNATOVICH
*Frank Laboratory of Neutron Physics of
Joint Institute for Nuclear Research
FLNP JINR, 141980, Dubna Moscow region, Russia*

> He surely meant the "Copenhagen Interpretation" of quantum mechanics. Extremely loosely speaking (our experts will surely flame me to hell and back) it says that the mathematics of QM is working and the whole bunch of philosophers (many world theory etc.) thus shall go fish.
>
> Hauke Reddmann
> fc3a501@rzaixsrv2.rrz.uni-hamburg.de
> GROUP sci.physics
> 1 Aug 1995 10:41:22 GMT

Abstract. Neutrons are well known entities used as a tool in condensed matter and nuclear physics research. Their properties are well known with great precision but only one problem has remained unsolved for more than 25 years — the problem of extralosses of ultracold neutrons in traps. These losses are small, but they are two orders of magnitude larger than the theoretically predicted ones. To solve this enigma an excursion into the fundamental principles of quantum mechanics was undertaken. The wave function of free particles was supposed to be a singular wave packet. Because of high energy components of the packet the neutron can penetrate deeply into the potential barrier and even overcome it by nontunneling way. It is shown here that these considerations do not contradict the essence of quantum mechanics and lead to some experimentally observable consequences: one is the weak temperature dependence of the loss coefficient and at the same time the second one is the strong temperature dependence of the UCN inelastic scattering on the walls. The investigation of these wavepackets shows that the uncertainty relations have nothing to do with the essence of quantum mechanics, and a particle can have precisely mathematically defined position and momentum simultaneously. It is shown that a different definition of a bound state can be provided.

1. Introduction

The problem of ultracold neutrons (UCN) is first recalled. A neutron is called ultracold (see the books [1]-[3]) if it can be stored in a bottle. It can be stored there when it's energy E is less than the repelling potential of the walls, i.e. if $E < E_{\lim} \equiv u = (\hbar^2/2m)4\pi N_0 b$, where m is the neutron mass. N_0 is the atomic density and b is the coherent neutron nucleus scattering amplitude. The velocity of such neutrons is of $v \leq v_{\lim} \approx 5$ m/s and the wave length is nearly 1000 Å. The interaction u varies from substance to substance, but the magnitudes of $E_{\lim} \approx 10^{-7}$ eV and $v_{\lim} \approx 5$ m/s are typical ones.

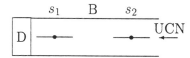

Figure 1. UCN storage experiment scheme: One opens the shutter s_2 (shutter s_1 is closed), fills the bottle B, then closes it, waits some time $t_{\exp}$, which is called the exposition time, then opens the shutter s_1 (s_2 is closed) and counts the number of UCN left in the bottle.

The UCN problem (anomaly) is related to the storage experiment, the scheme of which is shown on fig. 1. Here s_1 and s_2 denote two shutters that can be opened and closed. When s_1 is closed and s_2 is opened, neutrons fill the bottle. When both s_1 and s_2 are closed the UCN are trapped in the bottle. The time, during which they are trapped, is called "exposition time," $t_{\exp}$. And when the shutter s_1 is opened and s_2 is closed, the UCN left in the bottle after exposition are counted by the detector. In the experiment one measures the storage curve similar to the one shown in fig. 2. It shows in logarithmic scale the number of neutrons in the bottle survived after exposition time $t_{\exp}$. This curve can be represented by a function

$$N(t) = N_0 \exp(-t/\tau),$$

where τ is called "storage time." For monoline spectrum of stored neutrons τ is constant and it is determined by the own life time of the neutron τ_β with respect to β-decay and by losses in the walls:

$$1/\tau = 1/\tau_\beta + 1/\tau_l.$$

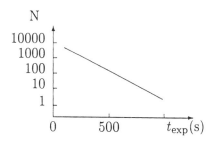

Figure 2. The storage curve: the number of neutrons, left in the bottle after exposition in closed bottle during the time t_{exp}. In an ideal case it is of an exponential type: $N(t_{exp}) = N_0 \exp(-t_{exp}/\tau)$, where τ is storage time.

The loss part $1/\tau_l$ is proportional to the probability of losses at single collision with the wall: $1/\tau_l \propto \eta$, where η is equal to the ratio: $\eta = b''/b'$ of imaginary and real parts of the scattering amplitude b. The imaginary part b'' can be represented as $b'' = \sigma_l/2\lambda$, where λ is the neutron wave length and σ_l is the loss cross section.

Let us enumerate the factors responsible for losses (τ_l).

1. Absorption cross section σ_a (own and impurities)
2. Inelastic scattering cross sections $\sigma_{inel}^{coh,inc}$ (own, impurities and vacuum gas)
3. Geometry: dimension, surface roughnesses, gaps.
4. Gravity: change of trajectories, variation of spectrum along the height.

Taking into account all these factors, we can say that at present for the coldest, purest bottle (Beryllium walls covered with oxygen) with the smallest self-absorption, we observe [4] *an additional loss cross section of 1 barn ($\eta \approx 3 \times 10^{-5}$) of unknown nature, which is of two-three order of magnitude higher than the theoretical one.* This is the main UCN problem or UCN anomaly. To explain the anomaly, it is necessary (if not doubting the experiments) to do something extraordinary, for instance, to impinge upon fundamental principles.

2. De Broglie's Wave Packets

To explain the anomaly (if it is of fundamental nature) one way is to suppose that the neutron is described by a wave packet containing high energy components. Then the part of the wave packet spectrum shared by these high energy components defines the probability of penetration through the potential barrier in a non tunneling fashion.

But for such an explanation to be viable, it is necessary to suggest that the high and low energy components of the wave packet are dependent. That means that the wave packet should be an immanent property of the neutron and should not spread. Now the question arises whether it is possible to construct a nonspreading wave packet in the framework of the existing quantum mechanical theory?

The answer to this question is affirmative for a free neutron, and the number of possible wave packets is infinite. But if we restrict ourselves only to spherically symmetrical case we come to two types of wave packets, describing a particle moving with velocity v. These were found by de Broglie [5].

One of them (non normalizable)

$$\Psi(\mathbf{r},t) = \exp(i\mathbf{v}\mathbf{r} - i\omega t)j_0(s|\mathbf{r} - \mathbf{v}t|), \tag{1}$$

is the solution of the homogeneous Shrödinger equation

$$(i\partial/\partial t + \Delta/2)\Psi(\mathbf{r},t) = 0,$$

and the other (normalizable) one

$$\psi(s,\mathbf{v},\mathbf{r},t) = c\exp(i\mathbf{v}\mathbf{r} - i\omega t)\frac{\exp(-s|\mathbf{r} - \mathbf{v}t|)}{|\mathbf{r} - \mathbf{v}t|}, \tag{2}$$

is the solution of the inhomogeneous equation

$$(i\partial/\partial t + \Delta/2)\psi(\mathbf{r},t) = -C(t)\delta(\mathbf{r} - \mathbf{r}(t)). \tag{3}$$

where s is a width of the wave packet, v is the neutron velocity, c in (2) is the normalization constant defined by the equality:

$$\int d^3r |\psi(s,\mathbf{v},\mathbf{r},t)|^2 = 1. \tag{4}$$

and $C(t)$ in (3) is $2\pi c \exp[i(v^2 - \omega)t]$. The frequency ω in (1) is: $\omega = (v^2 + s^2)/2$. It means, that energy is composed of kinetic part $v^2/2$ and thickness part $s^2/2$. The last one is needed to hold the packet to be nonspreading.

The frequency ω in (2) is: $\omega = (v^2 - s^2)/2$. It means, that the thickness part is formed at the expense of kinetic energy.

It is the width s that is important for the penetration of the neutron over a barrier and which would explain the UCN anomaly. The parameter s should be small for the wave function of UCN so that it simulates the reflection from the bottle walls close to the one obtained as for plane waves. But the spectrum due to this width should be wide enough to contain high energy components. Those constraints leads to the wave packet (2), which alone can meet these requirements.

The calculation of losses, described by the over barrier penetration, gives

$$s = 10^{-4}v.$$

It means that s depends on v. The higher the v, the narrower the wave packet, and finally the closer the wave embodies a corpuscular property. This is in line with the present day description of wave-particle duality.

3. Experimental Consequences of Such an Explanation

Such an explanation leads to some interesting consequences that can be experimentally verified. Let us consider the experiment in fig. 1 modified as is shown in fig. 3. The difference is now that around the storage bottle there are placed thermal neutrons counters. They count the neutrons created in the bottle during exposition time because of up scattering of UCN in collisions with the walls. The experiments were performed at different

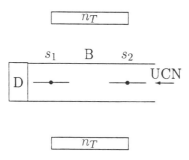

Figure 3. The scheme of storage experiment (fig. 1) can be modified by placing thermal neutrons counters n_T around the storage bottle.

temperatures of the bottle walls [4] had shown the different loss coefficient $\eta(T)$, extracted from storage times $\tau(T)$, but the dependence of $\eta(T)$ on temperature was very weak, as is shown by curve 1 in fig. 4. But at the same time if the $\eta(T)$ were extracted from thermal neutrons detectors counting it would show very steep dependence on T (curve 2 in fig. 4). This result looked quite paradoxical. The dependence of upscattered neutrons on temperature could be interpreted as a dependence of upscattering cross section on temperature. And if at $T = 300$ K almost all the losses can be ascribed to up scattering, then after decreasing the temperature and therefore expecting a vanishing of upscattering we should expect the vanishing of almost all losses and a strong decrease of the η, extracted from $\tau(T)$. But the result showed nothing like that. It looked like as if capture cross section depends on temperature in opposite way and balanced the decreasing of the up scattering with temperature. In our model it has however a very natural explanation. Indeed, the main part of η is due to over barrier transmission. But the neutrons, which entered the walls have three different fates: they can be up scattered or captured or they can be lucky enough to

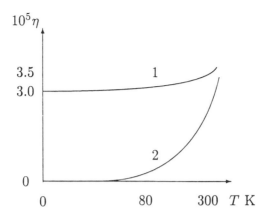

Figure 4. Dependence of η on temperature (curve 1), found from storage curves, and the one, found from counts of thermal neutron detectors n_T (curve 2) surrounding the storage bottle (fig. 3)

cross the other side of the wall. The cross sections for UCN are high, thus the neutrons that are lucky to cross the wall are few. Their main fate is to be up scattered or captured. But if capture cross section is much less than inelastic one, all the neutrons will be upscattered. At lower temperature the inelastic cross section vanishes, up scattering ceases and all the neutrons entered the wall penetrates deeper in it and at last they are captured. In the case of thin enough walls and small capture cross section the UCN can avoid capturing and can be measured on another side of the wall because of non tunneling transmission.

4. Prejudices and Incompleteness of Quantum Mechanics

4.1. UNCERTAINTY RELATIONS

With a wave packet of the form (1) in mind we can discuss some prejudices in quantum mechanics. These are concerned first of all with uncertainty relations.

The nonspreading wave packet (1) has precisely definite shape and velocity, and with respect to these properties it is legitimate to consider the possibility of a simultaneous definition of position and momentum. (The same point was advanced also by P.Marmet in the recent publication [6] .) The momentum of the wave packet is precisely defined because of precise definition of velocity v and of relation $p = mv$ with the strictly defined mass m of the considered particle.

The position should be defined for the quantum object as a whole, that means the wave function should be also included in the definition. The last is extended in space, thus we are to define position of the extended object.

For instance the position can be identified with the center of gravity of the wave function or with the maximum of its amplitude.

In the case of (1) the center of gravity coincides with the maximum of the amplitude. And it is seen that if this point is defined as the position of the particle, this position is precisely defined simultaneously with the momentum of the particle.

If we try to define position with the help of measurement, then even in classical physics we shall have uncertainty in the case of extended objects. If we are dealing with point object in classical physics, then we can define precisely its position with the help of measurements only if the particle is infinitely heavy, or if we prepare precisely defined test particles and deduce the previous position of the target particle from scattering measurements. But in this case the position point is again the matter of definition and such a definition is in the same way applicable to a quantum object with a wave function.

In conclusion we can say that the position and momentum of particle in quantum mechanics can be precisely defined simultaneously, so the quantum axiom of impossibility of such a definition is not correct.

Now the question arises about the meaning of the Heisenberg uncertainty relations. These relations relate the widths of an extended object with the width of it's Fourier image, and in that respect they are applicable both to quantum and classical object. For instance these relations are applicable to the classical electron with its electrostatic field.

In the case of wave packet (2) position of the particle can be identified with the singularity point in analogy with classical physics where the position of classical electron coincides with the singularity point of the Coulomb potential.

4.2. INCOMPLETENESS OF QUANTUM MECHANICS

We define incompleteness as follows. Suppose that we calculate some scattering amplitude. If with the same wave function of the incident particle we can calculate different outcomes of the experiment, and quantum mechanics (QM) does not supplies us with the rules for choosing the right outcome, we call such a theory incomplete.

Now we show that QM is indeed an incomplete theory. Consider a simple one-dimensional potential barrier. QM gives us a recipe in order to calculate reflection $|r(k)|^2$ and transmission $|t(k)|^2$ for an incident particle if it is described by a plane wave $\exp(ikx - i\omega t)$. But if our particle is described by a wave packet

$$\psi(x,t) = \int_{-\infty}^{+\infty} dk \Psi(k) \exp(ikx - i\omega t),$$

we can calculate the reflection coefficient in different ways. One way is

$$|R|^2 = \int_{-\infty}^{+\infty} dk |\Psi(k) R(k)|^2, \qquad |T|^2 = \int_{-\infty}^{+\infty} dk |\Psi(k) T(k)|^2, \qquad (5)$$

if we use normalization

$$\int_{-\infty}^{+\infty} dk |\Psi(k)|^2 = 1,$$

meaning we have only one particle. The expression (5) means that we calculate reflection coefficient for every plane wave and sum the coefficient over all components of the wave packet.

The other way is

$$|R|^2 = \int_{-\infty}^{+\infty} dk \, k |\Psi(k) R(k)|^2, \qquad |T|^2 = \int_{-\infty}^{+\infty} dk \, k |\Psi(k) T(k)|^2, \qquad (6)$$

and normalization should be

$$\int_{-\infty}^{+\infty} k \, dk |\Psi(k)|^2 = 1. \qquad (7)$$

It means that the incident flux is equal to unity.

Two definitions (5), (6) can give in principle different numbers. QM mechanics fails to tell which one is more correct. And it proves its incompleteness.

4.3. RECIPES OF QM

Now we can conclude, that QM mechanics is not a complete theory. But we should accept that it gives us very useful recipes on how to calculate scattering amplitudes and how to calculate energy levels for bound states. These recipes are different. For scattering you should use plane waves, and for energy levels - the other functions.

What is changed if we accept an immanent wave packet for a free particle? First, we should choose the rules for calculations of scattering amplitudes. For reflection from the potential barrier we choose the rule (6). And we see it does not contradict to present day QM.

Second, we should accept that the reflection does not change the wave packet. It also does not contradict to present day QM. Indeed, QM gives us a recipe on how to calculate scattering amplitudes for plane waves, but it does not predict which side will go that given particle. It is out the scope of QM. For that reason we can accept, that outgoing wave packet of the particle is identical with incoming one.

5. Bound States

QM gives us recipes on to how to calculate scattering amplitudes and levels of bound systems. Is this recipe unique? Is it possible to devise some other way to calculate the energy levels? The answer to this question is positive.

We shall show now a little bit different formalism, that leads to bound states too. Let us consider the simplest case of a particle in a spherical potential well. of depth V and radius R, We shall think of a particle placed somewhere at point $r = a < R$. The equation of motion will be like

$$(\Delta - u(r) + k^2)\psi = -c\delta(r-a), \qquad (8)$$

where c is some parameter related to normalization of the total wave- field. Substitution [7] of $\psi = (\mathcal{R}_0/r)Y_{00}$ into (8) yields

$$[d^2/dr^2 + k^2 - u(r)]\mathcal{R}_0 = -ca\delta(r-a).$$

In our simplest system the solution of the inhomogeneous equation can be represented as

$$\mathcal{R}_0(r) = A\sin(kr)\theta\,(0 < r < a) + B\cos(kr)\theta\,(a < r < R)$$

with $A = ca\cos(ka)/k$ and $B = ca\sin(ka)/k$. Matching this wave function to function $C\exp(-\kappa[r-R])$ outside the well, where $\kappa = \sqrt{V-k^2}$, we get a condition, defining a bound state:

$$k\mathrm{tg}(kR) = \kappa.$$

The constant c in (8) is determined with the help of the normalization condition

$$\int |\mathcal{R}_0|^2 dr = 1.$$

We have still some arbitrariness in the choice of position a in (8). But the energy level does not depend on this position. Only constant c can vary with variation of a. But it is very natural. Distribution of $c(a)$ can be interpreted as a probability to find particle at point a.

6. Conclusion

We have shown,

1. that QM is not a complete theory, because it does not permit to make a unique choice of scattering amplitudes between many different possibilities. If we impose some rules for the establishment of uniqueness, it will not contradict the existing QM.
2. that the rules for determination of bound states levels are also not unique. It is possible to devise the rules different from accepted ones and it will still not contradict existing QM.
3. the uncertainty relations, considered usually as a cornerstone of QM have in essence nothing to do with its ideology and particle can be described by a wavepacket with strictly mathematically defined momentum and position simultaneously.
4. description of free neutrons with de Broglie's singular wave packet leads to some interesting experimental consequences.

Of course the exercises with wave packets at present look like a game. A lot of work should be done before advantages of it become clear enough to abandon the commonly accepted formalism of QM. But even at this present stage of development the achievement of the new formalism consist in liberation of thinking from the dogmatic approach of the Copenhagen interpretation.

I have fished all that out of the mainstream of the ideas exposed by giants, whose fighting with "smoky dragoon" was a great encouragement, and today *I want to join my voice to all those who compliments Jean-Pierre Vigier on occasion of his jubilee.*

The research described in this publication was made possible in part by Grant No. J6P100 from the International Science Foundation an Russian Government. The author expresses his gratitude to organizing committee and personally to S. Jeffers for the invitation and support, and to I.Carron from Texas A&M University for his invaluable assistance.

References

1. Steyerl, A. (1977), Very low energy neutrons, in G.Höhler (ed.) *Neutron Physics, (Springer Tracts of Modern Physics)*, Springer-Verlag, Berlin, N.Y., **Vol. 80, Part 2**, pp. 57- 130.
2. Ignatovich, V.K. (1990) *The Physics of Ultracold Neutrons*, Clarendon Press, Oxford.
3. Golub, R., Richardson, D. and Lamoreaux, S.K. (1991) *Ultra-Cold Neutrons*, Adam Hilger, Bristol.
4. Alfimenkov, V.P., et al., (1992), Anomalous interaction of UCN with the Surface of Beryllium Bottles, *JETP Lett.* **55**, pp. 92-96.
5. Broglie L. de (1960) *Non-Linear Wave Mechanics: A Causal Interpretation*, Elsevier, Amsterdam.

6. Marmet, P. (1994) On the interpretation of Heisenberg's Uncertainty Relationships, *Phys. Essays* **7**, pp. 340-348.
7. Landau, L.D. and Lifshits, E.M. (1963) *Quantum Mechanics*, Gosizdat Fizmatlit, Moscow.

EXPERIMENTS TO TEST THE REALITY OF DE BROGLIE WAVES

J.R. Croca
Departamento de Física
Faculdade de Ciências - Universidade de Lisboa
Campo Grande, Ed. C1, 1700 Lisboa, Portugal
Email: croca@cc.fc.ul.pt

Abstract

Most experiments proposed to test the nature of the quantum waves are based on the so called collapse of the wave packet. Still there are open other possibilities, discussed here, leading to feasible experiments that may clarify this long standing problem.

1. Introduction

The nature of the quantum waves have been a major problem since the creation of quantum mechanics[1]. As is well known the Copenhagen interpretation, following Born's ideas, proposed that the quantum waves are probability waves without intrinsic physical reality. On the other side de Broglie[2] and many others always maintained that the quantum waves were indeed real physical waves. The first conceptual experiment[3] to test these ideas was proposed in 1980 and were soon followed by other more concrete experiments[4]. The proposed experiments were based on the collapse of the wave packet by the measure. The principles underlying the method can be summarized in the following way:

If the initial wave function ψ is divided in two ψ_1 and ψ_2, by means of a beamsplitter, or some other device, and a detector placed in the possible path of one of these waves, say ψ_2, localizes the particle, then the probability of finding it at the other path turns instantly to zero and consequently the other wave ψ_1 collapses into nothing. This conclusion results from the fact that the quantum waves are assumed to be a probability waves. If on the contrary one accepts that the quantum waves are real, then on must conclude that, since the measure did not affected physically the wave ψ_1 it will continue in existence. One concrete experiment, based on these ideas[5] was done by Mandel[6]. the result of the experiment was discussed by the author[7] and the conclusion is that due to the resolution time of the apparatus 10 ns versus the coherence length .1 ps of the utilized photon wave packet the experiment was not conclusive. Modification on the experiment were proposed, in the same work[7], to overcame the difficulties.

Koh and Sasaki[8] in 1988 proposed another approach to the problem. Their proposal is based in what they call *in flight reduction of the matter wave packet*. Basically the idea is that if one increases the length of the arms of the interferometer the blurring of the interference pattern would also increase till no interference is observable. These ideas[9] were later developed in the light of the causal theory of de Broglie.

According to de Broglie a quantum particle is a real physical entity composed of a wave plus a singularity. Under certain conditions, for instance by means of a beamsplitter, it is possible to split the wave in two, one with the singularity, the other without the singularity. The singularity remaining always undivided otherwise one should have a particle with half of its energy. Since the full wave and the empty wave are different they may interact in different ways with the medium. For instance, the full wave by means of its singularity is able to trigger a detector while the empty wave is not. On the other hand even when traveling in the free space the interaction with the zero point field may be different, Koh and Sasaki hypothesis. The empty wave may in this case loose intensity as it travels freely. This hypothesis was recently tested by Jeffers[10]. Unfortunately due to the fact that the used source was not monophotonic the experiment could not be conclusive. Since with these kind of sources there are always many photons presented in each bunch, when the packet is split there is no way of guaranteeing that along one arm of the interferometer follows an empty wave. Usually in each arm follows waves with singularity.

Recently[11] a method to test the different possibilities of interaction of these waves was proposed. This process avoids the shortcomings of the previous methods, such as the enormous length of the interferometer arms required for the interaction between the empty wave and the zero point field be of any significance.

2. The Experiment

The experiment as can be gathered from above is mainly based in the following assumptions taken from the causal theory of de Broglie:

1 - Quantum waves are real waves.
2 - A quantum particle is composed of a real wave plus a real singularity $\phi = \theta + \xi$, where ϕ is a solution of a non-linear Schrödinger master equation. Where θ represents the wave without significant energy and ξ the singularity, carrying practically all the particles energy.
3 - It is possible by means of a physical device to split the wave in two or more waves.
4 - The height energetic singularity remains always unsplit.
5 - It is not possible to have a singularity without the accompanying wave. De Broglie basic idea, for each quantum particle there is always an associated wave.
6 - It is possible to have a wave without singularity.
7 - Interference results from the superposition of the waves. The singularity being guided through a non-linear process by the resulting wave (guiding principle).
8 - It is possible to build monoparticle sources.

Since the empty wave θ is a real wave, then when it strikes a beamsplitter is divided in two real waves, whose intensity is related accordingly. By putting successive beam splitters, as shown in Fig.1, the intensity of the wave is reduced till practically nothing is left mixed up with the vacuum fluctuations.

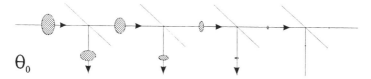

Fig.1. The intensity of the wave θ is reduced by the interaction with the sucessive beamsplitters.

The intensity reduction of the wave, as it crosses the n beamsplitters, can be described by

$$\theta = \theta_0 e^{-\mu n}, \qquad (1)$$

where μ represents the attenuating factor related with the amplitude transmission t by $\mu = -\log t$.

The behavior of the full wave ϕ (wave with the singularity) when crossing the successive beam splitters in light of de Broglie's ideas is different from the one of the wave with no singularity. Since assumption 6 prevents the accompanying wave to disappear completely, leaving only the singularity, it must be assumed that after a certain point the reduction of the intensity and concomitant loss of energy due to the interaction of the wave with the beamsplitter must, in some way, be compensated by energy supplied by the singularity. Naturally since according to de Broglie the ratio of energy between the singularity and the wave is so great the loss of energy of the singularity is for all practical effects negligible. Only on cosmic scale the loss of energy can be noticed given way to a mechanism to explain the redshift without big bang.

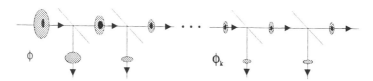

Fig.2. The amplitude of the wave accompanying the singularity decreases as it crosses the successive beamsplitters till a certain point, from this point on the amplitude remains for all practical instances constant.

In this conditions the interaction of the full wave with the beamsplitter may be described in two steps: When the full wave strikes on the first 50% beamsplitter the wave θ_0 would loose half of its amplitude $\theta = \theta_0/2$ in the second $\theta = \theta_0/4$, and so till the kth beamsplitter $\theta = \theta_0(1/2)^k$. From this point on having reached the minimum possible energy compatible with the existence of the quantum particle the wave starts regenerating itself at the expenses of the energy from the singularity, keeping the amplitude constant.

This process can be represented analytically by

$$\begin{cases} \theta = \theta_0 e^{-\mu n}, & n \leq k \\ \theta = \theta_0 e^{-\mu k}, & n > k. \end{cases} \quad (2)$$

which can be represented graphically, see Fig.2

2.1. The concrete experiment

Consider the Mach-Zehnder interferometer shown in Fig.3. Along both arms of the apparatus an equal number of beamsplitters are placed, to reduce the amplitude of the waves. Suppose that a single photon is send at a time from a monophotonic source.

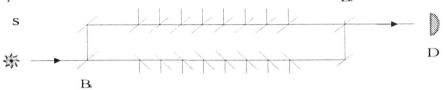

Fig.3. The experiment. A Mach-Zehnder interferometer with an equal number of beamsplitters along both arms. Here S is a monophotonic source, D a detector and B_s 50% beamsplitters.

According to usual interpretation of quantum mechanics the two probability waves going through paths one and two of the interferometer behave precisely in the same way and undergo the same amplitude reduction. Thus the prediction for the expected intensity at the detector will be

$$I_u = |\psi_1 + \psi_2|^2, \quad (3)$$

or

$$I_u \propto 1 + \cos \delta, \quad (4)$$

where ψ_1, ψ_2 are the waves along the arms 1 and 2 of the interferometer respectively, and δ is the relative phaseshift between the two waves. For this experiment the usual quantum mechanics predicts that, independently of the absolute intensity of the two overlapping waves, the visibility will be approximately equal to one.

Let us now see the expected visibility, for the same experimental set-up, when one assumes that the quantum waves are real.

According to the previous assumption, if the number of beamsplitters is less than k, the visibility is exactly the same as in the usual quantum theory, since the reduction in amplitude due to the beamsplitters affects both waves in the same way.

THE REALITY OF DE BROGLIE WAVES

If the number of beamsplitters n is greater than k, the singularity starts regenerating the accompanying wave in the traverse arm, while at the other interferometer arm the single empty wave continues loosing energy. For the case when the singularity arriving at the detector comes from the path one the two waves overlapping at the detector can be written: (Note that the wave is regenerated only in the arm with the singularity, but not on the other)

$$\begin{cases} \theta_1 = \theta_0 e^{-\mu k}, \\ \theta_2 = \theta_0 e^{-\mu n}. \end{cases} \qquad n \geq k \qquad (5)$$

Therefore, one obtains

$$I_c = |\theta_1 + \theta_2|^2 = |\theta_0|^2 (e^{-2\mu k} + e^{-2\mu n})(1 + \gamma \cos \delta), \qquad (6)$$

where the expected visibility γ is given by

$$\gamma = \text{sech}\mu(n-k), \qquad n \geq k. \qquad (7)$$

The prediction does not change if the singularity arrives at the detector by path two.

Then, on the basis of the previous assumptions if the number of beamsplitters n is less than k, the visibility will be one, but if is greater than k the visibility will decrease rapidly with $(n-k)$ as can be seen in Fig.4.

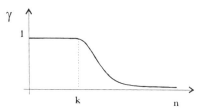

Fig.4. Plot for the predicted visibility assuming that the quantum waves are real.

In an actual interferometric experiment, as everybody knows, the experimental visibility always fails to reach 100%, even if predicted theoretically. This decrease in the expected fringe visibility is explained as a consequence of various reasons: The light is never monochromatic, imperfect alignment, instabilities in the interferometer, different temperature gradients, and other causes. So, it would not be correct to compare the theoretical results predicted by the usual quantum mechanics with the visibility predicted by relation (8) derived from the causal theory. In this situation and in order to achieve reliable experimental results the actual experiment could be carried out in the following way:

After having properly aligned the interferometer, in order to maximize the visibility γ_0 for an incident monophotonic beam of intensity I_0, one reduces the intensity to I_1 of the entering beam by placing a beamsplitter in front of the interferometer and measure the new visibility γ_1. By adding another beamsplitter the intensity is further reduced to I_2 to which corresponds a visibility γ_2. This process of adding beamsplitters continues till it gives origin to an experimental correspondence between intensities and visibilities $(I_0, \gamma_0), (I_1, \gamma_1), ..., (I_n, \gamma_n)$.

Now, after this like visibility calibration of the interferometer, the experiment is performed placing progressively the beamsplitters in the branches of the interferometer. Then one compares the visibilities obtained when the beamsplitters are placed inside the Mach-Zehnder interferometer, with the visibilities obtained in the calibration process.

There remains, of course, the problem of knowing the number of beamsplitters necessary so that the effect may manifest itself. Since there is no way to know a priori the value of the possible interactions only experimentally the question can eventually be solved.

References

1] F. Selleri, *Le Grand Dèbat de la Theorie Quantique*, (Flamarion, Paris, 1986)
2] L. de Broglie, *The Current Interpretation of the Wave Mechanics, A Critical Study*, (Elsevier, Amsterdam, 1964)
3] J. and M. Andrade e Silva, C.R. Acad. Sci. (Paris) 290(1980)501
4] F. Selleri, Found. Phys. 12 (1982) 1087; A. Garuccio, V. Rapisarda and J.P. Vigier, Phys. Lett. A 90 (1982) 17; J.R. Croca, in *Waves and Particles in Light and Matter*, Eds. A. Van der Merwe and A. Garuccio (Plenum Press, New York, 1994)
5] J.R. Croca, A. Garuccio, V.L. Lepore, and R.N. Moreira, Found. Phys. Lett. 3(1990)557
6] L.J. Wang, X.Y. Zou, and L. Mandel, Phys. Rev. Lett. 66(1991)1111
7] J.R. Croca, in *Waves and particles in Light and Matter*, Eds. Van der Merwe and A. Garuccio, (Plenum Press, New York, 1994)
8] Y. Koh and T. Sasaki, in *Microphysical Reality and Quantum Formalism*, eds. A. van der Merwe, G. Tarozzi and F. Selleri, (Kluwer Academic Publishers, 1988)
9] J.R. Croca, in *The Concept of Probability*, Eds. E.I. Bitsakis, and C.A. Nicolaides, (Kluwer Academic Publishers, 1989)
10] S. Jeffers and J. Sloann, Found. Phys. Lett. 333(1994)341
11] J.R. Croca, M. Ferrero, A. Garuccio, and V.L. Lepore, *An Experiment to Test the Reality of de Broglie Waves*, subm. for publ.

PRESELECTED QUANTUM OPTICAL CORRELATIONS

MLADEN PAVIČIĆ

Max-Planck-AG Nichtklassische Strahlung,
Humboldt University of Berlin, D-12484 Berlin, Germany,
and Department of Mathematics, GF, University of Zagreb,
*Kačićeva 26, POB 217, HR-41001 Zagreb, Croatia**

Abstract. Two previously discovered effects in the intensity interference: a spin–correlation between formerly unpolarized photons and a spin entanglement *at-a-distance* between photons that nowhere interacted, have been used for a proposal of a new *preselection experiment*. The experiment puts together two photons from two independent singlets and makes them interfere at an asymmetrical beam splitter. A coincidental detection of two photons emerging from different sides of the beam splitter *preselect* their pair–companion photons (which nowhere cross each other's path) into a nonmaximal singlet state. The quantum–mechanical nonlocality thus proves to be essentially a property of selection. This enables a loophole–free experimental disproof of local hidden–variable theories requiring detection efficiency as low as 67% and an exclusion of all nonlocal hidden–variable theories that rely on some kind of a physical entanglement by means of a common medium.

1. Introduction

In the past ten years the fourth order interference of photons has been given a considerable attention. [1–18] It proved to be a powerful tool for testing both local [2–12] and nonlocal [13, 14] hidden variable theories and it also revealed several new features of quantum phenomena. The interference turned out to be a genuine *quantum* phenomenon which does not have a proper classical counterpart as opposed to the interference of the 2nd

---
* Permanent address. Internet: mpavicic@dominis.phy.hr

order. [3] In particular, its visibility reaches 100% (in contradistinction to the classical 50%) and does not depend on the relative intensity of the incoming beams. [4]

The interference served us to recognize a beam splitter as a source of singlet photon states and as a device for spin correlated interferometry even when incoming photons are unpolarized. [8] We elaborate on this properties and introduce general beam splitter spin formalism for different input states in Sec. 2. In Sec. 3 we introduce the Bell inequalities in such a way to enable us to close the remaining loopholes in the Bell theorem. These properties and elaborations open the way for the preselection experiment presented in Sec. 4. The experiment is the first realization of the spin entanglement of independent photons which do not have any common history.

2. Interferometry with Asymmetrical Beam Splitters

In the next subsection we use the second quantization formalism in order to describe the fourth order interference of two incoming photons at a beam splitter without specifying their input state. In the subsequent two subsections we consider three kinds of input states and their outputs.

2.1. FORMALISM OF A BEAM SPLITTER AND ITS OUTPUT

In this subsection the only assumption we make about the input of a beam splitter is that it consists of two input beams falling on the beam splitter. The beams contain altogether two photons but one of them might be empty while the other might contain two photons (see Fig. 1). Let the state of incoming photons be $|\Psi\rangle$. Two particular special states will be specified in the next two subsections. The actions of beam-splitter BS, polarizers P1,P2 and detection D1,D2,D1$^\perp$,D2$^\perp$ are taken into account by the outgoing electric field operators which in the second quantization formalism one obtains in the following way.

The polarization is described by means of two orthogonal scalar field components. The scalar component of the stationary electric field operator in the plane wave interpretations will read: $\hat{E}_j(\mathbf{r}_j, t_j) = \hat{a}(\omega)e^{i\mathbf{k}_j \cdot \mathbf{r}_j - i\omega t_j}$, where ω_j is the frequency of incoming photons, $\mathbf{k}$ is the wave vector ($k = \omega/c$), $\mathbf{r}$ is the radius vector pointing at detector D, t is the time at which a photon is detected, $j = 1, 2$ refer to a particular outcoming photon in question, and the annihilation operators describe actions of detectors so as to act on the states as follows: $\hat{a}_{1x}|1_x\rangle_1 = |0_x\rangle_1$, $\hat{a}^\dagger_{1x}|0_x\rangle_1 = |1_x\rangle_1$, $\hat{a}_{1x}|0_x\rangle_1 = 0$, etc.

One can easily obtain [10] the electric outgoing field operators describing photons which pass through beam splitter BS and polarizers P1,P2 and are

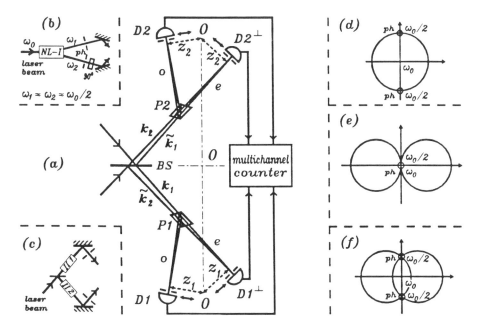

Figure 1. (a) Beam splitter fed by a down–converted photon pair coming out from a non–linear crystal of type I (b) (Sec. 2.2) and by two superposed down–converted photon pairs (c) (Sec. 2.4). Photon wave vectors form cones in space (d-f). For type I the same frequency cones coincide and with pinholes *ph* as in (d) we ideally always get either both photons together and parallelly polarized or none. Type II of the first kind (e) emit perpendicularly polarized photons; *ph* has them together with the pump beam (ω_0) which must be eliminated by a filter. Type II of the second kind (f), which serves as a source for the main experiment, emits a superposition of perpendicularly polarized photons through the shown *ph* because one cannot know which cones they belong to.

detected by detectors D1,D2:

$$\hat{E}_1 = (\hat{a}_{1x}t_x \cos\theta_1 + \hat{a}_{1y}t_y \sin\theta_1) e^{i\mathbf{k}_1 \cdot \mathbf{r}_1 - i\omega(t-t_1-\tau_1)}$$
$$+ i(\hat{a}_{2x}r_x \cos\theta_1 + \hat{a}_{2y}r_y \sin\theta_1) e^{i\tilde{\mathbf{k}}_2 \cdot \mathbf{r}_1 - i\omega(t-t_2-\tau_1)}, \quad (1)$$

$$\hat{E}_2 = (\hat{a}_{2x}t_x \cos\theta_2 + \hat{a}_{2y}t_y \sin\theta_2) e^{i\mathbf{k}_2 \cdot \mathbf{r}_2 - i\omega(t-t_2-\tau_2)}$$
$$+ i(\hat{a}_{1x}r_x \cos\theta_2 + \hat{a}_{1y}r_y \sin\theta_2) e^{i\tilde{\mathbf{k}}_1 \cdot \mathbf{r}_2 - i\omega(t-t_1-\tau_2)}, \quad (2)$$

where t_j is time delay after which photon $j = 1, 2$ reaches BS and τ_j is time delay between BS and Dj.

The electric outgoing field operator describing photons which pass through beam splitter BS and polarizers P1,P2 and are detected by detector D2$^\perp$ reads

$$\hat{E}_2^\perp = (-\hat{a}_{2x}t_x \sin\theta_2 + \hat{a}_{2y}t_y \cos\theta_2) e^{i\mathbf{k}_2\cdot\mathbf{r}_2 - i\omega(t-t_2-\tau_2^\perp)}$$
$$+ i(-\hat{a}_{1x}r_x \sin\theta_2 + \hat{a}_{1y}r_y \cos\theta_2) e^{i\tilde{\mathbf{k}}_1\cdot\mathbf{r}_2 - i\omega(t-t_1-\tau_2^\perp)}, \quad (3)$$

where $\tau_2^\perp$ is time delay between BS and D2$^\perp$.

The probability of joint detection of two ordinary photons (see Fig. 1: denoted by o) coming out from opposite sides of the beam splitter by detectors D1 and D2 is

$$P(\theta_1, \theta_2, \phi) = \langle\Psi|\hat{E}_2^\dagger \hat{E}_1^\dagger \hat{E}_1 \hat{E}_2|\Psi\rangle, \quad (4)$$

where $\phi = (\tilde{\mathbf{k}}_2 - \mathbf{k}_1)\cdot\mathbf{r}_1 + (\tilde{\mathbf{k}}_1 - \mathbf{k}_2)\cdot\mathbf{r}_2 = 2\pi(z_2 - z_1)/L$, where L is the spacing of the interference fringes. ϕ can be changed by moving the detectors transversely to the incident beams as indicated by '↔' in Fig. 1. θ_1 and θ_2 are the angles along which polarizers P1 and P2 are oriented with respect to a chosen fixed direction.

The probability of joint detection of two ordinary photons coming out from the same side of the beam splitter, e.g., the lower one, by the detector D1 (in the experiment we dispense with such a detection, but we do need the expression for the calculation) is

$$P(\theta_1 \times \theta_1) = \langle\Psi|\hat{E}_1^{\dagger 2} \hat{E}_1^2|\Psi\rangle. \quad (5)$$

And the probability of one ordinary and one extraordinary photon being detected by D1 and D2$^\perp$ (as enabled by birefringent polarizer P2) is given by

$$P(\theta_1, \theta_2^\perp, \phi) = \langle\Psi|\hat{E}_2^{\perp\dagger} \hat{E}_1^\dagger \hat{E}_1 \hat{E}_2^\perp|\Psi\rangle. \quad (6)$$

2.2. BEAM SPLITTER FED BY A DOWN-CONVERTED PHOTON PAIR

A laser beam of frequency ω_0 pumps a nonlinear crystal of the first type NL–I producing in it a down–converted pair of signal and idler photons of frequencies ω_1 and ω_2, respectively, which satisfy the energy conservation condition: $\omega_0 = \omega_1 + \omega_2$. By means of appropriately symmetrically (and as far away from the crystal as possible) positioned pinholes we select half–frequency sidebands so as to have $\omega_2 = \omega_1$ [see Fig. 1(d)]. Idler and signal photons coming out from the crystals do not have definite phases [15] with respect to each other and consequently one cannot have a second order interference. Photons emerging from a nonlinear crystal of the first type are parallelly polarized [16] and because we aim at a spin correlated photon state we use a polarization rotator for one of the beams. One achieves the best correlation with 90^0 rotation although for an asymmetrical beam splitter even 0^0 yields a nonvanishing probability of photons emerging from opposite sides of the beam splitter. [8] The state of the incoming polarized photons is thus given by $|\Psi\rangle = |1_x\rangle_1 |1_y\rangle_2$.

Assuming $\phi = 0$ (we can modify ϕ by moving detectors perpendicular to the light path as indicated by '$\leftrightarrow$' in Fig. 1) and introducing $s = t_x t_y$ and $r = \frac{r_x r_y}{t_x t_y}$ the probability of joint detection of two ordinary [see Fig. 1(a)] photons by detectors D1 and D2 reads

$$P(\theta_1, \theta_2) = \langle \hat{E}_2^\dagger \hat{E}_1^\dagger \hat{E}_1 \hat{E}_2 \rangle = \eta^2 s^2 (\cos\theta_1 \sin\theta_2 - r \sin\theta_1 \cos\theta_2)^2 \qquad (7)$$

where η is the (detection) efficiency. This probability tells us that the photons emerge from the beam splitter correlated in polarization whenever they emerge from two different sides of it.

The probability of one ordinary and one extraordinary photon being detected by D1 and D2$^\perp$ (as enabled by birefringent polarizer P2) is given by

$$P(\theta_1, \theta_2^\perp) = \eta^2 s^2 (\cos\theta_1 \cos\theta_2 + r \sin\theta_1 \sin\theta_2)^2 . \qquad (8)$$

On the other hand in case of a symmetrical beam splitter photons emerge from it unpolarized. To convince ourselves let us look at the singles–probability of detecting one photon by D1 and the other going through P2 and through either D2 or D2$^\perp$ without necessarily being detected by either of them [obtained by summing up Eqs. (7) and (8) and dividing them by η for only D1 detection] is

$$P(\theta_1, \infty) = \eta s^2 (\cos^2\theta_1 + r^2 \sin^2\theta_1), \qquad (9)$$

and analogously:

$$P(\infty, \theta_2) = \eta s^2 (\sin^2\theta_2 + r^2 \cos^2\theta_2) . \qquad (10)$$

We see that for $r = 1$, i.e., for a symmetrical beam splitter, Eq. (9) gives $P(\theta_1, \infty) = \eta/4$, i.e., the outgoing photon is unpolarized. In other words whenever photons emerge from different sides of a symmetrical beam splitter they emerge anticorrelated in polarization and unpolarized, i.e., they appear in a *singlet state*. It is therefore to be expected that unpolarized balanced incoming photons also appear correlated in polarization. That this is indeed the case we show in the next subsection.

And for a later use we give here the probability of both photons being detected at the same side of the beam splitter by, e.g., D1 assuming $t_y = t_x$

$$P(\theta_1 \times \theta_1) = \frac{\eta s^2}{2} sin^2(2\theta_1) . \qquad (11)$$

2.3. BEAM SPLITTER FED BY TWO UNPOLARIZED PHOTONS

Sources of unpolarized light can be cascade atom processes, but even better, photons emerging from the opposite sides of a symmetrical beam splitter (as we learned in the previous subsection) or photons emerging from two type II crystals of the second order [17] [from pinholes ph in Fig. 1(f)].

We obtain the general probability for unpolarized light, by calculating the mean value given by Eq. (4) for four input states $|1_x\rangle_1|1_x\rangle_2$, $|1_x\rangle_1|1_y\rangle_2$, $|1_y\rangle_1|1_x\rangle_2$, and $|1_y\rangle_1|1_y\rangle_2$ and adding them together. For $t_y = t_x$ they sum up to the following correlation probability:

$$P(\infty,\infty;\theta_1,\theta_2,\phi) = \frac{\eta^2 s^2}{4}[1 + r^2 - 2r\cos^2(\theta_1 - \theta_2)\cos\phi]. \qquad (12)$$

The correlation is maximal for a symmetric beam splitter for $\phi = 0$ in which case we obtain:

$$P(\infty,\infty,\theta_1,\theta_2) = \frac{1}{8}\sin^2(\theta_2 - \theta_1). \qquad (13)$$

2.4. BEAM SPLITTER FED BY TWO SUPERPOSED DOWN-CONVERTED PHOTON PAIRS

The main disadvantage of feeding a beam splitter by a down-converted photon pair is that 50% of photons emerge from it from the same side [8] [see Eq. (11)]. Namely, detectors still cannot (at least not efficiently enough) tell two photons from one and we therefore cannot *control* photons. Such a *control* of photons would however be possible if photons never appeared together from the same side of a beam splitter. This was achieved by Kwiat et al. [18] Their scheme is shown in Fig. 1(c). Two type II crystals of the first order down-convert two collinear and orthogonally polarized photons of the same average frequencies (half of the pumping beam frequency). The crystals are pumped by a 50:50 split laser beam (filtered out before reaching detectors) whose intensity is accommodated so as to give only one down-conversion at a chosen time-window. Since one cannot tell which crystal a down-converted pair is coming from, the state of the photons incoming at the beam splitter must be described by the following superposition

$$|\Psi\rangle = \frac{1}{\sqrt{2}}(|1_x\rangle_1|1_y\rangle_1 + f|1_x\rangle_2|1_y\rangle_2), \qquad (14)$$

where $0 \leq f \leq 1$ describes attenuation of the lower incoming beam.

The joint D1-D2 probability is given [as follows from Eq. (4)] by

$$P(\theta_1,\theta_2) = \frac{1}{2}[\cos\theta_1\sin\theta_2(t_x r_y + f\, t_y r_x \cos\phi)$$
$$+ \sin\theta_1\cos\theta_2(t_y r_x + f\, t_x r_y \cos\phi)]^2. \qquad (15)$$

The probability of both photons emerging from either the upper or the lower side of BS is for $\phi = 180°$ (+) and $\phi = 0°$ (−) given, respectively, by

$$P(\infty \times \infty) = (t_x t_y \pm f r_x r_y)^2 + (r_x r_y \pm f t_x t_y)^2. \qquad (16)$$

We see that for the crosstalk $t_y = r_x = 0$ and $\phi = 180°$ we obtain:

$$P(\theta_1, \theta_2, 180°) = \eta^2 (\cos\theta_1 \sin\theta_2 - f \sin\theta_1 \cos\theta_2)^2, \qquad (17)$$

which is functionally equivalent to Eq. (7) and has the advantage of giving a perfect *control* of photon as follows from Eq. (16) which yields $P(\infty \times \infty) = 0$. The main disadvantage of this solution, however, is that the crosstalk is very difficult to control in the laboratory [18]. Kwiat et al. wanted to go around this problem by choosing $\phi = 0°$ but this does not help either, as we show in the next section.

3. Bell inequalities

No Bell experiment carried out so far could conclusively disprove hidden-variable theories without additional assumptions. [19] Cascade photon pair experiments have to rely on the *no enhancement* assumption (made by Clauser and Horne [20]: *a subset of a total set of events gives the same statistics as the set itself*) because the directions of photons in the process (which is a three-body decay) are uncontrollable. The fourth order interference, on the other hand, provides directional photon correlation but it was believed that one has to discard 50% of counts which correspond to photons emerging from the same sides of a beam splitter. Therefore Kwiat *et al.* [18] and we [9, 11] devised two different schemes that, on the one hand, do not require discarding of counts and, on the other, reduce the required efficiency to (ideally) 67% which is 16% lower than the previously required efficiency of 83%. [21] We devised the preselection experiment which we present in Sec. 4 but which is, in effect, based on a recognition of a possibility to obtain nonmaximally entangled photons by means of coefficients of transmission and reflection as elaborated in Secs. 2.1 and 2.2. Kwiat *et al.*, on the other hand, used Eberhard's result [12] in order to make a proposal for a loophole–free Bell inequality experiment which we present below and in Sec. 2.4. Eberhard considered nonmaximally entangled photons and showed that for a violation of the Bell inequality only 67% efficiency is required. He obtained his result by employing an asymmetrical form of the Bell inequality for which he found angles of polarizers that violated this Bell inequality only after he set efficiency η to particular values and connected it to the background level.

To compare the two approaches let us first look at the usual Clauser–Horne form and then at Eberhard's form of the Bell inequality: $B \leq 0$. In the Clauser–Horne form B is defined so as to satisfy [see Eqs. (7), (9), and (10)]

$$\eta s^2 B \equiv P(\theta_1, \theta_2) - P(\theta_1, \theta'_2) + P(\theta'_1, \theta'_2) + P(\theta'_1, \theta_2) - P(\theta'_1) - P(\theta_2), \quad (18)$$

where $P(\theta'_1) = P(\theta'_1, \infty)$ and $P(\theta_2) = P(\infty, \theta_2)$, as given by Eqs. (9) and (10). B of the Eberhard's form is, in effect, defined so as to satisfy [see Eqs. (17) and the equations corresponding to Eqs. (8) and (9)—with $s = 1$ and $r = f$]

$$\eta s^2 B \equiv P(\theta_1, \theta_2) - P(\theta'_1, \theta'_2) - P(\theta_1, \theta'^{\perp}_2) - P(\theta'^{\perp}_1, \theta_2)$$
$$-(1 - \eta)[P(\theta_1) + P(\theta_2)], \quad (19)$$

where $(1 - \eta)P(\theta_1)$ is the probability of one photon being detected by D1 and the other reaching either D2 or D2$^{\perp}$ but not being detected by them due to their inefficiency. In other words, to be able to use either of the above forms we have to have a perfect *control* of all photons at BS. As we stressed in Sec. 2.4, Kwiat *et al*. do achieve this *control* but at the expense of the crosstalk. They also claim that the choice $\phi = 0°$ would preserve the *control* and at the same time dispense with the crosstalk. Unfortunately, this does not work what can be easily seen from Eq. (16). To obtain $P(\infty \times \infty) = 0$ one has to satisfy $r_x r_y = f t_x t_y$ and $t_x t_y = f r_x r_y$ what is however clearly impossible for $f < 1$. Thus, contrary to the claims of Kwiat *et al*. [18], the only way to make use of $f < 1$ for either $\phi = 0°$ or $\phi = 180°$ is the crosstalk $t_y = r_x = 0$ and this is apparently difficult to achive within a measurement. [18] It seems that the set–up is ideal for a loophole–free experiment with maximal non–product states, i.e., with $f = 1$ and $\eta > 83\%$ but that attenuation ($f < 1$) is not the best candidate for Bell's event–ready preselector [20]. We therefore propose another "event–ready set–up" which dispenses with variable f, enables a full *control* of photons, and offers a fundamental insight into the whole issue.

But before we dwell on the experiment itself let us first compare the two afore–introduced forms of the Bell inequality in two ways. First, we obtain $Max[B](r, \eta)$ surfaces (by a computer optimization of angles) for both forms (18) and (19). As we can see in Fig. 2, there is no difference between them although the maxima were achieved for different angles. (The differences are 10^{-5} in average, for 100 iterations used in numerical calculations of maxima.) The values above the $B = 0$ plane mean violations of the Bell inequality. For $r = 1$ we obtain $Max[B] = 0$ for $\eta = 0.828427$ in accordance with the result of Garg and Mermin.[21] For $r \to 0$ we obtain a violation of the Bell inequality for any efficiency greater then 66.75%. Secondly, we calculate η by setting $B = 0$, first, from Eq. (18) and then from Eq. (19). Again, [see Fig. 3] there is no difference between the two

PRESELECTED QUANTUM OPTICAL CORRELATIONS

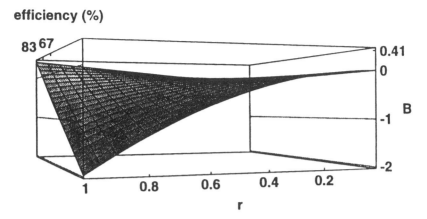

Figure 2. The surface $Max[B]$ [Eqs. (18) and (19)] for the optimal angles of the polarizers. Values above the $B = 0$ plane violate the Bell inequality $B \leq 0$.

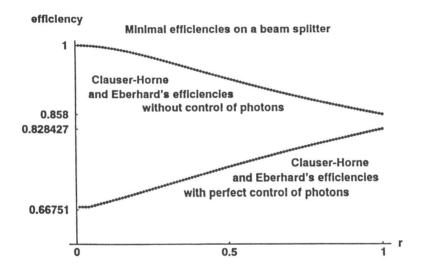

Figure 3. Lower plot: η's as obtained for $B = 0$ from Eqs. (18) and (19). Upper plot: η's as obtained for $B = r[\sin^2(2\theta'_1) + \sin^2(2\theta_2)]/2$ from Eqs. (7), (9), (10), and (18), and for $B = r[\sin^2(2\theta_1) + \sin^2(2\theta_2)]/2$ from Eqs. (7), (8), (9), (10), and (19).

forms. We can also see that there is no need to optimize the angles only *after* values for η have been fixed, contrary to the claim from Ref. [12]. The efficiencies for uncontrolled photons are shown as the upper curve in Fig. 3. Once again, there is no difference between the forms. On the other hand, we see that uncontrolled photons, i.e., the ones that may emerge from the same sides of BS as well, violate the Bell inequality—starting with 85.8%

efficiency—in opposition to the widespread belief that "unless the detector can differentiate one photon from two... no indisputable test of Bell's inequalities is possible." [18]

4. Preselection Experiment

A schematic representation of the experiment is shown in Fig. 4. Two independent sources $S1$ and $S2$ simultaneously emit two independent entangled pairs. Left photons from each pair fly towards detectors D1' or D1'$^\perp$ and D2' or D2'$^\perp$. Right photons from each pair interfere at an asymmetrical beam splitter which acts as an *event-ready* preselector and as a result the so preselected left photons, under particular conditions elaborated below, appear to be in a nonmaximal singlet state although the latter photons are completely independent and nowhere cross each other's path. There are several possible sources for such an entangled state of photons.

Atoms exhibiting cascade emission. The atoms of the two sources could be pumped to an upper level by two independent lasers. This level would decay by emitting two photons correlated in polarization in a triplet–like state. The independence of the two sources can be assured by slight differences in central frequency and drift of the two pump lasers. The sources have been elaborated by Pavičić and Summhammer and an experiment with such sources was estimated to be very difficult to carry out. [7]

Symmetrical beam splitters fed by two nonlinear crystals of type I. The experiment with such sources relies on quadruple recording obtained in the following way: whenever exactly two of the *preselection detectors* D1–D2$^\perp$ fire in coincidence (see Fig. 4) a gate for counters D1'–D2'$^\perp$ opens. In case only one or none of the so preselected D1'–D2'$^\perp$ detectors fires we discard the records. In case exactly two of four D1'–D2'$^\perp$ detectors fire, the corresponding counts contribute to our statistics. The sources have been elaborated by Pavičić [9] and an experiment with such sources was found to be rather demanding. Besides, the procedure of discarding data is a *no enhancement* assumption and the experiment cannot be considered a *loophole-free* one,

Symmetrical beam splitters in a crosstalk fed by two "superposed" crystals of type II. The experiment with such sources, emitting photons in a singlet like state, relies only on the *preselection detectors* D1–D2$^\perp$. When they fire in coincidence (see Fig. 4) a gate for counters D1'–D2'$^\perp$ opens. In case only one of the so preselected D1'–D2'$^\perp$ detectors fires we do *not* discard the records but use them to form frequencies approximating one–photon probabilities in the Bell inequalities formed by Eq. (18) and Eq. (19). The whole set–up

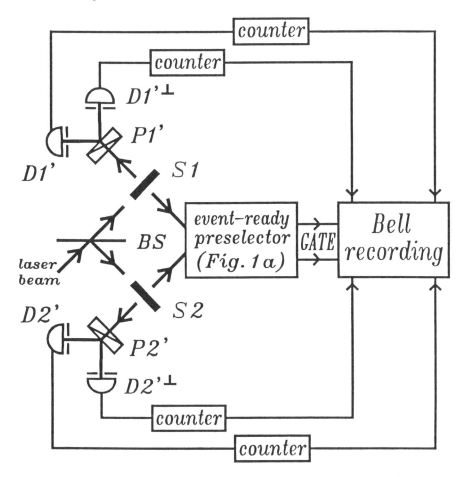

Figure 4. Proposed experiment. As the event–ready preselector, serves a beam splitter with detectors D1, D1$^\perp$, D2 and D2$^\perp$ as shown in Fig. 1. $S1$ and $S2$ are sources (type II crystals as explained in the text) emitting singlet–like photon pairs. As birefringent polarizers P1' and P2' may serve Wollaston prisms (which at the same time filter out the uv pumping beam).

would apparently be difficult to tune in because of the crosstalk [18] but this nevertheless seem to be a feasible loophole–free Bell experiment.

Crystals of type II of the second order. These crystals emit perpendicularly polarized photons into two intersecting cones as shown in Fig. 1(f). If we position pinholes ph at the intersections we shall obtain superposition of perpendicularly polarized photons, i.e., photons in a singlet–like state because one cannot know which cones they belong to. The sources seem to enable a rather feasible loophole–free Bell experiment. [17]

In what follows we shall adopt the latter two kinds of sources. Thus the state of photons immediately after leaving the sources $S1$ and $S2$ is given by a tensor product of two singlet-like states:

$$|\Psi\rangle = \frac{1}{\sqrt{2}}(|1_x\rangle_{1'}|1_y\rangle_1 - |1_y\rangle_{1'}|1_x\rangle_1) \otimes \frac{1}{\sqrt{2}}(|1_x\rangle_{2'}|1_y\rangle_2 - |1_y\rangle_{2'}|1_x\rangle_2). \quad (20)$$

The probability of detecting all four photons by detectors D1, D2, D1', and D2' is thus

$$P(\theta_{1'},\theta_{2'},\theta_1,\theta_2) = \langle\Psi|\hat{E}^\dagger_{2'}\hat{E}^\dagger_{1'}\hat{E}^\dagger_2\hat{E}^\dagger_1\hat{E}_1\hat{E}_2\hat{E}_{1'}\hat{E}_{2'}|\Psi\rangle$$
$$= \frac{1}{4}(A^2 + B^2 - 2AB\cos\phi), \quad (21)$$

where $\hat{E}_1, \hat{E}_2$, and ϕ are as given above, $\hat{E}_{j'} = (\hat{a}_{j'x}\cos\theta_{j'} + \hat{a}_{j'y}\sin\theta_{j'})\exp(-i\omega'_j t_{j'})$; $j = 1,2$; $A = Q(t)_{1'1}Q(t)_{2'2}$ and $B = Q(r)_{1'2}Q(r)_{2'1}$; here $Q(q)_{ij} = q_x\sin\theta_i\cos\theta_j - q_y\cos\theta_i\sin\theta_j$.

For $\phi = 0°$, $\theta_1 = 90°$, and $\theta_2 = 0°$ Eq. (21) yields (non)maximal singlet-like probability $P(\theta_{1'},\theta_{2'})$ given by Eq. (7) which permits a perfect control of photons 1' and 2'. This means that D1 and D2—while detecting coincidences—act as event-ready preselectors and with the help of a gate (see Fig. 4) we can extract those 1' and 2' photons that are in a non-maximal singlet state, take them miles away and carry out a loophole-free Bell experiment by means of P1', D1', D1'$^\perp$, P2', D2', and D2'$^\perp$ with only 67% efficiency in the limit $r \to 0$.

References

1. Paul, H., *Rev. Mod. Phys.* **57**, 209 (1986).
2. Ou, Z. Y., *Phys. Rev. A* **37**, 1607 (1988).
3. Ghosh, R. and Mandel, L., *Phys. Rev. Lett.* **59**, 1903 (1987).
4. Ou, Z. Y. and Mandel, L., *Phys. Rev. Lett.* **62**, 2941 (1989).
5. Campos, R. A., Saleh, B. E. A., and Teich, M. A., *Phys. Rev. A* **42**, 4127 (1990).
6. Żukowski, M., Zeilinger, A., Horne, M. A., and Ekert A. K., *Phys. Rev. Lett.* **71**, 4287 (1993).
7. Pavičić, M. and Summhammer, J., *Phys. Rev. Lett.* **73**, 3191 (1994).
8. Pavičić, M., *Phys. Rev. A* **50**, 3486 (1994).
9. Pavičić, M., *J. Opt. Soc. Am. B* **12**, 821 (1995).
10. Pavičić, M., *Int. J. Theor. Phys.*, **34**, 1653 (1995).
11. Pavičić, M., *Phys. Lett. A*, [to be published] (1995).
12. Eberhard, P. H., *Phys. Rev. A*, **47**, R747 (1993),
13. Wang, L. J., Zou, X. Y., and Mandel, L., *Phys. Rev. Lett.* **66**, 1111 (1991).
14. Zou, X. Y. et al., *Phys. Rev. Lett.* **68**, 3667 (1992).
15. C. K. Hong, Z. Y. Ou and L. Mandel, Phys. Rev. Lett. **59**, 2044 (1987).
16. Z. Y. Ou and L. Mandel, Phys. Rev. Lett. **61**, 50 (1988);
17. P. G. Kwiat *et al.*, Phys. Rev. Lett. [to be published] (1995);
18. P. G. Kwiat *et al.*, Phys. Rev. A, **49**, 3209 (1994).
19. E. Santos, Phys. Rev. Lett. **66**, 1388 (1991); **66**, 3227 (1991); **68**, 2702 (1992).
20. J. F. Clauser and M. A. Horne, Phys. Rev. D, **10**, 526 (1974).
21. A. Garg and N. D. Mermin, Phys. Rev. D **35**, 3831 (1987).

APPARENT CONTRADICTION IN EPR CORRELATIONS

LUIZ CARLOS RYFF[1]
University of Maryland Baltimore County,
Baltimore, MD 21228, U.S.A.

Different versions of an experiment which allows us to prove a Bell's theorem for two particles without using an inequality are discussed. The experiment could be used to test local realism with the introduction of additional assumptions based on direct experimental observation.

The first version of the experiment I wish to discuss is represented in Fig.1. It is a variant of an experiment proposed by Klyshko [1] (we could also have considered a variant of an experiment performed by Mandel and co-workers on induced coherence [2]). Two nonlinear crystals are coherently pumped by a laser, producing pairs of photons via spontaneous parametric down conversion. H, H_1,..., are beam splitters, and have real amplitude transmissivities T, T_1,..., and real amplitude reflectivities R, R_1,...; M are mirrors; and ϕ_i (i=1,2,3) are phase shifters. In the original experiment H_3 is a mirror, ϕ_3 and H_4 are removed, and H_1 and H_2 are 50%:50% beam splitters. I will consider a situation in which the probability of having both crystals emitting pairs of photons simultaneously is very small. For example, if H_1 is removed we can have detections at either site 1 or site 1', but never at both sites at the same time. The coincident detections at sites 1 and 2, 1 and 2', 1' and 2, and 1' and 2', respectively, are assumed to be events separated by a space like interval. Therefore, these events are not causally connected, and their sequence in time depends on the Lorentzian frame one uses to describe the experiment. Whenever it is impossible to know which crystal emitted the pair of photons, two photon interference is displayed. I will consider four different situations: (A) H_1 and H_4 are removed; (B) H_1 is in place and H_4 is removed; (C) H_1 is removed and H_4 is in place; (D) H_1 and H_4 are in place. We see that the probability of coincident detections occuring at 1' and 2 in situation (A), $P_A(1',2)=0$, since in this case only photons emitted by crystal I(II) can be detected at 1' (2). We can determine

---

[1]Permanent address: Universidade Federal do Rio de Janeiro, Instituto de Física, Caixa Postal 68528, 21945-970 Rio de Janeiro, RJ, Brazil

the expression for the probability amplitude of coincident detections occuring at sites 1 and 2' in the general case and then treat the different situations as different limiting cases. It is easy to see that [3,4]

$$A(1,2') = T\chi(e^{i\phi_1}iR_1)(T_2e^{i\phi_3}T_4) + iR\chi(T_1)[e^{i\phi_2}(iR_3iR_2e^{i\phi_3}T_4 + T_3iR_4)] \;, \tag{1}$$

where χ is the probability amplitude of an incident pumping photon being converted into a pair of photons via a parametric down conversion process. We can choose $T=R$ and renormalize the probabilities ($T \to 1$, $\chi \to 1$). In situation (B), $T_4=1$ and $R_4=0$. Thus, using (1) and choosing $R_1T_2 = T_1R_2R_3$, we obtain

$$A_B(1,2') = iT_1R_2R_3(e^{i\phi_1} - e^{i\phi_2})e^{i\phi_3} \;, \tag{2}$$

where $A_B(1,2')$ is the probability amplitude of coincident detections occuring at sites 1 and 2' in situation (B). From (2) we see that $\phi_1 = \phi_2 \to A_B(1,2') = 0$. That is, whenever detection at 1 occurs *and* no detection occurs at 3, we can infer that the other photon of the pair has been emitted by crystal II. Similarly, in situation (C), $T_1=1$ and $R_1=0$. Thus, using (1) and choosing $R_2R_3T_4 = T_3R_4$, we obtain

$$A_C(1,2') = -e^{i\phi_2}R_3R_2T_4(ie^{i\phi_3} + 1) \;, \tag{3}$$

From (3) we see that $\exp(i\phi_3) = i \to A_C(1,2') = 0$. That is, whenever detection at site 2' occurs, we can infer that the other photon of the pair has been emitted by crystal I. In situation (D), using (1) and the two previous conditions, $R_1T_2 = T_1R_2R_3$ and $R_2R_3T_4 = T_3R_4$, we obtain

$$A_D(1,2') = T_1R_2R_3T_4[ie^{i\phi_1}e^{i\phi_3} - e^{i\phi_2}(ie^{i\phi_3} + 1)] \;. \tag{4}$$

From (4) we see that the probability of coincident detections, $P_D(1,2') = |A_D(1,2')|^2$, will in general depend on the phase shifts. However, whenever it is possible to know which crystal emitted the pair of photons this is not so. For example, if $\exp(i\phi_3) = i$, the probability does not depend on ϕ_1. In this case, whenever a photon is detected at site 2' the probability of the other photon of the pair being detected at 1 (1') does not depend on ϕ_1, that is no interference is displayed. The reason as seen from (3), is that in this case we can infer that the photon detected at 1 (1') has been emitted by crystal I. Similarly, if $\phi_1 = \phi_2$, the probability does not depend on ϕ_3. In this case, whenever a photon is detected at site 1 the probability of the other photon of the pair being detected at 2' (2) does not depend on ϕ_3. In other words, no interference is displayed. The reason, as seen from (2), is that in this case, if no detection occurs at 3, it is possible to infer that the photon detected at 2' (2) has been emitted by crystal II. As we will now see, this leads to an apparent contradiction, from the usual point of view, and to a real contradiction, from the local realistic one.

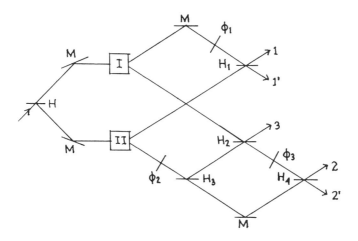

Figure 1

Let us imagine an infinite number of experiments like the one represented in Fig.1, with all possible different combinations of phase shifts and all possible outcomes. We can then select only those experiments in which $\exp(i\phi_3) = i$ and in which detection at 2' occurs (set S_2). In this case the photon detected at 1 (1') has apparently been emitted by crystal I. We can also select only those experiments in which $\phi_1 = \phi_2$ and in which detection at 1 occurs and no detection at 3 occurs (set S_1). In this case the photon detected at 2' (2) has apparently been emitted by crystal II. However, the experiments in which $\exp(i\phi_3) = i$ *and* $\phi_1 = \phi_2$ and in which coincident detections occur at sites 2' and 1 belong simultaneously to both sets. Therefore, apparently, each photon of the pair has been emitted by a different crystal.

To try to understand this result from the usual point of view we have to keep in mind that before the photons have been detected we only have potentialities, so to speak. The pair of photons is shared by the fields generated by the two crystals. It would be a meaningless question to ask which crystal "really" emitted the pair. This situation is similar to the situation in which light from two different lasers interfere [5]. However, in the experiment using two different lasers one single photon has apparently been emitted by both lasers at the same time, whilst in the case we are discussing each photon of the pair has apparently been emitted by a different crystal. On the other hand, from the point of view of local realism [6], we are facing a contradiction.

To see this we may consider a situation in which all the beam splitters are removed in Fig.1. In this case we can only have coincident detections at either sites 1 and 2, or sites 1' and 2'. In the first (second) case the pair has been emitted by crystal II (I). According to local realism, if the beam splitters had been in place things could not be different. But as we have just seen, when $\exp(i\phi_3)=i$ and $\phi_1=\phi_2$ and coincident detections at 1 and 2' occur we are allowed to conclude that each photon of the pair has been emitted by a different crystal, which leads to a conceptual contradiction. As we will now see, this leads to contradictory predictions too.

Assuming local realism, a Bell's theorem for two particles without using an inequality [7,8] can be proved as follows. Let us consider the situation in which a photon is detected at site 1 and no photon is detected at site 3. From (2) ($\phi_1=\phi_2$) we see that if H_4 had not been in place the other photon of the pair would have been detected at site 2. Hence, if H_4 and H_1 had not been in place, and a photon would thus have been detected at site 2, the other photon of the pair would have been detected at site 1. Now, if only H_1 had not been in place, and a photon would thus have been detected at site 1, we see from (3) [$\exp(i\phi_3=i)$] that the other photon would have been detected at site 2. But H_1 being in place or not should not interfere with the detection at 2 (2'). Therefore, whenever detection at 1 and no detection at 3 occurs, we are able to conclude that the other photon of the pair can only be detected at site 2. However, according to (4) we can have coincident detections at sites 1 and 2', which leads to a contradiction.

This reasoning can be cast in a more mathematical form. Let us assign the value i (I) for detections that occur at sites 1 and 2' (1' and 2). Thus, assuming there can be hidden variables states (HVS) λ of the photon pair which mimic the quantum mechanical results, we see that, when no detection at 3 occurs, we can only have: (B)

$a_P(\lambda)b_R(\lambda)=i,1$; (A) $a_R(\lambda)b_R(\lambda)=i,-1$; (C) $a_R(\lambda)b_P(\lambda)=i,1$, where $a_P(\lambda)$ [$b_R(\lambda)$] represents the result of a detection measurement performed at 1, 1' (2, 2') when H_1 (H_4) is in place (removed), and so on. Assuming locality, that is, that $a_R(\lambda)$ is the same in situations (A) and (C), for example, we see that $a_P(\lambda)=i\xrightarrow{B} b_R(\lambda)=1\xrightarrow{A} a_R(\lambda)=i\xrightarrow{C} b_P(\lambda)=1$. That is, $P_D(1,2')=0$ (local realism), in disagreement with the quantum mechanical prediction given by (4).

This apparent (real) contradiction, from the usual (local realistic) point of view, is not restricted to the situation in which two crystals are coherently pumped. Actually, it is an essential feature of experiments on EPR correlations. For example, let us consider the experiment represented in Fig.2. It is a variant of an experiment proposed by Horne, Shimony, and Zeilinger [9] (a variant of Franson's experiment [10] could also have been considered). Now we have only one crystal, but instead of only two pinholes to select the pairs of photons we have four. There are only two possibilities. Either one photon of the pair follows path A and the other follows path C, or one follows path D and the other follows path B. This can easily be verified by removing all the beam splitters. When the beam splitters are in place it may not be possible to know which of these two possibilities took place. As a consequence, they are simultaneously actualized, so to speak. On the other hand, as in the experiment represented in Fig.1 (in the original experiment H_2 and H_4 are removed, and H_1 and H_3 are 50%:50% beam splitters), there can be a situation in which there is an apparent conflict between the wave like and the nonlocal properties of photons. In this case, a third possibility, apparently in contradiction with the two previous ones, is actualized. For example, ϕ_3 can be chosen so that a photon following path A can never be detected at site 2 (this could be verified by blocking path D), and ϕ_1 and ϕ_2 can be chosen so that whenever a photon is detected at site 1 the other photon of the pair, in case it impinges on H_3, can only follow direction 3 (this could be verified by blocking the path with ϕ_3). In this case, whenever a photon is detected at site 2 we are able to infer that the other photon of the pair follows path B, and thus can be reflected at H_1 and be detected at 1. Whenever a photon is detected at site 1 and no detection at 3 occurs we are able to infer that the other photon of the pair follows the path with ϕ_3, and thus can be reflected at H_4 and be detected at 2. Therefore, whenever coincident detections occur at 1 and 2, *apparently* one of the photons of the pair follows path A and the other follows path B. However, no violation of the phase matching condition (momentum conservation) can be directly observed. Actually, our argument is based on this very condition. As in the first experiment discussed, this seemingly paradoxical result can be used to prove a Bell's theorem for two particles without using an inequality.

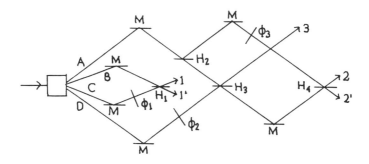

Figure 2

For the sake of completeness let us discuss the experiment represented in Fig.3, in which instead of being emitted with correlated directions, the photons of the pair are emitted with correlated polarizations. It is a variant of an experiment proposed by Zukowski and Pikacz [11] (in the original experiment H_3 and H_4 are removed, and H_1 and H_2 are 50%:50% beam splitters). The polarization entangled pair of photons could also be produced via a parametric down conversion process [12]. If all the beam splitters are removed, coincident detections can only be observed at sites 1' and 2 and at sites 3 and 2'. That is, the photons of the pair are either both transmitted or both reflected at the two channel polarizers. The half-wave plates ($\lambda/2$) change polarization parallel (perpendicular) to a into polarization perpendicular (parallel) to a, so that the transmitted and reflected beams can interfere when H_1 and H_2 are in place. When the beam splitters, excepting H_2, are in place ϕ_3 can be chosen so that whenever detection at site 1' occurs, v_2 can only be detected at site 2', that is, it follows the transmission channel. When the beam splitters, excepting H_4, are in place ϕ_1 and ϕ_2 can be chosen so that whenever detection at site 2' occurs and no detection occurs at site 3, v_1 can only be detected at site 1, that is, it follows the reflection channel. When all polarizers are in place we can have coincident detections at sites 1' and 2'. In this case, apparently one photon of the pair follows the transmission channel whilst the other follows the reflection channel, in disagreement with what was to be expected from the kind of polarization entanglement they are supposed to have. This version of our experiment can also be used to prove a Bell's theorem for two particles without using an inequality.

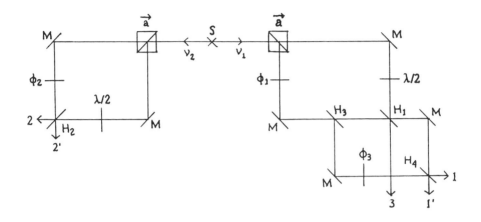

Figure 3

Let us now return to the experiment represented in Fig.1 to see how the introduction of two assumptions which are based on direct experimental observation allow us to obtain an inequality that could be used to test local realism in the case of real (i.e., non-ideal) situations. Let us initially consider situation (C) and select only those events in which detection at site 2' occurs. In this case, whenever a coincident detection at 1 occurs we know that the pair of photons has been emitted by crystal II. I will then assume that:

(i) If H_1 had been in place [situation (C) → situation (D)] the number of photons *emitted by crystal II* that would be coincidentally detected at 1 could not be greater than the number of photons coincidentally detected at 1 when H_1 is removed. (I will return to this point later.)

Therefore, the number of coincident detections at 1 and 2' in situation (D) *that correspond to the possibility in which the photon pair is emitted by crystal II* cannot be greater than $N_C(1,2')$, the number of coincident detections at 1 and 2' in situation (C).

Let us now consider situation (B) and select only those events in which detection at 1 and no detection at 3 occur. In this case, only the coincident detections at 1 and 2' can correspond to the possibility in which the photon pair is emitted by crystal I. I will then assume that:

(i) If H_4 had been in place [situation (B) $\longrightarrow$ situation (D)] the number of photons *emitted by crystal I* that would be coincidentally detected at 2' could not be greater than the number of photons coincidentally detected at 2' when H_4 is removed.

Therefore, the number of coincident detections at 1 and 2' in situation (D) *that correspond to the possibility in which the photon pair is emitted by crystal I* cannot be greater than $N_B(1,2')$, the number of coincident detections at 1 and 2' in situation (B). I will also assume that:

(ii) Coincident detections can only occur when the photons of the pair either (a) both are emitted by crystal I, or (b) both are emitted by crystal II.

Hence, from (i) and (ii) it follows that $N_D(1,2') \leq N_C(1,2')+N_B(1,2')$, or, in terms of probabilities,

$$P_D(1,2') \leq P_C(1,2')+P_B(1,2') \ . \ \text{(local realism)} \qquad (5)$$

In an ideal experiment we can choose the phase shifts so that $P_C(1,2')=P_B(1,2')=0$. In a real experiment these probabilities can be made quite small. Our reasoning can easily be adapted to the experiments represented in figures 2 and 3.

Let us examine assumptions (i) more closely. It was assumed, when changing from situation (C) [(B)] to situation (D), that the number of detections at site 1 (2') generated by photons emitted by crystal II (I) could not be increased by placing a beam splitter H_1 (H_4) in front of the detectors. Although this may appear to be a nonehancement assumption [6], this can be directly verified. For example, by blocking the path with ϕ_1 (ϕ_2) in situation (D). Nevertheless, it might actually still be argued, for example, that when H_1 (H_4) is in the position represented in Fig.1, in which case photons from two different directions impinge on it, its properties are modified, in such a way that photons coming from crystal II (I) become more "detectable" after impinging on H_1 (H_4) and being transmitted [13]. We also have to assume that photons coming from crystal II (I) become less detectable after impinging on H_1 (H_4) and being reflected, in such a way that the relations $P_D(1,2')+P_D(1',2')=P_C(1,2')+P_C(1',2')$ and $P_D(1,2)+P_D(1,2')=P_B(1,2)+P_B(1,2')$ remain valid. However, this sounds to me as a much too contrived supposition. Probably, some contrived supposition could also be introduced to argue that assumption (ii) might not be true. For example, one might assume that some times the photons are neither emitted by crystal I nor crystal II. Both crystals emit waves that are incapable of producing a detector click. However, they can interfere constructively at the beam splitters, then becoming capable of producing a detector click. In this case we could perform the following experiment to test this supposition. First we block the path from crystal I to H_1. The total number of detections at sites 1 and 1' would be $N_{II}=N_{II}(1)+N_{II}(1')$. We would then freed this path and block the path from crystal II to H_1. The total number of detections at sites 1 and 1' would be $N_I=N_I(1)+N_I(1')$. When both paths are freed the total number of detections $N= N_I+N_{II}$. On the other hand, according to the contrived supposition, we should have $N>N_I+N_{II}$.

To have only a rough estimation of the expected disagreement between the local realistic and the quantum mechanical predictions in a real experiment we can perform the following simple calculation. We can write $P_C(1,2')=(1/2)(1-R_3^2T_2^2)x$, where 1/2 is the probability of a photon following channel 1 (since we are only considering coincident detections we do not need to worry about the detectors efficiencies), the term inside parentheses is the probability of the other photon of the pair *not* following channel 3, and x is the probability of having a photon which impinges on H_4 following the "wrong" channel 2'. We can also write $P_B(1,2')=(1/2)(1-T_1^2T_3^2)x$, where 1/2 is the probability of a photon following channel 1, the term inside parentheses is the probability of the other photon of the pair *not* following channel 2, and I am assuming that the probability of having a photon which impinges on H_2 following the wrong channel is also x. Using (4) [with $\exp(i\phi_3)=i$] we see that (5) will be violated if $T_1^2R_2^2R_3^2T_4^2>(1/2)(2-R_3^2T_2^2-T_1^2T_3^2)x$. Choosing $T_2=R_2=T_3=R_3=(1/2)^{1/2}$, $T_1=T_4=(2/3)^{1/2}$, and $R_1=R_4=(1/3)^{1/2}$ (we must have $R_1T_2=T_1R_2R_3$ and $R_2R_3T_4=T_3R_4$) we see that we must have $x<0.157$. This sounds reasonable, since usually the phase shift in a Mach-Zehnder interferometer can be chosen so that 10% of the impinging photons follow the wrong channel.

References

[1] Klyshko, D.N.: Phys. Lett. A **132** (1988), 299.
[2] Zou, X.Y., Wang, L.J., and Mandel, L.: Phys.Rev.Lett. **67** (1991), 318.
[3] Greenberger, D.M., Horne, M.A., and Zeilinger, A.: Phys. Today **46** (1993), 22.
[4] Ryff, L.C.B.: Phys. Rev. A **51** (1995), 79.
[5] Pfleegor, R.L. and Mandel, L.: Phys. Rev. **159** (1967), 1084; J. Opt. Soc. Am. **58** (1968), 946.
[6] Clauser, J.F. and Shimony, A.: Rep. Prog. Phys. **41** (1978), 1881.
[7] Greenberger, D.M., Horne, M.A., Shimony, A, and Zeilinger, A.: Am. J. Phys. **58** (1990), 1131.
[8] Hardy, L.: Phys. Rev. Lett. **71** (1993), 1665.
[9] Horne, M.A., Shimony, A., and Zeilinger, A.: Phys. Rev. Lett. **62** (1989), 2209.
[10] Franson, J.D.: Phys. Rev. Lett. **62** (1989), 2205.
[11] Zukowski, M. and Pikacz, J.: Phys. Lett. A **127** (1988), 1.
[12] Alley, C.O. and Shih, Y.H.: Phys. Rev. Lett. **61** (1988), 2921.
[13] Ryff, L.C.B.: in *Advances in Fundamental Physics*, M. Barone and F. Selleri (eds.), Hadronic Press, Palm Harbor (1995), 369.

THE WAVE - PARTICLE DUALITY

EFTICHIOS BITSAKIS
Department of Philosophy, University of Ioannina
Department of Physics, University of Athens

Is light and more generally matter, constituted of waves, of particles, or of waves-particles? The debate concerning the nature of light began in the seventeenth century - at the moment of the publication of two books which laid the foundations of Optics: The book of Chr. Huygens, *Treatise of Light,* and the *Opticks* of I. Newton.

During the seventeenth century physicists studied mainly reflection, refraction, and double refraction. Newton, also studied the spectral decomposition of light and elaborated a theory of colours. Finally, phenomena of diffraction were already known at that epoch. On the basis of the existing data, two opposite conceptions or "paradigms" were elaborated concerning the nature of light: according to Huygens, light consists of waves. For Newton, on the contrary, light is constituted of material corpuscles. The history of Optics is, from this point of view, the history of the struggle between these two opposite paradigms. However, two centuries later Louis de Broglie tried to overcome the antithesis in his famous *Thèse* (1924): to achieve a unity of the opposites, by postulating the dual, wave-particle nature not only of the light, but of the totality of the constituents of matter. Yet, as is well known, the question: particles, waves, or wave-particles, has not received till now a definite answer. A critical presentation of the struggle of the above "paradigms" is not, therefore, of an exclusively historical interest.

Waves or particles?

The *Treatise of Light* of Christian Huygens has been published in 1690. Reflection, refraction and double refraction, "the strange refraction of the Iceland Crystal," were studied at the end of the seventeenth century. Huygens gave a geometrical form to his demonstrations, in conformity with the mechanistic world view prevailing at that time. For him, as for Descartes or Newton, "true Philosophy" conceives the causes of all natural phenomena in terms of mechanical motions. Otherwise one must "renounce all hopes of ever comprehending anything in Physics."[1]

It is inconceivable, Huygens maintained, to doubt that light consists of the motion of some sort of matter. Now, Huygens continues, "if one examines what their matter may be in which the movement coming from the luminous body is propagated, which I call Ethereal matter, one will see that it is not the same that serves for the propagation of

Sound." The ether, Huygens asserts, is a substance "as nearly approaching to perfect hardness and possessing a springiness as prompt as we choose." Finally, light spreads as sound does, by spherical surfaces and waves: "For I call them waves from their resemblance to those which are seen to be formed in water when a stone is thrown into it, and which presents a successive spreading as circles, though these arise from another cause, and are only in a flat surface."[2]

The theory of Huygens was a mechanistic one. However, considering light not as a substance, but as form of movement, Huygens was oblidged to introduce an hypothetical medium: the ether. We know the problems that this "enfant terrible" created to the physicists of the eighteenth and the nineteenth century. Newton, on the contrary, being a *shame faced* partisan of the atomic hypothesis, maintained in his *Opticks* (1704) that light consists of material particles: "Are not all Hypotheses erroneous, in which Light is supposed to consist of Pression, or Motion, propagated through a fluid Medium?" (Question 28). Huygens, as Newton notes, had not succeeded in explaining the "unusual" refraction of the Iceland Crystal by Pression or Motion. For him it is impossible to explain the double refraction "if light be nothing else than Pression; or motion propagated though Ether."[3] Consequently, Newton put the question: "Are not the Rays of Light very small bodies emitted from shining substances? For such bodies will pass through uniform Mediums in right Lines without bending into Shadow, which is the Nature of the Rays of the Light (Qu. 29).[3] In support of his theory, Newton invokes "the Authority of those the oldest and most celebrated Philophers of Greece and Phoenicia who made a Vacuum, and Atoms, and the Gravity of Atoms, the first Principle of their Philosophy."[4]

Faithful to his realism, Newton tried to explain the known in his time phenomena on the basis of an exclusively corpuscular conception of light. By this way he arrived to the following question: "Query 1: Do not Bodies act upon Light at a distance and their action bends its Rays; and it is not this action (cateris paribus) strongest in the least distance?"[5] We know today that General Relativity verified this anticipation, in a different epistemic frame.

The opposition between waves and particles is irreducible in the frame of formal logic. It seems equally irreducible in the case of the theories of light. However, as Louis de Broglie remarks, it would be impossible for the rigorous spirit of Newton not to notice that the phenomenon of coloration of thin slades that himself discovered, implied the existence of an element of periodicity, impossible to be explained by the single image of the corpuscles. In his "Theory of Access," as de Broglie notes, Newton arrived to define a "longueur d' access" which has a close similarity to what Fresnel named later "wave length" of a monochromatic ray of light.[6]

The known phenomena in the seventeenth century advocated in favour of the corpuscular theory. The discovery, on the contrary, of the inteference by Young, the polarisation by Malus etc, during the nineteenth century, were in favour of the wave theory. Auguste Fresnel (1788-1827) with his wave theory explained not only the rectilinear propagation of light, but also the refraction, the diffraction, the interference, the polarisation and the totality of the known phenomena. The wave theory of Huygens was the starting point of the theory of Fresnel. The old, abandoned "paradigm" dislogded the one which had disloged it one and a half century ago.

The electromagnetic theory of Maxwell has been considered as the definite solution of the dilemma: as a theory which cut the Gordian knot, in favour of the continuity. However, Hertz discovered in 1887 the photoelectric effect. This effect was incomprehensible in the frame of the wave theory. Thirteen years later Planck postulated the discontinuous emission and absorption of light. The famous formula: $E=h\nu$ indicated the eventual existence of a discontinuity within the continuous wave. Five years later Einstein explained the photoelectric effect, by accepting the existence of photons having an energy and a momentum equal to $h\nu/c$. The wave and the particle aspects seemed now intrinsically related. However, the riddle of the light seemed insoluble. A synthesis of the opposite paradigms, in spite of the indications of the formula of Planck and Einstein, was not yet perceptible on physical grounds.

Waves and Particles

The first who tried to overcome the antithesis and to create a synthesis of the opposite attributes not only of light but, more generally, of all material particles was, as is well known, L. de Broglie, in his famous *Thèse*. The ejection of the photoelectrons needs an energy $E=h\nu$, greater than the energy needed to tear away the electron from the surface of the metal. This fact indicates the existence of discontinuous quanta of energy. The theory of Maxwell, on the contrary, and a number of other optical phenomena, constituted the solid foundation of the wave conception. How could one overcome the antithesis? According to L. de Broglie, the objective of Wave Mechanics was to acheive a synthesis between the dynamics of the material point and the wave theory conceived in the manner of Fresnel.[7]

In his *Thèse* de Broglie writes: "The relation of the quantum would not have any sense, if the energy was distributed in a continuous way in space. However, we have seen that this is not the case. Consequently, it is possible to conceive that because of a great law of Nature, every bit of energy of proper mass m_0 is intrinsically related to a periodic phenomenon of frequence ν_0 in such a way that $h\nu_0=m_0c^2$, ν_0 being, evidently, measured in the system attached to the bit of the energy. This hypothesis is the basis of our system."[8]

Since 1923, as we have already noted, de Broglie considered the opposition between the waves and the particles as relative: not as an irreducible contradiction. The theory of quanta, de Broglie maintains, confirms this conclusion. Consequently, de Broglie tried to generalize for light and for massive particles as well, the wave-particle dualism implicit in the work of Planck and Einstein. The conception of de Broglie was, from its very beginning, a relativistic one: de Broglie postulated that to every massif particle a periodic phenomenon is associated: $E=h\nu=m_0c^2$ and also a wave lenght $\lambda=h/p$. So he notes in his *Thèse:* "My essential idea was to extend for every particle the coexistence of the waves and the particles discovered in 1905 by Einstein for the case of light and the photons. In conformity with the clear ideas of classical Physics, I tried to imagine a real physical wave transporting a very small object localized in space during the time."[9]

Let us recall now, how de Broglie arrived at a first formulation of the wave-particle duality in his *Thése*.

Louis de Broglie postulated that the equation of Planck, $E=h\nu$ possesses a general validity for photons and for massive particles as well. To every particle, considered as a bit of energy of proper mass m_0 corresponds an internal periodic phenomenon of frequency $\nu_0 = m_0 c^2 / h$ in the system of reference related to the particle. For another system of reference moving with a velocity $v=\beta c$ relative to the first, the frequency is

$$\nu = \nu_0 \sqrt{1 - \frac{v^2}{c^2}} = \frac{m_0 c^2}{h} \sqrt{1-\beta^2}$$

At the same time de Broglie postulated that a frequency

$$\nu' = \frac{E}{h} = \frac{m_0 c^2}{h\sqrt{1-\beta^2}}$$

is related with the particle. The two phenomena are in phase if the velocity of the wave is $V = c^2/v$. Consequently the wave length is given by the formula:

$$\lambda = \frac{\frac{h}{m_0 v}}{\sqrt{1-\beta^2}} = \frac{h}{p}$$

The wave-particle duality was generalized from the mathematical as well as from the physical point of view by L. de Broglie, via a synthesis of the variational principles related to corpuscles and to waves. In fact, with a profound intuition, de Broglie perceived an analogy between the principle of Fermat for geometrical optics and the principle of Maupertuis for the movement of material particles. By this way he achieved the synthesis of these different principles bringing to light the dual character of the quantum particles.

As is well known the wave equation has the form

$$\Delta \Psi = \frac{n^2}{c^2} = \frac{\partial^2 \Psi}{\partial t^2}$$

and its solution is

$$\Psi = a(M) e^{2\pi i (\nu t - \varphi(M))}$$

In the geometrical approximation the rays are perpendicular to the surfaces of equal phase ($\varphi(M) = c^{te}$)

According to the principle of Fermat, the variation of the phase between A and B is minimal:

$$\int_A^B \frac{d\Phi}{ds} ds = \text{min imal} \tag{1}$$

Let us imagine now the movement of a massive particle. According to the principle of Maupertuis, the variation of the action between A and B is minimal:

$$\int_A^B \frac{dS}{ds} ds = \text{minimal} \qquad (2)$$

$$\left(\frac{dS}{ds} = p = mv \right)$$

Or, it is possible to establish between (1) and (2) a formal analogy:

$$\int_A^B \frac{d\Phi}{ds} ds = k \int_A^B \frac{dS}{ds} ds$$

If we put:

$$k = \frac{1}{h} \text{ and } \Phi(M) = \frac{1}{h} S(M)$$

we have

$$\frac{d\Phi}{ds} = \frac{1}{\lambda} = \frac{1}{h} \frac{dS}{ds} = \frac{p}{h}$$

Consequently,

$$\lambda = \frac{h}{p} \qquad (3)$$

The well known formula (3) is the equation relating the wave length with the momentum of the particle.[10]

As L. de Broglie himself emphasized, the principle of Fermat applied to a wave is identical to the principle of Maupertuis applied to a moving particle. The possible trajectories of the particle are identical with the possible rays of the wave. And de Broglie concluded: "We think that this idea of a profound relation between the great principles of the Geometrical Optics and the Dynamics could be a precious guide for the realization of the synthesis of the waves and the quanta."[11]

The formal analogy between the two variational principles was the starting point of the profound intuition of de Broglie. However, de Broglie notes, someone who knows the history of sciences, could think that there is a hidden connection between the concepts of wave and particle in this "beautiful theory of Analytical Mechanics which has been developed by Hamilton and Jacobi and which corresponds to an ensemble of movements of corpuscles in a given field of force, the propagation of a wave in this region of the space, in the approximation of the geometrical Optics." This striking image, de Broglie notes, "permits to identify the trajectories of the corpuscles associated to a wave, with the "rays" of this wave defined by the geometrical Optics as othogonal curves on the surfaces of equal phase."[12] The theory of Hamilton-Jacobi, according to de Broglie, adds to the image of the propagation of the waves with their "rays" the image of a small localized accident intimately related to the wave and following one of these rays.[13]

The approach of de Broglie was a relativistic one. As he remarks, it is possible to relate the four-vector defined by the gradient of the phase of a monochromatic wave with the four-vector of energy-momentum of a particle, by introducing the constant of Planck h, and putting: $w = h\nu$, $p = h/\lambda$.

However, the debate concerning the nature of microparticles has not come to an end, and the ideas of de Broglie were forgotten for twenty five years.

Only particles - a corpuscular interpretation

Even today there is not a generally accepted solution of the problem of the wave-particle duality. The relevant debate goes back in the twenties. In fact, in the same year of the elaboration of the *Thèse* of de Broglie (1923), the American physicist W. Duane formulated a purely corpuscular explanation of the wave-like phenomena of microphysics. The pioneer work of Duane has been practically forgotten. Or, in the sixties, A. Landé brought to light the forgotten work of Duane and formulated, starting from his ideas, a corpuscular interpretation of quantum mechanics.

Landé puts the question: Why do particles in their statistical behavior obey wave-like laws? For him "duality" explains nothing. It is a purely descriptive term. Quantum particles are not waves: They have discrete charge, rest mass and spin. One can count electrons and electrons always behave exactly as particles ought to behave. A wave has intensity and phase. But, Landé remarks, where has single electron ever displayed an intensity and phase?

The mechanical reasons for the periodic diffraction patterns, Landé maintains, have been known since 1923, when Duane developed the first corpuscular interpretation of diffraction through a crystal, followed by Ehrenfest and Epstein's theory for screens with slits.

Landé, following the work of Duane, derived the laws of quantum mechanics from a non-quantal, probabilistic basis. His corpuscular interpretation was realistic, against the "dualistic doctrine" and its subjective interpretation. The objective of Landé was to deduce the quantum laws "from general non-quantal postulates of symmetry and invariance, imposed on the probability structure of mechanical events." His aim was to explain quantum principles: to show that they are consequences of more elementary principles known from pre-quantal physics.

The fault of Physics, Landé argues, consists in ignoring the rule of the quantized exchange of linear momentum (Duane's third quantum rule, 1923). This rule has been established "in perfect analogy to the quantum rule for the energy (Planck) and for the angular momentum (Sommerfeld-Wilson)." The third quantum rule yields, according to Landé, an explanation of all the wave-like phenomena, including diffraction and coherence.

Landé reconstructs the interpretation of Duane as follows: "The incident matter particles do not spread out as continuous matter waves, or manifest themselves as though they did. It is the crystal with its parallel lattice planes which is already spread out in space and reacts as one rigid mechanical body to the incident particles under the conservation laws of mechanics with the following restriction (Duanés quantum rule for linear momentum): A body periodic in space with linear periodicity of length L is thereby entitled *to change its linear momentum P parallel to L* in amounts $\Delta p = h/L$ [...] Hence, it can change its momentum perpendicular to the lattice planes in amounts

$\Delta p = h/l, 2h/l, 3h/l, ..., nh/l$. When these selective momenta are transferred from the crystal to the incident particles, the latter are reflected into exactly the same directions which can also be calculated according to Lawe-Bragg wave interference theory."[14]

On the basis of the quantum rule, Duane derived the relation

$$2p\sin\theta = nh/\lambda$$

identical with Bragg's wave rule, by virtue of translating the momentum p into a wave length $\lambda = h/p$. Duane's rule yields the same observed deflections directly, without appealing to a wave interlude.[15]

Let us take now two specific cases: The two-slit experiment and the phenomena of coherence.

How can one explain, Landé asks, the fact that the opening of a second slit blocks particles from places at which they formerly arrived? This is a crucial question. The solution of the problem, Landé maintains, is contained in the well-neglected investigation of Ehrenfest and Epstein representing a further development of Duane's theory. In Landé's own words: "A screen with one slit has periodic components of various lenghts L, composing a geometrical shape; a screen with two slits has a different set of L's. And since the several components L give rise to the impulse transfer $\Delta p = h/L$ respectively, the two cases of one and two slits yield different deflected angles with different intensities. The diffraction patterns produced statistically by deflected particles agree in both cases with those obtainable also via translation into waves and translation into particles."[16]

Now, concerning coherence. A frequent argument in favour of the wave theory, Landé notes, is the apparent impossibility of explaining the phenomena of coherence without using the interference of waves. Or, according to him, the quantum theory of particles, with the help of the third quantum rule, yields exactly the same observed results. Landé explains the phenomenon in particle's terms and finds the same results with those obtained by the wave theory.

More generally, Landé developed a formalism explaining the quantum phenomena on the basis of the statistical distribution of particles. The wave picture, Landé maintains, is an epiphenomenon produced by the laws and rules of symmetry which govern the interaction of particles with the experimental devices. Or, neither Duane nor his successors provided us with an explanation of the quantum rule $p = h/\lambda$ which they took for granted. And Landé concludes: "Since an explanation, that is the reduction to more elementary prostulates is now at hand, the wave like dualism has become an anachronism."[17]

Why was the pioneer work of Duane forgotten? Why does the same hold for the more elaborate theory of Landé? Before attempting an answer, we can put another question: Why has the wave theory of Huygens been abandoned for more than one century? The main reason was not the enormous prestige of Newton (Huygens was also a great physicist). The main reason of the victory of Newton is to be found in the internal laws of evolution of the scientific thought. In fact, the corpuscular theory was able to explain the phenomena in a more simple, intuitive way. More than that: the theory of Newton was more consistent with the dominant mechanistic world view, and the rising materialism of natural sciences. Newton's theory was abandoned later, when the phenomena of diffraction, interference and polarisation were studied and were explained by the wave theory.

Now, concerning quantum mechanics: the actual formalism, using new concepts, incompatible with classical physics, succeeded to explain an enormous number of phenomena. On the other side, it is well known that there is not a general concensus concerning the physical foundations of this discipline. That some of its fundamental concepts have not a clear physical counterpart. This fact explains the interminable debates on the foundations of quantum mechanics. Why has then the interpretation of Duane-Landé been forgotten? According to Landé: "The reason that Duane's third quantum rule was not immediately recognized as a way out of the duality paradox was historical: Duane proposed his statistical particle theory of diffraction of X-rays in support of the photon theory of light. At his time (1923) diffraction of electrons was not discovered. Moreover, the quantum rule $\Delta p = h/L$ for bodies of perdiodicity L in space seems artificial and conceived *ad hoc*. Since 1926, however, the three quantum rules have been victorious."[18]

However, even the more elaborate theory of Landé was not accepted by physicists. It is a matter of fact that Landé made a rigorous critique of the dominant interpretation and in particular of its vague concepts (wave, wave packet, reduction of the wave packet, etc), as well as of its subjectivism. On the other hand he contributed to the reinforcement of the realistic school and of modern atomism in particular. However, his ontology is mechanistic - he tries to explain quantum phenomena with pre-quantal concepts. Or, quantum mechanics is a new discipline, epistemically different from classical mechanics. It constitutes a new paradigm - its concepts are not reducible to their classical counterpart. Because of that, the corpuscular theory succeeds in explaining some simple quantal phenomena but its conceptual structure is inadequate for more complicated cases concerning the transformation of quantum particles, the physical meaning of the superposition principle, etc. The victory of the wave mechanics can be explained, in principle, by the internal laws of development of scientific knowledge.

However, if we reject the orthodox interpretation and if at the same time we are convinced that the corpuscular interpretation is inadequate, then we must put the question: Does the realist interpretation and in particular the interpretation of the school of de Broglie (de Broglie, Bohm, Vigier etc) offer a way out of the actual Tower of Babel?

Waves and Particles: The theory of the Double Solution

The objective of de Broglie, since 1923, was to create a synthesis between the dynamics of the material point and the wave theory taken in the spirit of Fresnel.[19] However, if one adopts a continuous solution without singularity, it would be impossible to take into consideration the atomic structure of matter. Or, de Broglie took his inspiration from relativity as well as from the ideas developed by Hamilton. Einstein, in particular, considered particles not as an element additional but external to the field, but as an entity incorporated in the structure of the field and constituting a kind of local inhomogeneity. And de Broglie writes: "Following the work of Mie and Einstein, I was endeavouring to represent the particle as a kind of local accident, a singularity within an extensive wave phenomenon. Accordingly, I introduced the u-solution of the Shrödinger equation, which differed from the normal Ψ-solution and permitted the inclusion of a singularity."[20]

In this spirit, de Broglie postulated the principle of the Double Solution: To every regular solution
$$\Psi = a e^{\frac{2\pi i}{h}\varphi}$$
of the wave equation of the wave mechanics, there corresponds a solution
$$u = f e^{\frac{2\pi i}{h}\varphi}$$
having the same phase φ but with an amplitude f involving a point singularity which is, in general, in motion. The u-wave represents the particle.[21] The theory of the double solution was formulated during the years 1926-1927.

However, de Broglie also elaborated the theory of the pilot wave, which is sometimes identified with that of the double solution. Yet, there is an essential difference between the two «models.» The pilot-wave model considers as dinstict realities the material point and the continuous wave represented by the Ψ function. It is the continuous wave that guides the movement of the particle.

The theory of the pilot wave has been the object of «nombreuses critiques» by Pauli and other participants at the Solvay Meeting of 1927. Finally, de Broglie arrived at the conclusion that it is impossible to elaborate a genuine, causal and objective interpretation of quantum mechanics by supposing that the particle is guided by the wave. Consequently de Broglie arrived at the conclusion that it was necessary to re-incorporate the particle into the wave phenomenon.[22]

In his *Thèse* and during the following years, de Broglie tried to create a genuine synthesis of the particle and the wave and to conserve the precise images of the physical realities attributed to these notions. Consequently he considered the particle as a very small region of high concentration of the field incorporated in an extended wave. He introduced a strong, localised non linearity in order to take into consideration the permanent and high concentration of energy in the interior of a wave described everywhere else by an equation essentially linear.

The u-wave is the true representation of the particle. The non linearity introduced into the wave must be very localised. «It should be significant only in the small region which constitutes the particle in the true sense of the word.» The singularity defining the particle must remain «inside of an extremely thin tube, the walls of which are formed by geodesics of an external field.»[23]

It is worthy to quote once more de Broglie. In a relatively recent (and forgotten) book (1961) he writes: «The complete representation of the wave with its very small singular region, must be made in the theory of the double solution with the help of a wave function u, of the form: $u = u_0 + v$ where u_0 is a function having very little values in the interior of the singular region and very small, negligible compared with v, in the exterior. It is then normal to accept that the regular function v representing the wave in every point of space with the exception of the very small singular region, obeys the linear equation used by Wave Mechanics and it closely resembles the wave Ψ. However, in the very small singular region, the function u obeys surely to a non linear equation because, according to an old remark of Einstein's only this linearity can ensure the guidance of the corpuscle, that is to say the high values of the field u, making the propagation of u_0 closely bound up with that of v.[24]

The corpuscle, incorporated into the wave, is «guided» by it. So, it would be possible to understand how, when the propagation of the wave is disturbed by an obstacle (boundary conditions) and is affected by inteference and diffration, the corpuscle is in preference directed towards the region of high intensity of the wave (bright fringes).[25]

If the non linearity is the real key of the corpuscular microphysics, then, L. de Broglie notes, we understand why the actual theory cannot describe the wave-particle duality. The actual linear theory puts out of sight the local, non linear accidents. Consequently it effaces the corpuscular structures.[26]

The theories of de Broglie, and in particular the theory of the double solution, incompatible with the dominant positivist ideology, were practically forgotten for twenty five years. So de Broglie was writing in 1953: «If the conceptions I set forth in 1927 were a day to revive from their ashes, this would be possible only under the fine form of the double solution and not under the mutilated and unacceptable form of the pilot-wave.»[27]

Waves and Particles: Further Developments

It was in 1952 that the «ressurection» of the theory of the double solution was realised. In fact, David Bohm, who ignored at that time the theory of de Broglie, published a deterministic version of the quantum mechanics, similar to that of L. de Broglie.[28]

During the following years David Bohm, J.P. Vigier et their collaborators, published a great number of papers in their effort to formulate a deterministic version of quantum mechanics. Bohm et Vigier introduced the concept of the «subquantum level», in order to explain the behaviour of the quantum particles.[29] L. de Broglie, from his side, considered the ideas of Bohm and Vigier as a development of the theory of the double solution. In his own words: «In consequence of a paper by Bohm and Vigier, I increasingly felt that this picture must be augmented by the introduction of the 'subquantum medium.'» According to Bohm et Vigier «the particle is subjected to constant random perturbations resulting from its continual interaction with a hidden, subjacent «subquantum medium» which is a kind of «hidden thermostat.»[30] The particle thus appears as a kind of granule carried by the flow associated with the propagation of the waves, but which constantly changes from one streamline to another as a result of collisions with the hidden particles of the subquantum level.»[31]

The concepts of sub-quantum level, of the «hidden thermostat,» of the ether of Einstein and Dirac introduced or used by Bohm and Vigier in conformity with the conceptions of the double solution, went far beyond the initial schema of de Broglie. First of all, they rejected the metaphysical idea of the vacuum related to ancient and classical atomism. At the same time they explained the movement of the particles in the quantum level as the result of hidden random interactions of the quantum particles with the subquantum medium. Finally, they considered the particles themselves as the «observable surface» of the more profound reality of the «vacuum» which, according to Bohm and Vigier is an immense reservoir of hidden energy. This conception concords with the classical idea of Dirac concerning the «ocean» of unobservable particles and with his prediction of the positron.

According to Louis de Broglie, these revolutionary ideas of Bohm and Vigier added to the primitive conceptions of the double solution a random element concerning the movement of the corpuscle in the midst of the wave.

During the fifties, J.P.Vigier and his collaborators (Tabayasi, Halbwachs, Hillion and Lochak) developed a new theory of elementary particles. Louis de Broglie published a book (1961) on this new theory. According to de Broglie, Vigier and his collaborators arrived to a concrete representation of the microphysical reality in conformity with the theory of the double solution: they considered the particle as a well localised region of the space, where a wave comparable to a droplet in rotation, moves via the quantification of the internal movements of the particles. Vigier and his collaborators, arrived at defining quantum numbers (isospin, strangeness, baryonic number) characteristic of every type of particle. By this way, they arrived to re-obtain the formula of Gell-Man. The new theory made a general classification of the known elementary particles and anticipated the existence of other ones.[32]

During the following years D. Bohm and his collaborators in London, J.P. Vigier and his collaborators in Paris, developed new ideas and new formalisms in a great number of papers in their effort to give a realistic and deterministic description of quantum phenomena. The concepts of the sub-quantum level, of the ether of Dirac-Einstein, of the particle as a singularity localized in space and time, were some of the fundamental concepts used by the collaborators of Louis de Broglie. However the new formalisms introduce an intrinsic non locality in quantum mechanics, incompatible with the relativistic locality.[33]

Concluding remarks

I tried to outline the efforts of de Broglie, Bohm, Vigier and their collaborators, to give a theoretical foundation to the wave-particle duality and, more generally, to formulate a realistic and deterministic conception of quantum mechanics, against the dominant interpretation of the Copenhagen School.

Now concerning the wave-particle duality: it is a matter of fact that it is impossible to explain quantum phenomena exclusively in terms of waves. On the other hand, the corpuscular theory of Duane and Landé cannot explain, as I already maintained, the great variety of the quantum realm. Consequently, a dialectics between waves and particles seems as the only way out of the actual situation. The School of de Broglie has till now created a rich theoretical corpus which has strong evidence in its favour, and which can constitute the starting point for further theoretical investigations. Thus, before concluding we will discuss some specific cases related to the problem of the wave-particle duality.

Let us first take the famous two-slit experiment. In the case of light considered in the frame of the electromagnetic theory, there is no problem: electromagnetic waves are considered as real and their interference is explained in classical terms. If, on the contrary, we consider that light is constituted of photons and if we consider also the case of massive particles, then an explanation of the classical type is not possible. In fact, if the photons were corpuscles the trajectory of the particle passing from the slit A would be independent of the ouverture or the closure of B.

However, as it is well known, the pattern of inteference is different in the two cases. Consequently, how does the particle «know» if the slit B is open or closed? It is evident that the opening or the closing of the slit affects the trajectory of the particle. How could this fact be explained?

The particle theory is incapable to give a convincing answer. On the other hand, it is impossible to represent the photon by a classical wave. If we use individual photons, then every photon interferes with itself (as it was anticipated by Dirac) and its impact is well localised. Consequently we must take into consideration the corpuscular aspect of our photon or of the massif particle. There is no other way out than to consider our systems as having a dual nature: to be wave-particles. In that case, if the distance of the slits A and B is smaller than the «diameter» of the region affected by the wave accompanying the particle, then it is possible to understand, at least in principle, the influence of the slit B on the particle passing through A, as well as the phenomenon of the interference of the particle with itself. It seems that the hypothesis of the wave particle duality is the only one compative with the two-slit experiment.

However a new question arises here: is it legitimate to attribute a trajectory to our particle? The orthodox answer is: no! But as it is well known, today it is possible to put two detectors C_1 and C_2 in an appropriate distance behind the slits A and B, and then we see that every particle passes either from slit A or from slit B. Consequently: our particle behaves like a wave and at the same time like a corpuscle. The theory must express this contradictory picture - the objective unity of these opposite attributes of the quantum particles. The theory of the double solution is a first response to the dilemma: waves or particles?[34]

Neutron interferometry offers new evidence in favour of the wave-particle hypothesis. The interference pattern taken in these experiments shows that «each neutron carries information about the physical situation in both widely separated beam paths.» However, the particle follows one of the two possible ways. Consequently «something» passes from the second path. In that case also we have to do with a phenomenon of self-interference of the single particle. The results of the neutron interferometry are incompatible both: with the wave and the corpuscular conception of quantum particles. It seems that we must go beyond our intuitive ideas and attribute an ontic status to the wave-particle duality.[35]

Or, what is this «something» which follows the second way, which does not carry a particle and which nevertheless provokes observable phenomena?[36] At this moment the «ghost waves» of Einstein make their enigmatic appearance. In fact, according to Einstein and de Broglie, waves without particles (ghost waves, ondes vides) have a real existence. F. Selleri brought anew to light this idea which is supported today by experimental evidence.[37]

The hypothesis of wave-particle duality, the double solution version and the existing experimental evidence are in favour of the following picture of the quantum particles: the central part of the particle, its «nucleus» carries the totality (or nearly) of its mass. This part represents the «corpuscle» which passes from one of the slits or the paths of the interferometer. This «nucleus» is surrounded by a wave intrinsically related to it. This picture concords with the fact that quantum particles have a well definite mass, charge, spin, and other quantum numbers. Also the structure of the quantum particles is today a fact and this fact is incompatible with the wave picture and with the concept of the wave packet. Also, particles are not geometrical points. They have a finite volume and this is also an argument against the simple wave picture. The particles structure, Enz notes, is seen as a clue to finding an answer to the question why are there unique given particle properties like discrete charge or mass. Finally, the field conception of Einstein is compatible with the existence of a structure in the small volume of high energy concentration.[38] The existence of individual photons, electrons, quarks etc, is not incompatible with the field conception of matter: the "separate" particle is a singularity of the field described by non-linear equations. This idea of Einstein is conformable with the conceptions of de Broglie and his School.[39]

If the preceding conceptions have a physical counterpart, then what is the meaning of the solution of the Schrödinger equation in the case of a free particle and, also, what is the meaning of the notion of the wave packet?

In fact, as it is well known, a free particle is represented by a plane wave of infinite lenght. Evidently this representation has nothing to do with the reality of the particle. On the other hand, the so called wave packet built as the sum of a great number of plane waves is a mathematical artifice permitting the localisation of the particle. In reality it is incompatible with the real structure of the quantum particles and with the fact that these particles possess definite mass, charge and other quantum numbers. Finally, the wave packet does not conserve a constant volume: it spreads and occupies, finally, the whole of the available space. For example, Auger electrons with initial uncertainty of position $\Delta x_0 = 10 Å$ will have, after one second, a size of 100 km![40] Also, M. Murascin notes that linearity implies the fact that the different contributions, having different wave numbers, act independently, so that the spreading of the wave packet becomes a serious problem in attempts to associate the wave packet with particles. With one Compton wave length, for example, the wave packet after one second will have a width of about 2×10^9 m.

It is true that in the non linear theories the existence of wave packets with constant shape is recognized. As P. Cornille maintains, in a non dualistic theory exposed by him in a recent paper, where the particle as a wave packet integrates the structure of the field, "a wave packet can propagate without spreading in the vacuum considered as dispersive, dissipative and inhomogeneuous medium containing scalar standing waves in the framework of a linear theory."[41]

The "father" of the concept of the wave packet was, as is well known, L. de Broglie. In the theory of the double solution, the velocity of the material point is equal to the group velocity of the plane waves. And de Broglie adds: "Ce fait remarquable fait penser que le point matériel doit être assimilé a un groupe d' ondes monochromatiques. Cette conception, qui avait été la mienne, a été reprise par M. Schrödinger et l' a conduit a# considérer le point matériel comme un 'Wellenpacket.' Au point de vue didactique, il est très utile d' employer cette image, mais il n' est pas sûr qu' elle corresponde a la réalité."[42]

The "spreading" is not the only problem related to the concept of the wave packet. Even if we accept the validity of the models of a non-spreading wave packet, this notion is incompatible with the real properties of the quantum particles and with their stability. The definite mass is one such attribute. On the other hand, the discontinuous charge, the spin, the parity and other quantum numbers are the external manifestations of structures, symmetries and movements non accessible in the actual level of approach of the microphysical reality. In the frame of a coherent statistical interpretation, it would be possible to interpret the concept of the wave particle as describing the movement of an ensemble of particles during the time. The spreading of the "wave packet" represents, in that case, the dispersion of the statistical ensemble.[43] As J.R. Croca writes, the spread of the "wave train" is a way of saying that as time increases, particles of a definite, constant, small size increase the distance between each other to a value that depends on the elapsed time and the respective range of velocities ΔV.[44]

The spreading and the incompatibility of the wave packet with the structure and the properties of the particles are not the unique problems related with this concept. In the non-statistical, single system interpretation of quantum mechanics, the so called "reduction" of the wave packet implies the existence of an action at a distance incompatible with relativity. This fact was emphasized already by Einstein in the Fifth Meeting of Solvay[45] with its well known thought experiment. Landé, from his point of view, writes: "I emphatically agree with Margenau that quantum theory deals with objective situations, and not with packets of expectations spreading out in space and suddenly collapsing as through a kind of telepathy."[46] The non-locality presupposed by the "reduction" of the "wave packet" was stressed also by L. de Broglie as early as in 1927: "Si l' onde Ψ est une 'réalité physique,' comment peut-on comprendre qu' une mesure faite dans une région de l' espace puisse modifier cette réalité physique dans d' autres regions de l' espace qui peuvent etre très eloignèes de la première?"[47]

However, the problem of a realistic interpretation of the superposition principle is more general than that of the wave packet. In fact, as it is well known, an initial pure ensemble is transformed in most cases into a mixture as a result of the "measurement". What is then the meaning of the formula:

$$\Psi = \sum_i c_i \Psi_i \qquad (1)$$

According to the dominant, single system interpretation, formula (1) represents a superposition of real waves. The measurement is then considered as a spectral analysis, a reduction of the wave packet, a projection, etc. Such an explanation is plausible in the case of superposition of real waves as, for example, electromagnetic or sound waves. The equation (1) on the contrary represents the possible states to be realized because of the interaction of the particle with the measuring apparatus. Consequently (1) is the measure of the potentialities of the statistical ensemble in given external conditions. This phenomenon is not a reduction or a collapse or a projection, but a local and causal transformation of the quantum system (in nature or in the laboratory). The equation (1) has meaning only for statistical ensembles of identical particles.[48]

In what precedes, I tried to outline the problem of the nature of the quantum particles and to formulate arguments in favour of the objective unity of the corpuscular and the wave aspect. However, the evidence in favour of the wave-particle duality is not conclusive. A new theory, embracing the totality of the known phenomena will differ, eventually, in its fundamental concepts and structure from the present linear theory. To quote Einstein: "I think it is not possible to get rid of the statistical character of the present quantum theory by merely adding something to the latter, without changing the fundamental concepts about the whole structure. Superposition principle and statistical interpretation are inseparably bound together. If one believes that the statistical interpretation should be avoided and replaced, it seems one cannot conceive a linear Schrödinger equation which implies by its linearity the principle of superposition of states."[49]

References

1. Huygens, Chr. (1962) *Treatise on Light*, Dover, New York, p. 3
2. Ibid., p. 4
3. Newton, I. (1952) *Opticks*, Dover, New York, p. 364
4. Ibid., p. 369
5. Ibid., p. 339
6. Broglie, L. de (1956) *Nouvelles Perspectives en Microphysique*, Albin Michel, Paris, pp. 168-169
7. Broglie, L. de (1963) *Recherches sur la Théorie des Quanta*, Masson, Paris, p. 29
8. Ibid., p. 21
9. Ibid., p. 63
10. Ibid., passim. See also, Yourgrau W. and Mandelstram S. (1960) *Variational Principles in Dynamics and Quantum Theory*, Pitnam, London
11. Broglie, L. de (1956), *Nouvelles Perspectives en Microphysique*, op. cit. p. 43
12. Ibid., pp. 171-172
13. Ibid., p. 207
14. Landé, A. (1965) *New Foundations of Quantum Mechanics*, Cambridge University Press, Cambridge, p. 5
15. Ibid., p. 61
16. Ibid., p. 10
17. Landé, A. (1960) *From Dualism to Unity in Quantum Physics*, Cambridge University Press, Cambridge, p. xii. See also id. *New Foundations of Quantum Mechanics*, op. cit. See also, Duane, N. (1923) *Proc. Nat. Acad. Sc.*, Wash. 9, 158
18. Landé, A. (1965), *New Foundations of Quantum Mechanics*, op. cit. p. 69
19. Broglie, L. de (1953) *La Physique Quantique restera-t-elle Indéterministe?* Gauthier-Villars, Paris, p. 29
20. Broglie, L. de (1964) *The Current Interpretation of Wave Mechanics*, Elsevier, Amsterdam, p. 44
21. Ibid., p. 36
22. Broglie, L. de (1953) *La Physique Quantique restera-t-elle Indéterministe?* op. cit., passim
23. Broglie, L. de (1964) *The Current Interpretation of Wave Mechanics*, op. cit., passim. Id., *Recherches sur la Théorie des Quanta*, op. cit., passim.
24. Broglie, L. de (1961) *Introduction a# la Nouvelle Théorie des Particules de M. Jean-Pierre Vigier et de ses Collaborateurs*, Gauthier-Villars, Paris, p. 94
25. Ibid., p. 93
26. Broglie, L. de (1956) *Nouvelles Perspectives en Microphysique*, op. cit. p. 163

27. Broglie, L. de (1953) *Le Physique Quantique restera-t-elle Indéterministe?* op. cit., p. 12
28. Bohm, D. (1952) *Phys. Rev.* 85, 166 and 85, 180
29. Bohm, D. (1954) *Phys. Rev.* 96, 208
30. See also, Bohm, D. (1957) A proposed explanation of quantum theory in terms of hidden variables at a sub-quantum mechanical level, in S. Corner (Ed.) *Observation and Interpretation*, Butterworths Publications, pp. 33-40
31. Broglie, L. de (1964) *The Current Interpretation of Wave Mechanics,* op. cit. pp. 41-43
32. Broglie, L. de (1961) *Introduction a# la Nouvelle Théorie...* op. cit. passim
33. See: 1) *Open Questions in Quantum Physics,* G. Tarozzi and A. van der Merwe (Eds), Reidel, Dordrecht, 1985. 2) *Microphysical Reality and Quantum Formalism,* A. van der Merwe et al (Eds), Kluwer Academic Publishers, Dordrecht 1988. 3) *The Concept of Probability,* E.I. Bitsakis and C.A. Nicolaides (Eds), Kluwer Academic Publishers, Dordrecht, 1989
34. See, Bitsakis, E (1991), The Physical Meaning of the Superposition Principle, *Physics Essays,* 1, 124-133
35. See, R. Rauch in *Proceedings Intern. Symposium of Found. of Quantum Mechanics,* Tokyo, 1983, Also J. Summhammer, The Physical Quantities in the Random Data of Newton Interferometry, in *The Concept of Probability,* E.I. Bitsakis and C.A. Nicolaides (Eds) Kluwer Academic Publishers, Dordrecht, 1959, pp. 207-219
36. Enz, U. (1993), Particle extension: the bridge between wave and particle aspects, in *Courants, amers écueils en Microphysique,* Fondation Louis de Broglie, Paris, pp. 131-144
37. See 1) F. Selleri, *Quantum Paradoxes and Physical Reality,* Kluwer Academic Publishers, Dordrecht, 1990. 2) M. Schmidt, Empty Waves and Particle Trajectories, in *Bell's Theorem and the Foundations of Modern Physics,* A. van der Merwe et al (Eds), World Scientific, Singapore, 1992, pp. 436-440. 3) F. Selleri, *Lett. Nuovo Cimento,* 1, 908 (1969). 4) Id., *Found. of Physics,* 12, 1087 (1982). 5) J.P. Croca, *Found. of Physics,* 17, 971, (1987)
38. See: 1) Bitsakis, E. (1991), The Physical Meaning of the Superposition Principle, *Physics Essays,* 4, 124-133. 2) Canals-Frau, D. (1991) About the Photon, *Physics Essays,* 4, 577-582. 3) Canals-Frau, D. (1993) A note About Bell's Theorem, *Physics Essays,* 6, 166-172. 4) Cornille P. (1993) Is the Physical World Built upon Waves, *Physics Essays,* 6, 289-307
39. Sachs, M. (1993), Quantum Mechanics from general relativity: a paradigm shift, in *Courants, amers écueils en Microphysique,* op. cit. pp. 393-400
40. Croca J.R. (1990), A Possible Experiment to Test the Nature of Spreading of Matter Wave Packets, *Physics Essays,* 3, 71-74. See also, Murascin M. 1(1995), Update of Mathematical Aesthetic Principles with Discussion of a More Viable Wave Packet Solution, *Physics Essays,* 8, 99-121
41. Cornille P., op. cit.
42. Broglie, L. de (1953), *La Physique Quantique restera-t-elle indéterministe?* op. cit. p. 32
43. Bitsakis, E. (1976), *Le Problème du Déterminisme en Physique* Thèse d' Etat, Paris. Id., *Physics Essays,* op. cit.
44. Croca J.R. (1990), op. cit.
45. Einstein, A. (1928), *In Electrons et Photons,* Gauthier-Villers, Paris
46. Landé, A. (1960), *From Dualism to Unity In Quantum Physics,* op. cit. p. 82
47. Broglie, L. de (1927), La Mécanique Ondulatoire et la structure atomique de la matiére et du rayonnement, *J. Phys. Rad,* série vi, t-8, no5, pp. 225-241
48. Bitsakis E (1988), Potential and real states in Quantum Mechanics, in *Microphysical Reality and Quantum Formalism,* op. cit. pp. 47-63
49. Cited by Sach M., in *Courants, Amers Ecueils en Microphysique,* op. cit. p. 396

QUANTUM MECHANICAL TUNNELING IN A CAUSAL INTERPRETATION

MICHAEL CLARKSON
*CRESS, York University,
North York, Ontario, Canada*

> "*I can safely say that nobody understands quantum mechanics.*"
> — R. P. Feynman

1. Quantum Mechanical Tunneling

Quantum Mechanical Tunneling is one of the uniquely quantum phenomena about which so much has been written, and so little is understood. One of the reasons for this is that it is very challenging to compare with experiment, as the tunneling coefficient is sensitive to the barrier shape near the top. Moreover, the experimental barrier shapes are not known to any degree of accuracy. Also, the tunneling coefficient depends on wave-packet shape, and theoretically the wave-packet shape can be rather arbitrary.

In our view, Quantum Mechanical Tunneling is second only to the "measurement problem" for the level of distortions, miscomprehension and obfuscation. It is perhaps an entirely unnecessary concept, as most tunneling states are quasi-bound decaying states. As we shall see, it is probably best treated in a Schrödinger formulation of quantum mechanics, which includes the Bohm-de Broglie type of mechanics. As is to be expected, quantum mechanical tunneling poses interesting questions to a "causal" theory.

The motivation for seeking a "causal" interpretation of quantum mechanics lay in the severe restrictions on causality imposed by the Copenhagen interpretation. The philosophical distaste that many such as Einstein and Schrödinger had for the abandonment of causality is well documented. Other motivations to search for a causal theory included:

- the incompleteness of quantum mechanics (EPR (Einstein et al., 1935)). Any description of underlying processes to the probabilistic description was missing.

- the inability of quantum mechanics to incorporate a description of the measurement process.

At the heart of much of the discussion lay the fundamental inability of the quantum mechanical description of nature to describe the measurement process. In the Copenhagen interpretation, an arbitrary line was established between the quantum system, and the classical measurement apparatus, which as Bohr emphasized, was of a fundamentally irreversible nature. Unfortunately the arbitrariness of the dividing line gives rise to the possibility of a *reductio ad absurdum* situation. This gave rise to a long debate on the role of the observer, where in desperation, the consciousness of the observer was introduced to break the time reversal invariance (Wigner, 1963).

The root of this aspect of the measurement problem clearly lies in the fact that the underlying theory to quantum (and classical) mechanics is based on an equation of motion that is time reversal invariant. Thus quantum theory will have great difficulties describing an irreversible process. This problem is a direct consequence of Poincaré's Theorem (Poincaré, 1893).

> For any closed system of a finite number of particles, whose motion is governed by the Liouville equation, a function of phase space will infinitely often assume its initial value.

Let us examine this a little more closely. Consider a Gibbs H-functional of the form

$$\eta(\rho_\Gamma) = \int\int \rho_\Gamma(\mathbf{p}, \mathbf{q}) \ln \rho_\Gamma(\mathbf{p}, \mathbf{q}) \; d\mathbf{p}\, d\mathbf{q} \geq 0$$

It is often overlooked that Hilbert introduced his mathematics to study exactly these types of problems in classical mechanics. We can reformulate the problem in *classical* Hilbert space (Koopman, 1931), in terms of an superoperator M in $\mathcal{H} = \mathcal{L}^2$ on phase space Γ

$$\Omega(\rho_\Gamma) = \int\int \rho_\Gamma(t) M \rho_\Gamma(t) \; d\mathbf{p}\, d\mathbf{q} \geq 0$$

As ρ_Γ is positive definite, M must be positive semi-definite $M \geq 0$, and if Ω is to be an H-functional, we require that it be a decreasing quantity $\frac{\partial M}{\partial t} \equiv D < 0$. Using the operator definition of the classical Liouvillian $\frac{\partial M}{\partial t} = iLM$, we consider the time derivative of Ω

$$\begin{aligned}\frac{\partial \Omega}{\partial t} &= \tfrac{\partial}{\partial t} \iint e^{-iLt}\rho_\Gamma(0) M e^{-iLt}\rho_\Gamma(0)\ d\mathbf{p}\,d\mathbf{q} \\ &= \iint e^{-iLt}\rho_\Gamma(0) D e^{-iLt}\rho_\Gamma(0)\ d\mathbf{p}\,d\mathbf{q}\end{aligned}$$

If we consider the initial distribution to be the microcanonical distribution $\rho_\Gamma(0) = \rho_{\text{can}}(0)$, at equilibrium

$$\begin{aligned}\rho_\Gamma(t) &= e^{-iLt}\rho_{\text{can}}(0) = \rho_{\text{can}}(0) \\ \frac{\partial \Omega}{\partial t} &= \iint \rho_{\text{can}}(0) D \rho_{\text{can}}(0)\ d\mathbf{p}\,d\mathbf{q} = 0\end{aligned}$$

If D is a multiplicative function of phase-space, and as $\rho_{\text{can}}(0) > 0$, the $D = 0$, in contradiction to our hypothesis that $D < 0$. Hence no entropy functional can exist in *classical or quantum* mechanics.

1.1. LYAPOUNOV FUNCTIONAL

We consider there to be 3 principal formulations of non-relativistic quantum mechanics:

1. Schrödinger (Hamiltonian)
2. Density Matrix (Liouvillian)
3. Path Integral (Lagrangian)

In the Liouville formulation, there is the possibility of trace invariant non-unitary transformations that does not exist in the Hamiltonian formulation using star-unitary Superoperators (Prigogine et al., 1973):

If $\mathcal{Q}(\mathcal{L})$ is a functional of the Liouvillian $\mathcal{L}$, then we define L-inversion as the result of replacing $\mathcal{L}$ by $-\mathcal{L}$

$$\mathcal{Q}(\mathcal{L})' \equiv \mathcal{Q}(-\mathcal{L})$$

We combine the L-inversion and adjoint to define star-Hermitian operation

$$\mathcal{A}^\star \equiv \mathcal{Q}^\dagger(-\mathcal{L}) = (\mathcal{Q}'(\mathcal{L}))^\dagger$$

Star-unitary transformations leave the trace invariant

$$\text{Tr}\left(\tilde{\mathcal{A}}^\dagger \tilde{\rho}\right) = \text{Tr}\left((\Lambda'\mathcal{A})^\dagger \Lambda\rho\right) = \text{Tr}\left(\mathcal{A}^\dagger \rho\right)$$

Misra gave conditions for existence of a Lyapounov functional, both for classical and quantum systems (Misra, 1978), which are cast in terms of spectral properties the Liouvillian, related to the ergodic properties. In general, it is necessary that it satisfies the condition of mixing, and sufficient that it satisfies a K-Flow.

The corollary to this is that as $[L, M] \neq 0$, if a Lyapounov superoperator M exists, it must fail to commute with at least some multiplicative operators. This means that the notion of a classical trajectory is phase space breaks down. This also provides a mechanism by which pure states become mixtures.

The purpose of the preceding digression was to note that treatment of irreversibility in quantum mechanics:

- Requires density matrices, not state-vectors.
- Implies a complementarity between trajectories and entropy.
- Implies an implicit nonlocality for irreversible systems.

Thus the different formulations of non-relativistic quantum mechanics are not equivalent, for an exact treatment of irreversible processes cannot be carried out in the Schrödinger formulation of quantum mechanics. However, the investigation of tunneling is traditionally carried out in a Schrödinger formulation, and unlike the Copenhagen interpretation, the causal theory provides a basis to discuss the interpretation.

2. Causal theory of quantum mechanics

The causal interpretation of quantum mechanics originated with de Broglie in the 1920's (de Broglie, 1926a; de Broglie, 1926b), based on the ideas of a pilot wave. He also called it the Theory of the Double Solution, and individual discrete particles are "piloted" by a guiding wave that satisfies Schrödinger's equation. No energy is assumed to be exchanged between the particle and the guiding wave, and the wave was assumed to be ubiquitous, and does not presuppose the existence of the particle.

This interpretation was criticized by Pauli at the 1927 Solvay conference (de Discussions de Cinquième Conseil de Physique, 1928). de Broglie was unable to respond satisfactorily. He himself gave up and later recanted, and the theory was largely abandoned for 25 years. During this period, von Neumann put forward a "proof" that no theory of the type put forward by de Broglie could be consistent with quantum mechanics (Von Neumann, 1955). Theories of this type came to be known as "hidden variable theories", because in contrast to the Copenhagen interpretation, these theories ascribed a definite position and momentum to a particle at any given instant of time, and postulated that although these quantities existed, they were hidden in the ensemble averages of by our lack of knowledge of the initial conditions.

The theory was in essence reformulated by Bohm in 1952 (Bohm, 1952b; Bohm, 1952c; Bohm, 1952a). His was a direct challenge to von Neumann's 'proof' of the impossibility of hidden variables (Von Neumann, 1955), which the hidden variables are none other than saying a particle's position exists. In fact the "proof" was flawed in that it assumed that the hidden variables are local, and the Bohm/de Broglie theory was a direct counter-proof. Bohm also answered Pauli's objection by reformulating the question in terms of wave-packets, not just plane waves (Bohm, 1952b).

Bohm's work was a direct challenge to the Copenhagen interpretation, which postulated that the types of questions that were accessible with the causal interpretation, could not even be asked. Bohm not only asked the questions, but in certain cases, gave the answers as well. Unfortunately, Bohm's troubles with the House Committee on Un-American Activities in 1952 forced him to leave the USA, and his theory was subsequently often treated as heresy, with strong political overtones, as a partial consequence.

It is important to note that Bohm's theory, especially in his reply to Pauli's objections, is strongly linked to the concept of wave-packets. Wave-packets may be mathematically well defined based on the superposition principle, but less so from an interpretational point of view, including the fact that wave-packets spread out with time. They are *not* simply not Gibbs ensembles of independent particles because of the superposition principle and Pauli's objection. In fact, in certain applications wave-packets are required to calculate the properties of individual particle events or trajectories.

2.1. WAVEFUNCTION SEPARATION

The causal interpretation makes use of the perfectly natural ansatz for a complex function:

$$\psi(x) = R(x)e^{iS(x)} \qquad \psi^* = R(x)e^{-iS(x)}$$

The Schrödinger equation for $\psi(x)$ and $\psi^*(x)$ are recast as a pair of equations:

- the Hamilton-Jacobi equation $\frac{\partial \rho}{\partial t} = -\nabla \left(\frac{\rho \nabla S}{m} \right)$
- the continuity equation $-\frac{\partial S}{\partial t} = \frac{(\nabla S)^2}{2m} + V - \frac{\hbar^2}{2m} \frac{\nabla^2 R}{R}$

By identification with the Hamilton-Jacobi equation, the particle (wave-packet) is guided with a velocity proportional to ∇S by the classical and quantum potential

$$V - \frac{\hbar^2}{2m} \frac{\nabla^2 R}{R}$$

There are nuances to the interpretations in the causal theory given by different contributors (de Broglie, Bohm, Vigier, Holland), and we note that it is not the only interpretation of quantum mechanics based on this ansatz: e.g. the hydrodynamical theories of Madelung and Nelson (Nelson, 1966).

2.2. THE DOUBLE SLIT EXPERIMENT INTERPRETATION

The causal interpretation provides a beautiful interpretation of the double slit experiment. The incident wave-packet is guided by the quantum potential, which is the non-local sum of influences from both slits. The particle rides along the quantum potential surface created by the wavefunction $\frac{\nabla^2 R}{R}$, and is detected most probably in one of the potential valleys [From Philipidis 1979 (Philipidis et al., 1979)]

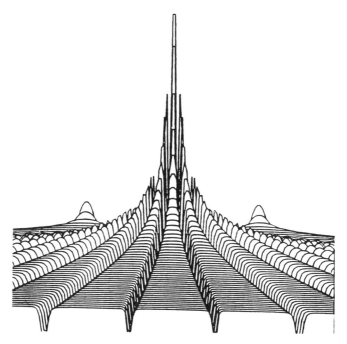

While the intuitive appeal of this interpretation is very strong, it also important to note that the form of the potential surface is a functional of the particle's mass. Thus the de Broglie wavelength, which is supposedly a property of the particle and its motion, generates the quantum potential surface by which it is guided.

The causal interpretation not only provides a beautiful interpretation of the double slit experiment, it also provides a prediction supplementary to orthodox quantum mechanics: the trajectories never cross the line of symmetry. We note that this is also predicted by Prosser's derivation based on the classical Poynting vector (Prosser, 1976)). To our knowledge, this prediction has never been tested.

However the interpretation is less obvious for multi-particle phenomenon, as the quantum potential acts in configuration space. It implies instantaneous action-at-a-distance for multi-particle correlated states (Bohm and Hiley, 1989). It also implies instantaneous action-at-a-distance for any change in the effects of a change in the boundary conditions of any given problem.

3. Tunneling

As we have mentioned, tunneling is a uniquely quantum phenomena. Conveniently it is a 1D single-particle phenomenon, which means that physical space = configuration space.

Quantum theory has very little to say in the way of predictions about tunneling. It only predicts only the transmission coefficient, based on the assumption of asymptotic plane waves. It also says that tunneling takes place without loss (or gain) of energy.

In the usual nomenclature of quantum mechanics, it treats the problem as a "stationary state," which is is start contrast to our notions of tunneling as a dynamical phenomena. The assumption in deriving the tunneling coefficient that a particle is represented by a plane wave implies that that the momentum of a particle in the barrier is imaginary. This leads to a wide variety of definitions for the barrier traversal times to choose from (Hauge and Stovneg, 1989), though most are of the order of $2mk/\hbar k(k_0^2 - k^2)$.

If we graft wave-packets onto quantum-theory as a model of the quantum particle, any use of non-monochromatic wave packets makes the transmission coefficient width dependent. This lack of agreement on even the definition of a tunneling time shows why we can't just look it up in a good book.[1]

3.1. HARD POTENTIAL BARRIERS

Most work on tunneling in a causal theory has treated hard potential barriers [From (Dewdney and Hiley, 1982; Dewdney, 1987)]

---

[1] See J. S. Bell, **Against 'Measurement'**, for an amusing yet very penetrating point of view on why one cannot answer most questions in quantum mechanics by just looking it up in a good book (Bell, 1989).

These results predict time delay in tunneling. It predicts diffraction fringes on the reflected wavepackets, the matter-wave equivalent of Wiener opticalfringes (Holland, 1992). It also predicts that tunneling only occurs for members of the ensemble of trajectories that are on the leading edge; in fact, determining the first reflected trajectory determines the tunneling coefficient.

It should be noted that these calculations require that they be done based on $\Psi(x)$ being a wave-packet (Holland, 1992, page 148); one is not free to image that wave-packets are a Gibbsian ensemble of independent trajectories. They also show an increase of momentum of about 30 % during the tunneling process.

Hard potential barriers shows spikes and holes in the quantum potential with the time development, which are often associated with step barriers. There are "diffraction fringes" in the tunneling coefficient, and it becomes non-monotonic with incident energy, as shown on the following page.

3.2. SECH SQUARED POTENTIAL BARRIERS

Inverse hyperbolic cosine squared barriers are "smooth" potential barriers that are normally free from the anomalous spikes and holes of the hard potential barrier, and in the quantum potential.

There are analytical expressions for the force and quantum potential:

$$i\hbar \frac{\partial \psi}{\partial t} = \frac{\hbar^2}{2m}\nabla^2\psi - \frac{V_0}{\cosh^2 x}\psi$$

and exact solutions in terms of the hypergeometric function ($^2F^1$)

$$\psi = c_1 \left(\cosh \tfrac{x}{a}\right)^{-2\lambda} F\left(z, z^*; \tfrac{1}{2}; -\sinh \tfrac{x^2}{a}\right) + $$
$$c_2 \left(\cosh \tfrac{x}{a}\right)^{-2\lambda} \sinh \tfrac{x}{a} F\left(z+\tfrac{1}{2}, z^*+\tfrac{1}{2}; \tfrac{3}{2}; -\sinh \tfrac{x^2}{a}\right)$$

where $\lambda = \tfrac{1}{4}\left(\sqrt{\tfrac{8\mu V_0 a^2}{\hbar^2}} - 1\right)$

and $z = -\lambda + \tfrac{ika}{2}$

QUANTUM MECHANICAL TUNNELING IN A CAUSAL INTERPRETATION

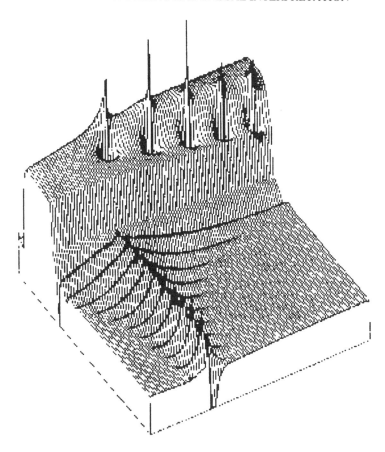

The transmission coefficient is monotonic, and given by

$$T = \frac{\sinh^2 \pi k a}{\sinh^2 \pi k a + \cosh^2\left(\frac{\pi}{2}\sqrt{\frac{8\mu V_0 a^2}{\hbar^2}-1}\right)}$$

We feel that this is a more realistic model to compare with experiments. Preliminary results also show a time delay due to tunneling. No spikes are seen due to continuous second derivative of the potential. Work is in progress to evaluate the tunneling coefficient and time delay as a function of incident energy, with wave-packets having a non-zero momentum spread.

3.3. COMPARISON WITH EXPERIMENT

Chiao at Berkeley has for the first time measured the delay time for a photon due to tunneling (Chiao et al., 1993; Chiao, 1993). He used a parametric down-converter crystal to create two coherent "wave-packets", one of which is directed along a path where it traverses a tunneling barrier. The parametric down-converter crystal create conjugate pairs at 702 nm. ($\pm$ 6 nm. rms) into 75% efficient single-photon counters gated at 500 ps. He then recombines the pair to constructively interfere, and measures the coincidence rate, or lack thereof.

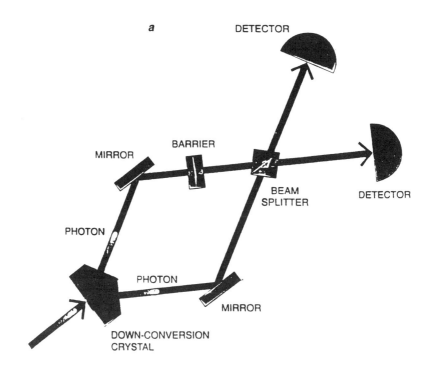

His results show that the the tunneling "delay" is negative! For a 1 micron band-gab barrier, the peak of the photon "wave-packet" appears 1.47 $\pm$0. 21 fs. *earlier*. There is an apparent tunneling velocity in the barrier is 1.7$\pm$0. 2c.

We know that energy individually conserved by tunneling process. Chiao explains this as being similar to a multi-photon pulse reshaping, although this is occuring for a single-photons.

4. Conclusions

After 60 years of debate on tunneling, for the first time the the delay time for a photon due to tunneling has been measured. This delay time appears to be in contradiction with the Bohm-de Broglie theory, in that there is no observed change in momentum of the photon. Although the objection might be raised that the calculations were carried out with a massive particle, we note that Bohmian mechanics for massless particles (photons) poses significant problems (Holland, 1992, page 541).

If there is a disagreement between experiment and theory, it does not necessarily invalidate the theory. To begin with, more realistic potential barriers need to be modeled. Also, more experimental data needed to refine the margin of error. We hope to calculate transmission coefficients in the causal theory over a wide range of potential and wave-packet shapes and sizes.

But a central part of the problem may lie in the fact that the Gaussian wave-packet is insufficient as a model of the photon. We re-iterate that in the Bohm-de Broglie theory, one cannot interpret the wave packet as being a Gibbsian ensemble of independent states, for the calculations with just a single trajectory give the wrong answer. This we are implicitly using the wave-packet as a model of the photon, which may be insufficient (Hunter and Wadlinger, 1988).

Acknowledgments

We gratefully acknowledge the liberal use of figures drawn from references (Philipidis et al., 1979), (Dewdney and Hiley, 1982) and (Holland, 1992).

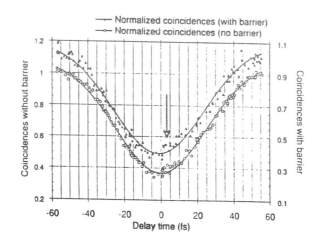

References

Bell, J. S. (1989). Against 'measurement'. In Miller, A. I., editor, *Sixty-two years of Uncertainty*, page 17. New York. Plenum.
Bohm, D. (1952a). Reply to a criticism of a causal re-interpretation of the quantum theory. *Physical Review*, 87:389.
Bohm, D. (1952b). A suggested intrepretation of the quantum theory in terms of "hidden" variables — I. *Physical Review*, 85:166. Reprinted in (Wheeler and Zurek, 1983).
Bohm, D. (1952c). A suggested intrepretation of the quantum theory in terms of "hidden" variables — II. *Physical Review*, 85:180.
Bohm, D. and Hiley, B. (1989). Non–locality and locality in the stochastic interpretation of quantum mechanics. *Physics Reports*, 172:93.
Brush, S. G. (1965). *Kinetic Theory*, volume 2. Pergamon, Oxford.
Chiao, R. Y. (1993). Measurement of the single-photon tunneling time. *Physical Review Letters*, 71:708.
Chiao, R. Y., Kwait, P. G., and Steinberg, . M. (1993). Faster than light? *Scientific American*, August:52.
de Broglie, L. (1926a). *C. R. Acad. Sci. Paris*, 183:447.
de Broglie, L. (1926b). *Nature*, 118:441.
de Discussions de Cinquième Conseil de Physique, R. (1928). *Electrons et Photons*. Gauthier-Villars, Paris.
Dewdney, C. (1987). Illustrations of the causal interpretation of one and two particle quantum mechanics. In Kostro, L., editor, *Problems in Quantum Physics; Gdansk 87*, page 62, Singapore. New World.
Dewdney, C. and Hiley, B. (1982). A quantum potential description of one-dimensional time-dependent scattering. *Foundations of Physics*, 12:27.
Einstein, A., Podolsky, B., and Rosen, N. (1935). Can quantum–mechanical description of physical reality be considered complete? *Physical Review*, 47:777. Reprinted in (Wheeler and Zurek, 1983).
Hauge, E. H. and Stovneg, J. A. (1989). Tunneling times: A critical review. *RMP*, 61(4):917.
Holland, P. R. (1992). *The Quantum Theory of Motion*. Cambridge University Press, Cambridge.
Hunter, G. and Wadlinger, R. (1988). Photons and neutrinos as electromagnetic solitons. *Physics Essays*, page 158.
Koopman, B. O. (1931). Hamiltonian systems and transformations in Hilbert space. *Proceedings of the National Academy of Sciences*, 17:315.
Misra, B. (1978). Nonequilibrium entropy, Lyapounov variables, and ergodic properties of classical systems. *Proceedings of the National Academy of Sciences*, 75:1627.
Nelson, E. (1966). *Physical Review*, 150:1079.
Philipidis, C., Dewdney, C., and Hiley, B. (1979). *Nuovo Cimento B*, 52:15.
Poincaré, H. (1893). Mechanism and experience. *Revue de Metaphysique et de Morale*, 1:534. Translated in (Brush, 1965).
Prigogine, I., George, C., Henin, F., and Rosenfeld, L. (1973). A unified formulation of dynamics and thermodynamics. *Chemica Scripta*, 4:5.
Prosser, R. D. (1976). The interpretation of diffraction and interference in terms of energy flow. *International Journal of Theoretical Physics*, 15:169.
Von Neumann, J. (1955). *Mathematical foundations of quantum mechanics*, volume 2 of *Investigations in physics*. Princeton University Press, Princeton, N.J. German edition, 1932.
Wheeler, J. A. and Zurek, W. H., editors (1983). *Quantum theory and measurement*. Princeton University Press, Princeton, N.J.
Wigner, E. P. (1963). The problem of measurement. *American Journal of Physics*, 31:6. Reprinted in (Wheeler and Zurek, 1983).

QUANTUM UNCERTAINTY, WAVE-PARTICLE DUALITY AND FUNDAMENTAL SYMMETRIES

PETER ROWLANDS
Physics Department, University of Liverpool,
Oliver Lodge Laboratory, Oxford Street, Liverpool, L69 3BX, UK

ABSTRACT

The nature of quantum uncertainty and the origin of wave-particle duality can be linked to fundamental symmetries between the parameters space, time, mass and charge, which determine that conservation and nonconservation, and continuity and discontinuity, remain *exactly opposite* properties, with precisely defined meanings. The built-in physical oppositions produced by these properties ensure that indeterminacy is inherent within the fundamental structure that underlies the whole of physics. In particular, it can be shown that no fundamental choice can be made, on any physical grounds, between wave and particle theories, between quantum mechanics and stochastic electrodynamics, and between the Einstein and Lorentz-Poincaré versions of relativity. Duality is absolute because fundamental physical differences between space and time, which emerge from basic symmetries, ensure that their mathematical combination in Minkowski space-time has no unique physical interpretation. In addition, fundamental links may be established between quantum uncertainty and conservation laws, and between irreversibility and causality, and new theorems may be derived to link the conservation laws with established physical symmetries.

1. The fundamental parameters of physics

It is natural, when confronted with profound problems like quantum mechanical uncertainty and wave-particle duality, to look for sophisticated solutions, but this does not seem to be nature's way of operating. Sophistication in the face of fundamental or universal facts seems to be evidence that we have not penetrated to a deep enough level in our understanding. The deepest and most fundamental level has also always been the simplest. But how can we reach that level? The answer seems to be that we should strip away all unnecessary assumptions and investigate directly the fundamental parameters of measurement. This might seem, at first, an impossible task, for the fundamental parameters, being simple, necessarily resist analysis. However, it might be possible to reach an understanding indirectly, if we can discover patterns of symmetry between them. In other words, we might be able to transfer problems which seem to be complicated when looked at in terms of mathematics or philosophy into much simpler ones explicable by deep-lying symmetries.

So what is the simplest and deepest level at which we can understand physics? What are the fundamental parameters of measurement? Obviously, they must include space and time, which have always been considered ultimately simple. And it seems almost equally clear that the only information which can be regarded as equally fundamental is that regarding the ultimate sources of the four known physical interactions. Two of these are known exactly, namely mass and electric charge. At the same time, although we don't know them as precisely, there must be source terms for the weak and strong forces, which, according to the rules of quantum electrodynamics, should be more like electric charge than like mass. Since Grand Unified Theories of particle physics suggest that, under ideal conditions, the three nongravitational forces would be identical in effect, I have found it convenient to refer to these sources as weak and strong 'charges', and to describe the three nongravitational sources under the collective label 'charge', in exactly the way that this concept is used when we talk about the process of 'charge conjugation' in particle physics. Space, time, mass and charge, then, are the parameters we shall take as the most fundamental concepts in physics, and in which we will make our search for fundamental symmetries. Understanding the fundamental nature of these quantities is not a philosophical issue. It leads directly to new physics, and even to new mathematics.

2. Conserved and nonconserved: mass and charge versus space and time

The conservation laws of mass and electric charge are among the most fundamental in physics, and, to the best of our current knowledge they appear to be true without exception. In all probablity, also, some type of conservation law applies to the other two nongravitational sources, manifesting itself in such aspects of fundamental particles as lepton and baryon conservation. Very significantly, the conservation laws of mass and charge are not merely global, conserving the total amounts of these quantities in the universe, but also *local*, conserving the particular amount of each quantity at a given place in a given time. Element of mass and charge have, in effect, *identities*, which are specific and permanent, and which they retain through all physical interactions, subject only to the fact that elements of charge may be annihilated by elements of the opposite sign.

As is well-known, of course, space and time are nonconserved quantities; but it is not always realised that nonconservation is the *exact opposite* of conservation and that it is just as definite a property. Nonconservation is, in fact, one of the most vital and significant of all physical properties, and it has many manifestions. Thus, just as the elements of mass and charge have individual and unchangeable identities, so those of space and time have no identity whatsoever, and this nonidentity *must be incorporated directly into physics*. Space and time, for example, are both translation symmetric: every element of space and time is exactly like every other, and must be made indistinguishable in all physical equations. And translation symmetry is not just a philosophical issue; it leads to two of the most significant laws of nature, for Noether's theorem tells us that the translation symmetries of time and space are precisely equivalent to the conservation laws of energy and linear momentum. Space, also, as a three-dimensional parameter, has rotation symmetry, meaning that there is no identity, either, for spatial *directions*. Space not only lacks a unique set of elements, but also a unique set of dimensions; and this, according to Noether's theorem, is equivalent to the conservation of angular momentum.

We could, in fact illustrate the exact oppositeness of conservation and nonconservation by defining the identity or uniqueness properties of mass and charge in terms of 'translation' *a*symmetries, translation asymmetry implying that one element of mass or charge cannot be 'translated to' or exchanged for any other within a system, however similar. This is precisely what we mean by local conservation.

But there are other manifestations of nonconservation in space and time besides translation and rotation symmetry. The whole structure of physics is founded on the fact that systems are defined by differential equations in which conserved quantities remain fixed while nonconserved quantities vary absolutely. We define conserved quantities only with respect to changes in nonconserved quantities. Quantities like energy, momentum, force or action remain constant, or zero, or a maximum or a minimum, because of the more fundamental conservation requirements involving mass and charge, while the space and time coordinates, expressed in terms of differentials, alter arbitrarily.

The absoluteness of nonconservation is illustrated by the *gauge invariance* which occurs in both classical and quantum physics. Here, electric and magnetic fields terms remain invariant while arbitrary changes are made in the vector and scalar potentials, or phase changes in the quantum mechanical wavefunction, as a result of translations (or rotations) in the space and time coordinates. In principle, gauge invariance implies that a system will remain conservative under arbitrary changes in the coordinates which do not involve changes in the values of conserved quantities such as charge, energy, momentum and angular momentum. We cannot specify an absolute phase or value of potential because we cannot fix values of coordinates which are subject to absolute and arbitrary change. And, even more significantly, this nonconservation must be local in exactly the same way as conservation is local, for, in the Yang-Mills principle used in particle physics, the arbitrary phase changes are specifically local, rather than global.

3. Real and imaginary: space and mass versus time and charge

Space and time then, are alike in respect of nonconservation, but they are by no means indistinguishable, and the mathematical combination which produces four-dimensional space-time in special relativity does not make them identical. In fact, this very combination is a source of one of the differences, for, while Pythagorean addition produces positive values for the squares of the three spatial dimensions, the squared value of time becomes negative, suggesting that time should be represented here by an imaginary number. This is often described as a 'convenient trick', but it is important to understand *why* it is convenient.

In the parallel representation of mass and charge, we have the intriguing fact, long known but never really explained, that forces between like masses are attractive, with negative sign, whereas forces between like electric charges are repulsive, with positive signs. Now, these force laws effectively square mass and charge terms, just as space and time terms are squared in Pythagorean addition. We may, therefore, choose to represent charges by imaginary numbers and masses by real ones – a procedure that would have just as much validity as using imaginary numbers in Minkowski space-time.

We might also observe that the other two forces – the strong and weak interactions – are like the electromagnetic in being repulsive for like particles, and so their sources could also be defined by imaginary numbers, if these could be distinguished from each other in some way. By good fortune, the mathematics required for such a situation is well-known and widely-used. This is the *quaternion* system, based on unity and i, j, k, the three square roots of -1. The real significance of quaternions is that they are unique. No other extension of ordinary complex algebra involving imaginary dimensions is possible: if we require a dimensional imaginary algebra (as this representation of source terms suggest we might) then we have only one possible choice: an algebra based on one real part and three imaginary. And the real part of the quaternion structure is also ready-made, for it allows us immediately to accommodate the parameter mass.

Using this mathematical structure, the three components of charge (say, ie, js, kw) begin to appear like the 'dimensions' of a single charge parameter, with their squared values used in the calculation of forces added, in the same way as the three parts of space, by Pythagorean addition; and space and time become a three real- and one imaginary-part system by *symmetry*. The requirements of algebra and symmetry, then, specify both the number of fundamental forces possible and also the number of space-time dimensions. A combination of the 4-vectors used in space-time and the quaternions used in mass-charge, has been found by the author to be identical, in principle, to the algebra used in the Dirac equation.[1-2]

But, even though charge may be a three-dimensional parameter like space, we should still expect some fundamental differences, since one parameter is conserved and the other is not. In fact, we should expect conservation in dimension for charge as well as in quantity: charge should exhibit rotation *a*symmetry. That is, we should expect separate conservation laws for the sources of the electromagnetic, weak and strong interactions, and no mechanism for interconversion. Particle theorists attempting Grand Unified Theories have been puzzled as to why the proton does not decay, but basic reasoning suggests that there may be a simple answer: the proton, which has a strong charge measured by its baryon number, cannot decay to products like the positron and neutral pion, which have none. Again, separate conservation laws for charges would easily indicate the reasons for separate laws of baryon and lepton conservation, baryons being the only particles with strong, as well as weak, components, and leptons being the only particles with weak, but no strong, components. (We should note here that the Weinberg-Salam electroweak unification only says that the forces under identical conditions are identical in effect, not that they have identical sources.)

There is yet another great advantage to an imaginary representation of charge. This stems the fact that equal representation must be given to positive and negative values of imaginary quantities. Neither positive nor negative values of imaginary numbers may be privileged in algebraic equations. In principle, every equation which has a positive solution also has an algebraically indistinguishable negative solution. Consequently, all charges must exist in both positive and negative states. This is the precise requirement for the existence of antiparticles, even for those particles, such as the neutron and neutrino, which have no electric charge, for such particles still have strong and/or weak charges whose signs may be changed under the process of 'charge conjugation'.

4. Divisible and indivisible: space and charge versus time and mass

There are yet more differences between space and time. A striking characteristic of space, for example, is that it is the only parameter which can be used in direct measurement – it is impossible to measure anything but space. 'Time'- measuring devices, in particular, all use some concept of repetition of a spatial interval, and require special conditions to be set up, whereas any object whatsoever can be used to measure space. Space, again, is reversible but time is not; indeed, time 'measurement' requires the reversibilty of space.

The reason for these differences seems to be that space, as used in all measurement applications and physical observations, is discrete. Mathematicians, of course, often try to define space in terms of a Cantorian continuum of real numbers, but this is not how we use it in practice. Space, as used, is always constructible in terms of some algorithmic process, and is therefore always countable; this is why it is the parameter used in measurement. In measurement, space is assumed to be discontinuous in both quantity and direction, meaning that it can be reversed or changed in orientation – a truly continuous quantity could not – and, without both these properties, measurement would be impossible. The whole process of measurement depends crucially on the divisibility of space, or creation of discontinuities within it. Absolute continuity cannot be visualised and any process used to describe it would deny continuity. True, the units or divisions of space, unlike those of charge, remain unfixed and indefinitely elastic, but this is because space, unlike charge, *remains nonconserved*; it cannot be fixed in any way. The elasticity of its grain size or indefinite recountability has nothing to do with Cantorian continuity. Differentiability is a property of nonconservation and can be defined by a discrete (Leibnizian) process as readily as a continuous (or Newtonian) one.

It is time, rather than space, which exhibits Cantorian or absolute continuity. Time cannot be reversed, precisely because it is absolutely continuous; any reversal of time would require some kind of discontinuity. For the same reason, time cannot be multidimensional. It is interesting that the ancient problem posed by Zeno's paradoxes disappears as soon as we accept that we can have discontinuities or divisibility in space, but not in time. The continuous, and therefore unmeasurable, nature of time seems also to be responsible for the fact that it is the independent variable in dynamical equations, while space is the dependent variable. Of course, we often read about a 'reversibility paradox', where time, according to the laws of physics is reversible in mathematical sign, when it is clearly not reversible in physical consequences. Time, however, as we have already said, is characterised by imaginary numbers, and these are not privileged according to sign. Hence, it is quite possible for time to have equal positive and negative mathematical solutions because it is imaginary (leading to a CPT rather than a CP theorem), but only one physical direction because it is continuous.

Exactly the same distinction, as between time and space, applies also to mass and charge. Mass appears to be an absolute continuum present in all systems and (as energy) at every point in space; this is why there is no negative mass and no mass 'dimensions', for either concept would necessarily require a break in the continuum. Charge, on the other hand, is divisible and observed in units (fixed because charge is conserved); it is also multidimensional. Divisibility, on this basis, seems to be the 'cause' of dimensionality. Although we cannnot easily prove this, we can at least see why *absolutely* continuous quantities cannot have more than one dimension.

5. A group of order 4

From the preceding analysis, it would seem that the properties of the four basic parameters are distributed between three sets of opposing paired categories: real / imaginary, conserved / nonconserved, divisible / indivisible, with each parameter paired off with a different partner in each of the categories:

TABLE 1. Properties of the fundamental parameters

parameter	properties		
space	real	nonconserved	divisible
time	imaginary	nonconserved	indivisible
mass	real	conserved	indivisible
charge	imaginary	conserved	divisible

This seems to be an exact symmetry: properties which match appear to be exactly identical, and properties which oppose to be in exact opposition. In a mathematical representation, this would be a noncyclic group of order 4, with any parameter as the identity element and each its own inverse. Using this group as a working hypothesis, we can investigate constraints on possible laws of physics which result from group properties. The application of a numerical relationship between the units of space and time in 4-vector space-time and those between the units of mass and charge in quaternion mass-charge, for instance, when applied to the direct and inverse relationships required for the elements of the group as a whole, suggest the necessity for at least four fundamental constants of the kind already known (including G, c and h, and one representing some Grand Unified value of charge). We can even investigate how the quaternion representation and the requirement of separate conservation for charges might determine which fundamental particle structures are possible.

Some new mathematical results can be generated by even more direct uses of the symmetries. Noether's theorem, as we have already stated, requires the translation symmetry of time to be linked to the conservation of energy, which is further linked by relativity to the conservation of mass. To put it another way, the nonconservation of time is responsible for the conservation of mass. This is a result we could have inferred from symmetry alone; and, by extending the analogy, we could link the conservation of the quantity of charge with the nonconservation, or translation symmetry, of space, which is already linked with the conservation of linear momentum. We could, therefore, propose a theorem in which the conservation of linear momentum is responsible for the conservation of the quantity of 'charge' (of any type), and, by the same kind of reasoning, we can make the conservation of *type* of charge linked to the rotation symmetry of space, and so to the conservation of angular momentum, as in the following scheme:

TABLE 2. Conserved quantities and linked symmetries

symmetry	conserved quantity	linked conservation
space translation	linear momentum	value of charge
time translation	energy	value of mass
space rotation	angular momentum	type of charge

Some special cases of these two general theorems are already known. The first incorporates the fact that the conservation of electric charge within a system is identical to invariance under transformations of the electrostatic potential by a constant representing phase changes – of the kind involved in the conservation of linear momentum. The second incorporates the link between spin and statistics, in which the spin angular momentum state of fermions and bosons depends on the respective presence or absence of a quantity of weak 'charge'.

6. Quantum mechnaical uncertainty

The application of real-imaginary 4-vectors and quaternions to a system of parameters based on group symmetry requires, as we have said, the existence of fundamental systems of units, and of algebraic laws by which they are related. This introduces the principle of *measurement*, in which fixed amounts of one quantity are set up against those of another. But it also brings us up against the first of the two major paradoxes of fundamental physics: that of quantum uncertainty. Quantum uncertainty is essentially an expression of the incompatibility of measurement and the definition of a 'conservative' *system*. The effects of the divisions between conserved and nonconserved parameters (defining the system), and real and imaginary ones (creating measurement), provide conflicting requirements for fundamental physics.

Conservative systems are defined so as to enable us to distinguish between conserved and nonconserved quantities. Effectively, we define a quantity, such as momentum, force, energy, action, or a function, such as the Hamiltonian and Lagrangian, by algebraic manipulation of the basic relations between units of measurement; and then we show that it behaves in such a way that the fundamental conserved quantities, mass and charge, remain unchanged while the nonconserved quantities, space and time, undergo continuous variation. A 'conservative' system so defined, however, would be incompatible with the principle of *measurement*. Systems and measurement cannot exist at the same time, though each is required by the fundamental symmetry. This is because measurement fixes the values of space and time, while nonconservation within a system requires them to remain unfixed.

The absolute measurement of nonconserved particle coordinates is an intrinsic impossibility, but quantum physics is required to overcome this difficulty without violating the principle of the conservative system; it effectively tells us, therefore, that physical measurement is incompatible with a system's exact definition. What Heisenberg uncertainty is telling us, then, is that a *physical* conservative system cannot be realised in practice because a 'measurement' fixes the values of space and time, quantities that, in a physical system, ought to be unfixed. In principle, a true system requires that $\Delta p = 0$, $\Delta E = 0$, $\Delta x = \infty$, $\Delta t = \infty$; changes of energy and momentum should never happen, changes of space and time should always happen. The immediate consequence of applying measurement, therefore, is that a fixed system cannot be conservative. In practice, we overcome this by deliberately making the system nonconservative, thus reducing the relationship between Δp and Δx or between ΔE and Δt to finite variations in each. Because the constant relating Δp and Δx, or ΔE and Δt, is very small in classical terms, we can make close approximations in classical physics to the ideal system, with the nonconservative aspects reduced to insignificance.

Quantum mechanics is not, thus, something of a different, probabilistic, nature imposed on an otherwise 'deterministic' system of classical physics; it is the logical result of applying measurement to a system in which the parameters space and time are *intrinsically* indeterminate. A deep understanding of the idea of nonconservation reveals that there is nothing fundamentally mysterious about the intrinsically random nature of quantum mechanics. "God playing dice" with the universe is merely a result of the absolutely symmetrical nature of the opposition of nonconservation and conservation in the fundamental parameters. As nonconserved quantities, space and time are necessarily translation and rotation symmetric, made up of units with no fixed identity, and described by values which are indeterminate within limits set only by the necessity of conserving mass and charge within a system. It is the application of the contradictory, but equally necessary, principle of measurement to such a system that forces on us the compromises of complementarity and Heisenberg uncertainty.

7. Wave-particle duality

The other great problem of contemporary physics is wave-particle duality. This is not explained by Heisenberg uncertainty, although the two may be linked by applying the de Broglie duality condition, $p = h/\lambda$, to the Heisenberg relation between momentum and position. Duality originates in an entirely separate symmetry: that between the continuous and the discontinuous parameters. Once again, the 4-vector combination of space and time – the 'Lorentz invariance' – and the related quaternion connection between mass and charge, both uniting real and imaginary quantities, is set up in opposition to a symmetry from another part of the parameter group.

Theories which incorporate Lorentz invariance as a fundamental component necessarily involve a mathematical combination of unlike physical quantities. Essentially, time and space are dissimilar in most respects – similar, in fact, only in their property of nonconservation. In particular, time is continuous, while space is discontinuous or discrete, and the same distinction occurs between charge and mass. These divisions, however, cannot be maintained when we combine them mathematically within the 4-vector or quaternion structures. Since unlike things cannot be combined by mathematical addition, we are obliged to make space timelike (and charge masslike) or time spacelike (and mass chargelike) to complete the process; and to combine all the quantities in this way requires making them either all continuous or all discrete. The choice between these alternative methods is responsible for duality.

However, as in the similar case of quantum uncertainty, we do not make all parameters discrete or all continuous by violating fundamental physical laws. The fundamental sharing of properties occurs in 'measurement' only, and not in ultimate 'reality'. When duality is combined with uncertainty, the system and measurement act together to restore the parameters to their true status.

The fundamental choice is available in many different forms: classical particles, Einstein-Minkowski relativity and quantum mechanics represent discrete options, classical waves, Lorentz-Poincaré relativity and stochastic electrodynamics continuous ones. The choice is completely arbitrary, because each option is fundamentally unphysical. Lorentz invariance is a mathematical procedure with no completely recoverable physical meaning, and the space-time combination is not, in fact, a true physical process. If we interpret it in wave terms, we obtain a classical wave theory incorporating the aether; if we choose a corpuscular explanation, we obtain special relativity combined with the Einstein process of signalling by lightquantum.[3]

Duality is absolute across the whole of physics, but it is shown most strikingly in the alternative theories of quantum and wave mechanics associated with Heisenberg and Schrödinger, which represent opposite extremes in both their definitions of the system and of measurement. Neither theory gives a complete description of reality in defining a mathematical system, and each requires completion with an ad hoc process of measurement. The Heisenberg formalism selects all the discrete options and is directly based on observables and real particles. The Schrödinger approach, on the other hand, chooses all the continuous ones, assuming a continuous, and therefore unmeasurable, interpretation of space and time in the concept of the wavefunction or state vector; particles such as electrons are delocalised and spread throughout space and time. Time, in this formalism, is no longer an observable quantity, information being derived from the wavefunction only by the application of momentum and position operators. In each case, a process of measurement, which is extrinsic to the system, restores the true attributes of the fundamental parameters which are lost in its definition.

The expressions in bold type in Table 3 represent the violations of fundamental conditions within each system, which must be corrected by the respective processes of measurement. The measurement processes introduce a virtual version of what each system excludes, thus providing a link between uncertainty and duality. In the Schrödinger measurement process, wavefunction collapse restores real localised particles in discrete space, both of which the system excludes, but only at the expense of knowledge of the wavelengths of the system. The Heisenberg measurement process, on the other hand, is made at the expense of causality, which the system retains; and, this time, measurement brings in nonlocality and the (real) vacuum.

Wavefunction collapse is, of course, not predicted within the Schrödinger *equation*. This is because the equation itself is not a true description of reality, allocating, as it does, a continuous nature to space. In the same way, the Heisenberg formalism is equally 'unreal' because it makes time discrete. The Schrödinger wavefunction is continuous, allowing no direct knowledge of time or position, and so denies causality of the discrete kind required by Einstein, until a virtual causality is introduced by the ad hoc process of measurement, when the observer introduces a particle-like discontinuity. Thus, the interpretation suggested by Born explains the squared wavefunction (squared because it links space with time in a process of measurement) as a probability amplitude when *we* (as observers) collapse the wavefunction. However, although position now becomes observable, the unobservable status of time remains. In the Heisenberg formulation, on the other hand, time is assumed to have a discrete structure, like space, and so can be brought into a meaningful uncertainty relation with energy.

TABLE 3. Comparison between Schrödinger and Heisenberg formulations of quantum mechanics

Schrödinger's wave mechanics

	The System		Measurement	
continuous	space charge momentum ang. mom.	virtual particles	restores discreteness of these	introduces localised particles
	time mass energy	real vacuum	not changed by measurement	

Heisenberg's quantum mechanics

	The System		Measurement	
discrete	space charge momentum ang. mom.	real particles	not changed by measurement	
	time mass energy	virtual vacuum	restores continuity of these	introduces nonlocalised vacuum

Much confusion has resulted from the fact that the two main formulations of quantum mechanics give the same basic results in application (FAPP, in John Bell's terminology), for their basic physical assumptions are nonetheless incompatible, and the axioms of one cannot be used to comment on those of the other. Thus the discrete time involved in the Heisenberg uncertainty relation has no meaning in wave mechanics, and cannot be carried over into the alternative theory. Its absence from the Schrödinger theory should, therefore, be the cause of no philosophical difficulty whatsoever, for that theory is 'correct' in assuming that time is continuous within the system, and so correctly leaves time continuous in applying the process of measurement. At no time, in fact, are the correct physical assumptions of either theory altered in the process of measurement.

The Schrödinger and Heisenberg models also apply the same options involving space and time to mass and charge, deviating from 'reality' by assuming, either that continuous mass is discrete, or that discrete charge is continuous. Heisenberg's quantum mechanics employs a discrete model of radiation, with discontinuous mass, and introduces the continuous vacuum and zero-point energy only as a result of measurement. This is why it is often assumed that the vacuum is a virtual concept that emerges only with the uncertainty principle. The alternative option, employed by

Schrödinger, treats the zero-point energy as real from the beginning. And this is the basis, not only of the Schrödinger theory, but also of stochastic electrodynamics and the Lorentz-Poincaré version of relativity, or the classical theory of the aether. Again, neither the discrete nor the continuous option gives an entirely true picture of reality, as each is limited by the processes of measurement and incomplete without it. There is undoubtedly a truly continuous real distribution of energy or mass in the vacuum, but matter, on the other hand (representing 'charged' particles), is discrete.

8. Irreversibility and causality

The standard (Copenhagen) interpretation of quantum mechanics assumes that measurement takes place at the interface between the 'world' and the 'measuring apparatus'. In our terminology, there is *measurement*, which fixes space and time; and a *system* (equivalent to the Copenhagen 'world') in which space and time vary absolutely. The fixing of space and time required by measurement violates the space-time properties of nonconservation, and it also violates time's additional property of continuity. Stopping time at any point to make a measurement makes it no longer continuous, and so irreversibility immediately manifests itself. In effect, measurement requires us to determine the arrow of time. The measurement process 'breaks the rules': continuous or indivisible time becomes countable, so its direction becomes manifest.

According to all known laws of physics, whether quantum or classical, time has two indistinguishable directions of mathematical symmetry, and we have found this to be characteristic of quantities determined by imaginary numbers. Physical irreversibility allows only one time direction, but, because of the mathematical indistinguishability of imaginary numbers of opposite sign, this can never be known in absolute terms. The laws of physics, in being constructed always for quantities involving the second power of time, prevent a mathematical realisation of time's direction.

In classical theories, the direction of time is associated with the concept of causality, and direct causality is characteristic of those theories which assume discrete time, such as Heisenberg's quantum mechanics and its classical analogue, special relativity. (We can hardly imagine Einstein's concept of signalling via the exchange of lightquanta without it.) Causality is the specific direction of time which appears as soon as we interrupt its flow, making it discontinuous and irreversible. Causality also requires a discontinuous mass, again as in Einstein's theory. Wave theories, however, (with their built-in gauge invariance) are not based on identifiable sequences of causally related discrete events.

In quantum mechanics, we have no need of causality until we actually make a measurement, but, in the Heisenberg theory, classical causality, though defined as part of the system, is lost, and continuity of time restored, with the uncertainty introduced with measurement. (In the Schrödinger theory, of course, causality does not appear at all.) The actions of a quantum system cannot be defined by classical causality, for measurable events cannot be separated out in a system whose definition cannot be kept apart from the process of measurement. Causality, therefore, is only known relatively when we interrupt the flow of time and discover it to be irreversible. In particular, we cannot discover an absolute causality corresponding to the absolute irreversibility of time required by the system. Causality is, in principle, an effect of measurement, where irreversibility is an effect of the system.

Irreversibility, as we have said, can only be discovered when time is made as discontinuous as space, and the squares of the two quantities are linked by Pythagorean addition. In the parallel case of mass and charge, squaring is described as 'interaction'; and so, by symmetry, interaction is linked with irreversibility. It is certainly the process of interaction, which is unique for any set of elements of mass and charge, which leads to irreversibility in physical terms, producing, for example, the wavefunction collapse in quantum mechanics. In effect, it is 'charged' particles, or sources of 'interactions', which provide the so-called 'apparatus' required for quantum mechanical 'measurements', irrespective of whether there are any 'conscious' observers. A further implication is that all interactions at the quantum mechanical level are irreversible.

Now, the continuity of mass has exactly the same cause as the irreversibility of time, and one concept, by symmetry, presupposes the existence of the other. Our lack of knowledge of the absolute direction of time is, therefore, precisely identical to the fundamental indeterminacy which would result from the interaction of infinitely-many bodies in a classical system – providing a source for a direct link between quantum mechanics and its analogue in stochastic electrodynamics. Determinacy, which is related to causality, does not exist within physical systems of any kind, even classical ones; it occurs only in the process of measurement. Indeterminacy is a characteristic of all physical systems. It has a special significance, however, in the quantum case, because, there, it is intrinsically inseparable from measurement. Quantum mechanics is the ultimate expression of the process in which all three divisions of the attributes of space, time, mass and charge – real / imaginary, conserved / nonconserved and continuous / noncontinuous – are combined and manifested, respectively, as Lorentz-invariance, indeterminacy and wave-particle duality.

9. References

1. Rowlands, P. (1994) An algebra combining vectors and quaternions: A comment on James D. Edmonds' paper. *Speculat. Sci. Technol.* **17**, 279-82.
2. Rowlands, P. (1995) Some interpretations of the Dirac algebra. *Speculat. Sci. Technol.* (in press).
3 Rowlands, P. (1992) *Waves versus Corpuscles: The Revolution that Never Was* (PD Publications, Liverpool).

ON THE CONTRADICTION BETWEEN QUANTUM MECHANICS AND RELATIVITY: A SUPERLUMINAL QUANTUM MORSE TELEGRAPH

A. GARUCCIO

Dipartimento di Fisica, Università di Bari
INFN, Sezione di Bari
Via Amendola 173, 70126 Bari, Italy
Department of Physics and Astronomy
University of Rochester
Rochester, N.Y. 14627

Abstract

An apparatus based on the intrinsic non-locality of quantum mechanics is described. This apparatus is able to transmit superluminal signals. The result is obtained using only the axioms of quantum mechanics and electromagnetism. In the apparatus a Michelson interferometer with a phase-conjugate mirror is used. The results of an experiment of Mandel, Wolf and co-authors on the interference produced on reflection at a phase conjugate mirror are reported and discussed.

1. Introduction

The incompatibility between quantum mechanics and any realistic local theory has been a matter of long debate and many experiments have been performed to discriminate between these two different approaches.

We can briefly summarize the situation in this way: Most physicists working on the subject are convinced that experiments have confirmed, without any reasonable doubt, the validity of quantum mechanical predictions and the intrinsic non-locality of this theory. However, they claim that this feature of quantum mechanics cannot give rise to any contradiction with relativity because it is not possible to use the correlated, non-local, quantum mechanical states to transmit signals faster than light.

Only a small number of physicists criticize these experiments on the basis of the important role played by some supplementary assumptions to claim violation of the predictions of Einstein Locality in this class of experiments. These researchers claim that the experimental results have not disproved Einstein Locality and that the issue of the compatibility between quantum mechanics and relativity has not yet been resolved.

Recently, there has been increased interest in exploiting the intrinsic nonlocality of quantum mechanics to study new phenomena like teleportation of a quantum state and quantum cryptography. In this letter we will prove that the intrinsic nonlocality of quantum mechanics leads to the possibility of superluminal communication and, therefore, contradicts the postulates of relativity. More precisely, we will describe an apparatus, based on quantum mechanics, which can permit binary signals to travel nearly instantaneously between distant points.

2. The superluminal quantum Morse telegraph (SQMT)

Let us consider a source S of correlated photon pairs. The two photons are emitted in opposite directions along the z axis and are correlated in polarization (Fig. 1). The state which describes the polarization of the pair is the "entangled" singlet state of positive parity

$$|\psi\rangle = \frac{1}{\sqrt{2}} \{|x_1\rangle|x_2\rangle + |y_1\rangle|y_2\rangle\} \quad (1)$$

where $|x_1\rangle, |y_1\rangle$ and $|x_2\rangle, |y_2\rangle$ are the two orthogonal polarization states along an arbitrary x-y frame for the first and second photons, respectively.

The state (1) is rotationally invariant along the z-axis; i.e. the mathematical structure of the state (1) remains unchanged if the polarization basis is rotated through an arbitrary angle around the z-axis.

Moreover, the state (1) has the property of preserving its mathematical structure as a symmetric entangled state even when a circular polarization basis is used.

Indeed, if we define

$$|R_1\rangle = \frac{1}{\sqrt{2}} \{|x_1\rangle + i|y_1\rangle\}, \quad |L_1\rangle = \frac{1}{\sqrt{2}} \{|x_1\rangle - i|y_1\rangle\},$$

$$|R_2\rangle = \frac{1}{\sqrt{2}} \{|x_2\rangle - i|y_2\rangle\}, \quad |L_2\rangle = \frac{1}{\sqrt{2}} \{|x_2\rangle + i|y_2\rangle\},$$

we can write (1) as

$$|\psi\rangle \frac{1}{\sqrt{2}} \{|R_1\rangle|R_2\rangle + |L_1\rangle|L_2\rangle\}. \quad (2)$$

The equivalence between these two representations has been tested experimentally [1,2,3].

Let us suppose now that the "polarization detection area" P is placed in the optical path of the first photon beam at a distance very far from the source S (for example one light-year). In this area an experimenter can choose whether to measure linear polarization or circular polarization. In the first case, he will detect the photon along one of the two channels of a linear polarization analyzer oriented along an arbitrary, but fixed, direction x. In the second case, he will insert the quarter-wave plate Q along the path of the beam with the optical axis at an angle of 45° with respect the direction x. The quarter-wave plate changes the phase relationships between the linear polarization components of the light, so that, for example, light which enters the waveplate with right-handed circular polarization will emerge linearly polarized along x, and light which enters with left-handed circular polarization will emerge with linear y-polarization. The new x and y components are then seperable as before by the linear polarization analyzer, and subsequent measurements of these components are equivalent to measuring circular projections of the initial polarization state.

In both the cases, the experimenter will perform a measurement on each impinging photon and, therefore, he will induce a quantum wave collapse.

If the experimenter measures linear polarization, the state of the photon pair will collapse into the mixture

$$50\% \quad |x_1\rangle|x_2\rangle$$
$$50\% \quad |y_1\rangle|y_2\rangle. \quad (3)$$

Vice versa, if the experimenter choices to measure circular polarization, the final state will be the mixture

$$50\% \quad |R_1\rangle|R_2\rangle$$
$$50\% \quad |L_1\rangle|L_2\rangle. \quad (4)$$

The two mixtures (3) and (4) are completely indistinguishable with a measurement of polarization performed only on the second photon, since the expectation value of any polarization operator for the second photon is always equal to zero for both the mixtures (3) and (4), and many authors have argued that this proves the impossibility of using the quantum nonlocality to transmit signals faster than light.

Now we will show that an interference apparatus based on phase-conjugation is able to distinguish the state (3) from (4) by performing measurements only on the second photon.

Let us suppose that the photon 2 impinges after a year from its production into the "interference detection area" I. In this area a phase-conjugation interferometer is constructed; it consists of a Michelson interferometer in which one metallic mirror is replaced by a phase conjugate mirror (P.C.M.). Let us first describe the behavior of this apparatus with an ideal PCM. In the next section we will describe the theory and the results of an experiment of interferometry with a real PCM.

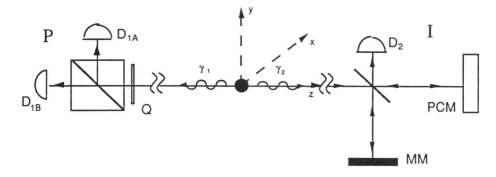

Fig.1. Outline of the superluminal quantum Morse Telegraph (SQMT). The first photon impinges on the polarization area P, while the second photon impinges on the interference area I. Q is a quarter-wave plate that can be inserted or removed from beam 1 to detect linear polarization or circular polarization, respectively. In the area I one metallic mirror of a Michelson interferometer has been substituted for a phase-conjugate mirror.

A PCM is a non linear medium that performs a complex conjugation on the spatial part of the complex amplitude of an impinging electromagnetic field. The effect is to reflect the impinging wave back in the direction of propagation while inverting the phase. Consequently, the PCM maintains the polarization state of the incident wave, more precisely, a linear polarized wave is reflected as a linear polarized wave, while a circular polarized wave is reflected with identical circular polarization[4]. This behavior of a PCM is different from that of a conventional mirror which inverts the circular polarization state of the light, and has an interesting consequence on the visibility of the interference in the phase-conjugate interferometer. If the impinging light is linearly polarized, for example in the x-direction, the splitting of the beam by the beamsplitter BS and the reflection of the produced beams on the metallic and phase-conjugate mirrors do not produce any variation of the polarization. So, when the two beams overlap at the beamsplitter, an interference effect will occur with an oscillation of the amplitude which is a function of the optical path of the two beams. Conversely, if the impinging beam has, for example, right-handed circular polarization $|R_2\rangle$, the beam reflected from the metallic mirror has left-handed circular polarization, while the beam reflected from the PCM maintains its right circular polarization. So, when they overlap at the beamsplitter BS, they will not interfere, since they are in two orthogonal states of polarization.

In conclusion, the phase-conjugate interferometer is able to distinguish states of circular polarization from states of linear polarization. This means that if the experimenter in P measures linear polarization on the first beam, the state vector (1) is collapsed in the mixture (3) and the experimenter in the interference area I sees, at the same time, an interference effect on the second beam. Vice versa, if the experimenter in P measures circular polarization, the state (2) collapses in the mixture (4) and the experimenter in I cannot see any interference effect.

If the two experimenters agreed in the past to interpret the non-detection of the interference as a "dot" and the detection of the interference as a "line", the experimenter in P has the possibility of transmitting messages in the Morse language with a velocity that is greater than the velocity of the light.

Of course, the experimenter in the interference area I needs to collect a given quantity of photons in order to distinguish interference from non-interference, so the detection of a signal (dot or line) requires a time t. If the photons spend a time T arriving from the source S to the detection areas, then the velocity of transmission of the Morse signal is

$$v = 2cT/t, \qquad (5)$$

where c is the velocity of light. The velocity v can be greater than c, if T/t is greater than 0.5.

In order to make a numerical evaluation of t, let us suppose that n is the rate of the emitted pairs per second, and that the mirrors of the interferometer are positioned to obtain constructive interference in the photodetector D_2. When linear polarized photons impinge on the interferometer giving rise to the interference effect, the photons detected in D_2 during the measurement time t will be $N_L = nt$. When circularly polarized photons impinge without producing interference, the detected photons will be $N_C = nt/2$. If we require that the two numbers differ at least 5 standard deviations and we assume that $\sigma_L = \sqrt{N_L}$, the measuring time t must the inequality

$$\frac{N_L - N_C}{\sqrt{nt}} \geq 5, \quad <=> \quad t > \frac{100}{n} \quad .$$

If, for example, n = 10,000 pairs/sec, the measuring time for a single signal will be t = 0.001 sec and the minimum distance between the two experimenters at which the quantum telegraph operates as a superluminal apparatus is 300 Km.

3. Interference produced on reflection at a PCM

The interference produced on reflection at phase-conjugate mirror has been studied in detail both theoretically and experimentally [5,6,7]. We will summarize here the main results in order to apply it to our apparatus.

Let us consider (Fig.2) an electromagnetic wave with wave vector **k** incident upon the PCM and frequency $\omega/2\pi$

$$\mathbf{E}^{(i)}(\mathbf{r},t) = \varepsilon\, A^{(i)} \exp[i(\mathbf{k}\cdot\mathbf{r} - \omega t)]. \qquad (6)$$

where ε is the complex unit polarization vector satisfying the condition $\varepsilon^* \cdot \varepsilon = 1$.

Let us suppose the PCM is pumped with two beams of identical frequency $\omega/2\pi$ (Fig.2). The reflected wave leaving the PCM is given by

$$E^{(r)}(r,t) = \mu\, \varepsilon^*\, A^{(i)*} \exp[i(-k \cdot r - \omega t)]. \qquad (7)$$

where μ is the complex reflectivity of the mirror and depends on the intensity of the pumping beams, the strength of the coupling between pumping beams via the nonlinear susceptibility, and the length of the PCM. In general μ is less than 1, but it can be equal to or greater than 1 under well defined conditions[5].

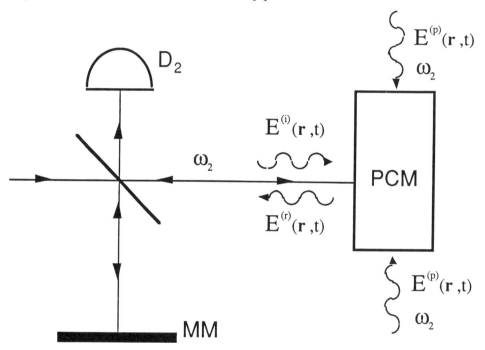

Fig. 2. The modified Michelson interferometer. One of two metallic mirrors is replaced with the phase-conjugate mirror PCM. This mirror is pumped by two optical beams of the same frequency as the impinging wave and with opposite direction.

The superposition of the incident and reflected fields result in a total field

$$E(\mathbf{r},t) = E^{(i)}(\mathbf{r},t) + E^{(r)}(\mathbf{r},t) =$$
$$= (\varepsilon\, A^{(i)} \exp[i\mathbf{k}\cdot\mathbf{r}] + \mu\, \varepsilon^*\, A^{(i)*} \exp[-i\mathbf{k}\cdot\mathbf{r}]) \exp[-i\omega t]. \quad (8)$$

and in a total light intensity

$$I(\mathbf{r},t) = |E(\mathbf{r},t)|^2 =$$
$$|\varepsilon|^2 |A^{(i)}|^2 + |\mu|^2 |\varepsilon|^2 |A^{(i)}|^2 + \mu^* \varepsilon^2 A^{(i)2} \exp[-2i\mathbf{k}\cdot\mathbf{r}] + c.c.. \quad (9)$$

Since $|\varepsilon|^2 = 1$, and assuming

$$\mu = |\mu| \exp[i\phi],$$
$$A^{(i)} = |A^{(i)}| \exp[i\alpha],$$
$$\varepsilon^2 = |\varepsilon^2| \exp[i\delta],$$

we have

$$I(\mathbf{r},t) = |\varepsilon|^2 |A^{(i)}|^2 (1 + |\mu|^2) + 2 |\mu| |\varepsilon^2| |A^{(i)}|^2 \cos(2\mathbf{k}\cdot\mathbf{r} - \phi + \delta + 2\alpha), \quad (10)$$

and the visibility of the interference

$$V = \frac{2\,|\mu|\,|\varepsilon^2|}{1 + |\mu|^2}$$

exhibits a maximum when $|\mu| = 1$ and $|\varepsilon^2| = 1$.

If we make $|\mu| = 1$, and if the impinging wave is x-linearly [y-linearly] polarized, we have $\varepsilon = \mathbf{i}$ [$\varepsilon = \mathbf{j}$] (**i** and **j** are the vectors of the x-axis and y-axis, respectively), and $\delta = 0$. Then, the scalar product ε^2 is equal to 1, and the visibility of the interference V is equal to 1.

Vice versa if the impinging light is right-handed [left-handed] circularly polarized, the polarization vector is $\varepsilon = (\mathbf{i} + i\mathbf{j})/\sqrt{2}$ [$\varepsilon = (\mathbf{i} - i\mathbf{j})/\sqrt{2}$], the scalar product ε^2 is equal to zero, and consequently $V = 0$.

Mandel, Wolf, and co-authors[6,7] confirmed these predictions with an experiment (Fig. 3) in which a PCM is inserted in a Michelson interferometer. In the experiment a phase shifter was introduced in different positions in order to vary the phase of the incident and pumping waves. The displacement of the position of the interference pattern was determined from measurements of the light intensity impinging on the photodetector PD.

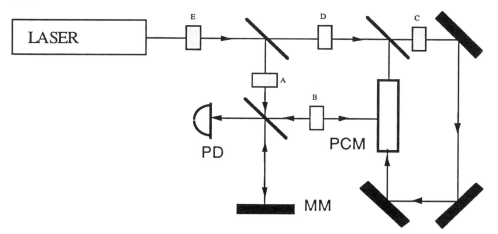

Fig.3. Outline of the experimental setup to verify the existence of interference produced at a reflection at a phase-conjugate mirror. A phase shifter was introduced at one of five different positions A-E, and the position of the interference patter was determined with respect an arbitrary reference point by the photodetector PD. The predictions of formula (10) were completely verified.

4. Conclusions

The experiment of Mandel, Wolf, and co-authors proves that the Michelson interferometer with a phase-conjugate mirror is able to verify the interference between a linearly polarized impinging wave and its conjugate wave in the case of real PCM. Therefore this interferometer can be used as receiver in a superluminal quantum Morse telegraph, and this result opens the way to an experimental test of the contradiction between quantum mechanics and theory of relativity.

It is worth noting that in our reasoning the assumption of the existence of a physical reality has never been used. In this sense this result is not a different version of the paradox of Einstein-Podolsky-Rosen[8], but a new paradox of quantum mechanics. If the existence of superluminal signals can be proved experimentally, the complete non-locality of this theory will be confirmed and we will be compelled to reject the relativistic postulate of nonexistence of superluminal signals.

Acknowledgments

The author wishes to thank Prof. L. Mandel for the hospitality shown in his laboratory and the discussions that were the starting point for this paper. Dr. C. H. Monken , D. Branning, and J.R. Torgerson also provided useful discussions. The author is indebted to D. Branning and J.R. Torgerson for the careful reading of the manuscript. The author would like to acknowledge the Administration Council of the University of Bari and the Italian Istituto Nazionale di Fisica Nucleare for supporting his research work at the University of Rochester.

References

1] Clauser, J.F. *Nuovo Cimento* **338**, 740 (1976)

2] Duncan, A.J., Kleinpoppen, H., and Sheikh, Z.A. in: *Bell's theorem and the Foundations of Modern Physics, Edited by A. van der Merwe, F. Selleri, and G. Tarozzi p.* 161 (Word Scientific, Singapore, 1992).

3] Torgerson, J.R., Branning, D., Monken, C.H., and Mandel, L. *Phys. Rev A* 51, 4400 (1995)

4] Pepper, D. M. Nonlinear optical phase coniugation in *Laser Handbook Vol $*, edited by M.L. Stitch and M. Bass, Nort- Holland , Amsterdam (1985)

5] Boyd, R.W., Habashy, T.M., Mandel, L., Nieto-Vesperinas, M., and Wolf, E. *J.Opt. Soc. Am.* **4B**, 1260 (1987)

6] Jacobs, A.A., Tompkin, W.R., Boyd, R.W., and Wolf, E. *J.Opt. Soc. Am.* **4B**, 1266 (1987)

7] Boyd, R.W., Habashy, T.M., Jacobs, A.A., Mandel, L., Nieto-Vesperinas, M., Tompkin, W.R., and Wolf, E. *Opt. Lett.* 12 **42**, (1987)

8] Einstein, A., Podolsky, B., and Rosen, N. *Phys. Rev.* **47**:777 (1935).

INCOMPATIBILITY BETWEEN EINSTEIN'S RELATIVITY AND LORENTZ'S EQUATIONS

PAUL MARMET
*Physics Department,
University of Ottawa,
Ottawa, Ontario,
Canada K1N 6N5*

Abstract

Lorentz has developed a set of equations so that one could transform the parameters x, v and t from one frame of reference into a new one, taking into account the invariance of the velocity of light. Almost at the same time, Einstein proposed a physical model and synchronization methods to demonstrate the relativity of simultaneity. We show that Einstein's prediction of relativity of simultaneity is not compatible with the Lorentz's equations.

1 Einstein's Relativity

Einstein[1] predicted several phenomena that have been clearly verified experimentally. Clock retardation is an important one. However, at least one consequence predicted by relativity has never been proved experimentally. In particular, when two events appear simultaneous in one frame of reference, it is predicted that they do not necessarily appear simultaneous in another frame of reference. In other words, Einstein claims relativity of simultaneity. This prediction is not a fundamental hypothesis in relativity; it is only a predicted consequence that has never been confirmed experimentally.

Surprisingly, that conclusion is not compatible with the Lorentz transformations which are the basis of relativity. Several papers[2-5] have been written pointing out the incompatibility of Einstein's description.

1.1 EINSTEIN'S DEMONSTRATION

Using Einstein's own example let us calculate the time interval between the arrival of light from two events taking place simultaneously in A and B, when observed at the final location R_f located at the middle of the train as shown in Figure 1.

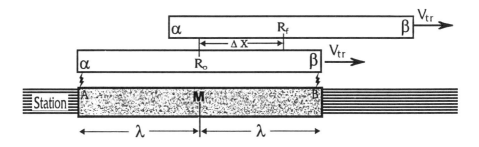

Figure 1

This represents the station, the initial and final train location.

Einstein[1] does not use the independent reference points α and β on the train. He refers only to points A and B where the events are produced so that α is indistinguishable from A and β is indistinguishable from B. The length of the moving train is such that α coincides with A, then β coincides with B.

The observer M at the middle of the station always uses the Station Frame of Reference and the train traveler always uses the Train Frame of Reference. We will identify here the Station Frame of Reference by "Sta" and the Train Frame of Reference by "Train". This choice is preferred to the one often used in mathematics. In relativity, Einstein considers that there is no absolute frame of reference, therefore it is not appropriate to refer to a stationary and a moving frame of reference. For this reason, we will give names to frames of reference such as Station Frame of Reference and Train Frame of Reference.

Fig. 1 contains the following parameters: λ=half length A→M of the station in "Sta" (equal to M→B), R=observer's location at the train center, M=station center and V_{tr}=train velocity with respect to the station.

At t=0, two events take place at A and B respectively at the moment the initial position of the observer (in R_o) passes exactly in front of M (Fig. 1). Then, the distance of the middle of the train R_o from source A is:

$$(A \rightarrow R_o)(\text{"Sta"}) = \lambda \, (\text{"Sta"}) \tag{1}$$

Since the observer moves at velocity V_{tr}, the distance of the observer R from A as a function of time t is (in "Sta"):

$$(A \rightarrow R_f)(\text{"Sta"}) = \lambda \, (\text{"Sta"}) + V_{tr} \, t_\ell \, (\text{"Sta"}) \tag{2}$$

We define t_ℓ ("Sta") as the time for light to travel the distance across the left-hand side of Fig. 1 before reaching the train observer in R_f as seen by the observer on the station. On the left-hand side of the experiment, we have:

$$t_\ell(\text{"Sta"}) = (A \rightarrow R_f)(\text{"Sta"})/c \tag{3}$$

Putting Eq. 2 into Eq. 3, we find:

$$t_\ell(\text{"Sta"}) = \left\{ \frac{\lambda(\text{"Sta"}) + V_{tr} \, t_\ell(\text{"Sta"})}{c} \right\} \tag{4}$$

$$t_\ell(\text{"Sta"}) \left\{ 1 - (V_{tr}/c) \right\} = \frac{\lambda(\text{"Sta"})}{c} \tag{5}$$

$$t_\ell(\text{"Sta"}) = \left\{ \frac{\lambda(\text{"Sta"})/c}{(c - V_{tr})/c} \right\} \tag{6}$$

We find:

$$t_\ell(\text{"Sta"}) = \frac{\lambda(\text{"Sta"})}{(c - V_{tr})} \tag{7}$$

Eq. 7 is exactly the same equation as the one published by Einstein[1] in 1905. Eq. 7 gives the time t_ℓ taken by light to travel from A to R_f in the Station Frame of Reference ("Sta"). It is not in the Train Frame of Reference ("Train") as Einstein concluded.

In a similar way, it can be easily shown that the time $t_\curvearrowright$("Sta") (on the right-hand side of the experiment) for a photon starting in B (then located in front of β) to reach the center R_f of the moving train is:

$$t_\curvearrowright(\text{"Sta"}) = \frac{\lambda(\text{"Sta"})}{(c + V_{tr})} \tag{8}$$

This Eq. 8 is the same as the one given by Einstein. It is also in frame "Sta" instead of in frame "Train". We see that Eqs. 7 and 8 give different times as seen from "Sta". That difference is interpreted by Einstein to imply that light does not arrive simultaneously at the train center R_f which is in the "Train". Einstein[1] concludes: *"Observers moving with the moving rod would thus find that the two clocks were not synchronous"*. However, Einstein[1] also writes that it is "measured in the stationary system".

Einstein[1] and others[7,8] conclude that the train observer in R_f sees the flashes as non simultaneous. However, it is in frame "Sta" that the pulses are seen not to arrive simultaneously in R_f when issued from A and B. Since the observer R_f is not in frame "Sta", the demonstration of non simultaneity is not valid for the observer in the train frame of reference.

1.2 THE WEAKNESS OF THE EXPRESSION "EVENT" IN RELATIVITY

In the study of simultaneity discussed above, we have used terms so that the discussion could be compatible with previous articles. For example, Einstein[1] uses the word *"event"* several times when he describes the experiment. Einstein writes: "... *to connect in time series of events* ..." or "... *can determine the time values of events* ..." etc. This subtle difference between the time of an *"event"* and the moment when a "source of light is emitting a pulse of light" is fundamentally important. In Einstein's experiment, fundamental information is missing when he refers to *"events"* instead of moments when light is emitted or received. When there is an event, everybody understands that it takes place at a given location and at a given time. However, it is clear that nobody will ask the question: "What is the velocity of the event?" Scientifically, an "event" is a sort of "handicapped" expression that contains the information about location and time but is deficient in information about the velocity. For an event, the values of the coordinates x and time t exist, but the derivative $v = dx/dt$ does not exist. To be more specific, Einstein considers that an event taking place in A is indistinguishable from an event taking place in α. However, α is moving with respect to A. In that case, one cannot use the same velocity v for A as for α in the Lorentz transformations.

Instead of mentioning an *"event"*, we can consider the moment when an electric discharge generates a pulse of radiation between two moving bodies coming momentarily in contact. This is a complete physical description since the emitting atoms located in α and in A possess locations and velocities which are necessary for the application of the Lorentz transformations. Einstein ignored the fact that the event between the station and the train implies two different velocities: the train velocity and the station velocity. Therefore it is necessary to consider two pairs of clocks α and A and also β and B to take into account the two different velocities at each location. Since an event has no velocity, it was impossible for Einstein to apply completely the Lorentz's transformation.

1.3 EINSTEIN'S CLOCK SYNCHRONIZATION

In Einstein's experiment, clocks A and B are first synchronized. Clocks A and α are synchronized by definition since they are adjacent, and Einstein represents them as a unique clock. The same synchronization exists by definition between clocks B and β. Then Einstein[1] brings his hypothesis #2: *"If the clock at A synchronizes with the clock at B, and also with the clock at C, the clocks at B and C also synchronize with each other."*

Let us examine Fig. 1. Since A is synchronized with B and B with β, Einstein's statement means that β is then synchronized with A. Let us verify the compatibility of that statement with the Lorentz transformations.

1.4 LORENTZ TRANSFORMATION

Lorentz found relationships between different frames of reference that are compatible with a constant velocity of light in different frames of reference. Therefore, any synchronization must be compatible with the Lorentz equations. Using the principle of invariance and substituting Lorentz's notation with primes by "Train" and the notation with non primes by "Sta", the Lorentz transformations become:

$$t(\text{"Train"}) = \gamma \left\{ t - vx/c^2 \right\}(\text{"Sta"}) \quad \text{and} \quad x(\text{"Train"}) = \gamma \left\{ x - vt \right\}(\text{"Sta"}) \tag{9}$$

where:
$$\gamma = \left\{ 1 / \sqrt{1 - (v/c)^2} \right\} \tag{10}$$

Where v represents the relative velocity between systems.

2 Clock Synchronization

Einstein described a synchronization method between clocks. However, they could not be tested for their compatibility with the Lorentz transformations since they were not known to Einstein.

2.1 SYNCHRONIZATION OF A WITH B

Einstein's first synchronization method is related to the synchronization of clock A with clock B. Clock A is in the same frame of reference as clock B. Therefore, their relative velocity v is zero. Since time $t_A(\text{"Sta"})$ is set to zero, using the Lorentz Eq. 9, we find that time $t_B(\text{"Sta"})$ is also zero. We have replaced (in Eq. 9) $t(\text{"Train"})$ by $t_B(\text{"Sta"})$, $t(\text{"Sta"})$ by $t_A(\text{"Sta"})$ and v by $v_{AB}=0$. This shows the compatibility of synchronization of clock A with B with the Lorentz transformations. Therefore:

$$t_B(\text{"Sta"}) = t_A(\text{"Sta"}) \tag{11}$$

2.2 SYNCHRONIZATION OF α WITH A

In order to verify the synchronization of α with A, we have to consider the relative velocity $v=V_{tr}$. However, since α is adjacent to A, the distance x is zero. Since we set the clock $t_A(\text{"Sta"})$ to zero, Eq. 9 gives $t_\alpha(\text{"Train"})=0$:

$$t_A(\text{"Sta"}) = t_\alpha(\text{"Train"}) \tag{12}$$

In the case of synchronization of β with B, we find in the vicinity of B the relation:

$$t_B(\text{"Sta"}) = t_\beta(\text{"Train"}) \tag{13}$$

2.3 SYNCHRONIZATION OF β WITH A

Einstein uses his hypothesis #2 recalled above in section 1.3 in order to synchronize β with A. Let us test that hypothesis against the Lorentz equation. What is the time on a clock facing B located on the train moving at a velocity v, as seen from A (Fig. 1). The distance of that clock from A is equal to 2λ. The velocity v of the train is equal to V_{tr}. From the Lorentz equation (Eq. 9) the time on that clock as seen by the observer in A is:

$$t_{(\text{facing B})}(\text{"Train"}) = \gamma \left\{ t_A - V_{tr} 2\lambda/c^2 \right\} (\text{"Sta"}) \qquad (14)$$

Similarly, the time shown by a clock facing A as seen by observer B, is, according to the Lorentz transformation:

$$t_{(\text{facing A})}(\text{"Train"}) = \gamma \left\{ t_B + V_{tr} 2\lambda/c^2 \right\} (\text{"Sta"}) \qquad (15)$$

Here in Eq. 15, V_{tr} is negative because the relative distance decreases with time. This negative sign is required since there can be no absolute positive direction nor negative direction in relativity. Only the relative distance has a physical meaning. We see that in this case, time on the moving clock (facing A) observed from a remote point of observation (B) is not equal to zero if the time in B (and in A) is zero. This is in contradiction with Einstein's hypothesis #2 mentioned above stating that if clock A synchronizes with clock B and also with C, the clocks B and C also synchronize with each other.

This means that this Einstein's synchronization hypothesis is not compatible with the Lorentz equations and therefore is not compatible with the constancy of the velocity of light. Consequently, all conclusions deduced from that hypothesis are not compatible with the constant velocity of light.

2.4 CLOCKS CHARACTERISTICS

The synchronization of all clock mechanisms must be compatible with the Lorentz equations. First, each clock must have been built with a mechanism that maintains an identical rate. This means that two clocks located in the same frame of reference must continuously show the passage of identical units of time as a function of time. Atomic clocks provide stability and excellent reliability. The second operation consists of synchronizing the clocks. This means establishing an initial moment $t=0$ for which both clocks are adjusted to the same setting.

In Newtonian mechanics, it is understood that after synchronization, a pair of clocks always stays synchronized as a function of time. However, in relativity, when a clock is brought to a different velocity, the Lorentz equations show that the time shown by that clock will be different. The Lorentz equations show that this is a necessary consequence of the constancy of the velocity of light.

2.5 LORENTZ'S CLOCK SYNCHRONIZATION

In relativity, we use clocks that have previously been designed to show identical time when they are observed in the same frame of reference.

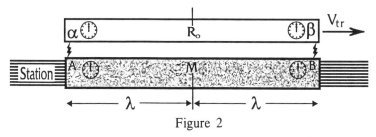

Figure 2

Synchronization between clocks on the station and on the train.

On Fig. 2, let us first synchronize clock B with respect to clock A, located some distance away in the same frame of reference. This can be done by asking observer A to send a pulse of light toward B and having B to reflect light back to A. The pulses of light between A and α and between B and β are produced by an electric discharge between the train and the station at those points. By measuring the time interval for light to make the return trip between A and B, the two clocks A and B in frame "Sta" can be synchronized. We have shown above that this method of synchronization used by Einstein is compatible with the Lorentz transformations. Therefore:

$$t_A(\text{"Sta"}) = t_B(\text{"Sta"}) \tag{16}$$

A different method must be used to synchronize clock α which is moving with respect to A. A sudden electric discharge between α ("Train") and A ("Sta") (see Fig. 2) is observed by both observers α and A which are adjacent. At the moment of an electric discharge between α and A, observer α sets his own clock to the same setting (time equals zero) as the one seen on clock A. We have verified above (section 1.3) that the foregoing synchronization is compatible with the Lorentz equations. On the other end of the train, another observer in β also synchronizes his clock at the moment β makes a sudden electric discharge with B. Observer β sets his clock to the same setting as the one he observes on clock B.

We have seen above that although clock A is synchronized with clock B, and also that clock B is synchronized with β, the Lorentz equation gives that clock β is not synchronized with clock A (and α is not synchronized with B).

Let us calculate the time in β as seen by the observer in A in the case of figures 1 and 2. At time t=0("Sta"), the distance x("Sta") between A and β is equal to 2λ and the velocity v is $+V_{tr}$ (V_{tr} is positive since dx/dt increases with time). Equation 10 shows then that when the observer in A observes a time equal to zero, written t_A("Sta")=0, then time on clock β, written $t_β(A)$("Train") observed by A is:

$$t_\beta(A)(\text{"Train"}) = \gamma \left[\frac{-2\lambda V_{tr}}{c^2} \right] \quad (17)$$

where (A) is the point where the observations are made, and the index β of t (Eq. 17) is the clock observed.

Then clocks A and β must show a difference of time when observed from A in order to stay compatible with the Lorentz transformations. Similarly, the Lorentz equations show that clock α does not show the same time as A when observed from B. When time in B equals zero, as observed from B, the distance x("Sta") between clocks α and B equals 2λ and the relative velocity v equals $-V_{tr}$ (minus sign of V_{tr} because dx/dt decreases). From Eq. 9, the time on clock α as seen from B ($t_α(B)$("Train")) is:

$$t_\alpha(B)(\text{"Train"}) = \gamma \left[\frac{+2\lambda V_{tr}}{c^2} \right] \quad (18)$$

For the observer located in M, in the middle of the station, the Lorentz Equation 9 predicts that the time on clock α observed from M ($t_α(M)$)("Train"), at time t(M)("Sta")=0 is:

$$t_\alpha(M)(\text{"Train"}) = \gamma \left[\frac{+\lambda V_{tr}}{c^2} \right] \quad (19)$$

For the observer in M, the time on clock β observed from M

$(t_\beta(M))(\text{"Train"})$, at time $t(M)(\text{"Sta"})=0$ is:

$$t_\beta(M)(\text{"Train"}) = \gamma \left[\frac{-\lambda V_{tr}}{c^2}\right] \quad (20)$$

For observer M the difference of time between clock α and clock β at $t(M)=0$, given from Eq. 19 and 20 is:

$$t_\beta(M)(\text{"Train"}) - t_\alpha(M)(\text{"Train"}) = \gamma \left[\frac{-2\lambda V_{tr}}{c^2}\right] \quad (21)$$

3 Physical Interpretation

Equation 21 deserves more explanation. This equation gives the difference of time between clocks α and β measured by an observer on the "Sta". In fact, that time difference is independent of any location on the "Sta". For example, an observer located in A sees clock α showing the same time as clock A; also, an observer in B sees the same time on clock β since they are at distance zero. However, when clocks B and β meet each other, the observer in A, finds that clocks β and B do not show the same time. We have seen above that this difference in time is absolutely necessary to be compatible with the Lorentz transformations and the constancy of the velocity of light.

3.1 TRAVEL TIME ("Train").

In order to compare Einstein's predictions on simultaneity with Lorentz transformations let us compare the time taken by light to travel from A to R_f with the time taken for light to travel from B to R_f, as seen by R in the Train Frame of Reference("Train"). Let us use the Lorentz transformations.

We have established that the proper value of the velocity of light is constant in any frame of reference. We can compare the time $t_\ell(\text{"Train"})$ taken by light to cross the distance on the left-hand side of the diagram $x_\ell(\text{"Train"})$ (from A to R_f) with the time $t_n(\text{"Train"})$ to cross the right-hand side $x_n(\text{"Train"})$ (from B to R_f) by dividing those two distances by c. Therefore we have:

$$t_\ell(\text{"Train"}) = x_\ell(\text{"Train"})/c \quad \text{and} \quad t_n(\text{"Train"}) = x_n(\text{"Train"})/c \quad (22)$$

From Eq. 9, we have:
$$x_\ell(\text{"Train"}) = \gamma(x_o - vt)(\text{"Sta"}) \quad (23)$$
where x_o is the distance traveled by light.

3.2 TIME TO MOVE FROM A TO R_f.

The distance traveled by light represented by x_o is the distance $x_\ell(A \to R_f)(\text{"Sta"})$ in Eq. 2. The time for light to travel the left-hand side is called t_ℓ. We have:
$$x_o = (\lambda + V_{tr} t_\ell)(\text{"Sta"}) \quad (24)$$
and v is the train velocity V_{tr} in the Station Frame of Reference.
Substituting Eq. 24 in 23, we find:
$$x_\ell(\text{"Train"}) = \gamma\left\{(\lambda + V_{tr}t_\ell)(\text{"Sta"}) - V_{tr}t_\ell(\text{"Sta"})\right\} = \gamma\left\{\lambda(\text{"Sta"})\right\} \quad (25)$$

3.3 TIME TO TRAVEL FROM B TO R_f.

On the right hand side of the experiment, the distance traveled by light $x_\pi(B \to R_f)(\text{"Sta"})$ represented by x_o is:
$$x_o = (\lambda - V_{tr} t_\pi)(\text{"Sta"}) \quad (26)$$
where $t_\pi(\text{"Sta"})$ is the time for light to travel from B to R_f.

However, on the right-hand side, we have seen that the velocity V_{tr} is negative since the distance between the source and the observer is decreasing with time. Therefore, substituting the values of v for $-V_{tr}$ and the corresponding values of t and x_o we obtain for the right-hand side, using Eqs. 26 and 9:
$$x_\pi(\text{"Train"}) = \gamma\left\{(\lambda - V_{tr}t_\pi)(\text{"Sta"}) - (-V_{tr}t_\pi)(\text{"Sta"})\right\} = \gamma\left\{\lambda(\text{"Sta"})\right\} \quad (27)$$

Consequently, following the full Lorentz transformation, we find that the two distances $x_\ell(\text{"Train"})$ and $x_\pi(\text{"Train"})$ (Eq. 25 and 27) are identical in frame "Train". Substituting Eq. 25 and 27 into Eq. 22, we find:
$$t_\ell(\text{"Train"}) = t_\pi(\text{"Train"}) \quad (28)$$

Therefore, the time taken by light on the left hand side ("Train") is identical to the time taken by light on the right hand side ("Train"). The two pulses arrive simultaneously in R_f (as seen in frame "Train"). One must conclude that the two pulses of light emitted simultaneously at the ends of the station ("Sta") also arrive simultaneously for the train observer ("Train"), contrary to Einstein's hypothesis. Several consequences must be reexamined[9].

From this demonstration, we must conclude that the arguments used by Einstein on the relativity of simultaneity are not compatible with the Lorentz's transformations and consequently not compatible with a constant velocity of light.

The author thanks the Natural Sciences and Engineering Research Council for a grant received for a related subject which created the conditions suitable for this research.

References

1- Einstein, A. (1905) *On the Electrodynamics of Moving bodies*, "Zur Elektrodynamik bewegter Körper", Annalen der Physik, **17**. From: Lorentz, H.A., Einstein, A., Minkowski, H., and Weyl, H. (1923) "The Principle of Relativity", Dover Pub. Inc.

2- Phipps, T.E. (1986) *Heretical Verities: Mathematical Themes in Physical Description*, Ed. Classic Non-Fiction Library, Urbana, Illinois. Page 88-89, .

3- Phipps, T.E., (1991) *Proper Time Synchronization* in Foundation of Physics, Vol. 21, No: 9, .

4- Phipps, T.E., (1993) *On Hertz's Invariant Form of Maxwell's Equations* Physics Essays, **6**, No: 2, Page 246-256,

5- Van Flandern, T., (1993) *Dark Matter, Missing Planets & New Comets: Paradoxes Resolved, Origins Illuminated*, North Atlantic Books, Berkeley,

6- Einstein, A., (1920) *Relativity: The Special and the General Theory*, Methuen & Co. Ltd., London.

7- Born, M., (1962) *Einstein's Theory of Relativity*, Dover Pub., N.Y.

8- Dukas, H. and Hoffmann, B., (1972) *Albert Einstein, Creator and Rebel* Ed. The Viking Press, New York,

9- Marmet, P., (1993) *Absurdities in Modern Physics: A Solution*, Ed. Les Éditions du Nordir, c/o R. Yergeau, 165 Waller Ottawa, Ont. K1N6N5

ON SELF-INTERACTION AND (NON-)PERTURBATIVE QUANTUM GRT

H.-H. v. BORZESZKOWSKI
Technische Universität Berlin, Institut für Theoretische Physik,
„Gravitationsprojekt", An der Sternwarte 16,
D-14482 Potsdam, FR Germany,
Tel: (+331) 7499326, Fax. (+331) 7499203

In this paper, we discuss some aspects of perturbative general relativity. First, this will reveal reasons for the the failure of perturbative standard approaches and second show that there are more suitable perturbative methods. The latter signal in particular that in (non-perturbative) quantized general relativity one has to expect quantum limitations unknown from special-relativistic quantum-field theory. The method used is based on the (nonlinear) Fierz-Pauli equations with self-interaction and the so-called EIH method.

1. Introduction

A great deal of effort is invested in the study of non-perturbative methods in order to found a consistent quantum theory of general relativity (quantum GRT). In particular, Ashtekar's connection representation of GRT together with the related loop variable calculus is at present often considered as a hopeful candidate for such a formalism. This search is mainly motivated by the fact that the used perturbative approaches do not lead to a unitary, renormalizable theory. The conjecture is that these approaches are responsible for this failure because they assume that at small distances where gravity dominates a smooth given background is assumed [1].

In view of these arguments it seems to be unnecessary to waste further effort on perturbative approaches to quntum GRT. Otherwise, in this field till now there were mainly used methods which were developed to solve problems of quantun electrodynamics. Therefore, they do not seem to be appropriate to be applied to GRT and it should be interesting to rediscuss this matter within the framework of more suitable perturbative methods, i. e., of methods that do not assume a given smooth background. In this paper, we shall do this by using the approximation method developed by Einstein, Infeld, and Hoffmann [2] (cf. also [3], [4]) and by Fock [5]. (For the sake of brevity, let us call it EIH method.)

The plan of the paper is as follows. In Sec. 2, the method usually used and some objections raised against it will be outlined. In Sec. 3, then we shall consider the whole matter in the light of the EIH method.

2. The Standard Perturbative Method and Its Criticism

In the case of non-linear equations the only viable approximation methods are that ones of a small parameter, where it depends on the problem under considerration which parameter is used. In approaches to quantum GRT one generally starts with the *ansatz*

$$g_{\mu\nu} = g^0_{\mu\nu} + \kappa h_{\mu\nu}$$

$$(h_{\mu\nu} = h^{(1)}_{\mu\nu} + \kappa h^{(2)}_{\mu\nu} + ...)$$

(1)

where $g^0_{\mu\nu}$ is assumed to be a classical background metric which in particular can be equal to the Minkowski metric, and κ denotes the square root of Einstein's gravitational constant $8\pi G/c^4$ (G is Newton's gravitational constant, and c is the vacuum velocity of light).[1] On the quantum level, the term $\kappa h_{\mu\nu}$ is considered as weak quantum perturbation.

In particular, Feynman [7] tried to arrive at quantum GRT by exploiting the *ansatz* (1), where he assumed $g^0_{\mu\nu} = \eta_{\mu\nu}$. Inserting it in the Lagrangian for gravity coupled to scalar matter this leads to a series of terms which in [7] is discussed according to the rules known from quantum electrodynamics, i. e., in terms of Feynman's tree graphs describing effects like gravitational Compton scattering, *bremsstrahlung* etc. In higher-order approximation also radiative corrections (closed loops) and hard renormalization problems arise.

To establish a covariant background method of quantization the Feynman approach was modified by assuming an arbitrarily curved classical background metric $g_{\mu\nu}$ [8]. Inserting then (1) in the Einstein equation (for the sake of simplicity we here confine ourselves to the vacuum case)

$$R_{\alpha\beta}(g^0_{\mu\nu} + \kappa h_{\mu\nu}) = R_{\alpha\beta}(g^0_{\mu\nu}) + \kappa R^{(1)}_{\alpha\beta}(h_{\mu\nu}) + \kappa^2 R^{(2)}_{\alpha\beta}(h_{\mu\nu}) + ... = 0,$$

(2)

where (the upright line denotes the covariant derivative with respect to the background metric):

$$R_{\alpha\beta}(g^0_{\mu\nu}) = \text{the Ricci tensor of the background},$$

$$R^{(1)}_{\alpha\beta}(h_{\mu\nu}) = \frac{1}{2}g^{\rho\tau}(h_{\rho\tau|\alpha\beta} + h_{\alpha\beta|\rho\tau} - h_{\tau\alpha|\beta\rho} - h_{\tau\beta|\alpha\rho}),$$

(3)

---

[1] The special case, where $g^0_{\mu\nu} = \eta_{\mu\nu}$ is assumed, first was considered by Einstein [6] in order to calculate gravitational waves as solution of the first-order approximation of the gravitational equations.

$$R^{(2)}_{\alpha\beta}(h_{\mu\nu}) = -\frac{1}{2}\left[\frac{1}{2}h^{\rho\tau}{}_{|\beta}h_{\rho\tau|\alpha} + h^{\tau|\rho}_{\beta}(h_{\tau\alpha|\rho} - h_{\rho\alpha|\tau})\right.$$

$$+ h^{\rho\tau}(h_{\rho\tau|\alpha\beta} + h_{\alpha\beta|\rho\tau} - h_{\tau\alpha|\beta\rho} - h_{\tau\beta|\alpha\rho})$$

$$\left. - (h^{\rho\tau}{}_{|\rho} - \frac{1}{2}h^{\tau})(h_{\tau\alpha|\beta} + h_{\tau\beta|\alpha} - h_{\alpha\beta|\tau})\right],$$

etc.

where in [8] the implications of this expansion series were disdicussed under the assumption

$$R_{\alpha\beta}(g^0_{\mu\nu}) = 0, \qquad (4)$$

to which we shall return below.

As was mentioned repeatedly (cf., e. g., [1], [9]), the following arguments speak against such a perturbative background method.

(a) The practical failure of this approach: Einstein's GRT is perturbatively non-renormalizable at two loop for pure gravity and at one loop for gravity interacting with matter. Changing now the theory by adding higher-derivative terms to the action integral of GRT the ultraviolet behavior of the quantized theory is improved (e. g., the theory can be made renormalizable this way). However doing this one comes to a theory which is not unitary.

(b) The unsatisfactory theoretical assumption forming the basis of this approach (possibly it is the source of the practical failure): It is a fixed background structure $g^0_{\mu\nu}$ assumed. On the one hand, this additional mathematical element seems to be necessary if one wants to formulate quantum GRT in analogy to usual quantum field theory. But otherwise this leads to a bimetric structure which contradicts the spirit of the basic principles of GRT, namely the principles of general relativity and strong equivalence. For instance, in the case $g^0_{\mu\nu} = \eta_{\mu\nu}$, this means that for the gravitational quantum perturbations the causal structure of the Minkowski space is the correct one, while the motion of usual matter is governed by $g_{\mu\nu}$.

These arguments do show that the *above-outlined* method is not suitable for solving fundamental problems like that one of quantum GRT. But do they really force us to drop all perturbative methods in GRT, as, for instance, is argued in [1]? Our point is that this question cannot be answered as long as one only refers to the method mentioned above because it does not regard the non-linear nature of GRT sufficiently. Therefore, in the next section a more suitable approximation method will be discussed. But before doing this, in the remainder of this section still some remarks that summarize a criticism of the above method presented previously [10].

For this purpose, let us consider a two-parameter approximation which was used to consider high-frequency waves in GRT in [11] and [12]). In this approximation method one assumes again (1), where

$$|g^0_{\mu\nu}| = O(1), \quad |h_{\mu\nu}| = O(1), \quad \kappa < 1. \tag{5}$$

But now, besides κ, a second smallness parameter is introduced by the assumption

$$\partial |g^0_{\mu\nu}| = \frac{|g^0_{\mu\nu}|}{L}, \quad \partial ||h_{\mu\nu}| = \frac{|h_{\mu\nu}|}{\lambda}, \quad \lambda < L, \tag{6}$$

where L and λ are characterisic lengths over which $g^0_{\mu\nu}$ and $h_{\mu\nu}$ change significantly. The ratio λ/L is a second small parameter so that now the orders of magnitude of the different terms in (2) are governed by the relation between κ and λ/L.

Because the first three terms in (2) are of the following orders of magnitude,

$$R_{\alpha\beta}(g^0_{\mu\nu}) = O(L^{-2}), \quad \kappa R^{(1)}_{\alpha\beta} = O(\kappa \lambda^{-2}), \quad \kappa^2 R^{(2)}_{\alpha\beta} = O(\kappa^2 \lambda^{-2}), \tag{7}$$

only for $\lambda/L > \kappa$, i. e., in the case that the field $h_{\mu\nu}$ is slowly changing one has to satisfy (2) by setting term by term equal to zero so that in particular Eq. (4) is satisfied. For $\lambda/L \leq \kappa$ one however gets $R_{\alpha\beta}(g^0_{\mu\nu}) \leq \kappa^2 R^{(2)}_{\alpha\beta}$ such that one cannot realize linearized Einstein equations describing gravitational (quantum) perturbations propagating in front of a background satisfying Eq. (4).

This shows that the above-considered covariant approximation procedure is limited to a low-frequency region. Therefore, it can be used in some special cases, but it is not justified to consider it as a rigerous approximation method.

3. Nonlinear Fierz-Pauli Equations and the EIH Approximation Method

At the beginning let us compare the motion of a number N of electrically charged particles under the influence of the electromagnetic field produced by all charged particles with the motion of (electrically neutral) particles under the influence of their common gravitational field.

The electromagnetic case can be described as follows:[2]
(i) The equation of motion of the Ath particle is given by

$$m_o \frac{d^2 \xi^{(A)}_\alpha}{ds^2} = e F_{\alpha\beta}(x) \left[\frac{d\xi^{(A)\beta}}{ds} \right], \tag{8a}$$

---

[2] In the electromagnetic case we follow Barut's representation [13], where we slightly change the notation in order to make it compatible with that one of the gravitational case. For the latter case, cf.[4].

where $\xi_\alpha^{(A)}$ is the world-line of the Ath particle in Minkowski space, s an invariant time parameter, and $F_{\alpha\beta}$ is the total electromagnetic field; the bracket says that the term has to be taken on the world-line, $x = \xi$.

(ii) The electromagnetic field $F_{\alpha\beta}$ obeys Maxwell's equations

$$F_{\alpha\beta}{}^{,\beta}(x) = \sum_{B=1}^{N} e^{(B)} \frac{d\xi_\alpha^{(B)}}{ds} \delta(x_\alpha - \xi_\alpha^{(B)}). \tag{9a}$$

In the gravitational case the corresponding two groups of equations take the following form [4]:

(i') The equation of motion of the Ath particle under the influence of the total gravitational field pruced by all the particles reads

$$\frac{d^2 \xi_{(A)}^\alpha}{ds_A^2} + \Gamma_{\mu\nu}^{(A)\alpha}(x) \left[\frac{d\xi_{(A)}^\mu}{ds_A} \frac{d\xi_{(A)}^\nu}{ds_A} \right] = 0, \tag{8b}$$

where $ds_A = (g_{\alpha\beta} d\xi_{(A)}^\alpha \xi_{(A)}^\beta)^{1/2}$ is an invariant parameter along the world-line of the Ath particle.

(ii') The total gravitational field is given by Einstein's equations

$$E^{\alpha\beta} = -\frac{8\pi G}{c^4} \sum_{B=1}^{N} m_0^{(B)} \int_{-\infty}^{\infty} ds_B \delta_{(4)}(x - \xi_{(B)}) \frac{d\xi_{(B)}^\alpha}{ds_B} \frac{d\xi_{(B)}^\beta}{ds_B}. \tag{9b}$$

($E^{\alpha\beta}$ is the Einstein tensor and the right-hand side is proportional to the energy-momentum tensor of the N point particles.)

Before turning to the point most interesting in our context one remark on the infinities arising in both cases. Eq. (8a) is highly nonlinear because, due to the term B=A in (9a), $F_{\alpha\beta}$ in (8a) depends on $\xi^{(A)}$. This provides a self-field of the Ath particle that becomes infinite at $x = \xi^{(A)}$. As was shown in [13], and [14], one can separate an infinite inertial part from this self-field which may be removed by mass renormalization, while the remaining part gives rise to observable effects like anomalous magnetic moment, Lamb shift etc. - In the gravitational case (8b), (9b) one meets insofar a similar situation as there also occur terms which become infinite at $x = \xi^{(A)}$. Infeld and Plebanski [4] tried to solve this problem by defining the so-called good delta funktion $\delta_{(4)}$. This seems to be a suitable method because the renormalization procedure of electrodynamics does not work here, as will be seen in the following discussion.

Now, as far as the equations (8a) and (9a) are concerned, they can be solved by the following procedure: One solves Eq. (9a) for an arbitrary particle motion and inserts the solution in Eq. (8a) so that one finally obtains an integro-differential equation whose solution gives us the motion of the charged particles.

This procedure can also be applied when the charged particles, electrons say, are described by the Dirac field $\psi(x)$. Then, Eqs. (8a) and (9a) must be replaced by:

$$(-i\gamma^\mu \partial_\mu - m_0)\psi(x) = e\gamma^\alpha \psi(x) A_\alpha(x), \tag{10a}$$

$$F_{\alpha\beta}{}^{,\beta}(x) = e\bar{\psi}(x)\gamma_\alpha\psi(x) .$$
(11a)

In the Lorentz gauge $A^\alpha{}_{,\alpha} = 0$ we can again eliminate A^α by inserting the solution of (11a) given by the expression,

$$A_\alpha = \int dy D(x-y)\bar{\psi}(y)\gamma_\alpha\psi(y),$$
(12)

where $D(x-y)$ is the causal Green's function, and iserting this into (10a).

Now, when one passes to the gravitational case then one finds a quite different situation because then the two equatios (8b) and (9b) are not independent of each other (cf., [4]). Because of the contracted Bianchi identity

$$\nabla_\beta E^{\alpha\beta} = 0,$$
(13)

Eq.(8b) is a consequence of Eq. (9b) so that (8b) must be considered as condition of integrability of (9b). Therefore, Infeld and Plebanski [4] stated the following: One cannot solve the field equations with arbitrary motion, i.e., one cannot find the g's as functionals of of the source appearing on the right-hand side of the equations (9b).

The same is of course true when, as in the electromagnetic case, one replaces the classical equations of motion (8b) by quantum matter equations. Then, instead of the equations (8b) and (9b), one has to consider the equations:

$$(-i\gamma^\mu \nabla_\mu - m_0)\psi(x) = 0,$$
(10b)

$$E^{\alpha\beta} = -\frac{8\pi G}{c^4} T^{\alpha\beta}_{Dirac},$$
(11b)

where (10b) is the covariantly written Dirac equation and $T^{\alpha\beta}_{Dirac}$ the energy-momentum tensor of the Dirac matter.

Since the relation

$$\nabla_\beta T^{\alpha\beta}_{Dirac} = 0$$
(14)

(following from (10b) and representing thus a part of the dynamics described by (10b)) is also a consequence of (11b) the dynamical equations (14) have to be considered as integrability condition of (11b). Therefore, one cannot solve (11b) for arbitrary „motions of the electrons". In other words, one cannot expect to find a Green's function by means of which Eqs. (10b) and (11b) can be reduced to an integro-differential equation for the motion of Dirac matter.

Thus, it needs an approximation method that takes this typical feature of GRT into consideration. Such method is the EIH method.

To describe some aspects of this method and some of its implications for quantum GRT we first will rewrite Einstein's equation in the „nonlinear Fierz-Pauli form".[3]

---

[3] In the remainder of this section we mainly follow the paper [15].

Assuming two metric tensors $g^*_{\mu\nu}$ and $g_{\mu\nu}$ such that

$$g^*_{\mu\nu} = g_{\mu\nu} + \gamma_{\mu\nu} \tag{15}$$

and

$$g^{*\mu\nu} = g^{\mu\nu} + \gamma^{\mu\nu} \quad \text{(with } g^*_{\mu\lambda} g^{*\lambda\nu} = \delta^\nu_\mu \text{)} \tag{16}$$

then according to Eisenhart [16] the Riemann tensor $R^{*\alpha}{}_{\beta\gamma\delta}$ is given by

$$R^{*\alpha}{}_{\beta\gamma\delta} = -\Gamma^{*\alpha}{}_{\beta\gamma,\delta} + \Gamma^{*\alpha}{}_{\beta\delta,\gamma} - \Gamma^{*\lambda}{}_{\beta\gamma}\Gamma^{*\alpha}{}_{\lambda\delta} + \Gamma^{*\lambda}{}_{\beta\delta}\Gamma^{*\alpha}{}_{\lambda\gamma}$$

$$= R^{\alpha}{}_{\beta\gamma\delta} - \Delta^{\alpha}{}_{\beta\gamma|\delta} + \Delta^{\alpha}{}_{\beta\delta|\gamma} - \Delta^{\lambda}{}_{\beta\gamma}\Delta^{\alpha}{}_{\lambda\delta} + \Delta^{\lambda}{}_{\beta\delta}\Delta^{\alpha}{}_{\lambda\gamma} \tag{17}$$

$$= R^{\alpha}{}_{\beta\gamma\delta} + \Delta^{\alpha}{}_{\beta\gamma\delta}$$

where $\Delta^{\alpha}{}_{\beta\gamma\delta}$ takes the form

$$\Delta^{\alpha}{}_{\beta\gamma} = g^{*\alpha\lambda}(-\gamma_{\beta\gamma|\lambda} + \gamma_{\gamma\lambda|\beta} + \gamma_{\lambda\beta|\gamma}) \tag{18}$$

(the upright line denotes again the covariant derivative wih respect to the background now denoted by $g_{\mu\nu}$).

Linearizing the above expressions by setting $\gamma_{\mu\nu} = \delta g_{\mu\nu}$ where $\delta g_{\mu\nu}$ is a small perturbation of $g_{\mu\nu}$, one obtains Palatini's formulas for $g^*_{\mu\nu}$, $\Gamma^{*\alpha}{}_{\beta\gamma}$, $R^{*\alpha}{}_{\beta\gamma\delta}$, etc. Then the „linearized" Einstein-Hilbert Lagrange density reads

$$\mathbf{L}^* = \sqrt{-g^*}\, g^{*\mu\nu} R^*_{\mu\nu} = \sqrt{-g^*}\, g^{*\mu\nu}(R_{\mu\nu} + \delta\Gamma^{\lambda}{}_{\mu\nu|\lambda} - \delta\Gamma^{\lambda}{}_{\mu\lambda|\nu})$$
$$\approx \sqrt{-g}\, g^{\mu\nu}(R_{\mu\nu} + \delta\Gamma^{\lambda}{}_{\mu\nu|\lambda} - \delta\Gamma^{\lambda}{}_{\mu\lambda|\nu}) \tag{19}$$

Assuming (19) as point of departure its Palatini variation with respect to the variables $\Gamma^{*\alpha}{}_{\beta\gamma}$, $\Delta^{\alpha}{}_{\beta\gamma} = \delta\Gamma^{\alpha}{}_{\beta\gamma}$, and $g^{*\mu\nu}$ yields finally the „nonlinear Fierz-Pauli equations" (for details, see [15]):

$$\frac{1}{2}(g^{\mu\nu}\gamma_{\alpha\beta|\mu\nu} + \gamma_{|\alpha\beta} - \gamma_{\alpha\lambda|\beta}{}^{\lambda} - \gamma_{\beta\lambda|\alpha}{}^{\lambda})$$
$$-\frac{1}{2}g_{\alpha\beta}(g^{\mu\nu}\gamma_{|\mu\nu} - \gamma_{\mu\nu|}{}^{\mu\nu}) = -R_{\alpha\beta} + \frac{1}{2}g_{\alpha\beta}g^{\mu\nu}R_{\mu\nu} + \frac{8\pi G}{c^4}T_{\alpha\beta}. \tag{20}$$

with the nonlinear term $E_{\alpha\beta}(g) = R_{\alpha\beta} - (1/2)g_{\alpha\beta}g^{\mu\nu}R_{\mu\nu}$ (We denote in (20) $\delta g_{\mu\nu}$ again by $\gamma_{\mu\nu}$ and set $\gamma = g^{\mu\nu}\gamma_{\mu\nu}$.)

For further discussion and comparison with the method used of Sec. 2 first let us again assume the *ansatz* (15) with $g_{\mu\nu} = \eta_{\mu\nu}$. Then in the first-order approximation, up to the additional condition arising here in the corresponding approximation of (14), one finds for $\gamma_{\alpha\beta} = h_{\alpha\beta}^{(1)}$ the same equation as in Sec. 2, namely the linaer Fierz-Pauli equation. In this approximation, for the vacuum case one can calculate plane waves etc.

Passing now to the second-order level one gets the equations (20) with $\gamma_{\alpha\beta} = h_{\alpha\beta}^{(2)}$ and $g_{\alpha\beta} = \eta_{\alpha\beta} + \kappa h_{\alpha\beta}^{(1)}$, i, e., nonlinear Fierz-Pauli equations in which the Einstein tensor formed from $g_{\alpha\beta} = \eta_{\mu\nu} + \kappa h_{\alpha\beta}^{(1)}$ occurs as source term, and additionally again the corresponding approximation of Eq.(14) as integrability condition. On this level, $\eta_{\alpha\beta}$ and $h_{\alpha\beta}^{(1)}$ lose their seperate physical meaning, in particular also the first-order plane waves and the first-order Feynman graphs. And this repeats in each order of approximation.

One encounters nearly the same situation when one starts with $g_{\mu\nu} \neq \eta_{\mu\nu}$ (the only difference is that now already the first-order approximation provides a nonlinear Fierz-Pauli equation). Therefore, one can generally state:

(1) There is no series of Feynman graphs because the used iterative procedure allows only to interpret the highest order calculation.

(2) The Green's function cannot be calculated because the dynamics of matter cannot be considered as arbitrary.

Finally, from this point of view a further remark on the often used first-order formalism which starts from the equation:

$$R_{\alpha\beta}^{(1)}(h_{\mu\nu}) = \frac{1}{2} g^{0\rho\tau}(h_{\rho\tau|\alpha\beta} + h_{\alpha\beta|\rho\tau} - h_{\tau\alpha|\beta\rho} - h_{\tau\beta|\alpha\rho}) = 0 \qquad (21)$$

They follow only then from (20) when one assumes that the background $g_{\mu\nu} \neq \eta_{\mu\nu}$ is a solution of Einstein's vacuum equations. But - as was argued in [15] - to assume this means to overdetermine $g^*_{\mu\nu}$. If one otherwise starts with an arbitrary $g_{\mu\nu}$, for instance, with $g_{\mu\nu} = \eta_{\mu\nu}$, and uses the above-described approximation procedure then in general one will in no approximation arrive at Eq. (21) because then the term $E_{\mu\nu}(g_{\alpha\beta})$ which is a known function of the known h_l ($l < n$) has also to appear. (To justify (21) it remains only to hope for the stroke of luck that in some approximation of the gravitational equations (in general with matter term) accidentally the approximate solution of the problem is identical with an exact vacuum solution. But even this would not help because the next order-calculation would destroy the results obtained this way.)

4. Conclusion

The typical features of GRT force us to apply to it, instead of the approximation method outlined in Sec.2, the iteration method of Sec.3. This method however signals that GRT is essentially classical in the sense that there are no significant quantum effects. Indeed, the EIH method says that one can perform approximate calculations on an arbitrarily high approximation level n. However, because only the nth order describing the propagation of the perturbation $h^{(n)}$ on the background metric $g + h^{(1)} + ... + h^{(n-1)}$ can be physically interpreted (but not the orders $l < n$) neither tree graphs nor loops will occur.

On the quantum level, this circumstance is reflected by Rosenfeld's inequalities [17] which can be derived as follows (cf., for the following [18]). By taking the average of Einstein's equations

$$R_{\alpha\beta} = -\kappa(T_{\alpha\beta} - \frac{1}{2}g_{\alpha\beta}T) \qquad (22)$$

over the volume $L_o^3 \cong (\Delta x)^3$ of the measurement region (given by the extent of the measurement body) one gets for the uncertainties (or inaccuracies) ΔR and $\Delta \Gamma$ of the quantities $R_{\alpha\beta}$ and $T_{\alpha\beta}$ the relation[4]

$$\Delta R \approx \kappa \Delta T \approx \kappa \Delta E (\Delta x)^{-3} \qquad (23)$$

and thus for the uncertainties Δg of g_{ik},

$$\Delta g \approx \Delta R(\Delta x)^2 \approx \kappa \Delta E (\Delta x)^{-1}, \qquad (24)$$

where ΔE is the uncertainty of E. Hence, Heisenberg's fourth uncertainty relation

$$\Delta E \Delta t \geq \hbar \qquad (25)$$

and Bohr's condition of a classical measurement body[5]

$$\Delta x \geq c \Delta t \qquad (26)$$

yield

$$\Delta g (\Delta x)^2 \approx \kappa \Delta E \Delta x \geq \kappa c \Delta E \Delta t \geq \kappa \hbar c \qquad (27)$$

and thus the Rosenfeld inequalitiy

$$\Delta g_{ik} \Delta x^i \Delta x^k \geq \kappa \hbar c, \qquad (28)$$

where $(\kappa \hbar c)^{1/2}$ is Planck's fundamental length.[6]

Eq. (28) states a relation between the uncertainties of the basic quantity $g_{\alpha\beta}$ of Riemannian geometry (and, consequently, of the Christoffel connection and the Riemann-Christoffel curvature tensor) and the position in space-time. This relation holds true for each method of quantization if gravity is determined by Einstein's equations. For, it is a

---

[4] Here the $\approx$ signs relate the usual (stochastic) measurement errors, while $\geq$ denotes inequalities arising from fundamental relations of general relativity, quantum field, and measurement theories.

[5] In the Rosenfeld paper his condition is shown to be imposed in order to guarantee optimal measurements.

[6] As the derivation shows, (28) is a consequence of the (strong) equivalence of passive and active gravitational masses to the inertial mass (and thus to energy) of bodies. This equivalence underlies Einstein's purely metric theory of general relativity. In virtue of this, the energy-momentum tensor of matter is the source of gravitational fields, and only therefore the above line of arguments can be given.

simple consequence of these equations and of quantum mechanics applied to the matter source of Einsteinian gravity. Therefore, (28) must be compatible with each quantization procedure applied to such fields.

If one assumes Einstein's condition that the Lorentz signature of space-time has to be maintained, i. e., that. $\Delta g \leq 1$, then Rosenfeld's relation (28) says that $(\Delta x) \geq \hbar \kappa c$, i. e., that Planck's length is the smallest measurable distance. According to Rosenfeld, relation (28) (possibly) shows that, at distances of the order of magnitude of Planck's length, the quantum fluctuations induced by the measuring procedure are so violent that the quantum gravity region is not accessible to measurement. This interpretation is especially plausible when one follows Rosenfeld's own "measurement-theoretical" derivation of (28).

5. References

1. Ashtekar, A. (1991) *Non-Perturbative Canonical Gravity*, World Scientific, Singapore etc.
2. Einstein, A., Infeld, L. and Hoffmann, B. (1938) The gravitational equations and the problem of motion, *Ann. Math.* **39**, 65.
3. Einstein, A. and Infeld, L. (1949) On the motion of particles in general relativity theory, *Can. J. Math.* **1**, 209.
4. Infeld, L. and Plebanski, J. (1960) *Motion and Relativity*, Pergamon Press, New York.
5. Fock, V. A. (1959) *The Theory of Space, Time, and Gravitation* , Pergamon Press, New York.
6. Einstein, A. (1916) Näherungsweise Integration der Feldgleichungen der Gravitation, *Sitzungsber. Preuss. Akad. Wiss.*, pp.688-696; (1918) Über Gravitationswellen, *Sitzungsber. Preuss. Akad. Wiss.*, pp. 154-167.
7. Feynman, R. P. (1963) Quantum theory of gravitation, *Acta Phys. Pol.* **XXIV**, 697.
8. DeWitt, B. S. (1979) Quantum gravity: The new synthesis, in S. W. Hawking and W. Israel (eds.), *General Relativity: An Einstein Centenary Survey*, Cambridge U. P., Cambridge, and the literature cited therein.
9. Isham, C. J. (1991) *Conceptual and Geometrical Problems in Quantum Gravity*, Lectures presented at the 1992 Schladming Winter School, Imperial/ TP/ 90-91/14. 12, 633.
10. v. Borzeszkowski, H.-H. (1982) On high-frequency covariant background quantization, *Found. Phys.* **12**, 633; (1990) Remarks on the physical reality of gravitons, *Found. Phys.* **20**, 435.
11. Brill, D. R. and Hartle, J. B. (1964) Method of self-consistent field in general relativity and its application to the gravitational geon, *Phys. Rev.* **B 135**. 271.
12. Isaacson, R. A. (1968) Gravitational radiation in the limit of high frequency, *Phys. Rev.* **166** , 1263 and 1272.
13. Barut, A. O. (1979) Nonlinear problems in classical and in quantumelectrodynamics, in A. F. Ranada (ed.) *Nonlinear Problems in Theoretical Physics* (Lecture Notes in Physics 98) Springer-Verlag, Berlin etc.
14. Barut, A. O. (1974) Electrodynamics in terms of retarded field, *Phys. Rev.* **10**, 3335.
15 v. Borzeszkowski, H.-H. and Treder, H.-J. (1994) Einstein equations and Fierz-Pauli equations with self-interaction in quantum gravity, *Found Phys.* **24**, 949.
16. Eisenhart, L. P. (1927) *Non-Riemannian Geometry*, Amer. Math. Soc. Colloquium, New York.
17. Rosenfeld, L. (1966) Quanten und Gravitation, in H.-J. Treder (ed.), *Entstehung, Entwicklung und Perspektiven der Einsteinschen Gravitationstheorie*, Akademie-Verlag, Berlin.
18. v. Borzeszkowski, H.-H. and Treder, H.-J. (1994) Classical gravity and quantum matter fields in unified field theory, *Gen. Rel. Grav.* (in press).

CLASSICAL PHYSICS FOUNDATIONS FOR QUANTUM PHYSICS

LLOYD MOTZ
Department. of Astronomy
Columbia University
New York, NY 10027

DAVID W. KRAFT
School of Science & Engineering
University of Bridgeport
Bridgeport, CT 06601

Abstract

We employ classical principles to justify various quantum concepts. Among these are (1) the Planck formula for the energy of a photon, (2) the de Broglie relation, (3) the Schrödinger energy and momentum operators, (4) the position-momentum commutator relation and (5) the Schrödinger wave function. Thus for the Planck formula we require only basic thermodynamics, the Doppler effect and the assumption that photons exist. The de Broglie relation results from the quantization of action combined with the hypothesis of the existence of de Broglie waves and the properties of standing waves. Items (3) and (4) are direct consequences of classical definitions of action in terms of energy and/or momentum, the existence of a unit of action, and the use of the function of action $\varphi(S) = \exp(iS/\hbar)$, where S is the action. The wave function stems, in addition, from the inability to assign a particle to a precise classical path having a definite action; instead, we must have a bundle of paths, of varying action, which differ from each other by no more than the unit of action. The sum over these paths yields the wave function.

1. Introduction

Quantum physics contains concepts and procedures that seem arbitrary and are often justified only because they provide correct answers. Using fundamental classical principles we provide herein simple deductions of several quantum physics results. These include the Planck formula for the energy of a photon [1,2], the de Broglie relation [2], the Schrödinger momentum and energy operators, the position-momentum commutator relation, and the Schrödinger wave function [3].

2. Photon Energy

Consider an adiabatically enclosed cylinder of length L and of cross sectional area A, in which is a movable piston with a perfectly reflecting surface. Let the energy density of black-body radiation in this enclosure be u so that $U = uAL$ is the total energy of this radiation and $p = u/3$ is the radiation pressure. If we now push the piston slowly an infinitesimal distance $dx = v\,dt$, we do an amount of work $pA\,dx$ and the total energy of the gas increases by $pA\,dx = uA\,dx/3 = U\,dx/3L$. If there are N photons present and ε is their average energy, then the average energy increase for each photon is

$$d\varepsilon = \frac{U\,dx}{3NL}$$
$$= \frac{\varepsilon\,dx}{3L}$$

so that

$$\frac{d\varepsilon}{\varepsilon} = \frac{dx}{3L}. \tag{1}$$

Consider now the change in the frequency of a photon when it is reflected from the moving piston. From the Doppler effect we see that for each reflection, the relative change in the frequency of a photon that is incident normally on the piston is $2v/c$, where v is the constant speed of the piston and c is the photon speed. The number of reflections in dt is $c\,dt/2L$ so that the total relative change in the frequency of the photon in time dt is $dv/v = (2v/c) \times (c\,dt/2L) = v\,dt/L = dx/L$. However, since the motion of the photons of the black-body radiation is random, a photon will possess a component of motion at right angles to the piston face for only one-third of the time. Hence the relative change in frequency will be

$$\frac{dv}{v} = \frac{dx}{3L}. \tag{2}$$

Comparison of Eqs. (1) and (2) yields

$$\frac{d\varepsilon}{\varepsilon} = \frac{dv}{v} \tag{3}$$

so that

$$\varepsilon = \text{constant} \times v. \tag{4}$$

CLASSICAL PHYSICS FOUNDATIONS FOR QUANTUM PHYSICS

Thus if we accept basic thermodynamics, the Doppler effect and the existence of photons, we are forced to conclude that the energy of a photon is proportional to its frequency. The constant, which has the dimensions of action, is the same for all frequencies and is therefore a universal constant. We may take this as the unit of action.

The above derivation is based on the average energy of a photon and on the average change in its frequency, but this does not invalidate the final result since one can show from general thermodynamic arguments that the nature of blackbody radiation is not altered by an adiabatic compression or expansion [1]. Thus all photons must undergo the same relative change in their energy and frequency if the radiation is to remain blackbody radiation. It is also worth noting that the same result is obtained for a beam of monochromatic radiation of frequency v traveling back and forth parallel to L. Such a beam exerts a pressure on the reflecting surface of the piston that is exactly equal to the energy density of the beam. The derivation proceeds as before except for the numerical factor in Eqs. (1) and (2), and one again obtains Eq. (3), which leads to the final result, Eq. (4).

Finally, we note that the above derivation is the reverse of what Einstein did in his famous 1905 paper [4] where he assumed the relation $E = hv$ and deduced the existence of photons. Here we assume the existence of photons and then derive the energy formula

3. De Broglie Relation

To derive the de Broglie relation, we consider a particle with speed v moving back and forth in an enclosure of length L. The action, S, associated with this motion is the product of the particle's momentum and displacement; thus, for a round trip parallel to the length, $S = 2Lmv$. We now quantize the action according to $S = nh$, where h is the unit of action and n is a positive integer; thus

$$2Lmv = nh, \quad n = 1, 2, 3, \dots . \tag{5}$$

The number of nodes, q, in the standing de Broglie wave associated with the motion equals one plus the number of half-wavelengths in the length L:

$$q = 1 + \frac{2L}{\lambda}. \tag{6}$$

The minimum number of nodes, $q = 2$, occurs for $\lambda = 2L$. If we associate this case with the lowest level of action, $n = 1$, we see that the relation between the quantum number and the number of nodes is

$$n = q - 1. \tag{7}$$

Thus each half de Broglie wavelength represents a unit of action. Eliminating q between Eqs. (6) and (7) yields

$$n = \frac{2L}{\lambda} \tag{8}$$

which, when substituted into Eq. (5), yields

$$\lambda = \frac{h}{mv}. \tag{9}$$

Thus the de Broglie relation follows from the quantization of action, combined with the hypothesis of the existence of de Broglie waves and their interpretation as standing waves in an enclosure.

4. Schrödinger Momentum and Energy Operators

Consider a particle with motion characterized by an action function $S(r,p)$ where **r** is the position vector and **p** is the momentum. For the action associated with a differential displacement **dr**, we write

$$dS = \mathbf{p} \cdot d\mathbf{r}$$
$$= p_x dx + p_y dy + p_z dz.$$

However we also have

$$dS = \frac{\partial S}{\partial x} dx + \frac{\partial S}{\partial y} dy + \frac{\partial S}{\partial z} dz,$$

so that

$$p_x = \frac{\partial S}{\partial x}, \quad p_y = \frac{\partial S}{\partial y}, \quad p_z = \frac{\partial S}{\partial z}. \tag{10}$$

To be valid, an operator equation must hold for an arbitrary well-behaved function. For this purpose we take a function of the action of the form

$$\varphi(S) = \exp(iS/\hbar). \tag{11}$$

Upon differentiating Eq. (11) and using Eq. (10), we obtain

CLASSICAL PHYSICS FOUNDATIONS FOR QUANTUM PHYSICS 409

$$\frac{\partial}{\partial x}\varphi(S) = \frac{i}{\hbar}\exp(iS/\hbar)\frac{\partial S}{\partial x}$$
$$= \frac{i}{\hbar}\varphi(S)p_x.$$
(12)

Rewriting Eq. (12), we have

$$p_x \varphi(S) = \frac{\hbar}{i}\frac{\partial}{\partial x}\varphi(S)$$

so that we establish the correspondence

$$p_x \Leftrightarrow \frac{\hbar}{i}\frac{\partial}{\partial x},$$
(13)

and similarly for p_y and p_z.

To obtain the energy operator, we consider a particle with motion characterized by Hamilton's principal function [5]

$$S(\mathbf{r},\mathbf{p},t) = \mathbf{p}\cdot\mathbf{r} - Et$$
(14)

where E is the energy. We write the differential of this function as

$$dS = \mathbf{p}\cdot d\mathbf{r} - E\,dt,$$

from which we note that

$$\frac{\partial S}{\partial t} = -E.$$
(15)

Again consider the function $\varphi(S)$ given in Eq. (11). Taking the partial derivative with respect to time yields

$$\frac{\partial}{\partial t}\varphi(S) = \frac{i}{\hbar}\exp(iS/\hbar)\frac{\partial S}{\partial t}$$
$$= \frac{i}{\hbar}\varphi(S) \times (-E),$$
(16)

where we have used Eq. (15). Rewriting Eq. (16), we have

$$E\varphi(S) = i\hbar \frac{\partial}{\partial t} \varphi(S),$$

so that we establish the correspondence

$$E \Leftrightarrow i\hbar \frac{\partial}{\partial t}. \qquad (17)$$

Thus the forms for the quantum mechanical operators for momentum and energy are each obtained from a classical action function.

5. Position-momentum Commutator Relation.

Consider a particle moving along a path for which the action is well defined and is characterized by a function $\varphi(S)$. We form the commutator $[q,p] = qp - pq$ where q is the position and p is its conjugate momentum. Letting this quantity operate on $\varphi(S)$ and using Eq. (13) for the form of the operator for p yields

$$\begin{aligned}(qp-pq)\varphi(S) &= q\frac{\hbar}{i}\frac{\partial}{\partial q}\varphi(S) - \frac{\hbar}{i}\frac{\partial}{\partial q}q\varphi(S) \\ &= \frac{\hbar}{i}\left[q\frac{d\varphi}{dS}\frac{\partial S}{\partial q} - \varphi(S) - q\frac{d\varphi}{dS}\frac{\partial S}{\partial q}\right] \\ &= i\hbar\varphi(S),\end{aligned}$$

from which we see that

$$qp - pq \Leftrightarrow i\hbar. \qquad (18)$$

Thus the familiar quantum mechanical commutator of q and p is a consequence of having an action function for a single path for which S is well defined. Note that for this purpose the function of the action need not be of the form $\exp(iS/\hbar)$ used earlier.

6. Wave Function

A function of action such as $\varphi(S) = \exp(iS/\hbar)$ cannot be a quantum mechanical wave function for, since $\varphi^*\varphi = 1$, the particle would be limited to a single path, i.e. to a definite action. However, since the action is quantized, we do not know the action (path) to within h. Therefore we must have a function that depicts a bundle of paths in which no two paths differ in action by more than h. Such a function can be represented by a linear combination of functions of the form φ(S), i.e., a sum over actions (paths)

$$\psi = \sum_j c_j \exp(iS_j/\hbar). \quad (19)$$

The particle has now been localized to within a group of paths (actions) and if we interpret the expansion coefficients as probability amplitudes, the function ψ possesses the requisites for a quantum mechanical wave function. In a more general approach, Motz [3] introduced a density of paths per unit action interval and, replacing the sum in Eq. (19) by an integral, wrote a wave function of the form

$$\psi = \int \exp(iS/\hbar)\rho \, dS,$$

where ρ is the density function. Expressing ρ and S as functions of position then leads to a wave function which may be compared with Feynman's formulation [6].

The transition from the concept of an uniquely defined classical path to that of a bundle of trajectories also leads to a physical interpretation of the de Broglie wavelength [3]. Consider the picture we obtain by assigning to the particle a bundle of trajectories in the form of a tube whose diameter must be of the order of magnitude of the diameter of the particle. For simplicity, consider only circular orbits and let p be the magnitude of the particle momentum. If the diameter of the tube is Δr, the difference in the actions associated with the inner and outer radii of the tube is pΔr. According to the picture developed above, this must then be equal to h. Hence pΔr = h, or Δr = h/p, which is just the de Broglie wavelength of the particle. Thus the de Broglie wavelength represents the extension we must assign to a particle when we describe its motion in terms of a bundle of classical trajectories.

References

1. Motz, L.: A simple thermodynamic derivation of the quantum formula ε = hν, *American Journal of Physics* **41** (1973), 1016-1017.
2. Motz, L. and Kraft, D.W.: Derivation of the Planck formula and the de Broglie relation from physical principles, submitted to *The Physics Teacher*.
3. Motz, L.: Gauge invariance and the Hamilton-Jacobi equation, *Il Nuovo Cimento* **69 B** (1970), 94-104.
4. Einstein, A.: Concerning a heuristic point of view about the creation and transformation of light (transl. by Motz, L.), in Boorse, H. and Motz, L. (1966) *The World of the Atom*, Basic, New York, Vol. 1, p. 544.
5. See, for example, Goldstein, H. (1980) *Classical Mechanics*, 2nd ed., Academic, Reading, p. 484.
6. Feynman, R.P.: Space-time approach to non-relativistic quantum mechanics, *Reviews of Modern Physics* **20** (1948), 367-387. See also Feynman, R.P. and Hibbs, A.R. (1965) *Quantum Mechanics and Path Integrals*, McGraw-Hill, New York.

INERTIAL TRANSFORMATIONS: A REVIEW

F. Selleri

Università di Bari - Dipart. di Fisica
INFN - Sezione di Bari
Via Amendola 173
I-70126 Bari

Abstract. Recently published space-time transformations between inertial systems ("inertial transformations") are reviewed. They are based on three assumptions: (1) The two-way velocity of light is c in all inertial systems and in all directions; (2) Time dilation effects take place with the usual relativistic factor; (3) Clocks are synchronized in the way chosen by nature itself, *e.g.* in the Sagnac effect. They form a group, and agree with the available experimental evidence in spite of the implied noninvariance of the one-way velocity of light. Energy and momentum are defined consistently. Formally they equal the usual relativistic expressions only in a privileged frame, but numerically they do so in all inertial frames. All the experimental data concerning kinematics of high energy particle physics are thus explained also within this new framework.

1. INTRODUCTION

It is well known that the one-way velocity of light has never been measured accurately. The difficulty is that in order to measure it one needs synchronized clocks, but in order to synchronize clocks one must know the one-way velocity of light, so that the logical situation becomes circular. For this reason all the historical laboratory experiments (Fizeau, Foucault, Michelson, etc.) measured the two-way velocity of light. Since such measurements are obviously possible with just one clock, the synchronization problem did not arise. Only the famous experiments on the occultations of Jupiter's satellites (Römer, 1676) and on the aberration of light (Bradley, 1728) were one-way measurements, even if not very precise ones. In building the Special Relativity Theory (SRT) Einstein[1] postulated that the velocity of light has always the same value c in all directions and relative to all inertial frames.

Most contemporary authors of papers on relativity seem to believe that the one-way velocity of light is not measurable as a matter of principle. It is fortunately still possible to disagree, and to believe instead that, owing to serious practical difficulties, it has never been measured up to the present time. After all light goes from a point to another in a well-defined way, and it would be very strange if its true velocity were forever inaccessible to us.

As a consequence of this difficult empirical situation there is a good agreement among physicists at least on the conclusion that the constancy of the one-way velocity of light is a useful convention and that it is not necessarily dictated by objective properties of the natural world. For example Mansouri & Sexl observed[2] that when clocks are synchronized according to the Einstein procedure the equality of the velocity of light in two opposite directions is a consequence of synchronization and cannot really be checked with experiments.

Much earlier Reichenbach[3] had come to similar conclusions, and in 1979 Jammer[4] stressed again that Einstein's procedure is not necessary, and that different synchronizations could be used without ever leading to any conflict with experience.

Obviously, if the invariance of the velocity of light is only a convention, it is interesting to develop alternative theories based on different stipulations. Probably for this reason in recent times there has been a flourishing of new ideas, very often opposed to the relativism of the existing theory[5]-[21]. The task of the present paper is to review the results obtained by the present author in some recent papers[22]-[25].

In order to understand what is implied in these alternatives it is important to stress that the Lorentz transformations are necessary consequences of the relativity principle. This well known fact can be rephrased as follows: One can always choose two Cartesian co-ordinate systems in the inertial reference frames S and S_0 and assume: (1) that space is homogeneous and isotropic, and that time is homogeneous; (2) that relative to S_0 the velocity of light is the same in all directions, so that Einstein's sychronization can be applied in this frame and the velocity v of S relative to S_0 can be measured; (3) that the origins of S and S_0 coincide at $t = t_0 = 0$; (4) that planes (x_0, y_0) and (x, y) coincide at all times t_0; that also planes (x_0, z_0) and (x, z) coincide at all times t_0; but that planes (y_0, z_0) and (y, z) coincide at time $t_0 = 0$ only. It then follows[22] that the transformation law from S_0 to S has necessarily the form

$$\begin{cases} x &= f_1(x_0 - vt_0) \\ y &= g_2 y_0 \\ z &= g_2 z_0 \\ t &= e_1 x_0 + e_4 t_0 \end{cases} \quad (1)$$

where the four coefficients f_1, g_2, e_4, and e_1 can depend on v. A fifth coefficient, still present in Ref.[22], can be eliminated by invoking rotational invariance around the x axis. If at this point one assumes the validity of the relativity principle (including invariance of light velocity) the previous transformations reduce <u>necessarily</u> to the Lorentz ones. In other words, if one considers a four-dimensional space in which the coefficients f_1, g_2, e_4, and e_1 are represented as Cartesian co-ordinates, one can say that for a given value of v all coefficients are completely fixed by the relativity postulate, and therefore represented by a geometrical point. In this space there is no finite area representing relativity, only a structureless unprotected point. All other points lead to the logical negation of the relativity principle. It can therefore be concluded that any shift, either small or large, of any coefficient away from its relativistic value implies necessarily the existence of a privileged frame[22].

2. DIFFICULTIES WITH RELATIVISTIC REALITY

Galilei gave the first modern formulation of the relativity principle by discussing elementary experiments on a moving ship. He would probably have developed a different idea if every object had been invested by a very intense flux of radiation coming from the direction of motion of the ship and if, as a consequence, every form of life had been instantaneously wiped out. Fortunately our ships are not exposed to such effects, but there exist conceptually similar situations. In fact there are perfectly conceivable frames of reference in which the light coming from stars and galaxies toward which these frames are travelling is Doppler shifted towards violet to such an extent, that it is composed of ultrahard gamma rays: in these frames all forms of life and all material aggregations (instruments) would immediately be destroyed. In what sense, then, these frames should be considered equivalent to those moving with small cosmic velocity, if not even in principle it is possible to admit that physical experiments can be carried out with an apparatus at rest in them?

In the 2.7 °K cosmic background radiation an anisotropy has been detected which is probably due to the motion of the solar sistem in space (towards the constellation Leo with the velocity of about 300 km/s). Some authors considered the anisotropy of the background radiation as a proof of lack of validity in nature of the relativity principle. For example Bondi[27], who discussed a "Clear conflict between cosmology and ordinary physics." Or Bergmann[27], who stated: "The principle of relativity would hold only for certain types of experiments (those excluding interaction with the background radiation, for instance), or provided experiments are not refined beyond a certain degree of accuracy or sensitivity."

The SRT leads to major difficulties if used for understanding how objective reality has to be described. To begin with, *what one sees* cannot be considered real in the present, because by looking at distant objects one does not see them as they are now, but as they were when the light now entering our instruments left them. Also, it is not reasonable to attribute reality to the future, because common sense tells us that it does not yet exist and that it is at least partly undetermined. For these reasons a reasonable conception of reality seems to be the following: <u>all that exists now, here and elsewhere</u>.

Let us adopt a Minkowski diagram having space in abscissae and time in ordinates. At time $t=0$ an observer U_0 located in the origin of an inertial system S must regard as being objectively real down to the smallest detail all events in space. In a bidimensional diagram space is represented as x axis, whose equation is $t=0$, and which therefore contains all events simultaneous with the instantaneous presence of U_0 in the origin at $t=0$.

If we consider, however, another inertial reference frame S', its axes x' and t' are represented in the Minkowski diagram as straight lines in the plane (x, ct) because of the linearity of the Lorentz transformations. The observer U_0' at rest in S' must attribute reality to all events happening at his present time $t'=0$. These events are of course different from those constituting the reality of U_0. According to the relativity principle it does not make any sense to ask which of the two observers is right: they are both right. So all the events on the x' axis, whose equation is $t'=0$ can be considered just as real as those on the x-axis. The reality line of the observer U_0' has an inclination in time with respect to that of U_0 and also passes through the origin of the Minkowski diagram. Therefore U_0' will attribute reality to events in U_0's future.

The meaning of the given argument can be better understood by assuming that in S there are different observers $U_1, U_2, ... U_n, ...$ placed at rest in different points $x_1, x_2, ...x_n, ..$ of the x axis, all provided with clocks synchronized according to the Einstein procedure. These observers are all equivalent in their description of reality, since time $t=0$ is the same for all of them, and reality consists of the events represented on the x axis; naturally they are also equivalent to the observer U_0 situated in $x=0$ considered before. It is now clear that the reality line of U_0' passes through the personal future of some of the observers at rest in S [those situated in points having positive (negative) x if S' moves with velocity in direction $+x$ $(-x)$], e.g., through the future of U_1.

Infinitely many reality lines pass through every point above the x axis of the diagram (ct, x), each such line representing the (relativistic) reality of some legitimate inertial observer. Passing from two to four dimensions we can conclude that all of space-time (ct, x, y, z) is real, despite the different perception humans have of it. In other words my future should be real, i.e. fixed in the tiniest details, despite its looking to me as largely undetermined, unshaped, presently unreal.

Karl Popper, in his autobiography[28], is critical of hyperdeterminism. He recalls a discussion with Einstein in Princeton (1950): "I tried to persuade him to give up his determinism, which amounted to the view that the world was a four-dimensional Parmenidean block universe in which change was a human illusion, or very nearly so. (He agreed that this had been his view, and while discussing it I called him 'Parmenides')". Popper's identification is justified, since for both Einstein and Parmenides the subjective impression of evolution is pure appearence. Popper found this description of reality unacceptable, and it is difficult to disagree with him.

Coming now to general relativity one can observe that also in its case the validity of the relativity principle is far from obvious. Newton was convinced that an 'absolute' space exists and had produced the nice example of the water in the rotating bucket for showing that absolute effects can be produced. In his 1916 paper on general relativity Einstein[29] gave the example of two deformable spheres A and B placed in the interstellar space "at so great a distance from each other and from all other masses that only those gravitational forces need to be taken into account which arise from the interaction of different parts of the same body". At a certain moment one of them (let us say A) is set in rapid rotation around the line joining the centres of A and B. From a strictly rotational pint of view one could say that relativity holds, because the observer in B says "A is rotating", but also the observer on A could say that he sees B rotating. The situation changes however if one considers the deformation of A due to the centrifugal forces, because in this case both observers must agree that it is due to the rotational motion of A with respect to empty space. Since no deformation arises for B, it must be concluded that not all aspects of rotation can be considered to be relative, and thus that some are absolute.

At this point Einstein adds that inertial forces and gravitational forces both give rise to accelerations which are exactly the same for all bodies: therefore, he says, fictitius forces must be of gravitational origin. This is of course the equivalence principle, Einstein's physical reformulation of the more abstract Mach's principle, a beautiful physical idea. To pull it on the side of relativity seems rather doubtful, however.

Einstein implies essentially the following: if the Universe had been at rest in a reference frame different from the one in which it actually stays, then the fictitius forces would behave differently, and the reference frames which today we call inertial would not be such anymore, while other frames that today we consider accelerated would become inertial. In other words the fact that the Universe stays in a frame rather than in another one, according to Einstein does not imply a breakdown of the relativity principle. Here it is difficult to follow him, because his reasoning gives clear priority to the reference frames (which are useful human constructions) over the concrete reality of the whole Universe. It can well seem dangerous to reason in such a way. But if something has to be modified there remains only Newton's idea that privileged frames do exist.

3. NATURE'S CHOICE OF SYNCHRONIZATION

It was shown in[22] that the most general transformation laws of the general type (1) between two inertial frames S_0 and S satisfying the conditions of constant two-way velocity of light and of time dilation according to the usual relativistic factor are:

$$\begin{cases} x = \dfrac{x_0 - \beta c t_0}{R(\beta)} \\ y = y_0 \\ z = z_0 \\ t = R(\beta) t_0 + e_1 (x_0 - \beta c t_0) \end{cases} \quad (2)$$

where e_1 is a function of velocity v, $\beta = v/c$, and

$$R(\beta) = \sqrt{1-\beta^2} \qquad (3)$$

A term containing x_0 and y_0 in the right hand side of the fourth Eq. (2) has been eliminated by invoking rotational invariance around the x axis, which is parallel to velocity. Length contraction by the usual factor is also a consequence of (2). The velocity of light compatible with (2) was shown to be:

$$\frac{1}{\tilde{c}(\theta)} = \frac{1}{c} + \left[\frac{\beta}{c} + e_1 R(\beta)\right]\cos\theta \qquad (4)$$

where θ is the angle between the direction of propagation of light and the absolute velocity $\vec{v}$ of S. The transformations (2) represent the complete set of theories equivalent to STR: if e_1 is varied, different elements of this set are obtained, which are all equivalent as far as the explanation of experimental results is concerned. The Lorentz transformation is recovered as a particular case with

$$e_1 = -\frac{1}{c}\frac{\beta}{R(\beta)} \qquad (5)$$

Different values of e_1 are obtained from different clock-synchronisation conventions. In all cases but that of STR such theories do not lead to the validity of the relativity principle, but imply the existence of a privileged frame[24].

The simplest possibility in (2) is obviously $e_1 = 0$. A simple way of justifying this choice is to apply the so-called absolute synchronisation[2], by setting all clocks of S to time $t = 0$ when the passing clock at rest in the privileged system S_0 shows the time $t_0 = 0$.

There are however much stronger reasons for adopting a space-independent transformation of time. The assumed indifference of the physical reality concerning clock synchronisation exists only insofar as one neglects accelerations: when these come into play every inertial system exists, so to say, only for a vanishingly small time interval and it is physically impossible in the accelerated frame to adopt any time-consuming procedure for the synchronisation of distant clocks (such as Einstein's procedure). Yet physical events take place and synchronisation is somehow fixed by nature itself: we will see how this happens next.

Our choice of synchronization ("absolute", according to Mansouri and Sexl[2]) is made by considering accelerations. This would perhaps not please purists, but one can stress that the normally accepted relationship between SRT and accelerated systems is far from negligible. Accelerations are for instance essential in the so-called twin paradox, which is a prediction of SRT made before the general theory was even conceived. Very large accelerations enter in the experiment on the lifetime of muons circulating in the CERN Muon Storage Ring[30] which is considered the most precise empirical test of the time-dilation phenomenon. The acceleration of Earth is essential for perceiving the retardation/anticipation of the eclipses of Jupiter's satellites. The same can be said about the detection of stellar aberration. And so on.

Accelerations are instead avoided when they seem to generate some difficulties within the existing theory. This is perhaps an understandable reaction initially, but it is not acceptable in the long run.

A simple accelerated system is a rotating disk, and the Sagnac effect[31] is well known to take place in such conditions: a monochromatic light source placed on the disk emits two coherent beams of light in opposite directions. These travel along a circumference concentrical with the disk, until they reunite in a point M and interfere, after a 2π propagation. The result can be achieved by forcing the light to propagate tangentially to the internal surface of a cylindrical mirror. The positioning of the interference figure dipends on the disk rotational velocity ("Sagnac effect"). Some textbooks deduce the Sagnac formula (our Eq. (10) below) in the laboratory, but say nothing about the description given by an observer placed on the rotating platform. By reviewing the argument of Ref.[24] we will see that SRT predicts a null effect on the platform, while our approach based on the inertial transformations gives the right answer. For simplicity we consider a laboratory at rest in the privileged frame.

<u>Sagnac effect in the laboratory</u>. Light propagating in the rotational direction of the disk must cover a distance larger than the disk circumference length L by a quantity $x = v t_{01}$ equaling the shift of M during the time t_{01} taken by light to reach the interference region. Therefore

$$L + x = c t_{01} \quad ; \quad x = v t_{01} \tag{6}$$

whence:

$$t_{01} = \frac{L}{c - v} \tag{7}$$

Light propagating in the direction opposite to that of rotation must instead cover a distance smaller than the disk circumference length L by a quantity $y = v t_{02}$ equaling the shift of M during the time t_{02} taken by light to reach the interference region. Therefore

$$L - y = c t_{02} \quad ; \quad y = v t_{02} \tag{8}$$

whence

$$t_{02} = \frac{L}{c + v} \tag{9}$$

The time difference Δt_0 between the two propagations is the parameter fixing the phase difference in the considered interference point. From (7) and (9) it follows

$$\Delta t_0 = t_{01} - t_{02} = \frac{2L}{c} \frac{\beta}{1 - \beta^2} = \frac{2L_0}{c} \frac{\beta}{R(\beta)} \tag{10}$$

Obviously $L = L_0 R(\beta)$ is the disk circumference length reduced in the laboratory by the usual relativistic factor (3) if L_0 is the rest length of the same disk. The consistency of Eq. (10) with experimental data has been checked in many experiments.

<u>Sagnac effect on the disk</u>. Every small portion of the circumference of the rotating platform can be considered to be instantaneously at rest in a moving inertial frame of reference locally "tangent" to the disk. Therefore Eq. (4) applies for the velocity of light on the disk. Only the cases of light moving parallel and antiparallel to the local absolute velocity must be considered. It follows from (4) that the inverse velocity of light for these two cases is respectively given by:

$$\begin{cases} \dfrac{1}{\tilde{c}(0)} = \dfrac{1}{c} + \left[\dfrac{\beta}{c} + e_1 R(\beta)\right] \\ \dfrac{1}{\tilde{c}(\pi)} = \dfrac{1}{c} - \left[\dfrac{\beta}{c} + e_1 R(\beta)\right] \end{cases} \quad (11)$$

The time difference on the disk is given by

$$\Delta t = t_1 - t_2 = \dfrac{L_0}{\tilde{c}(0)} - \dfrac{L_0}{\tilde{c}(\pi)} \quad (12)$$

Substituting (11) in (12) one gets

$$\Delta t = \Delta t_0 R(\beta)\left[1 + \dfrac{c e_1 R(\beta)}{\beta}\right] \quad (13)$$

where Δt_0 is given by (10) and $R(\beta)$ is the usual factor describing the dilation of time intervals in a moving frame. As one can see only the value $e_1 = 0$ leads to physical agreement with (10), while the prediction (5) of SRT gives instead $\Delta t = 0$. Therefore we reach the important conclusion that of all theories having different values of e_1 only one ($e_1 = 0$) gives a rational description of the Sagnac effect on the rotating platform. In the case of $e_1 \neq 0$ the predicted time difference on the platform disagrees with the prediction (10) in the laboratory.

Finally we review some arguments of Ref.[23] which, like the Sagnac effect, point to the superiority of the inertial transformations. Two identical spaceships **A** and **B** are initially at rest on the x_0 axis of the (privileged) inertial system S_0 at a distance d_0 from one another. Their clocks are synchronised with those of S_0. At time $t_0 = 0$ they start accelerating in the $+x_0$ direction, and they do so in the same identical way having the same velocity $v(t_0)$ at all times t_0 of S_0, until at $t_0 = \bar{t}_0$ they reach a preassigned velocity parallel to $+x_0$. For all $t_0 \geq \bar{t}_0$ the spaceships can be considered at rest in a different inertial system S, which they concretely constitute.

It can been shown that the transformation relating S_0 and S is necessarily the inertial one, if no clock synchronisation is applied correcting what nature itself generated during the acceleration of the two spaceships. Since **A** and **B** accelerate exactly in the same way, their clocks will accumulate exactly the same delay with respect to those at rest in S_0. Such delay will arise both from instantaneous velocities, and from the cosmic gravitational field which is present in accelerated frames due to the equivalence principle. Motion is the same for **A** and **B** and all effects of motion will necessarily coincide, in particular time delay. Therefore two events simultaneous in S_0 will be such also in S, even if they take place in different points of space. Clearly we have a case of absolute simultaneity and the condition $e_1 = 0$ must hold in (2), reducing these transformations to their inertial form, given below in Eq. (5).

Another convincing argument showing that the condition $e_1 = 0$ gives the most natural description of physical reality is the following. Suppose that our spaceships have passengers P_A and P_B, who are twins. Of course in principle nothing can stop them from re-synchronising their clocks once they have finished accelerating and the two spaceships are at rest in S. If they do so, however, they find in general that they have different biological ages at the same (re-synchronised) S time, even if they started the space trip at exactly the same S_0 time and with the same velocity, as stipulated above. Everything is regular, instead, if they do not operate any asymmetrical modification of the time shown by their clocks.

4. THE INERTIAL TRANSFORMATIONS

In the previous section we showed quite generally that the condition $e_1 = 0$ is the right choice of synchronization. This implies that from all positions in S_0 the time in S will be seen to be the same, and therefore that no position dependent time-lag factor will be present in the transformation of time:

$$\begin{cases} x = \dfrac{x_0 - \beta c t_0}{R(\beta)} \\ y = y_0 \\ z = z_0 \\ t = R(\beta) t_0 \end{cases} \quad (14)$$

The velocity of light deducible from a theory based on (14) can easily be found by putting $e_1 = 0$ in (4):

$$\frac{1}{\tilde{c}(\theta)} = \frac{1 + \beta \cos\theta}{c} \quad (15)$$

The transformation (18) can be inverted and gives:

$$\begin{cases} x_0 = R(\beta)\left[x + \dfrac{\beta c}{R^2(\beta)}t\right] \\ y_0 = y \\ z_0 = z \\ t_0 = \dfrac{1}{R(\beta)}t \end{cases} \qquad (16)$$

Note that there is a formal difference between (14) and (16). The latter implies, for example, that the origin of S_0 (satisfying $x_0 = y_0 = z_0 = 0$) is described in S by $y = z = 0$ and by

$$x = -\dfrac{\beta c}{1-\beta^2}t$$

This origin is thus seen to move with speed $\beta c/(1-\beta^2)$, which can exceed c, but cannot be superluminal. In fact a light pulse seen from S to propagate in the same direction as S_0 has $\theta=\pi$, and thus [using (15)] has speed $\tilde{c}(\pi) = c/(1-\beta)$, which satisfies

$$\dfrac{c}{1-\beta} \geq \dfrac{c\beta}{1-\beta^2}$$

One of the typical features of these transformations is obviously the presence of velocities which can grow without limit *when they are relative to moving systems* having absolute velocities βc near to c. Absolute velocities can instead never exceed c[25]. In STR one is used to relative velocities that are always equal and opposite, but this symmetry is a consequence of the particular synchronisation used and cannot be expected to hold more generally.

Consider now a third inertial system S' moving with velocity $\beta'c$ and its transformation from S_0, which of course is

$$\begin{cases} x' = \dfrac{x_0 - \beta' c t_0}{R(\beta')} \\ y' = y_0 \\ z' = z_0 \\ t' = R(\beta')t_0 \end{cases} \qquad (17)$$

where $R(\beta')$ is given by (3) with β' replacing β. By eliminating the S_0 variables from (17) and (16) one obtains the transformation between the two moving systems S and S':

$$\begin{cases} x' = \dfrac{R(\beta)}{R(\beta')}\left[x - \dfrac{\beta'-\beta}{R^2(\beta)}ct\right] \\ y' = y \\ z' = z \\ t' = \dfrac{R(\beta')}{R(\beta)}t \end{cases} \qquad (18)$$

A transformation having the form (14) has once been written down by Tangherlini[32], while (16) and (18) were obtained for the first time in Ref.[22]: we call (14)-(16)-(18) "inertial transformations". In its most general form (22) an inertial transformation depends on two velocities (v and v'). When one of them is zero, either S or S' coincides with the privileged system S_0 and the transformation (18) becomes either (14) or (16).

A feature characterising the transformations (14)-(16)-(18) is the existence of <u>absolute simultaneity</u>: two events taking place in different geometrical points of S at the same t are judged to be simultaneous also in S' (and vice versa), this property being consequence of the absence of space variables in the transformation of time. Of course the existence of absolute simultaneity does not imply that time is absolute: on the contrary, the β-dependent factor in the transformation of time gives rise to time-dilation phenomena similar to those of STR. <u>Time dilation</u> in another sense is however also absolute: a clock at rest in S is seen from S_0 to run slower, but a clock at rest in S_0 is seen from S to run <u>faster</u> so that both observers will agree that motion relative to S_0 slows down the pace of clocks. Quantitatively one has for both situations:

$$\Delta t = \sqrt{1-\beta^2}\Delta t_0 \qquad (19)$$

where Δt and Δt_0 are the time intervals between any two given events as measured with clocks at rest in S and in S_0, respectively. The difference with respect to STR is however more apparent than real, if an experimental confirmation of time dilation is considered: a meaningful comparison of rates implies that a clock T_0 at rest in S_0 must be compared with clocks at rest in different points of S, and the result is therefore dependent on the convention adopted for synchronising the latter clocks. Nevertheless the conceptual consequences of the new situation are very positive because a theory based on absolute simultaneity does not have the ugly "Einstein-Parmenides" paradox that we reviewed in section 2. All observers agree on what must be considered the present reality, or, in other words, in a Minkowski diagram the x axes of all moving observers coincide.

Absolute length contraction can also be deduced from (14)-(16). A rod at rest on the x axis of S between the points with co-ordinates x_2 and x_1 is seen in S_0 to have end points x_{02} and x_{01} at a common time t_0, where from (14):

$$x_2 = \frac{x_{02} - vt_0}{\sqrt{1-\beta^2}} \quad ; \quad x_1 = \frac{x_{01} - vt_0}{\sqrt{1-\beta^2}} \tag{20}$$

From this one obtains

$$x_2 - x_1 = \frac{1}{\sqrt{1-\beta^2}}(x_{02} - x_{01}) \tag{21}$$

The reasoning can be inverted by considering the rod at rest in S and observed from S_0, and using the transformation (16). One gets then, after a few simple steps:

$$x_{02} - x_{01} = \sqrt{1-\beta^2}(x_2 - x_1) \tag{22}$$

which could also be obtained by inverting (21). The two results are thus mathematically equivalent and lead to the conclusion (with which both observers agree) that motion relative to S_0 leads to contraction. This is obviously an absolute effect, but again the discrepancy with the STR is due to the different conventions concerning clock synchronisation: the length of a moving rod can only be obtained by marking the <u>simultaneous</u> positions of its end points, and is therefore dependent on the very definition of simultaneity of distant events.

We conclude the section by recalling that the inertial transformations form a group[24]. In fact let $\Omega(\beta, \beta')$ be the transformation (18), dependent on the two dimensionless absolute velocities β and β', and let $I \equiv \{\Omega(\beta, \beta')\}$ be the set of all such transformations. Two elements of I differ from one another only for the value of one or both velocities β and β'. The Tangherlini transformation (14) is $\Omega(0, \beta)$; its inverse (16) is $\Omega(\beta, 0)$, so that they both belong to I. It follows that:

[1] The identical transformation is an element of I because for $\beta = \beta' = 0$ one gets $\Omega(0, 0) \in I$.

[2] The inverse transformation of $\Omega(\beta, \beta')$ is obtained by inverting (18) and turns out to be $\Omega(\beta', \beta) \in I$. Therefore the inverse of a transformation is obtained by interchanging the two absolute velocities β and β'.

[3] The product of two inertial transformations is obtained as follows: consider the transformation $\Omega(\beta', \beta'')$ from S' to S''. By combining it with $\Omega(\beta, \beta')$ one obtains $\Omega(\beta, \beta'') \in I$. This can be written

$$\Omega(\beta, \beta')\Omega(\beta', \beta'') = \Omega(\beta, \beta'') \tag{23}$$

This is the multiplication law of inertial transformations: as one can see the common velocity disappears from the product.

[4] The associative law of the multiplication can also be established. Consider four inertial frames S, S', S'' and S''' and the following transformations

$$\Omega(\beta, \beta') : S \Rightarrow S' \quad ; \quad \Omega(\beta', \beta'') : S' \Rightarrow S'' \quad ; \quad \Omega(\beta'', \beta''') : S'' \Rightarrow S'''$$

From (23) one easily gets:

$$[\Omega(\beta, \beta')\Omega(\beta', \beta'')]\Omega(\beta'', \beta''') = \Omega(\beta, \beta'')\Omega(\beta'', \beta''') = \Omega(\beta, \beta''')$$

and

$$\Omega(\beta, \beta')[\Omega(\beta', \beta'')\Omega(\beta'', \beta''')] = \Omega(\beta, \beta')\Omega(\beta', \beta''') = \Omega(\beta, \beta''')$$

so that the associative law is satisfied.

5. OPTICAL INTERFERENCE EXPERIMENTS

Inertial and Lorentz transformations give identical predictions concerning optical interference experiments[24]. Consider a laboratory at rest in an inertial frame of reference S moving with absolute velocity $\vec{v}$ and suppose that an experiment on the interference of light is performed with instruments at rest in this laboratory. A light ray is divided in two (coherent) parts by a semitransparent mirror placed in point P. The first part propagates along the broken path $P\text{-}A_1\text{-}A_2$ ... $A_{m-1}\text{-}Q$, where suitably oriented reflecting mirrors are placed in the intermediate points, the second one along the similar path $P\text{-}B_1\text{-}B_2$ ... $B_{n-1}\text{-}Q$. Finally the two parts come to overlap in Q where they interfere. The point Q can be any point of an extended interference figure.

The vectors $\vec{\ell}_{ai}$ (having modulus ℓ_{ai}), with $i = 1,2,...m$, are defined to coincide with the rectilinear segments described by light and oriented in the direction of propagation, from P towards Q. Similarly on the second path define the vectors $\vec{\ell}_{bj}$ (having modulus ℓ_{bj}), with $j = 1,2,...n$. The interference in Q is determined by the time delay ΔT between the two rays. The prediction of SRT is calculated as follows:

$$\Delta T = T_B - T_A = \frac{L_B - L_A}{c} \tag{24}$$

where

$$L_A = \sum_{i=1}^{m} \ell_{ai} \quad ; \quad L_B = \sum_{j=1}^{n} \ell_{bj} \tag{25}$$

Consider next the same quantity ΔT by starting from the inertial transformations, according to which the inverse velocity of a light ray propagating in S in a direction forming an angle θ with the absolute velocity $\vec{v}$ of S is given by (15). In such a case the time delay between the light rays that followed the two different paths is:

$$\Delta T = \sum_{j=1}^{n} \frac{\ell_{bj}}{\tilde{c}(\theta_{bj})} - \sum_{i=1}^{m} \frac{\ell_{ai}}{\tilde{c}(\theta_{ai})} \tag{26}$$

where θ_{ai} (θ_{bj}) is the angle between $\vec{\ell}_{ai}$ and $\vec{v}$ ($\vec{\ell}_{ai}$ and $\vec{v}$). By inserting (15) in (26) one gets:

$$\Delta T = \sum_{j=1}^{n} \ell_{bj} \frac{1+\beta \cos\theta_{bj}}{c} - \sum_{i=1}^{m} \ell_{ai} \frac{1+\beta \cos\theta_{ai}}{c} =$$

$$= \frac{L_B - L_A}{c} + \frac{1}{c^2} \sum_{j=1}^{n} \vec{\ell}_{bj} \cdot \vec{v} - \frac{1}{c^2} \sum_{i=1}^{m} \vec{\ell}_{ai} \cdot \vec{v} = \qquad (27)$$

$$= \frac{L_B - L_A}{c}$$

The last step is a consequence of

$$\sum_{j=1}^{n} \vec{\ell}_{bj} = \sum_{i=1}^{m} \vec{\ell}_{ai} \qquad (28)$$

because the two sides of (28) are separately equal to the vector joining points P and Q. As one can see the results (24) and (27) coincide. Therefore a theory based on the inertial transformations can explain the results of all the interferometric experiments carried out up to now (Michelson-Morley[33], Kennedy-Thorndike[34], Q. Majorana[35], ...) as well as the SRT does.

6. ENERGY AND MOMENTUM IN THE PRIVILEGED FRAME

In this section and in the following ones we review the results obtained in Ref.[25] concerning energy and momentum conservation. We will now show that there are necessary expressions of energy and momentum (referred to the fundamental frame) for a particle having rest mass m and velocity $\vec{u}_0$ if one assumes the following two physical conditions to hold:

1. <u>Mass-energy equivalence</u>, which can be written

$$E(u_0) = m(u_0)c^2 \qquad (29)$$

where $m(u_0)$ is the velocity-dependent mass and $E(u_0)$ is the total energy, including the kinetic energy $T(u_0)$ and the energy equivalent of the rest mass mc^2:

$$E(u_0) = T(u_0) + mc^2 \qquad (30)$$

Mass energy equivalence could be formulated by saying that mass is a totally redundant ingredient of physics, and that it can be completely eliminated. For historical reasons it is however better to be more conservative, and say only that every physical property of energy is shared by mass multiplied by c^2, and that every function of mass can be adequately performed by energy divided by c^2.

2. <u>Momentum as "quantity of motion"</u>. The interpretation of momentum as quantity of motion means that we must multiply the particle velocity by the amount of matter having that velocity, or by the total mass:

$$\vec{p}(u_0) = m(u_0)\vec{u}_0 \qquad (31)$$

A force $\vec{F}$ displacing a particle by $\vec{ds}$ generates a variation dE of the particle energy which satisfies

$$\vec{F} \cdot \vec{ds} = dE \qquad (32)$$

Using Newton's law and dividing by dt_0 this can be written

$$\frac{d\vec{p}}{dt_0} \cdot \vec{u}_0 = \frac{dE}{dt_0} \qquad (33)$$

From (29) and (31) we see that energy and momentum depend on time only because they are functions of velocity. Therefore (33) becomes

$$\left(\frac{\partial \vec{p}}{\partial u_{0x}} \dot{u}_{0x} + \frac{\partial \vec{p}}{\partial u_{0y}} \dot{u}_{0y} + \frac{\partial \vec{p}}{\partial u_{0z}} \dot{u}_{0z} \right) \cdot \vec{u}_0 = \frac{\partial E}{\partial u_{0x}} \dot{u}_{0x} + \frac{\partial E}{\partial u_{0y}} \dot{u}_{0y} + \frac{\partial E}{\partial u_{0z}} \dot{u}_{0z} \qquad (34)$$

where a dot indicates a time (t_0) derivative. Eq. (34) must hold for arbitrary values of the acceleration. Therefore we get

$$\frac{\partial \vec{p}}{\partial u_{0x}} \vec{u}_0 = \frac{\partial E}{\partial u_{0x}} \qquad (35)$$

and other two similar Eq.s for the y and z components. Given (29) and (31) Eq. (35) can be transformed into an equation for $m(u_0)$, which turns out to be:

$$\frac{dm(u_0)}{du_0} \left[c^2 - u_0^2 \right] = m(u_0) u_0 \qquad (36)$$

This differential equation can easily be solved and gives

$$m(u_0) = \frac{m}{\sqrt{1 - u_0^2/c^2}} \qquad (37)$$

where $m = m(0)$. From (29) and (31) it follows immediately

$$E(u_0) = \frac{mc^2}{\sqrt{1 - u_0^2/c^2}} \quad ; \quad \vec{p}(u_0) = \frac{m\vec{u}_0}{\sqrt{1 - u_0^2/c^2}} \qquad (38)$$

The kinetic energy is now obtainable from (30) and is:

$$T(u_0) = mc^2 \left[\frac{1}{\sqrt{1-u_0^2/c^2}} - 1 \right] \qquad (39)$$

We thus see that mass-energy equivalence and the interpretation of momentum as quantity of motion lead to the usual (relativistic) expressions for energy and momentum in the privileged frame. This does not force us to conclude that (38) and (39) hold with respect to all inertial frames. Only when the principle of relativity is assumed must the general validity of these expressions be granted, while we will see in the following section how they must be modified in our theory when calculated with respect to moving frames.

7. ENERGY AND MOMENTUM IN ALL INERTIAL FRAMES

The inertial transformation for space and time from a moving inertial frame S to the privileged frame S_0 are given by (16) which in differential form becomes:

$$\begin{cases} dx_0 = R\left[dx + \dfrac{v}{R^2}dt\right] \\ dy_0 = dy \\ dz_0 = dz \\ dt_0 = \dfrac{1}{R}dt \end{cases} \qquad (40)$$

Considering dx_0, dy_0, dz_0 in S_0 (dx, dy, dz in S) to be the displacements of a particle taking place in time dt_0 (in time dt) and dividing the first three Eq.s (40) by the fourth one, we get the absolute transformations of velocities:

$$u_{0x} = u_x R^2 + v \quad ; \quad u_{0y} = u_y R \quad ; \quad u_{0z} = u_z R \qquad (41)$$

where u_{0x}, u_{0y}, u_{0z} and u_x, u_y, u_z are the particle velocity components in the two frames. The following two consequences of Eq.s (41) will be useful:

$$1 - \frac{\vec{u}_0 \cdot \vec{v}}{c^2} = R^2 \left[1 - \frac{\vec{u} \cdot \vec{v}}{c^2}\right] \qquad (42)$$

and

$$\sqrt{1 - \frac{u_0^2}{c^2}} = R\left[\left(1 - \frac{\vec{u} \cdot \vec{v}}{c^2}\right)^2 - \frac{u^2}{c^2}\right]^{1/2} \qquad (43)$$

Consider now again the three Eq.s (41) multiplied by m, add to them the identity $mc^2 = mc^2$, and divide the four equations side by side by (43). The result is

INERTIAL TRANSFORMATIONS

$$\begin{cases} p_{0x} = R\left[p_x + \dfrac{v/c^2}{R^2}E\right] \\ p_{0y} = p_y \\ p_{0z} = p_z \\ E_0 = \dfrac{1}{R}E \end{cases} \qquad (44)$$

if E_0 and $\vec{p}_0$ are now energy and momentum in the fundamental frame, and if we define energy E and momentum $\vec{p}$ in the moving frame as:

$$E = \frac{mc^2}{\sqrt{\left(1 - \vec{u}\cdot\vec{v}/c^2\right)^2 - u^2/c^2}} \quad ; \quad \vec{p} = \frac{m\vec{u}}{\sqrt{\left(1 - \vec{u}\cdot\vec{v}/c^2\right)^2 - u^2/c^2}} \qquad (45)$$

It is amusing to notice that Eq.s (45) imply for energy and momentum exactly the same transformation laws holding for space and time: in fact the analogy between (16) and (44) is complete if only dimensionally homogeneous quantities [(x, y, z, ct) and $(p_x, p_y, p_z, E/c)$] are considered.

The reality condition for the square roots in (45) can easily be shown to imply

$$u \leq \frac{c}{1 + (v/c)\cos\theta} \qquad (46)$$

As seen in (15) the right hand side of this inequality is the velocity of light relative to the moving frame at an angle θ with respect to its absolute velocity. We thus see that (46) is the condition (in the moving frame) that no material object should move faster than light. This must obviously be satisfied because the reality of the square roots in (38) implies that it holds relative to the privileged frame: if a physical system is faster or slower than another one cannot be a matter of points of view, otherwise the first system could be seen overtaking the second one from some inertial frame, and not doing so from another one, which is absurd.

8. INELASTIC PARTICLE COLLISIONS

Let us consider the inelastic collision of two particles producing $n-2$ particles in the final state:

$$P_1 + P_2 \Rightarrow P_3 + P_4 + \ldots + P_n \qquad (47)$$

and let $m_1, m_2, m_3, \ldots, m_n$ be their rest masses, and $\vec{u}_{01}, \vec{u}_{02}, \vec{u}_{03}, \ldots \vec{u}_{0n}$ their absolute velocities. In order to evaluate the effect of Earth motion in theories different from the SRT let us assume that the target particle is also in motion with respect to the privileged frame. The conservation laws of energy and momentum are then

$$\begin{cases} \dfrac{m_1 c^2}{\sqrt{1-u_{01}^2/c^2}} + \dfrac{m_2 c^2}{\sqrt{1-u_{02}^2/c^2}} = \sum_{i=3}^{n} \dfrac{m_i c^2}{\sqrt{1-u_{0i}^2/c^2}} \\ \dfrac{m_1 u_{01}}{\sqrt{1-u_{01}^2/c^2}} \cos\theta_1 + \dfrac{m_2 u_{02}}{\sqrt{1-u_{02}^2/c^2}} = \sum_{i=3}^{n} \dfrac{m_i u_{0i}}{\sqrt{1-u_{0i}^2/c^2}} \cos\theta_i \\ \dfrac{m_1 u_{01}}{\sqrt{1-u_{01}^2/c^2}} \sin\theta_1 = \sum_{i=3}^{n} \dfrac{m_i u_{0i}}{\sqrt{1-u_{0i}^2/c^2}} \sin\theta_i \end{cases} \quad (48)$$

where θ_i is the angle formed between the i-th final particle momentum and P_1 while θ_1 is the angle between P_1 and P_2, always in the privileged frame S_0. The previous equations take a simpler form if one puts:

$$\varepsilon_k \equiv \dfrac{u_{0k}/c}{\sqrt{1-u_{0k}^2/c^2}} \qquad (k=1,2,3,...n) \quad (49)$$

whence

$$\sqrt{1+\varepsilon_k^2} = \dfrac{1}{\sqrt{1-u_{0k}^2/c^2}} \quad (50)$$

In fact Eq.s (48) become

$$\begin{cases} m_1\sqrt{1+\varepsilon_1^2} + m_2\sqrt{1+\varepsilon_2^2} = \sum_{i=3}^{n} m_i\sqrt{1+\varepsilon_i^2} \\ m_1\varepsilon_1\cos\theta_1 + m_2\varepsilon_2 = \sum_{i=3}^{n} m_i\varepsilon_i\cos\theta_i \\ m_1\varepsilon_1\sin\theta_1 = \sum_{i=3}^{n} m_i\varepsilon_i\sin\theta_i \end{cases} \quad (51)$$

By squaring and adding the second and third equations it follows:

$$m_1^2\varepsilon_1^2 + m_2^2\varepsilon_2^2 + 2m_1 m_2 \varepsilon_1 \varepsilon_2 \cos\theta_1 = \sum_{i,j=3}^{n} m_i m_j \varepsilon_i \varepsilon_j \cos(\theta_i - \theta_j)$$

By subtracting the latter to the square of the first Eq. (51) one gets:

$$m_1^2 + m_2^2 + 2m_1 m_2 \left[\sqrt{1+\varepsilon_1^2}\sqrt{1+\varepsilon_2^2} - \varepsilon_1\varepsilon_2\cos\theta_1\right] = \\ = \sum_{i,j=3}^{n} m_i m_j \left[\sqrt{1+\varepsilon_i^2}\sqrt{1+\varepsilon_j^2} - \varepsilon_i\varepsilon_j\cos(\theta_i - \theta_j)\right] \quad (52)$$

It will next be shown that the threshold condition can be obtained from Eq. (52). The term within square brackets in the left hand side can be written

$$Q \equiv \sqrt{1+\varepsilon_1^2}\sqrt{1+\varepsilon_2^2} - \varepsilon_1\varepsilon_2\cos\theta_1 = \frac{1 - \vec{u}_{01}\cdot\vec{u}_{02}/c^2}{\sqrt{1-u_{01}^2/c^2}\sqrt{1-u_{02}^2/c^2}} \quad (53)$$

Now, the partial derivative of Q with respect to u_{01} can easily be shown to be

$$\frac{\partial Q}{\partial u_{01}} = \Gamma(u_{01} - u_{02}\cos\theta_1)$$

with a positive Γ. It is easy to show that in all physically interesting cases the right hand side of the previous equation is positive, so that Q is a growing function of u_{01}. The result is obvious if the cosine in the latter expression is negative (or zero). If the cosine is positive, we can observe that collisions are possible only when the component of the velocity of particle 1 along the velocity of particle 2 is larger than the latter, a condition requiring $u_{01}\cos\theta_1 - u_{02} > 0$. Then also $u_{01}\cos\theta_1 - u_{02}\cos^2\theta_1 > 0$, which implies $u_{01} - u_{02}\cos\theta_1 > 0$.

The smallest value of Q is thus obtained by giving u_{01} its lowest possible value compatible with the concrete taking place of the inelastic reaction. Therefore we are interested in the smallest possible values of all $(n-2)^2$ square brackets in the right hand side of (52). These are given by

$$\begin{cases} \varepsilon_3 = \varepsilon_4 = \ldots = \varepsilon_n \equiv \varepsilon \\ \theta_3 = \theta_4 = \ldots = \theta_n \equiv \theta \end{cases} \quad (54)$$

and lead to the value one of all brackets. Thanks to (54) the system (51) becomes

$$\begin{cases} m_1\sqrt{1+\overline{\varepsilon}_1^2} + m_2\sqrt{1+\varepsilon_2^2} = M\sqrt{1+\varepsilon^2} \\ m_1\overline{\varepsilon}_1\cos\theta_1 + m_2\varepsilon_2 = M\varepsilon\cos\theta \\ m_1\overline{\varepsilon}_1\sin\theta_1 = M\varepsilon\sin\theta \end{cases} \quad (55)$$

These three equations fix $\overline{\varepsilon}_1$ (which gives the energetic threshold) and the other two unknown quantities ε and θ. At threshold Eq. (52) becomes

$$m_1^2 + m_2^2 + 2m_1m_2\overline{Q} = M^2 \quad (56)$$

where $\overline{Q}$ is the threshold value of Q for fixed θ_1 and ε_2. From now on the bar will be avoided, but it will remain understood that u_{01} is calculated at threshold. Next we apply the previous results to the situation where the target particle, moving with respect to the privileged system, is instead at rest in a laboratory forming a different inertial system, i.e. on Earth. If $\vec{v}$ is the instantaneous absolute velocity of our planet, then we must write $\vec{u}_{02} = \vec{v}$ in Eq. (53). From (53) and (56) we see that the threshold condition becomes

$$\frac{1 - \vec{u}_{01} \cdot \vec{v} / c^2}{\sqrt{1 - u_{01}^2 / c^2} \sqrt{1 - v^2 / c^2}} = \frac{M^2 - m_1^2 - m_2^2}{2 m_1 m_2} \quad (57)$$

If we transform u_{01} to the same inertial frame by using Eq.s (42) and (43), Eq. (57) becomes:

$$\frac{1 - \vec{u}_1 \cdot \vec{v} / c^2}{\left[\left(1 - \vec{u}_1 \cdot \vec{v} / c^2\right)^2 - u_1^2 / c^2\right]^{1/2}} = \frac{M^2 - m_1^2 - m_2^2}{2 m_1 m_2} \quad (58)$$

The physical meaning of this result will become clear in the coming section.

9. REDEFINITION OF ENERGY

If we multiply (58) by $m_1 c^2$ and remember the definitions (45) of energy and momentum, we get the threshold condition in the form

$$E_1 - \vec{p}_1 \cdot \vec{v} = \frac{M^2 - m_1^2 - m_2^2}{2 m_2} c^2 \quad (59)$$

There is an unexpected term $\vec{p}_1 \cdot \vec{v}$ in (59). It is tempting to attribute it to an ether wind, given that ether (at rest in S_0) would have a velocity proportional to $-\vec{v}$ relative to S. One could then assume that accompanying the ether wind ($\vec{p}_1$ parallel to $-\vec{v}$) adds the extra-energy $\vec{p}_1 \cdot (-\vec{v})$ to the energy of the particle, while opposing the ether wind ($\vec{p}_1$ parallel to $\vec{v}$) subtracts the same extra-energy.

It is anyway useful to define the "effective energy":

$$\tilde{E} = E - \vec{p} \cdot \vec{v} \quad (60)$$

which should of course bear an index "1" when referred to particle P_1. The "effective" energy is thus the total energy actually available to a particle. In the privileged frame, where the ether wind does not exist ($\vec{v} = 0$), the effective energy coincides with the definition (38) of energy. Remembering also that $\vec{p} \cdot \vec{v} = p_x v$ the transformations (44) can be transformed thus:

$$\begin{cases} p_{0x} = \dfrac{p_x + (v / c^2) \tilde{E}}{R} \\ p_{0y} = p_y \\ p_{0z} = p_z \\ E_0 = \dfrac{\tilde{E} + v p_x}{R} \end{cases} \quad (61)$$

These are exactly the (inverse) Lorentz transformations for the quantities $(\vec{p}, \tilde{E})$. This result does not imply that we are reverting to the SRT, first of all because the Eq.s (45) hold, and secondly because in our theory space and time obey the inertial (and not the Lorentz) trasformations.

Let us recapitulate. In the privileged system S_0 the velocity of light is isotropic, but in the system S, which can be any other inertial frame, this is not so. Our previous identification of S with an Earth-based laboratory was made of course for an easier comparison with the empirical evidence about inelastic thresholds, particle masses, collision kinematics, and so on. There are however good reasons for believing that in different inertial frames all that evidence would remain the same. Therefore we assume that the transformations (61) between S_0 and S hold unmodified for all possible S. But (61) is a Lorentz transformation (LT) and it is well known that:

(1) The inverse of a LT is also a LT;

(2) If LTs hold between S_0 and S and between S_0 and S', then a LT holds also between S and S'.

We thus conclude that (61) is the universal transformation of $(\vec{p}, \tilde{E})$ between any two inertial frames. Given the energy and the momentum of a particle in S_0 the numerical values of (effective) energy and momentum predicted in our approach and in SRT for a different frame are exactly the same because both theories use the LTs, and these fix uniquely the transformed energy and momentum in terms of their S_0 values.

Therefore our Eq. (59) becomes

$$\tilde{E}_1 = \frac{M^2 - m_1^2 - m_2^2}{2 m_2} c^2 \qquad (62)$$

This result shows that the practical use of the inertial transformations leads necessarily to the validity in an Earth-based laboratory of that threshold condition which received so many precise confirmations in high energy experiments.

The dynamical equivalence of the two theories is however not limited to threshold conditions, but is totally general, given that we have shown that energy and momentum must be defined in such a way as to coincide numerically for all particles and in all inertial frames once they coincide in the fundamental frame. Therefore the kinematics of high energy procesess, the determination of particle masses, and so on, do not require a different analysis from the one successfully carried out by particle physicists up to the present time.

10. CONCLUSIONS

There remain at least two fundamental questions to study, before a theory based on the inertial transformations can be coinsidered reasonably complete:

(1) Maxwell's equations must be reformulated. They will take a more general form, dependent on the absolute velocity of the inertial frame with respect to which they are considered, and will assume the usual form only in the privileged frame. This work is underway and will probably generate a good agreement with experiments, in view of the results presented in section 5 concerning optical interference.

(2) Also the general theory needs modifications because the ds^2 cannot be considered anymore invariant, given that the inertial transformations predict a frame-dependent one-way velocity of light. It will be necessary to show that the modified theory makes the same successful predictions as the general theory of relativity. Fortunately it is possible to be optimistic also in this connection, given that the research carried out by different authors in the last decades has produced theories structurally different from the general theory of relativity which lead nevertheless to the same successful predictions of the latter theory.

REFERENCES

(1) A. Einstein, Ann. der Physik, 17, 891 (1905).

(2) R. Mansouri e R. Sexl, General Relat. and Gravitation, 8, 497 (1977).

(3) H. Reichenbach, *The Philosophy of Space &Time* (Dover Publ., New York, 1958).

(4) M. Jammer, "Some fundamental problems in the special theory of relativity", in *Problems in the Foundations of Physics*, G. Toraldo di Francia, ed., pp. 202-236 (Società Italiana di Fisica, Bologna, and North Holland, Amsterdam, 1979).

(5) R.H. Dishington, "Cause and effect in special relativity", in: *Advances in Fundamental Physics*, pp. 187-202, ed. by: M. Barone & F. Selleri, Hadronic Press, Palm Harbor (1995).

(6) H.E. Wilhelm, Apeiron, 15, 1 (1993); Phys. Essays, 6, 420 (1993).

(7) P. Cornille, Phys. Essays, 5, 262 (1992).

(8) F. Winterberg, Z. Naturforsch. 44a, 1145 (1989).

(9) J. Lévy, Phys. Essays, 6, 241 (1993).

(10) H. Aspden, "Tests of photon theory in terms of precision measurement", in: *Problems in Quantum Physics: Gdansk '87*, p. 353, ed. by: L. Kostro et al., World Scientific, Singapore (1988).

(11) J.P. Wesley, Found. Phys. Lett., 7, 493 (1994).

(12) H.C. Hayden, Galilean Electrodyn., 2, 63 (1991).

(13) C.I. Mocanu, "Hertzian extension of Einstein's special relativity to non-uniform motions", in: *Advances in Fundamental Physics*, p. 217, ed. by: M. Barone & F. Selleri, Hadronic Press, Palm Harbor (1995).

(14) S.V.M.Clube, "Cosmological redshifts and the law of corresponding states", in: *Frontiers of Fundamental Physics*, p. 107, ed. by: M. Barone and F. Selleri, Plenum, New York/London (1994).

(15) G. Galeczki and P. Marquardt, Nuovo Cimento, **109 B**, 1331 (1994).

(16) P.F. Browne, Found. Phys., **6**, 457 (1976).

(17) G. Pocci and T. Sjödin, Nuovo Cimento, **63 B**, 601 (1981).

(18) S.J. Prokhovnik, *Light in Einstein's Universe*, Reidel, Dordrecht (1995).

(19) J.S. Bell, "How to teach special relativity", in: J.S. Bell, *Speakable and Unspeakable in Quantum Mechanics*, Cambridge Univ. Press (1987).

(20) T.E. Phipps, Jr., Phys. Essays, **2**, 180 (1989).

(21) G.Cavalleri and C.Bernasconi, Nuovo Cimento, **104 B**, 545 (1989).

(22) F. Selleri, "Theories equivalent to special relativity" in *Frontiers of Fundamental Physics*, pp. 181-192, M. Barone &F. Selleri, eds. (Plenum, London/New York, 1994)

(23) F. Selleri, "Space, time, and their transformations", in print on *Chin. Jour. Eng. Electronics*.

(24) F. Selleri, "Noninvariant one-way velocity of light", submitted to Found. Phys.

(25) F. Selleri,."Noninvariant one-way velocity of light and particle collisions", submitted to Found. Phys. Letters.

(26) H. Bondi, Observatory, **82**, 133 (1962).

(27) P.G. Bergmann, Found. Phys. **1**, 17-22 (1971).

(28) K. Popper, *Unended quest, An intellectual biography*, (Fontana/Collins, Glasgow, 1976), p. 129.

(29) A. Einstein, Ann. der Physik, **49**, 769 (1916).

(30) J. Bailey et al., Nature, **268**, 301-305 (1977).

(31) M.G. Sagnac, Comptes Ren.., **157**, 708-710 (1913); ibid, **157**, 1410-1413 (1913).

(32) F.R. Tangherlini, Nuovo Cim. Suppl. **20**, 351 (1961).

(33) A.A.Michelson and E.W. Morley, Amer. Journ. of Science, **34**, 333 (1887).

(34) R.J. Kennedy and E.M.Thorndike, Phys.Rev. **42**, 400-418 (1932).

(35) Q. Majorana, Phil. Mag. **37**, 145-150 (1919).

THE LORENTZ INVARIANCE REVISITED

M. SURDIN
CFR Laboratoire Mixte CNRS-CEA
Gif-sur-Yvette 91190
France

1°. The importance of Lorentz Invariance (L.I.) is exemplified by Louis de Broglie's thesis (1). De Broglie notes : « ... that the periodic phenomenon relative to a moving body appears to an observer at rest as slowed down in the ratio of 1 to $\sqrt{1-\beta^2}$ ($\beta = v/c$). Thus, the frequency measured by an observer at rest is

$$v_1 = v_0\sqrt{1-\beta^2} = \frac{m_0 c^2}{h} \cdot \sqrt{1-\beta^2}$$

On the other hand, the energy of a mobile of rest mass m_0 is

$$\frac{m_0 \cdot c^2}{\sqrt{1-\beta^2}},$$

the corresponding frequency according to the quantum relation is

$$v = \frac{1}{h} \cdot \frac{m_0 \cdot c^2}{\sqrt{1-\beta^2}}$$

The two frequencies v_1 and v are essentially different since the factor $\sqrt{1-\beta^2}$ is not included in the same way ».

To resolve this difficulty de Broglie advanced the seminal concept of phase concordance :
« the periodic phenominon relative to a mobile, the frequency of which for an observer at rest is $v_1 = \frac{m_0 c^2}{h} \cdot \sqrt{1-\beta^2}$ appears to this observer as being constantly in phase with a wave of frequency $v = \frac{1}{h} \cdot \frac{m_0 \cdot c^2}{\sqrt{1-\beta^2}}$ propagating in the same direction as the mobile, with velocity $V = c/\beta$ ».

The de Broglie wave was born.
Inspired by the preceding the Lorentz Invariance of several « Equations of Physics » will be examined. The following criterion will be used :
<u>An equation will be considered as L.I. if its two members have an identical Lorentz invariance.</u>

2°   The fundamental invariants are :
- c, the « limiting relativistic velocity », is invariant by definition.
- h, Planck's constant; de Broglie considers implicitly that h is invariant.
- e, the elemetary electric charge is invariant. According to Jackson (2) this invariance is an experimental fact.
- k, Boltzmann's constant, is an invariant (3).

The following relations will be used hereafter :

$$L = L_0\sqrt{1-\beta^2} \tag{2.1}$$

where L is the length of a rod measured in the moving frame and L_0 is the length measured in the rest frame. Length contraction.

$$T = T_0 / \sqrt{1-\beta^2} \tag{2.2}$$

The slowing down of clocks.

$$M = M_0 / \sqrt{1-\beta^2} \tag{2.3}$$

where M_0 is the rest mass.

Finally :

$$\Theta = \Theta_0 \sqrt{1-\beta^2} \tag{2.4}$$

where Θ is the temperature measured in the moving frame and Θ_0 is the temperature measured in the rest frame. Eq(2.4) is favored by Planck, von Laüe, Einstein and de Broglie. However, another relation was considered. For a detailed discussion see (4).

In what follows one often encounters the product $M.L$. Thus

$$M.L = \frac{M_0}{\sqrt{1-\beta^2}} \cdot L_0\sqrt{1-\beta^2} = M_0.L_0 \tag{2.5}$$

Hence the product $M.L$ is L.I.

3°   In the framework of Newtonian Mechanics the energy conservation is considered. Here the Coulomb case is examined :

$$\tfrac{1}{2} m.(dr/dt)^2 = e^2/r \tag{3.1}$$

It is easily seen that eq(3.1) is not $M.L$.

However, a similar equation, where dr/dt is replaced by c, is L.I. : viz :

$$mc^2 = e^2/r \tag{3.2}$$

This equation usually gives the classical radius of the electron, where m is the mass and e is the charge of the electron.

Here as in the following case one observes that replacing v by c obtains a L.I. equation.

Thus, the equation relative to the wavelength of the de Broglie wave

$$\lambda = \frac{h}{m.v} \tag{3.3}$$

is not L.I.. However, the equation relative to Compton's wave for an electron of mass m

$$\lambda_c = h/mc \tag{3.4}$$

is a L.I..

Consider Schrödinger's equation

$$i\hbar \frac{\partial}{\partial t}\Psi + \left(\frac{\hbar^2}{2m}\nabla^2 + e^2/r\right)\Psi = E\Psi \qquad (3.5)$$

with the solution

$$E_0 = -\tfrac{1}{2}me^4/\hbar^2 \qquad (3.6)$$

Eq(3.5) is not L.I.. However, the time independent equation

$$\left(\frac{\hbar^2}{2m}\cdot\nabla^2 + e^2/r\right)\Psi = E\Psi \qquad (3.7)$$

is L.I..

The Klein-Gordon equation

$$\left[\Box + \left(\frac{mc^2}{h}\right)^2\right]\Psi = 0 \qquad (3.8)$$

where

$$\Box = 1/c^2 \cdot \partial^2/\partial t^2 - \nabla^2 \qquad (3.9)$$

is not L.I.. On the other hand, the time independant $(\partial/\partial t = 0)$ equation is L.I..

Similarly, the Hamilton-Jacoby relativistic equation (5), viz :

$$\left(\frac{\partial S}{\partial x}\right)^2 + \left(\frac{\partial S}{\partial y}\right)^2 + \left(\frac{\partial S}{\partial z}\right)^2 - \frac{1}{c^2}\left(\frac{\partial S}{\partial t}\right)^2 = m^2 c^2 \qquad (3.10)$$

where S is the action, is not L.I.. However, in this case as well, the time independant $(\partial/\partial t = 0)$ equation is L.I..

Heisenberg's relation

$$\Delta p.\Delta x \cong \hbar \qquad (3.11)$$

is not L.I..

4° Consider now relations whereby the gravitational constant intervenes. A group of four equations having the same form is :

$$\Delta\varphi = \frac{6\pi GM}{a(1-\varepsilon^2)} \qquad (4.1)$$

where $\Delta\varphi$ is the advance of the perihelion of planets; a is the semi-major axis of the ellipse and ε the excentricity.

$$\alpha = \frac{4GM}{c^2.R} \qquad (4.2)$$

where α is the bending of a light ray passing near a spherical object of mass M and radius R.

The gravitational red-shift

$$\Delta\nu/\nu = 1/c^2\left[\frac{GM}{R+\Delta R} - \frac{GM}{R}\right] = -\frac{GM\Delta R}{R^2} \qquad (4.3)$$

where M is the mass of a stellar object of radius R and ΔR is the radial distance between the emitter and the receiver of an e.m. radiation of frenquency ν.

Finally, for a spherical « black hole » of mass M and radius R one has

$$c^2 = \frac{2GM}{R} \tag{4.4}$$

The four equations would be L.I. if

$$G = G_0\left(1 - \beta^2\right) \tag{4.5}$$

In that case the first relation of the Large Numbers Hypothsis of Dirac's, which gives the ratio of the electric and gravitational energies for a hydrogen atom

$$e^2 / (Gm_H \cdot m) \cong 0.25 \times 10^{40} \tag{4.6}$$

where e is the elementary electric charge, m_H the mass of the hydrogen atom and m the mass of the electron, is L.I..

Furthermore, Planck's lentgh

$$L_P = \left(G\hbar/c^3\right)^{1/2} = G^{1/2}\left(\hbar/c^3\right)^{1/2} \tag{4.7}$$

is a L.I..

5° Relations whereby a constant magnetic field enters.
Consider Larmor's frequency

$$\omega = \frac{eH}{mc} \tag{5.1}$$

where m is the mass of the electron and H a constant magnetic field. If this equation were to be L.I. H should be L.I..

A spherical body of mass M, radius R, rotating around one of its diameters at an angular velocity Ω creates a magnetic field (6).

$$\langle H^2 \rangle = 1/8 \cdot \frac{GM^2\Omega^2}{c^2 \cdot R^2} \tag{5.2}$$

If G satisfies eq(4.5), eq(5.2) is L.I. if H is L.I..

In a study of layered superconductors of high critical temperature (7) the superconducting state disappears for a critical magnetic field

$$H_c(\Theta) = H_c(0)\left(1 - \Theta/\Theta_c\right) \tag{5.3}$$

One has

$$-\frac{\partial H_c(\Theta)}{\partial \Theta} = \Gamma = \frac{H_c(0)}{\Theta_c} = \frac{3mck}{e\hbar} \tag{5.4}$$

where m is the mass of the electron.

Eq(5.4) is particularly interesting since Γ depends only on fundamental quantities, it is a universal constant. If one considers eq(2.4) one finds that $H_c(0)$ is L.I..

Bohr's magneton

$$\mu = e\hbar/2mc = \frac{1}{4\pi}(e\lambda_e) \tag{5.5}$$

where λ_c is Compton's wavelentgh is L.I. since μ is expressed by a product of a magnetic field by a length.

6° Relations concerning electromagnetic radiation.

The energy
$$E = h\nu \tag{6.1}$$
may be considered as that of a plane monochromatic wave. The enregy density is then
$$E/V = \frac{h\nu_0\sqrt{1-\beta^2}}{V_0 \cdot \sqrt{1-\beta^2}} \tag{6.2}$$
is L.I..

The equation
$$V = V_0\sqrt{1-\beta^2} \tag{6.3}$$
is due to the fact that the contraction of an elementary cube is in the direction of the motion (5).

The fact that the energy density (eq(6.2)) is L.I. is compatible with the fact that H is L.I.. The energy density Δ, potential or kinetic, is L.I..

For a transverse wave propagating in vacuum the energy density, using suitable units, is
$$\Delta = \frac{1}{4\pi}\langle\varepsilon^2\rangle \tag{6.4}$$
where ε is the amplitude of the electric field and $\langle\varepsilon^2\rangle$ is the mean square of this field. This relation is L.I..

Furthermore, the average radiation pressure for a wave propagating in vacuum is
$$P = \frac{1}{2\pi}\langle\varepsilon^2\rangle \tag{6.5}$$
is also L.I..

7° The application of the above criterion to several « equations of physics » obtains consistent results. However, the conclusions are in disagreement with results obtained in Special Relativity (SR) in the case where the equations comprise operators such as $\frac{1}{c}\cdot\partial/\partial t$ or $\frac{1}{c^2}\cdot\partial^2/\partial t^2$, as for example in the Klein-Gordon eq(3.8) or that of Hamilton-Jacoby eq(3.10). These equations are L.I. in SR but not according to the criterion advanced above.

One may, however, consider the following notion : the velocity c is a velocity of reference; the waves in general, as well as the e.m. waves propagate in vacuum at a velocity v_g very near, but smaller than c. This idea is not a new one, de Broglie considered it time and again (8), (9).

Following Einstein, one may generalise this concept by advancing the following principle : A wave transporting energy propagates at a velocity smaller than c.

Such a wave is detectable. On the other hand, a wave which does not transport energy is not detectable directly. Such a wave may propagate at a velocity higher than c, for example a phase wave.

Under these conditions the operators $\frac{1}{c}\cdot\partial/\partial t$ and $\frac{1}{c^2}\cdot\partial^2/\partial t^2$ should be replaced by $\frac{1}{v_g}\cdot\partial/\partial t$ or $\frac{1}{v_g^2}\cdot\partial^2/\partial t^2$ and, thus eq(3.8) and (3.10) become L.I. according to the criterion above.

The clocks considered in SR necessarily send detectable signals, the propagation velocity of these signals, v_g, is the group velocity, $v_g < c$. As v_g is very near c, the replacement of c by v_g should not introduce in practice any modification in the formulae. Lorentz transformations are the ideal case where v_g is replaced by c. It would be, however, interesting to resume the arguments, according to the above principle, which led to the derivation of Lorentz transformations.

References

1. de Broglie, L. (1925) Recherches sur la théorie des quantas, *Ann. Phys.* Série 10 TIII Masson Paris.
Reproduced (1992) : *Ann. Fond. Louis-de-Broglie*, 17, 1-109.

2. Jackson, J. B. (1975) *Classical Electrodynamics*, J. Wiley & Sons.

3. de Broglie, L. (1968) Thermodynamique Relativiste et Mécanique Relativiste, *Ann. Inst. Henri-Poincaré* 9, 89-108.

4. Ter Haar, D. and Wiegeland, H. (1971) Thermodynamics and Statistical Mechanics in the Special Theory of Relativity, *Physics Reports* 1C, 31-54.

5. Landau, L. and Lifchitz, E. (1965) *La théorie du Champ*. Ed. La Paix, Moscow

6. Surdin, M. (1980) Le Champ Magnétique des Corps Tournants, *Ann. Fond. Louis-de-Broglie* 5, 127-143.

7. Surdin, M. (1994) Essai de Réponse à quelques questions, *Ann. Fond. Louis-de-Broglie* 19, 173-200.

8. de Broglie, L. (1988) *Louis de Broglie que nous avons connu*, Ann. Fond. Louis-de-Broglie.

9. de Broglie, L. (1941) *Problèmes de Propagations Guidées des Ondes Electomagnétiques*. Gauthier-Villars Paris.

A NEW TYPE OF MASSIVE SPIN-ONE BOSON:
AND ITS RELATION WITH MAXWELL EQUATIONS

D. V. AHLUWALIA
Los Alamos Meson Physics Facility, P-25
H-846, Los Alamos National Laboratory
Los Alamos, NM 87545 (USA)
and
The Los Alamos Institute for Scientific and Social Studies
P.O.Box 1364, Los Alamos, NM 87544 (USA)

E-mail: *av@p25hp.lanl.gov*

1. Introduction

The text book understanding of quantum field theory states that a fermion and its associated antifermion have *opposite* relative intrinsic parity; and that a boson and its associated antiboson carry *same* relative intrinsic parity. No particles that do not fall within this understanding, coupled with the fact that no quantum field theory was known to exist that contradicted this canonical wisdom, has led to almost complete neglect of the fact that long ago Wigner [1] [1] argued that space-time symmetries, and that these manifest in quantum constructs as projective representations, do allow for theories where a fermion and its associated antifermion have *same* relative intrinsic parity; and that a boson and its associated antiboson carry *opposite* relative intrinsic parity. The only scholarly footnote to the canonical wisdom in the widely read literature, see for example Refs. [7], that refers

---

[1] Part of the work in Ref. [1] was done in collaboration with V. Bargmann and A. S. Wightman, as Wigner [1] points out and we have noted in Ref. [2]. Also similar work was done previously and independently, as we learned later while writing [3], by L. L. Foldy and B. P. Nigam [4, 5]. So, whenever we refer to Wigner-type bosons or theory, the reader may wish to keep this historical note in mind. However, it must be noted that the work of Ref. [1] is the most comprehensive and is within a much deeper and fundamental framework. *I thank Zurab K. Silagadze for kindly bringing to my attention the work of Foldy and Nigam on the subject* [6]. *Also see Acknowledgements.*

to Wigner's observations is Appendix C of Weinberg's latest monograph on the quantum theory of fields [8].

Recently [2], purely by accident (see Acknowledgements), in an attempt to understand an old work of Weinberg [9] and to investigate the possible kinematical origin for the violation of P, CP, and other discrete symmetries [3], a Wigner-type quantum field theory was constructed for a spin-one boson. This new theory made it clear that *boson-antiboson relative intrinsic parity is not (necessarily and) uniquely determined by the representation space in which the theory is constructed.* For example, the $(1/2, 1/2)$ representation space describes spin-one particles in which a particle and its antiparticle carry *same* relative intrinsic parity. [2] On the other hand [2], the $(1,0) \oplus (0,1)$ representation space not only allows for the construction of a theory in which boson-antiboson relative intrinsic parity is same ("Weinberg's theory" [9]) but also supports a construct in which boson-antiboson relative intrinsic parity is opposite ("our Wigner-type theory" [2]). This is a surprising result, particularly in view of the canonical wisdom that states that there is no physical distinction between the two representation spaces, i.e., $(1/2, 1/2)$ and $(1,0) \oplus (0,1)$, for the description of spin-one particles. [3] For an eloquent and forceful argument putting forward the canonical wisdom, one may refer to Weinberg's paper [10] presented at the fiftieth anniversary of the famous 1939 Wigner's paper [11] on the unitary representations of the inhomogeneous Lorentz group. In fact Weinberg [12] himself was surprised at the new theory and wrote "When I saw your paper I did not believe the result. ... I suspect that this is why you are getting such a surprising result. In all my work, I assume that states of given mass and spin are non-degenerate."

The Wigner-type quantum theory of fields is a very natural consequence of space-time symmetries and the fundamental structure of quantum mechanics. The underlying physical principles for the standard quantum theory of fields and Wigner-type theory are the same — standard quantum theory is one of the classes of the Wigner's general classification [1]. Our recent construction of a Wigner-type theory was purely accidental, as already noted, and it was indeed a very generous and knowledgeable referee from *Phys. Rev. Lett.* who brought Wigner's 1962 work to my attention (see acknowledgements). *Now* that such a theory (i.e., the theory for a

---

[2] At this time (September 1995) it is not known if $(1/2, 1/2)$ representation space too can support a theory of the Wigner type.

[3] The canonical wisdom translated to the spin-one case under consideration can now be stated as: There is no physical distinction between the usual (See footnote[2]) $(1/2, 1/2)$ representation space description and the Weinberg's construct [9] in the $(1,0) \oplus (0,1)$ representation space.

Wigner-type boson) exists and no bosons [4] exist (as far as the nature has revealed its secrets to us so far) that are of Wigner type leads us to the conclusion that *either, in some future experiment (or as a theoretical inevitability in a much broader theoretical framework), we shall discover the spin-one Wigner boson that the new theory describes; or there is some deep underlying reason, yet to be discovered, that prohibits the existence of the spin-one Wigner boson as a physical reality.*

Assuming that no subtle error of any significance exists in our or Weinberg's work presented in Refs. [2, 8, 9], the above conclusion is inescapable. Given this situation, we take the liberty of reviewing at this conference on "The Present Status of Quantum Theory of Light: A Symposium to Honour Jean-Pierre Vigier" the theory of the new type of boson and show that in the massless limit the theory suggests some very definite and fundamental modifications to Maxwell equations.

2. A New Type of Spin-One Boson

Consideration of the space-time symmetries $\Rightarrow (1,0)$ and $(0,1)$ matter fields Lorentz transform ("boost") in the following fashion: [5]

$$(1,0): \quad \phi_R(\vec{p}) = \exp\left(+\vec{J}\cdot\vec{\varphi}\right)\phi_R(\vec{0}), \tag{1}$$

$$(0,1): \quad \phi_L(\vec{p}) = \exp\left(-\vec{J}\cdot\vec{\varphi}\right)\phi_L(\vec{0}). \tag{2}$$

$\vec{J} = 3 \times 3$ angular momentum matrices with J_z diagonal.
$\vec{\varphi} =$ *the boost parameter* defined as

$$\cosh(\varphi) = \frac{E}{m}, \quad \sinh(\varphi) = \frac{|\vec{p}|}{m}, \quad \hat{\varphi} = \frac{\vec{p}}{|\vec{p}|}. \tag{3}$$

$\vec{p} =$ the three-momentum of the particle (of mass m).
Note: No "i" in the argument of exponentials that appear in the above equations.
Reason: $\vec{K}$, the generator of the boost, $= \pm i\vec{J}$. The plus sign for the $(0,1)$-, and minus sign for $(1,0)$-, matter fields.
Under parity: $(1,0) \rightleftharpoons (0,1)$. Parity covariance $\Rightarrow$ we introduce the $(1,0) \oplus (0,1)$ representation space spinor

$$\psi(\vec{p}) = \begin{pmatrix} \phi_R(\vec{p}) \\ \phi_L(\vec{p}) \end{pmatrix}. \tag{4}$$

---

[4] Even though in this presentation we confine our attention to spin one, the results that we report are valid for bosons of spin one and greater. See Ref. [2] for details.

[5] Our notations and conventions are closest to those found in Ryder's book [7] on quantum field theory. The reader not familiar with representations of the Lorentz group will find a very readable discussion in Chapter 2 of Ryder's book.

As the reader will soon discover, and as I learned from the generalization [13] of the work of Ryder [7], the wave equation satisfied by this $(1,0)\oplus(0,1)$ spinor is determined by the boost properties of the $\phi_R(\vec{p})$ and $\phi_L(\vec{p})$, given in Eqs. (1) and (2), *and* the algebraic relation between these fields at zero momentum. In the past it had been argued (see p. 44 of Ryder's book on quantum field theory [7]), and it does not matter which specific spin in the $(j,0)\oplus(0,j)$ representation space is under consideration, that "when a particle is at rest, one cannot define its spin as either left- or right-handed, so" for zero momentum "$\phi_R(\vec{0}) = \phi_L(\vec{0})$." That this is not true for the $(1/2,0)\oplus(0,1/2)$ Dirac field is manifest when we look at the explicit form of zero-momentum spinors. [6] In canonical representation the argument is made in Ref. [2] for the $(j,0)\oplus(0,j)$ representation space; and in Weyl (also called chiral) representation [the representation in which $\psi(\vec{p})$ is written in Eq. (4)] the reader should carefully follow Weinberg's arguments on pp. 220-224 of his text [8] to obtain spin-1/2 zero-momentum spinors (given in Eqs. 5.5.35 and 5.5.36 of [8]) to arrive at the same conclusion as us for $j = 1/2$.

Given Eqs. (1) and (2), the spinors at momentum $\vec{p}$, $\psi(\vec{p})$, are known if we specify the zero-momentum spinors $\psi(\vec{0})$. From the work of Ref. [2] and the detailed study on the closely related subject contained in [8] it is clear that it is the choice of the zero-momentum spinors, i.e., specification of $\psi(\vec{0})$, that must determine the parity, and other related structure, of the theory. [7] In Ref. [2] we postulated that in the *canonical representation* [8] the six-dimensional $(1,0)\oplus(0,1)$ representation space for a particle at rest can be spanned by the six zero-momentum spinors:

Canonical Representation:

---

[6] This observation was first made in the Summer of 1991 by C. Burgard [14] while trying to understand Ryder's *ab initio* derivation [7] of the Dirac equation.

[7] This we assert, not as a theorem, but as an essentially unavoidable conclusion made in an attempt to reconcile the apparently contradicting conclusions of Refs. [2] and [8, 9]. In principle, of course, there remains the possibility that something subtle is being missed by the authors of Ref. [2] or/and the author of [8].

[8] Canonical representation is defined as:

$$\psi_{\text{canonical}}(\vec{p}) = \frac{1}{\sqrt{2}} \begin{pmatrix} I & I \\ I & -I \end{pmatrix} \psi(\vec{p}) = \frac{1}{\sqrt{2}} \begin{pmatrix} \phi_R(\vec{0}) + \phi_L(\vec{0}) \\ \phi_R(\vec{0}) - \phi_L(\vec{0}) \end{pmatrix}. \quad (5)$$

The $\psi(\vec{p})$ in the middle of the above equation refers to the Weyl representation spinor of Eq. (4) and $I = 3 \times 3$ identity matrix.

$$u_{+1}(\vec{0}) = \begin{pmatrix} m \\ 0 \\ 0 \\ 0 \\ 0 \\ 0 \end{pmatrix}, \quad u_0(\vec{0}) = \begin{pmatrix} 0 \\ m \\ 0 \\ 0 \\ 0 \\ 0 \end{pmatrix}, \quad u_{-1}(\vec{0}) = \begin{pmatrix} 0 \\ 0 \\ m \\ 0 \\ 0 \\ 0 \end{pmatrix}, \quad (6)$$

$$v_{+1}(\vec{0}) = \begin{pmatrix} 0 \\ 0 \\ 0 \\ m \\ 0 \\ 0 \end{pmatrix}, \quad v_0(\vec{0}) = \begin{pmatrix} 0 \\ 0 \\ 0 \\ 0 \\ m \\ 0 \end{pmatrix}, \quad v_{-1}(\vec{0}) = \begin{pmatrix} 0 \\ 0 \\ 0 \\ 0 \\ 0 \\ m \end{pmatrix}. \quad (7)$$

The $\sigma = 0, \pm 1$ on $u_\sigma(\vec{0})$ and $v_\sigma(\vec{0})$ carry the meaning of the projection of spin on the z-axis, and the possibility of relative phases between various spinors (not important for the present considerations) are left implicit. On studying the C, P, and T properties of the associated wave equation, it turns out that the u- and v-spinors are related by the operation of Charge conjugation and carry opposite relative intrinsic parities [2].

Setting $\vec{p} = \vec{0}$ in Eq. (5), of footnote [8], and comparing the resulting equation with Eqs. (6) and (7) we find that for the u-spinors $\phi_R(\vec{0}) = +\phi_L(\vec{0})$ and for the v-spinors $\phi_R(\vec{0}) = -\phi_L(\vec{0})$.

To sum up, therefore, we have the needed algebraic relation between $\phi_R(\vec{p})$ and $\phi_L(\vec{p})$ at zero momentum:

$$\phi_R(\vec{0}) = \wp_{u,v} \, \phi_L(\vec{0}), \quad (8)$$

with $\wp_{u,v} = +1$ for the u-spinors and $\wp_{u,v} = -1$ for the v-spinors.

The three $u_\sigma(\vec{p})$ and the three $v_\sigma(\vec{p})$ spinors, with $\sigma = 0, \pm 1$ representing the three spinorial degrees of freedom, are obtained by applying the $(1,0) \oplus (0,1)$ boost implicit in definition (4) and the $(1,0)$ and $(0,1)$ boosts given in Eqs. (1) and (2). That is, [9]

$$u_\sigma(\vec{p}) = \begin{pmatrix} \exp(+\vec{J} \cdot \vec{\varphi}) & 0 \\ 0 & \exp(-\vec{J} \cdot \vec{\varphi}) \end{pmatrix} u_\sigma(\vec{0}), \quad (9)$$

$$v_\sigma(\vec{p}) = \begin{pmatrix} \exp(+\vec{J} \cdot \vec{\varphi}) & 0 \\ 0 & \exp(-\vec{J} \cdot \vec{\varphi}) \end{pmatrix} v_\sigma(\vec{0}); \quad (10)$$

with $\sigma = 0, \pm 1$.

---

[9] Note: we stay in the Weyl representation unless specifically indicated otherwise. For example, Eqs. (6) and (7) are written in canonical representation and this is explicitly noted right above these expressions.

The reader will note that the six $(1,0) \oplus (0,1)$ spinors thus obtained follow purely from the projective representations of the Lorentz group and the associated boosts, and the $(1,0) \oplus (0,1)$ fields operator for the quantum description of these particles follows from the canonical arguments of translational invariance etc. [for details refer to any recent book on quantum theory of fields, such as [8]]

$$\Psi(x) = \sum_{\sigma=0,\pm 1} \int \frac{d^3p}{(2\pi)^3} \frac{1}{2\omega_{\vec{p}}} \left[u_\sigma(\vec{p}) \, a_\sigma(\vec{p}) \, e^{-ip\cdot x} + v_\sigma(\vec{p}) \, b_\sigma^\dagger(\vec{p}) \, e^{ip\cdot x} \right] , \quad (11)$$

where $\omega_{\vec{p}} = \sqrt{m^2 + \vec{p}^2}$; and $[a_\sigma(\vec{p}), a_{\sigma'}^\dagger(\vec{p}')]_\pm = \delta_{\sigma\sigma'}\delta(\vec{p}-\vec{p}')$; etc. and we leave implicit the possibility that the $b^\dagger$ may carry a hidden numerical factor.

It is now a straightforward mathematical exercise to couple the right- and left-handed fields to obtain the *free-field wave equation* for the $(1,0) \oplus (0,1)$ spinors. Using Eq. (8) on the right-hand side of Eq. (1) to re-express $\phi_R(\vec{0})$ in terms of $\phi_L(\vec{0})$ and then using Eq. (2) to replace $\phi_L(\vec{0})$ by $\exp(\vec{J}\cdot\vec{\varphi})\phi_L(\vec{p})$, and executing a similar exercise beginning with the right-hand side of Eq. (2), we obtain two coupled equations for $\phi_R(\vec{p})$ and $\phi_L(\vec{p})$. These two equations are then transformed into a single wave equation for the $(1,0) \oplus (0,1)$ spinor (4). This wave equation for the $(1,0) \oplus (0,1)$ spinor reads: [10]

$$\left(\gamma_{\mu\nu} p^\mu p^\nu - \wp_{u,v} m^2 I \right) \psi(\vec{p}) = 0 , \quad (12)$$

with

$$\gamma_{\mu\nu} p^\mu p^\nu = \begin{pmatrix} 0 & B + 2(\vec{J}\cdot\vec{p})p^0 \\ B - 2(\vec{J}\cdot\vec{p})p^0 & 0 \end{pmatrix} , \quad (13)$$

where $B = \eta_{\mu\nu} p^\mu p^\nu + 2(\vec{J}\cdot\vec{p})(\vec{J}\cdot\vec{p})$. $\quad (14)$

Here, $\eta_{\mu\nu}$ is the flat space-time metric with the diagonal $(1,-1,-1,-1)$. The "0" on the diagonal represents a 3×3 block of zeros. The off-diagonal terms are the 3×3 block matrices.

From Eq. (13) we read off the following "gamma matrices" (not to be confused with the Dirac "gamma matrices"):

$$\gamma_{00} = \begin{pmatrix} 0 & I \\ I & 0 \end{pmatrix}, \gamma_{i0} = \gamma_{0i} = \begin{pmatrix} 0 & J_i \\ -J_i & 0 \end{pmatrix}, \quad (15)$$

$$\gamma_{ji} = \gamma_{ij} = \begin{pmatrix} 0 & I \\ I & 0 \end{pmatrix} \eta_{ij} + \begin{pmatrix} 0 & \{J_i, J_j\} \\ \{J_i, J_j\} & 0 \end{pmatrix}, \quad (16)$$

---

[10] The procedure is in fact valid for *any* spin in the $(j,0) \oplus (0,j)$ representation space. In particular, when applied for $j = 1/2$ one obtains the well-known Dirac equation.

i and j run over a spacial index $1, 2, 3$.

For the $(1,0) \oplus (0,1)$ representation space $\psi(x) \equiv \psi(\vec{p}) \exp(-i\wp_{u,v}\, p \cdot x)$, and Eq. (12) requires:

$$\left(\gamma_{\mu\nu}\partial^\mu \partial^\nu + \wp_{u,v}\, m^2\, I\right)\psi(x) = 0. \tag{17}$$

This wave equation is identical to Steven Weinberg's equation for the $(1,0) \oplus (0,1)$ representation space in all aspects except an important factor of $\wp_{u,v}$ attached to the mass term. It is this factor that leads to a fundamentally different, i.e., Wigner type, CPT structure in our theory. The CPT analysis of the theory is presented in Ref. [2] and is found to be very intricately related to the $\wp_{u,v}$ factor of our theory. The most important result that emerges from this analysis, and we simply quote it here, is that *the operations of Charge conjugation and Parity anticommute in a quantum field theory built upon wave equation (17) and field operator (11) and thus results in the underlying spin-one boson and antiboson carrying opposite relative intrinsic parities.* To the best of our knowledge the mathematical construction of Ref. [2], and further discussed here in its physical content, is the first explicit and non-trivial example of a quantum theory of the Wigner type. We look forward to further work on physical and mathematical implications of such a construct for future fundamental works in quantum field theory and possible experimental discovery of the Wigner-type bosons.

So far we have essentially emphasized the differences of our construct and that of Weinberg. We now discuss the aspects that are same (or similar) in both theories. The dispersion relations associated with both theories are determined by setting the determinant of $(\gamma_{\mu\nu} p^\mu p^\nu - \wp_{u,v} m^2 I)$ (for our theory), and the determinant of $(\gamma_{\mu\nu} p^\mu p^\nu - m^2 I)$ (for Weinberg's theory), equal to zero. The resulting equation is a 12th-order polynomial in $(E, |\vec{p}|, m)$ and results in the dispersion relations summarized in Table I.

TABLE 1. Dispersion relations $E = E(p, m)$ associated with Eq. (17) and similar equation of Weinberg. $p \equiv |\vec{p}|$

Dispersion Relation	Multiplicity	Interpretation
$E = +\sqrt{p^2 + m^2}$	3	Causal, "particle" $u_{\pm 1}(\vec{p}),\ u_0(\vec{p})$
$E = -\sqrt{p^2 + m^2}$	3	Causal, "antiparticle" $v_{\pm 1}(\vec{p}),\ v_0(\vec{p})$
$E = +\sqrt{p^2 - m^2}$	3	Acausal, Tachyonic
$E = -\sqrt{p^2 - m^2}$	3	Acausal, Tachyonic

There are two observations that we wish to make in regard to the tachyonic solutions:

1. In the $m \to 0$ limit all dispersion relations are non-tachyonic.
2. For $m \neq 0$, the the tachyonic solutions may be reinterpreted, on introducing a quartic self-coupling. The "*negative* mass squared" term can then be interpreted as in the simplest versions of field theories with broken symmetry. Such an analysis [15] shows that the resulting theory describes four particles: two charged particles of mass m (of Wigner type in our theory and usual type in Weinberg's theory), a (Goldstone-like) spin-one massless (Majorana-like) particle, and a massive (Majorana-like) spin-one particle of mass $\sqrt{2}|m|$.

In any case, at this stage negative mass squared may be considered to provide the sought-after physical origin of the "$-m^2$" that appears in the simplest versions of field theories with broken symmetry. Or, in absence of this re-interpretation one may argue that these solutions violate the original input on the mass parameter via equation (3), and hence may be considered physically inadmissible (without perhaps interactions that make it possible to re-interpret the "$-m^2$ as above!). While we do not suspect that usual interactions can induce transitions between the $E = \pm\sqrt{\vec{p}^2 + m^2}$ and $E = \pm\sqrt{\vec{p}^2 - m^2}$ sectors, we do not know of any proof on the subject.

3. The Massless Limit and Maxwell Equations

That the massless limit is well behaved for the representation space under consideration was shown in the sixties by Weinberg [9]. So to obtain the massless limit of the the theory under consideration, one may wish to set $m = 0$ in (12), or equivalently in (17), and argue that $\eta_{\mu\nu}p^\mu p^\nu \phi_R(\vec{p}) = 0$ and $\eta_{\mu\nu}p^\mu p^\nu \phi_L(\vec{p}) = 0$ for a massless particle, to obtain:

$$2\vec{J} \cdot \vec{p}\left(\vec{J} \cdot \vec{p} + p^0 I\right) \phi_R(\vec{p}) = 0, \tag{18}$$

$$2\vec{J} \cdot \vec{p}\left(\vec{J} \cdot \vec{p} - p^0 I\right) \phi_L(\vec{p}) = 0; \tag{19}$$

and since it is known (see, for example, [13]) that

$$\left(\vec{J} \cdot \vec{p} + p^0 I\right) \phi_R(\vec{p}) = 0 \tag{20}$$

$$\left(\vec{J} \cdot \vec{p} - p^0 I\right) \phi_L(\vec{p}) = 0, \tag{21}$$

are indeed free Maxwell equations one may claim that we have obtained Maxwell equations in the limit $m = 0$ of our, or that of Weinberg's, theory. **No**, this is not so. The reason, as we already noted a few years ago [16], is that Maxwell equations (20) and (21) do *not* follow from Equations (18) and (19) *because* the matrix $2\vec{J}\cdot\vec{p}$ is *non-invertible*. The Determinant $(2\vec{J}\cdot\vec{p})$ identically vanishes.

The analysis of the massless limit is not yet complete, but at this stage one can already note that all solutions of (Maxwell) Eqs. (20) and (21) are solutions of Eqs. (18) and (19), but there may exist solutions that do not satisfy Maxwell equations but still are solutions of Eqs. (18) and (19). Another point to note is that (Maxwell) Eqs. (20) and (21) are first order in space-time derivatives. Eqs. (18) and (19) are of second order in space-time derivatives. Hence at least *additional boundary conditions are to be satisfied. Any departures, therefore, from Maxwell equations will only be expected, or are most likely, for phenomenon that involve strong fields and/or strongly varying fields.* [11]

To complete the story we note that in the beginning of this section we assumed "$\eta_{\mu\nu}p^\mu p^\nu \phi_R(\vec{p}) = 0$ and $\eta_{\mu\nu}p^\mu p^\nu \phi_L(\vec{p}) = 0$ for a massless particle." There is no justification to invoke this assumption *a priori*. Therefore, the Maxwell equations (20) and (21) should, rigorously speaking, be replaced, instead of (18) and (19), by:

$$\left(\eta_{\mu\nu}p^\mu p^\nu I + 2\vec{J}\cdot\vec{p}\left(\vec{J}\cdot\vec{p}+p^0 I\right)\right)\phi_R(\vec{p}) = 0, \qquad (22)$$

$$\left(\eta_{\mu\nu}p^\mu p^\nu I + 2\vec{J}\cdot\vec{p}\left(\vec{J}\cdot\vec{p}-p^0 I\right)\right)\phi_L(\vec{p}) = 0. \qquad (23)$$

In reference to the above equations and Eqs. (18), (19), (20), and (21), it seems important to observe that these equations have solutions only if the appropriate dispersion-relation determining determinant vanishes. These determinants are:

$$\text{Determinant}\left(\vec{J}\cdot\vec{p}\pm p^0 I\right) = \mp E\left(\vec{p}^{\,2} - E^2\right) \qquad (24)$$

$$\text{Determinant}\left(2\vec{J}\cdot\vec{p}\left(\vec{J}\cdot\vec{p}\pm p^0 I\right)\right) = (\text{identically})\ 0, \qquad (25)$$

$$\text{Determinant}\left(\eta_{\mu\nu}p^\mu p^\nu I + 2\vec{J}\cdot\vec{p}\left(\vec{J}\cdot\vec{p}\pm p^0 I\right)\right) = -(\vec{p}^{\,2} - E^2)^3. \qquad (26)$$

A comparison of the dispersion relations implied by setting the above determinants to zero with the dispersion relations for $m \neq 0$ case, tabulated in Table I, again indicates Eqs. (22) and (23) as the sole candidate for the massless limit of our (or, Weinberg's) theory.

So, to sum up our analysis of the massless limit of our (or that of Weinberg's theory), we conclude that: *Present theoretical arguments suggest that in strong fields, or high-frequency phenomenon, Maxwell equations*

---

[11] What sets the scale that determines "strong" in the above statement? This question requires a precise answer, and we wish to take up this subject in the future. But for the moment we shall assume that a definite scale can be defined, or at least that the question can be answered in a specific experimental set up.

may not be an adequate description of nature. Whether this is so can only be decided by experiment(s). Similar conclusions, in an apparently very different framework, have been independently arrived at by M. Evans [17] and communicated to the author.

4. Concluding Remarks

First, we showed that in the $(1,0) \oplus (0,1)$ representation space there exist not one but two theories for charged particles.[12] In the Weinberg construct, the boson and its antiboson carry same relative intrinsic parity, whereas in our construct the relative intrinsic parities of the boson and its antiboson are opposite. These results originate from the commutativity of the operations of Charge conjugation and Parity in Weinberg's theory, and from the anti-commutativity of the operations of Charge conjugation and Parity in our theory. We thus claim that we have constructed a first non-trivial quantum theory of fields for the Wigner-type particles.[13] Second, the massless limit of both theories seems formally identical and suggests a fundamental modification of Maxwell equations. At its simplest level, the modification to Maxwell equations enters via additional boundary condition(s).

References

1. E. P. Wigner, in *Group Theoretical Concepts and Methods in Elementary Particle Physics Physics - Lectures of the Istanbul Summer School of Theoretical Physics, 1962*, edited by F. Gürsey; (Gordon and Breach, 1964). Also see: Z. K. Silagadze, Yad. Fiz. **55**, 707 (1992) [*Sov. J. Nucl. Phys.* **55**, 392 (1992)].
2. D. V. Ahluwalia, M. B. Johnson, and T. Goldman, Phys. Lett. B **316**, 102 (1993).
3. D. V. Ahluwalia, "Theory of Neutral Particles: McLennan-Case Construct for Neutrino, its Generalization, and a Fundamentally New Wave Equation," *Int. J. Mod. Phys. A* (in press). LANL HEP archive: hep-th/9409134.
4. L. L. Foldy, Phys. Rev. **102**, 568 (1956).
5. B. P. Nigam and L. L. Foldy, Phys. Rev. **102**, 1410 (1956). Typographical error: In the right hand side of Eq. (37) of this reference $U_s(\theta)$ should be corrected to read $U_s(\theta = 0)$.
6. Z. K. Silagadze, private communication (August, 1994)
7. L. H. Ryder, *Quantum Field Theory*; G. Sterman, *An Introduction to Quantum Field Theory* (Cambridge University Press, Cambridge University Press, Cambridge, U.K., 1993); M. Kaku, *Quantum Field Theory: A Modern Introduction* (Oxford University Press, Oxford, U.K., 1993); C. Itzykson and J.-B. Zuber *Quantum Field Theory*

---

[12] Considerations of particles that are self-charge conjugate leads to a yet another theory in the $(1,0) \oplus (0,1)$ representation space. This construction appears in Ref. [3].

[13] Despite the fact that the $\wp_{u,v}$ factor appears only in the mass term one cannot claim that Weinberg's theory and our theory have the same physical content in the massless limit. The differences may arise from how the zero momentum $(1,0) \oplus (0,1)$ spinors are chosen in the two theories. This certainly is true for the differences in the two theories for massive particles. This aspect requires further study. Rigorously speaking, the "rest spinors" and "zero-momentum spinors" must be distinguished while speaking of the massless limit. For massless particles there are no rest spinors.

(McGraw-Hill Inc., U.S.A, 1980); and B. Hatfield, *Quantum Field Theory of Point Particles and Strings*, (Addison-Wesley Publishing Company, California, U.S.A., 1992).
8. S. Weinberg, *The Quantum Theory of Fields, Vol. I (Foundations)* (Cambridge University Press, Cambridge, U.K., 1995).
9. S. Weinberg, Phys. Rev. **133**, B1318 (1964); **134**, B882 (1964). Also see, H. Joos, Forts. Phys. **10**, 65 (1962).
10. S. Weinberg, Nucl. Phys. B (Proc. Suppl.), **6**, 67 (1989).
11. E. P. Wigner, Ann. of Math. **40**, 149 (1939).
12. S. Weinberg, private communication (October, 1993); also see Appendix C of his text cited here as Ref. [8].
13. D. V. Ahluwalia, Texas A & M University, Ph. D thesis (unpublished, 1991); Abstract in: Dissertation Abstracts International B **52**, 4792-B (1992).
14. C. Burgard, private communication (Summer, 1991).
15. D. V. Ahluwalia and T. Goldman, Mod. Phys. Lett. A **8**, 2623 (1993).
16. D. V. Ahluwalia and D. J. Ernst, Mod. Phys. Lett. A **7**, 1967 (1992).
17. M. Evans, private communication (September, 1995).

Acknowledgements

This is an unusually long acknowledgement for an anonymous referee who has inspired and taught me in many ways. So I begin with a few personal comments in the nature of an introduction. I am deeply aware of the unusual nature of this acknowledgement and the criticism that it may draw, but to keep its contents in my files will be unfair to my fellow students of physics for many reasons.

In part, the purpose of this long addenda in the form of an acknowledgement to the manuscript is to document and acknowledge the significant contributions of this anonymous referee from *Phys. Rev. Lett.* in the construction of the Wigner-type quantum theory of fields. That the construction of a Wigner-type quantum theory of fields was purely accidental will also become apparent in the process of reading the two reports (from the same referee) that follow. To fully appreciate the impact of the referee on our work the reader should read Ref. [2] concurrently. While the physics these reports contain is important, it is equally important how that physics weaves with history and personal affection. Having talked about phases, projective representations, and given proofs of some important theorems, and having asked many questions, the referee suddenly seems overtaken by his affection for the man from whom he must have learned much of all this and writes "I am not professor Wigner (he has been 90 this month; let him in peace)."

I treat these reports as little monographs and these little monographs have provided me much guidance in the construction of the theory that I reviewed here. These reports are exceptional in their clarity, unsurpassed in their generosity, and exhibit a deep affection that their author holds for physics and the giants in his field. I reproduce these reports here not only to document the contributions that this referee made to my work, but also in the hope that future generations will be inspired and guided by the content and style of these reports when they write their reviews for the manuscripts of their colleagues. I remain deeply thankful to so unusual a referee. The very existence of this referee gives me hope for the future of physics in these difficult cultural times when so little of support exists for fundamental science and it is so difficult to find a true mentor in the classic sense of this word. For the roughly six-month period during which I worked on the revision of the manuscript the referee became my mentor, and perhaps my collaborators too gained from the knowledge and good advice of the referee.

The text of the two reports that appears here is the unedited exact version, and all spellings are left as in the original to preserve complete flavour of the original reports. For example at one place we have "Majorana articles" (instead of Majorana particles); at other particles appears as "particules." Similarly all punctuation, spacing after commas, italicized letters and boldface letters are reproduced as they appear in the original.[14]

First Report:

This paper cannot be published in any physical journal. Indeed it presents an interesting idea but this idea is not new: for instance the possibility to have boson anti-boson relative parity is presented (with an equivalent, but different representation of the Lorentz group for particle states) in the E.P. Wigner (the physicist who has introduced parity in quantum mechanics in 1928) contribution to the volume: *Group Theoretical concepts and methods in Elementary particle Physics*. F. Gürsey editor, Gordon and Breach, New-York 1964.

In this paper Wigner shows that to have opposite relative parity for boson anti-boson one has to pay a price: the doubling of the number of states, i.e. a new type of degeneracy (referred to as "Wigner type" by the knowledgeable physicists). So the author should refer to this Wigner paper and since he does not claim a doubling of the usual number of states he has to explain where is the discrepancy with Wigner. If he succeeds that will be very interesting to publish it. Unhappily there is not enough information in the manuscript to know if the author may succeed. For instance the author should write explicitly the Charge Conjugation operator $U(C)$ in his field theory.

Here are comments and questions for helping the author:

---

[14] No other referee report was received by the editors of *Phys. Rev. Lett.* and despite a strong recommendation from the referee to publish the paper the manuscript was rejected by the editors. The manuscript, with the sole addition of an acknowledgement to the anonymous referee, was then submitted to *Phys. Lett. B* where it was independently reviewed and accepted without any revisions.

1) In the traditional notation $(j,0) \oplus (0,j)$ for the finite dimensional reducible representation of the connected Lorentz group, the parity operation exchanges these inequivalent irreducible components. So to use equation (2) (taken from your thesis) you have to give more precisions on what you have done: the parity operator cannot be block diagonal in the direct sum; probably you have already taken a symmetric and antisymmetric combinations of these two subspaces which carry inequivalent representations of the connected Lorentz group? The reader needs to know in detail. Redactional details: your equation (3) contains no information as long as you do not give t', x' as function of t, x ! Why do you call spinors the vectors $(1,0$ and $(0,1)$ which correspond to $\vec{E} \pm i\vec{B}$ in electromagnetism (they were introduced by Helmoltz).

2) You do quote Wigner paper of 1939 on the inhomogenuous Lorentz group, but it is irrelevant here (e.g. in this beautiful and very important mathematical paper, Wigner studied the unitary representations of the full group, including time reversal, although he did know that in physics time reversal has to be represented by an antiunitary operator: see his paper of 1932 in Göttingen Nachrichten). The important fact to remember from this 1939 paper (or from Wigner's earlier book) is that in quantum physics, the relativity group is realized by a projective representation, i.e. a representation up to a factor. The phase of the unitary operator representing a discrete operation is therefore arbitrary; and you are completly right to emphasize after your equation (6) that there is a *"global phase factor"*: to neglect it, as you suggest, might be throwing the baby out with the water of the bath tub. The way out is simple: independently of the arbitrary phase of the parity operator, **the relative phase ϵ between particle and anti particle states is not arbitrary and is naturally defined as:**

$$U(P)U(C) = \epsilon U(C)U(P) \qquad (1)$$

It is a simple exercise to prove $\epsilon^2 = 1$ by using associativity of the group law. Indeed multiplying (1) on the right you obtain $U(P)U(C)^2 = \epsilon U(C)U(P)U(C)$ and using (1) again $U(P)U(C)^2 = \epsilon^2 U(C)^2 U(P) = \epsilon^2 U(P)U(C)^2$.

In your theory, you should define explicilty the charge conjugation operator $U(C)$. Does it commute or anticommute with $U(P)$?

3) In the frame of quantum field theory do not forget that operators which correspond to physical identity, as for instance those representing C^2, P^2, $(CP)^2$, T^2, $(CPT)^2$ etc... are not a multiple of the identity operator, but their value depends on the superselection sector on which they act. The values of the square of the antiunitary operators are well fixed on each superselection sector; (this is not the case of the unitary operators such as P^2 for the non vacuum sectors, but the relative value between different sectors might be well defined). The proof uses again the associativity of

the group law. Let $V = UK$ be an antiunitary operator: in a basis it is the product of a unitary operator and of the complex conjugation K. We assume that the restriction to a superselection sector of the square of this antiunitary operator is a multiple of the identity: $V^2 = \omega I$; since V^2 is unitary, ω is a phase: $\overline{\omega}\omega = 1$. Then $V^3 = \omega V = V\omega = \overline{\omega} V$; so $\omega = \overline{\omega} = \pm 1$. For instance, in usual quantum field theory, without the "Wigner type" degeneracy, $V(CPT)^2 = 1$ on the bosonic sector and -1 on the fermionic sector. This implies that $V(CPT)^2\psi = -\psi V(CPT)^2$ for any fermion field ψ.

4) There is arbitrariness in defining parity of a single field without interaction, but this arbitrariness can be completely reduced with the interactions. Indeed to measure the parity of a particle you must interact (by a parity conserving interaction!) with the particle, either to produce it or by the study of its spontaneous decay (the case interesting for you?). Do you wish that these particles have electromagnetic interactions: write their electric current if they have an electric charge. If not they can still interact by an induced magnetic moment and, for spin 1 particle, a quadrupole moment. Or do you expect these mesons with absolutly no electromagnetic interaction?

5) Finally a serious study will lead you to rise the fundamental question of CTP. What do you expect about the CTP behaviour of your particles? That is the type of question you must be able to answer if you want to make an interesting contribution to present particle physics. Indeed if you want their interaction to violate CTP invariance, you have to reconstruct all present day physics!

I am not professor Wigner (he has been 90 this month; let him in peace). But I wished to help you; that is why I wrote this long report asking you many questions.

P.S. While writing this report I have no access to a collection of Physical Reviews of the sixties. In one of them, if my memory is faithful, you will find a paper by T.D. Lee and G.C. Wick which deals with the discrete symmetries C, P, T in quantum field theory and give the values of the square of their representing operators on different superselection rules sectors.

Second Report

The first named author has appreciated my exceptionally long report. He has read and well assimilated the literature I suggested. Congratulations!

This very new version of the manuscript has now three authors and carries a very well chosen title. Indeed Bargmann, Wightman and Wigner had studied, this subject forty years ago, in an unpublished book (several chapters were distributed as preprints). The authors explain well the scope of their paper. They have made a thorough construction of the field theory of a non usual Wigner type; that is completely new and all given references are relevant. *This paper should be published.*

However the authors have missed an important point: in quantum theories, symmetry groups are implemented through **projective** representations. As the authors rightly write (page 7 and footnote 2) one can ignore an overall phase factor (except in Majorana theory), but a crucial argument of parity should not be obscured by conventions. In my report, I recalled the **proof** that for the operators C, P, T and their products, there are signs of ϵ independent from conventions: on each superselection rule sector, *the value ϵI of the square of the antiunitary operators: T^2, $(PT)^2$, $(CT)^2$, $(CPT)^2$ and the sign in $PC = \epsilon CP$* which indicates the commutation or the anticommutation of P and C; of course, these last two possibilities correspond respectively to same or opposite parity for particles and antiparticles. To stress this important point, the authors should show the anticommutation of P and C in the theory they develop and compute the value of the squares of the four antiunitary operators on the bosonic and fermionic states. Those characterize the Wigner's types.

In his letter presenting this new version, the first named author points out that for zero mass particles the situation is different (specialy for Majorana articles). He is completly right and I am looking forward for the announced paper. I want to bring to his attention that this was understood in Tiomno thesis (princeton, around 1947-48), unpublished I believe. Either Tiomno (in Brazil) or Wightman, in Princeton, could give him more details.

Implications of Extended Models of the Electron for Particle Theory and Cosmology

E. J. Sternglass
University of Pittsburgh
Pittsburgh, PA. 15261

Abstract

Although quantum theory has been very successful in the mathematical description of molecules, atoms and atomic nuclei, it has failed to resolve the conceptual problem posed by the wave-particle duality of fundamental entities such as the electron and photon. As long as these entities are regarded as infinitely small points the problem of the origin of mass, the reason for the observed wide variety of particles, and the connection between electromagnetic, nuclear and gravitational interactions cannot be resolved. Moreover, since in the present standard model of an expanding universe its initial size must have been of the order of the observed nuclear particles or smaller, such questions as to the nature of dark matter and the origin of the large-scale structures and motions in the very early universe can also not be answered without a more detailed understanding of the wave-nature of particles. It will be argued in the present paper that the crisis in fundamental physical theory and cosmology can be resolved along the lines originally suggested by Bohm, Weinstein, Vigier and others according to which the electron is regarded as a stable, rotating, extended structure describable in hydrodynamic terms similar to a vortex tube that can be excited to high states of internal motion such as envisioned more recently by superstring theory so as to account for the massses of mesons and other massive particles. Just as in the case of superstring theory, the existence of charges can then be explained as arising from the scission of a closed string or vortex ring under the action of zero-point vibrations as in the analogous case of the toroidal geons investigated by Wheeler, each of the ends becoming the extended source of an electric field contributing a fractional charge of one-half, held together to give integral charges by a strong local Poincare-type cohesive or gravitational force as suggested by Motz. Such a strong local space-curvature or gravitational force may explain the quantum potential of Bohm, since a high local curvature constant is found to be a necessary consequence of the evolution of matter in a series of scission processes from a Lemaitre-type 'primeval atom' consisting of a highly excited relativistic state of the electron - positron system originally developed to provide a model for the neutral pi-meson and subsequently found to provide molecular-type models for all hadrons.

1. Introduction

Despite the fact that the descriptions of atomic structure and the intensities of the spectral lines they emitted using the wave equation developed by Schroedinger and the matrix mechanics of Heisenberg and Born were extremely successful, they did not resolve the conceptual problems of how to understand the wave-particle duality of matter and light. Instead, under the powerful influence of Heisenberg, Bohr and Born, the inability to resolve this puzzle led to the view that it was in principle impossible to arrive at detailed space-time or causal descriptions of phenomena on the atomic or particle scale, that one must be satisfied with accepting the existing mathematical formalism of quantum theory as providing as complete an account of physical phenomena as is possible, and that efforts to arrive at a deeper conceptual understanding by means of visualizable models on the particle scale were inherently doomed to failure.

2. Historical Background

Before describing a possible solution to the wave-particle duality problem, it is of interest to briefly examine the question as to why the view found such wide acceptance in the early part of this century, in sharp contrast with the more optimistic view held in the 19th century that phenomena on every scale were amenable to being understood in terms of detailed, visualizable geometric and dynamical models that have indeed continued to be enormously successful in the fields of chemistry and molecular biology to this day.

As discussed in more detail elsewhere [1], it seems that one important factor in the argument against models of a geometric and dynamic form as used in classical physics was due to a cyclical, generationally based revolt against a period of highly rational synthesis that animated individuals in many fields beginning at the turn of the century. As the historian of ideas Karl Joel described it, such periods of synthesis, when apparently disparate phenomena were related or unified, have historically been followed by periods of analysis, during which many new phenomena were discovered that seemed to resist efforts at unification and description in terms of rational, clearly visualizable models. These periods seem to last for about three generations or roughly a century, and this cyclical pattern was traced by Joel over many centuries with particular clarity in the development of scientific ideas. Thus, in the 17th century, Descartes provided a great synthesis in the description of the world in terms of an all-pervading aether that provided for interactions by direct contact and unified geometry with arithmetic and Newton related the action of gravity on earth with that in the heavens while Boyle described heat as a form of motion of the constituents of matter. In sharp contrast, the following century saw the discovery of many new and apparently unrelated chemical substances and a proliferation of different "effluvia" and unrelated forces acting instantaneously in the effort to understand heat, magnetism and electricity

This period was followed by the great synthesis of the 19th century that saw a unification of all chemical elements as composed of hydrogen atoms by Dalton and the establishment of a connection between electric, magnetic and optical phenomena by Faraday and Maxwell which revived the Cartesian ideas of action by contact in the form of the concept of a field. This period also saw the revival of the 17th century theory of heat as a form of motion, as well as the Cartesian idea that matter was nothing but a form of localized vortex motion in an ideal fluid aether by Helmholtz, Riemann, and Clifford and developed extensively by Kelvin, Maxwell, Thomson , Larmor and others. This effort ultimately led to Einstein's General Theory that explained gravity as the manifestations of local distortions in a flat Euclidean space-time substrate produced by the presence of matter as the crowning achievement of classical physics. Moreover, toward the end of this period, Lorentz was able to explain many of the optical properties of matter in terms of his theory of electrons with charges interacting via an ideal structureless aether, and Poincare developed the idea that all matter, being essentially electrical in nature, obeyed the law of relativity in that no particle of matter or ray of light could ever be seen as exceeding the speed of light by different observers in uniform motion relative to each other. Thus, all the laws of nature would be the same for each observer, a concept subsequently arrived at on more general, purely kinematical grounds by Einstein in his theory of special relativity as discussed by Whittaker [2].

Seen in the light of Joel's ideas, the 20th century was by contrast a cyclical return to a period of analysis, dominated by the discovery of many different types of forces that appeared to be unrelated to electromagnetic phenomena such as the strong nuclear force and the weak nuclear force governing radioactive decay discovered at the end of the 19th century, followed by the discovery of hundreds of nuclear particles of short life and seemingly unrelated masses beginning in the 1930s that appeared to make a unification of physical theory a hopeless enterprise. Moreover, the abandonment of hope for a great synthesis along the lines so successful for the classical physics of the 19th century by a new generation of scientists born around the beginning of the 20th century was paralleled by a rise in anti-rationalism and anti-religious movements in the form of Stalinism and fascism and , a return to mysticism such as astrology, as well as a widening fear and mistrust of science that made wars ever more destructive and finally threatened to destroy life on this globe with the development of the atomic bomb.

But in addition to this general cyclical change in the "Zeitgeist" or spirit of the world at the beginning of the 20th century that influenced the thinking of scientists such as Einstein discussed by Feuer [3], the particular sequence of scientific discoveries also contributed greatly to the widespread abandonment of causal, space-time models for atomic phenomena in the 20th century. Thus, as described in detail by Whittaker [2], the experiments of Michelson and others between 1881 and 1905 that failed to detect any evidence for the existence of an aether, combined with the failure to develop simple aether models to explain the transverse nature of electromagnetic waves, eventually led to the wide acceptance of Einstein's conclusion that there was no aether. Moreover, the fact that the experiments of Millikan , Compton and others completely confirmed Einstein's radical suggestion that radiation appeared to have particle-like properties for which no aether was needed essentially ended all attempts to describe matter in terms of vortices by the mid-1920s when Einstein was awarded the Nobel prize for his theory of the photo-electric effect. At the same time, evidence for a large physical extent of the region over which coherence of light can be observed, extending over distances of many meters, made it seem impossible to arrive at pictures or models of photons that would simultaneously explain both their point particle-like properties and their properties as wave-like phenomena in the aether [4].

These conceptual problems were even further aggravated by the discovery of the neutron in 1932. Although Rutherford, who had postulated its existence in the eary 1920s, believed that the neutron was composed of a proton and an electron in a closely bound state, a severe dificulty in constructing such a model arose when it became clear that the neutron had a spin or angular momentum of one half the Planck unit h. Since both the proton and the electron had the same spin of $h/2$, only a spin 0 or a spin h was allowed by the vector addition rules that had been established for spin. As a result, Heisenberg developed a theory of nuclear particles in which the neutron and proton were regarded as the same entity, but with a new quantum number that was +1 for the neutron and -1 for the proton. Since this purely formal approach was found by Fermi in 1934 to lead to a successful theory of radioactive beta-decay and thus also gave theoretical support for the existence of yet another particle not related to electromagnetic phenomena, namely the neutrino, any further attempts to formulate detailed space-time models of nuclear particles along classical lines appeared to be not only unnecessary but impossible.

Combined with the enormous success of Heisenberg and Born's purely statistical interpretation of the deBroglie matter waves, any hope of describing phenomena on the atomic and nuclear particle scale in terms of detailed space-time or causal models appeared in vain, and both de Broglie and Schroedinger gave up their efforts to understand the wave-function associated with matter particles like the electron in terms of real, physical entities such as wave packets related in some way to the electromagnetic field around an electron. Ironically, it was Lorentz who had delivered one of the most serious objections to all such attempts when in a letter to Schroedinger he pointed out that a wave-packet was inherently unstable and would quickly run-apart.

3. Revival of Causal Space-Time Models

Not until after the end of World War II did any efforts begin to overcome the difficulties of developing space-time models on the sub-atomic scale, or to find another interpretation of the deBroglie waves than the purely probabalistic one. Part of the impetus for this effort was the discovery of a whole series of new, unstable particles in the course of cosmic ray research, beginning with the discovery of the positively charged anti-particle of the electron, or the positron, followed by that of the muon, whose mass was some two hundred times larger than that of the electron in the early 1930s. But it was the discovery of the pi-meson with a mass of about 270 electron masses in 1947 when physicists were able to end their wartime activities that triggered the renewed efforts. This particle turned out to possess a strong interaction with protons and neutrons, and was therefore of great theeoretical interest in understanding the nature of nuclear forces, unlike the muon which behaved as if it were just a heavy or excited form of the electron.

In fact one of the first efforts to develop a model for the electron as an extended entity capable of quantized internal excitation which might explain the meson masses was published by Bohm and Weinstein in 1948 [5], using a rigid, non-relativistic, extended shell- structure for the electron, an approach that they hoped would at the same time overcome the problem of the infinities encountered in dealing with a point-charge in quantum electrodynamics. They concluded that such a classical model when quantized could lead to the existence of excited states whose energies were integral multiples of $\pi (1/\alpha)\, m_0\, c^2$ or $n\, 400\, m_0\, c^2$ where α is the fine-structure constant $\hbar c/e^2 = 137.036$ where h is Planck's constant divided by 2π, m_0 is the rest-mass of the electron and c is the velocity of light, thus giving a mass of the order of the π meson. As will be discussed below, this turned out to be an amazingly prescient result.

Four years later, Bohm published a paper on a new interpretation of the wave-function as a physically real quantity associated with each fundamental particle, along the lines initially advocated by de Broglie and Schroedinger. This was related to Bohm's earlier effort to describe the electron as an extended entity that can be described in causal space-time terms rather than as a point particle whose position and momentum can only be determined statistically [6]. As he put it, his aim was not to exclude the possibility of the description of atomic and nuclear particle phenomena which would involve additional elements or parameters permitting a detailed causal and continuous description of all processes on the quantum level in precise terms, elements that could be called "hidden variables" until they became accessible to experimental observation. These might include substructures of known particles, their coordinates and momenta, as experimental capabilities advanced to explore structures below the size of the electron of about 10^{-13} cm.

The principal new idea he advanced was to introduce the interpretation of the wave-function as giving rise to a scalar "quantum potential" that was proportional to the second derivative of its real amplitute or its curvature, and inversely proportional to its magnitude R. This can therefore exert a force on the particle in a way that is analogous to that of known fields, such as an electromagnetic field, a meson field, or a gravitational field. As will be shown below, this, too, now appears to to have been enormously prescient.

This "quantum field" represented by the wave-function ψ had the property that it does not radiate as the electromagnetic field does, so that it simply changes its form while its integrated intensity remains constant. Moreover, he showed that it produces important effects only at very small distances of the order of the classical electron size or shell-radius, so that its effects are negligible in the atomic domain. Also, it allows a particle to stand still because electromagnetic force is balanced by the "quantum mechanical force" produced by the ψ - field. But just what the physical origin of this ψ - field was remained a mystery.

Nevertheless, this concept appeared to have the abilty to explain such puzzling phenomena as the well-known double-slit experiment where a double-slit interference pattern is produced even when only one electron at a time is fired at the screen. The conceptual puzzle is explained by Bohm in terms of an objectively real ψ – field that allows one to regard the process in causal terms, where the field that is associated with the electron passing through one slit extends out to the position of the other slit, so that in effect the electron passes through the two slits at the same time, which would be absurd for a point particle conceived as a small hard ball directed at a solid screen [8].

Pursuing the effort to find a mathematical description of an extended electron, Bohm joined forces with Vigier to develop a model involving relativistic fluid masses which are kept together by appropriate internal tensions that tend to hold these masses together in some stable form [9]. Interestingly, one of the stable forms they mention is a rotating torus or vortex ring discussed by Poincare in a 1885 paper [10], which was in effect a revival of the idea of Helmholtz so widely accepted as a promising model for the ultimate constituents of matter in the 19th century before the discovery of the electron by J. J. Thomson in 1897. Moreover, they suggested that such a "molecular" model could eventually hope to explain the physical origin of spin. Again, as will be discussed below, this appears now to have been remarkably prescient.

4. The Relativistic Electron Pair Model

Encouraged by the views of deBroglie, Einstein, Bohm, and Vigier, it seemed worthwhile to see whether detailed space-time models of the neutron and other heavy particles might be possible in which an extended electron and positron, taken to be nothing but stable sources and sinks of an electromagnetic field [11], could play the role of basic entities from which all the other, heavier particles known to decay to electrons, positrons and radiation quanta could be constructed.

The first and most unexpected result was related to the occurrence of a minimum approach distance between two charges if one adopts a definition of the force between two moving charges suggested by Einstein in his famous 1905 paper on "The Electrodynamics of Moving Bodies" in which his kinematic theory of special relativity is developed. It turns out if one uses the symetrical expression for the force exerted on a moving charge proposed by Einstein, namely that which is measured by an observer at rest relative to either one of the charges, one is automatically led to the existence of a minimum possible approach distance between two elementary charges rotating about each other in force equilibrium which is of the order of the classical electron shell diameter $d_{cl} = e^2/m_o c^2$ when the electron's mass and charge are used. Thus, the existence of such a minimum approach distance occurs naturally, regardless of any detailed assumptions as to the internal structure of the electron or knowledge of the tensile forces that hold it together, consistent with the assumption that the electron is a finite sized entity, yet behaves mathemaically in all its inteactions as if it were an infinitely small point.

Following a suggestion of Feynman in 1960 during a visit to the California Institute of Technology it turned out that a system composed of an electron and a positron in a highly relativistic Bohr-type orbit using the relativistic Coulomb force law discussed by Einstein and quantized along the lines suggested by Bohm and Weinstein [5] gives a series of excited states whose mass is almost exactly that worked out by these authors in their 1948 paper.

With the simple relativistic Coulomb force law that holds for the case of the motion at a right angle to the direction of motion, $\gamma_{12} e^2 / r^2_{12}$, where γ_{12} is the Lorentz contraction factor $(1-\beta_{12}^2)^{-1/2}$ and $\beta_{12} = v_{12}/c$, the relativistic centrifugal force $\gamma_{12} m_o \beta^2 c^2 / r_{12}$ is leads to a limiting value of $r_{12} = e^2/m_o c^2$ except for a small correction factor of 1/4 when $\gamma_{12} >> 1$ [12]. Quantization of the orbital angular momentum in units of $\hbar$ leads both to the normal, low energy positronium states and the high energy states n $(2/\alpha) m_o c^2 = n\ 274\ m_o c^2$ which for n =1 is very close to the pion masses. Correcting for the spin-spin and spin - orbit forces due to the magnetic moments leads to a mass of 263 $m_o c^2$, very close to the best present observed value of 264.068 $m_o c^2$ for the neutral pion

As described in detail in subsequent papers [13,14] and recently summarized in a review article [15], this basic relativistic positronium model has been extended to describe the charged pions as excited states of the muon, accounting for the observed masses and lifetimes to a surprising degree. Moreover, the forces between the pair systems turn out to be stronger than the non-relativistic Coulomb interaction by roughly the factor γ_{12} or 274, but with a short range since this force depends critically on the relative orientation and phase of the motions in each pair, thus giving rise to a Yukawa type force consistent with the solution of the Klein-Gordon equation appropriate to bosons of integral spin formed by these pair systems, and explaining the relation between the electromagnetic and the strong nuclear force.

One of the interesting results of the extremely relativistic motions of the pair system is that the Sommerfeld precession becomes so large that it takes up half of the total angular momentum $\hbar$. Thus, for an observer in the precessing frame, orbital angular momenta are quantized in units of $\hbar/2$. Furthermore, because of the high mass of the electrons and positrons in the relativistic states, their source size is contracted by a factor γ_{12} or 274, with the rsult that the rate of radiation is greatly reduced. As a consequence, the decay of the charged pion, in which the extra electron carries a 1/2 unit of h decaying to a charged mu-meson with the emission of a quantum of radiation that has to carry away one half a unit $\hbar$ is very slow, of the order of 10^{-6} seconds compared with the emission of two gamma rays in the case of the neutral spin 0 pion, which takes only about 10^{-16} seconds. Likewise, the subsequent decay of the charged muon core consisting of a central spin 1 pair system with spins opposed by giving rise to two spin 1/2 neutrinos of opposite helicity accompanied by the extra electron is also very slow, explaining the "weakness" of the beta decay processes along the lines originally developed by Fermi, who formulated his theory of beta decay by analogy to electromagnetic processes but with a weak coupling constant whose value could only be determined by a fit to the empirical results.

Thus the relativistic pair model automatically explains the difference in the strengths of the electromagnetic, the strong and the weak force, without the need for any new and unknown particles or interactions.

The same basic positronium-like structures also describes the so-called pion resonance states as excited rotational states of molecular-like systems cosisting of two, three and four electron-positron systems. Finally, it was found that four such pair-systems arranged in two set of pairs that each have the mass of a K-meson can be bound by an electron exchanged between them to give a highly stable proton, whose observed excited states again fit the same type of molecular rotational states as the pion resonances [14].
The proton structure thus consists of three dynamic components in agreement with the results of SU3 symmetry models, whose masses are of the same order as the quark masses in the standard model, except that they are integrally charged as suggested by Han and Nambu [16]. Thus, the essence of the standard model that regards pions and baryons to be composed of quarks and antiquarks forming "charmonium" or positronium-like substructures of all nuclear particles is preserved, and since no free, fractionally charged quarks have ever been observed despite decade-long searches, it seems simplest to accept Han and Nambu's suggestion that they may indeed be integrally charged, with spin 1/2 and point-like effective size and interactions, just as is found for the massive relativistic electron and positron in the pair model.

A further test of the hypothesis that all matter is composed of finite-sized electrons and positrons occurred in 1974 with the discovery of a totally unexpected narrow new resonance in the collision of high energy electrons and positrons, the J/ψ meson with a mass of 3097 Mev . Since highly excited states of the relativistic electron pair system were predicted that are integral multiples of the pion mass, it was very encouraging when it turned out that the 22nd level of the charged pion of mass $273.5 \, m_o c^2$ calculated for the model gave a mass of $6017 \, m_o c^2$ or 3075 Mev, and the 23rd state of the neutral pion of theoretical mass 263.1 Mev gave a mass of $6051.3 \, m_o c^2$, corresponding to 3092.2 Mev , extremely close to the observed J/ψ value. This resonance was subsequently interpreted as consisting of a charmed quark -antiquark pair each with spin 1/2 analogous to positronium, thus further supporting the basic idea of integrally charged quarks being involved.

5. Model for the Origin of the Universe

Encouraged by this discovery of highly excited and relatively long-lived states of the type already foreseen by Bohm and Weinstein in 1948, and with the growing evidence for a Big Bang model of the universe that according to Lemaitre started from an extremely massive, dense "primeval atom" , it seemed possible that this initial state of the universe may have been a highly excited state of the electron pair system from which all matter particles in the universe originated in a series of some 260 division processes by two, as originally suggested by Lemaitre [17]

As described in detail elsewhere [15, 18, 19, 20] , this led to a model for the evolution of matter in a series of repeated internal pair production processes occurring in 27 major stages of ten divisions by two each during which no radiation is emitted, ending with a phase-transition from individual pairs to baryons, pions and leptons at the moment of the Big Bang. On the assumption that the greatest possible density of the initial state was the Planck density $\rho_{Pl} = c^5/\hbar G^2$ as suggested by Ginzburg [21] and that the effective volume V_{ef} in which the energy was concentrated was that occupied by the rotating electron-positron system field with an outer radius of 1/4 of the Compton wavelength of the electron, one can arrive at a mass of the universe M_U that turns out to be in good agreement with what one calculates using the Dirac large number relation $(e^2 / m_o^2 G)^2 = M_u/m_o$. Its value turns out to be 1.7362×10^{85} m_o or 9.4551×10^{81} proton masses M_p, compared with present observational estimates of about 10^{80} M_p, in reasonable agreement with the fact that only about 1% of the total mass is at present in visible form. Thus, the hypothesis that matter is composed of extended electrons and positrons can be regarded to have passed yet another test, not only giving a reasonable value for M_u but also providing an explanation of the physical origin of the Dirac large number relation.

Moreover, because the process of internal pair-production does not involve the emission of radiation, the massive pairs that are the seeds of galaxies and stars are in effect a form of "cold dark matter", most of which is still trapped in the centers of the largest cosmological structures [19], leading to the delayed ejection of massive objects that are initially dark and point-like, and then explode in a form of Mini-Big Bang that we see as quasars.

6. Implications for the Physical Nature of the Wave-Function

Perhaps of greatest interest for an understanding of the wave-particle duality of the electron and the nature of the quantum potential is the conclusion arrived at in the examination of the decay of the Lemaitre atom that if the pairs are to divide so as to create the particles of the universe, the local value of G within each pair has to increase from its Newtonian value. This is necessitated by the fact that the effective volume V_{ef} is given by $(\hbar e^4 / m_o^3 c^5)$ so that if the fundamental atomic constants $\hbar$, e, m_o and c are assumed not to vary as the masses of the pairs M_{ee} decline so that the volume V_{ef} remains constant and the critical density for pair production M_{ee} / V_{ef} is to continue to have the form $c^5/\hbar G^2$ then G has to become $G_{ee} (M_{ee}) = (e^2/m_o^2)(m_o/M_{ee})^{1/2}$. This has the important consequence that when M_{ee} reaches a value of the order of the electron mass, $G_{ee} (M_{ee})$ must have increased by the factor $(M_u / m_o)^{1/2}$ or by 4.167×10^{42} to (e^2 / m_o^2), which means that the local gravitational curvature force has risen to a value equal to that of the Coulomb force.

But that is the magnitude of the tensile force needed to hold a classical shell electron as originally studied by Bohm and Weinstein [5] in equilibrium against the repulsive Coulomb force, and it is within a factor of $(1 / \alpha)$ of the value of the local gravitational force arrived at by Motz in his model for an extended electron [22]. Thus, the condition that allows the universe to come into being, starting with a single massive electron pair system, necessarily leads to a local space-curvature or gravitational force of the magnitude needed both to stabilize the electron and positron internally and to balance the pair system against the centrifugal force associated with the large precession. Now according to the hydrodynamic model studied by Bohm, Vigier and others, a spinning vortex ring model is one of the structures believed to be stable. If one therefore accepts the inherent stability of vortex rings as first derived by Helmholtz, then one can regard the contracted electromagnetic field between the relativistic electron and positron as just such a quantized vortex ring, but one that was cut in half, twisted 180° and held together by the low pressure in the interior of the vortex tubes.

Thus, the ends of the tubes are the region where lines of electric force identified with lines of vorticity emerge and disappear, explaining the physical origin of charges as the sources and sinks of lines of vorticity, while the magnetic field is identified with the ideal fluid circulating about the lines or tubes of the electric field, resembling a helium superfluid that is now known to sustain quantized vortex rings indefinitely. As a result, the quantization of charge is explained by the quantization of vorticity, or the existence of a finite, minimum value of the electronic charge is a consequence of the quantization of angular momentum. Ironically, this is something that Einstein suspected as early as 1909, but at that time he had just eliminated the fluid aether in his theories of radiation quanta and special relativity, and without a medium there are no vortexes. It was a decision that changed the course of physical theory for nearly a century, since without the universal medium or reference frame of Descartes and Newton, there could also be no way to understand rotation by an initial pair of fundamental particles keeping a pair of charges in force equilibrium in an otherwise emp\ty universe.

7. Implications for the Nature of Photons and the Origin of Space Curvature

But if spinning half - vortices in the aether are involved in models of the fundamental particles, and these charges annihilate into radiation quanta, then photons and neutrinos that carry no charge cannot be anything else than spinning *closed* vortex rings moving at the speed of light in empty space. This would explain the fact that high energy electrons produce pulses of electromagnetic fields as they pass by an atom that are indistinguishable from photons. And the wave-function giving rise to the quantum potential of Bohm that is associated with matter can now be considered a manifestation of the intense distortion of normally flat space-time produced by the locally high value of the gravitational constant in the neighborhood of the basic entities. It leads to a Schwarzshild radius equal to the Compton wavelength, creating a region that is impenetrable or "hard" like "ponderable matter" within the electron and causing it to act as a point entity in all interactions, so that it seems that impulses can travel finite distances across an electron or photon with infinite velocity as in a rigid body.

It follows that the gravitational interaction can be regarded as produced by the internal motions of self-stabilized, spinning vortex rings in an ideal fluid possessing both wave and particle-like properties, producing a distortion of space analogous to that seen on the surface of a liquid by half of a vortex ring. It is in fact a phenomenon familiar to anyone who has ever watched the dimples produced by a spoon idly drawn along the surface of a cup of coffee, an example used by Helmholtz at the end of his epoch - making paper published nearly 140 years ago. Since vortex rings at large separations attract each other according to an inverse square law, it now sems that both Descartes and Newton were correct in their beliefs about the nature of matter and gravity and the existence of an aether. Moreover, since Lorentz showed that if all forces are related to the electromagnetic force, it is possible to explain the inability to detect uniform motion relative to the aether in terms of a real contraction of objects and a slowing down of clocks.

The high local value of the gravitational constant within the spinning electron and positron also resolves another major problem in contemporary physical theory. This is the problem of the small size of the "strings" in superstring theory, presently identified with the Planck length using the value of Newton's gravitational constant. But when the local value within the electron now regarded as the fundmental constiuent of all nuclear particles is associated with a local value that is 10^{42} times larger, the Planck length $(\hbar G/c^3)^{1/2} = 1.616 \times 10^{-33}$ cm increases by the square root of this number, bringing it to the size of the electron Compton wavelength. This is therefore the size of the fourth, curled-up inner space of superstring theory, and the remaining five spaces or degrees of freedom can be identified with those needed to explain the excited states of baryons. Thus, the unification of all forces of nature aimed for by superstring theory can be realized wth the assumption that superstrings are stable, quantized vortex rings that can be excited to very high states of vibration and rotation, and which produce an intense local space curvature, just as in the case of stable quantized vortex rings that are observed in superfluid helium. These rings do not radiate either when spinning or when describing orbits that are geodesics in the locally highly curved space. It is a world of smoke-rings that last s forever, since a rotating universe will neither fly apart nor collapse into a fiery singularity.

Thus, it seems that the belief of Bohm and Vigier that there is a substructure of nuclear particles that can only be understood by constructing causal space-time models analogous to those used in classical hydrodynamics has been vindicated by superstring theory. Moreover, it is indeed necessary to have "hidden variables" involving the coordinates and angular momenta of sub-nuclear entities requiring added inner dimensions or degrees of freedom not allowed by the Copenhagen interpretation of quantum mechanics. Thus the belief that it should be possible to explain the various forces of nature in a unitary manner, long maintained by Einstein in the face of widespread opposition, was increasingly shared by a small number of physicists like deBroglie, Bohm and Vigier who were influenced by Einstein's courage and perseverance in believing in the comprehensibility of nature to the end of his life. This hope now appears to have been fully justified by the latest developments in physical theory , laboratory experiments showing action at a distance within the entities of matter and light, and astronomical observations with a new generation of powerful telecopes able to reach the most distant galaxies as a new century with a cyclical return to a period of synthesis and unification is about to begin.

8. References

1. Sternglass, E, J. *Cornell Review*, Spring (1951)
2. Whittaker, E, T. :*A History of the Theories of Aether and Electricity*, Vol.2, Chpt.2 Thomas Nelson and Sons, London, 1951.
3. Feuer, L. S..: *Einstein and the Generations of Science*, Basic Books, New York,1974.
4. Whittaker , E. T.: Vol. 2, p.94-95.
5. Bohm, D. and Weinstein, M.: *Physical Review* **74** (1948)1789
6. Bohm , D. : Physical Review **85** (1952), 166.
8. Sternglass, E. J. in *Horizons of a Philosopher*, eds J. Frank, H.Minkowski and E.J. Sternglass, E. J. Brill, Leiden 1963, p. 422.
9. Bohm, D. and Vigier, J.-P.:*Physical Review* **109** (1958)1882.
10. Poincare, H.: *Acta Mathematica* **7** (1885)259.
11. Sternglass, E. J.: *Comptes Rendus* **246** (1958)1386.

12. ----------------- *Physical Review* **123** (1961) 391.
13. ----------------- *Nuovo Cimento* **35** (1965)227.
14. ----------------- *Proceedigs 2nd Conference on Resonant Particles*, Ohio Universit, Athens, Ohio. (1965), edit.B. A. Munir, p..33.
15. ----------------- in *Frontiers of Fundamental Physics* , eds.. M. Barone and F. Selleri , Plenum Press N.Y. 1994, p..59.
16. Han , M. and Nambu,Y. *Physical Review B* **139** (1965)1006.
17. Lemaitre, G.: *Nature* **128** (1931) 701.
18. Sternglass, E.J. : Lett. Nuovo Cimento **41** (1984)203.
19. ------------------ in *Testing the AGN paradigm* , *AIP Conference Proceedings* **254** (1992)105
20. Sternglass, E. J. in *Dark Matter* , *AIP Conference Proceedings* **336** (1995) 513.
21. Ginzburg, V. L. *Comments Astron. Astrophys.* **3** (1971)7.
22. Motz, L. *Nuovo Cimento* **26** (1962) 672

THE TAKABAYASI MOVING FRAME, FROM THE A POTENTIAL TO THE Z BOSON

Roger BOUDET
Université de Provence
Pl. V. Hugo, 13331 Marseille, France

1. Introduction

During the years 1950, an important part of the works of the physicists of the "*Louis de Broglie's school*" (in particular O. Costa de Beauregard[1], F. Halbwachs [2], G. Jakobi, G. Lochak [3], T. Takabayasi [4], J.P. Vigier [5]) has been devoted to an intrinsic presentation of the electron Dirac theory. In this presentation, the abstract formalism of the Dirac spinors was replaced by the use of quantities and equations independant of all galilean frame of the Minkowski spacetime $M = R^{1,3}$ (here $R^{p,n-p}$ means the signature of a R^n euclidean space). So was introduced, in particular, the "Takabayasi frame" which is a frame of four orthonormal spacetime vectors v, n_1, n_2, n_3, defined at each poin x of M, in such a way that v is a timelike vector, the spacetime velocity of the Dirac particle (colinear to the spacetime current $j = \rho v$ where $\rho > 0$ is the probability density), and that the bivector (antisymmetric tensor of rank 2) $(\hbar c/2)n_1 \wedge n_2$ represents the intrinsic angular momentum, or spin.

In 1961 a geometrical interpretation of the $U(1)$ gauge, already suggested in [3], was precisely described by F. Halbwachs, J.M. Souriau and J.P. Vigier [6]: the couple (n_1, n_2) determines a plane $P(x)$ (the "spin plane"). A change of the electromagnetic gauge is nothing else but the replacement of the couple (n_1, n_2) by another couple (n'_1, n'_2) defining the same plane $P(x)$ and making, inside $P(x)$, with the first couple, an angle $\varphi(x)$. The parameter $\varphi/2$ is the "phase" χ of the Dirac theory. The gradient $(\hbar c/2)\partial_\mu \varphi$ is a "gauge photon".

Using the formalism of the Real Algebra of Spacetime, introduced in

Quantum Mechanics by D. Hestenes [7], I have completed in [8] the two above physical properties of the plane $P(x)$ by a third one. One can write $p_\mu = (\hbar c/2)\omega_\mu - eA_\mu$, where p_μ is the momentum-energy spacetime vector, A_μ is the electromagnetic potential acting on the electron and $\omega_\mu = (\partial_\mu n_1).n_2 = -n_1.(\partial_\mu n_2)$ is a spacetime vector, representing the infinitesimal rotation on itself of the "spin plane" $P(x)$. This result is a part of a general property I have established in [9]: if one excepts the role of the Yvon-Takabayasi angle (see below), the Tetrode momentum-energy tensor of the electron is the product of $\rho\hbar c/2$ by a tensor expressing the infinitesimal rotation on itself of the partial frame $\{v, n_1, n_2\}$.

So, one can foresee the interest of a geometrical interpretation of Quantum Mechanics, based on the properties of moving frames of spacetime. In the electron theory, three entities (spin, electromagnetic gauge, momentum-energy) without any apparent bond in the abstract formalism of the Dirac spinors, are unified in a relation with a spacetime plane. Also, one can imagine that, what we call energy, corresponds to movements of frames in spacetime. That may be perhaps a first step towards the unification of Quantum Mechanics and the General Relativity.

The construction of the Takabayasi frame can be made evident by the following expression $\Psi = \psi U$ of the Dirac wave function, established, for the first time, to my knowledge, in [3]. In this expression, U is some "spinor unity" and ψ may be written in the form $\psi = (\rho e^{i\beta})^{1/2} R$. The real scalar β is the Yvon-Takabayasi angle, whose geometrical interpretation is simple (it allows one to write, in a unic way, all non isotropic bivector of M as the sum of two supplementary simple bivectors), but the physical role is still mysterious (this angle does not intervene directly in the gauge transformations). The matrix $\underline{i}$ verifies $\underline{i}^2 = -1$, but corresponds to the real unit pseudo-scalar (antisymmetric tensor of rank 4) of M. At least, the matrix R represents a Lorentz rotation which transforms at each point x the laboratory galilean frame into the Takabayasi frame, and in particular the (x_1, x_2) plane into the "spin plane" $P(x)$.

This result was found again in 1967, in a quite independant way, by D. Hestenes [7], who brought an important additional algebraic property: ψ is a Hamilton biquaternion, i.e., an element of the Clifford algebra associated with $R^{3,0}$ or, as well, by means of an important connexion between the bivectors of $M = R^{1,3}$ and the vectors of $R^{3,0}$ (leading for example to the expression $F = \vec{E} + \underline{i}\vec{H}$ of the electromagnetic field), as the elements of the even Clifford sub-algebra of $R^{1,3}$ (see [10]).

2. The Pauli and Dirac spinors, and the doublet of Dirac spinors as Hamilton quaternions and biquaternions. The biquaternionic interpretation of SU(2)×U(1)

The $\sigma_k, \gamma_\mu, \tau_k$ matrices act, respectively, on doublets of complex numbers which define the Pauli spinors, on doublets of Pauli spinors which define the Dirac spinors, and on doublets of left Dirac spinors. The role of these objects will be study in the frame of E^3. Their relations with M will be established in the next paragraph.

1. We recall that a Hamilton quaternion is associated with the three dimensional euclidean space $E^3 = R^{3,0}$ and may be written in cartesian coordinates x, y, z

$$q = w + ix + jy + kz \tag{1}$$

where i, j, k verify $i^2 = j^2 = k^2 = -1$, and $ij = -ji$, $jk = -kj$, $ki = -ik$, and $i = -jk$, $j = -ki$, $k = -ij$. The field of the Hamilton quaternions is the even Clifford sub-algebra associated with $R^{3,0}$.

We emphasize that, from the point of view of the inventors themselves of the complex and hypercomplex numbers, Argand [11], Grassmann (see [12]), Hamilton [13], Clifford (see [10]), these "numbers" are in reality real geometric entities. So, i, j, k are real bivectors, i.e. oriented portions of real planes. If $\{\vec{i}, \vec{j}, \vec{k}\}$ is an orthonormal frame of E^3, in such a way that $\vec{i}^2 = \vec{j}^2 = \vec{k}^2 = 1$, one can write $i = \vec{j} \wedge \vec{k} = \vec{j}\vec{k}$, $j = \vec{k} \wedge \vec{i} = \vec{k}\vec{i}$, $k = \vec{i} \wedge \vec{j} = \vec{i}\vec{j}$.

One can write too, $i = \underline{i}\vec{i}, j = \underline{i}\vec{j}, k = \underline{i}\vec{k}$, where $\underline{i}$ satisfies the relation $\underline{i}^2 = -1$ but is in reality the real unit pseudo-scalar of E^3, $\underline{i} = \vec{i} \wedge \vec{j} \wedge \vec{k} = \vec{i}\vec{j}\vec{k}$, and one can put q in the form $q = w + \underline{i}\vec{U}$, i.e. as the sum of a real scalar and a real bivector. (Note that the associative operations on vectors are to be considered in the Clifford algebra associated with $R^{3,0}$).

Using $i = -jk$, one can deduce from (1),

$$q = w + kz - j(-y + kx) = u_1 - ju_2 \tag{2}$$

and associate with q, in the form of a column vector, a doublet $\xi = (u_1, u_2)$ of "complex numbers" $u_1 = w + i'z$, $u_2 = -y + i'x$, where the "imaginary" number $i' = \sqrt{-1}$ is to be identified here with the bivector $k = \vec{i} \wedge \vec{j}$. This doublet constitutes the Pauli spinor (see [14], [15]). As a confirmation, if $q = ix + jy + kz$ and $q^2 = 0$, the "numbers" u_1, u_2 satisfy the system proposed by Elie Cartan([16], eq. (1), p. 54)

$$u_1 z + u_2(x - i'y) = 0$$
$$u_1(x + i'y) - u_2 z = 0$$

as being the source of the definition of spinors, because it makes apparent the Pauli matrices $\sigma_1, \sigma_2, \sigma_3$.

Nota. The solution $u_1 = \pm[(x - i'y)/2]^{1/2}$, $u_2 = \pm[(-x - i'y)/2]^{1/2}$, that E. Cartan has considered for his system, was the source of beautiful works of projective geometry, but has for the while, to my knowlewge, no equivalent in Quantum Mechanics. This solution, and the fact that this system may be associated with an isotropic straight line of E^3, which is an imaginary object, has been probably one of the reasons why, if one could associate a clear signifiance to the Pauli matrices, for example in the definition of the generators of the $SU(2)$ group, the entities on which they act, the Pauli spinors, have appeared as rather mysterious until the publication of [14] and [15]. (In my opinion, based on talks with my master, P. Vincensini, to whom his own master, E. Cartan, had initially proposed the redaction of [16], E. Cartan has never surmised that the Pauli and Dirac spinors were real objects. The belief in the basic complexity of Nature still remains in works of mathematicians as R. Penrose, or J. Rzewuski, despite the indisputable contribution of D. Hestenes to the proof of the reality of the objects used in Quantum Mechanics).

One can write [15]

$$\vec{i}q\vec{k} = -iqk = jk(u_1 - ju_2)k = u_2 - ju_1 \quad \Leftrightarrow (u_2, u_1) = \sigma_1 \xi$$
$$\vec{j}q\vec{k} = -jqk = -j(u_1 - ju_2)k = -ku_2 - jku_1 \quad \Leftrightarrow (-i'u_2, i'u_1) = \sigma_2 \xi$$
$$\vec{k}q\vec{k} = -kqk = -k(u_1 - ju_2)k = u_1 - j(-u_2) \quad \Leftrightarrow (u_1, -u_2) = \sigma_3 \xi$$

That explains why the matrices σ_k obey the same relations as orthonormal vectors of E^3, but when they act on a spinor Pauli ξ, as a consequence, the corresponding Hamilton quaternion q is multiplied on the left by $\vec{k}$. In the same way $i'\xi$ means $i'\xi = (ku_1, ku_2) = (u_1 k, u_2 k)$ and corresponds to the transformation of q into $qk = q\vec{i}\vec{k}$.

2. A biquaternion (or "complex") quaternion) is of the form

$$Q = q_1 + \underline{i}q_2, \quad q_k = w_k + \underline{i}\vec{U}_k \tag{3}$$

So a biquaternion is the sum of a scalar, a vector, a bivector and a pseudo-scalar of E^3, i.e. in eq. (3), $w_1, -\vec{U}_2, \underline{i}\vec{U}_1, \underline{i}w_2$. The ring of the biquaternions is the Clifford algebra associated with $R^{3,0}$.

Because $\underline{i}q = q\underline{i}\vec{k}^2 = (qk)\vec{k}$, and for reasons we will explicite in par. 3, one can write

$$Q = q_1 + q_2'\vec{k}, \quad q_2' = q_2 k \tag{4}$$

and one can associate with Q the doublet $\Psi = (\xi_1, \xi_2')$ of the Pauli spinors associated with the quaternions q_1, q_2'. Ψ is a four complex components Dirac spinor.

Note that $i'\Psi = (i'\xi_1, i'\xi_2')$ corresponds to Qk. The fact that $\exp(i'\varphi)\Psi$ means $Q\exp(k\varphi)$ and corresponds to a rotation in a real plane of E^3, instead to some abstract operation belonging to the $U(1)$ group, is of a fundamental importance for the comprehension of the gauge theories. This particularity of spinors has been pointed out for the first time, to my knowlege, by D. Hestenes in [7], and has allowed him to find again, by an algebraic way, the result deduced in [3] and [6] from tensorial considerations.

3. For the construction of the $SU(2) \times U(1)$ group we are interested only by what we will call a left biquaternion.

Let us define the following biquaternion, which is an idempotent

$$u = \frac{1}{2}(1 + \vec{k}) \quad \Rightarrow \vec{k}u = u\vec{k} = u, \quad u^2 = u \tag{5}$$

We will say that the left biquaternion associated with a biquaternion Q is the biquaternion $Q^L = Qu$.

We propose here to associate with the following composition of two left biquaternions Q_1^L, Q_2^L

$$\phi = Q_1^L + Q_2^L \vec{i} = Q_1 u + Q_2 u \vec{i} \tag{6}$$

(which is itself a biquaternion) a doublet $\Phi = (Q_1^L, Q_2^L)$ that we will call a left doublet.

One can write

$$\begin{aligned}
Q\vec{i} &= Q_1^L \vec{i} + Q_2^L \vec{i}^2 = Q_2^L + Q_1^L \vec{i} & \Leftrightarrow (Q_2^L, Q_1^L) = \tau_1 \Psi \\
Q\vec{j} &= -Q_1^L k\vec{i} + Q_2^L \vec{i}^2 k = Q_2^L k + (-Q_1^L k)\vec{i} & \Leftrightarrow (Q_2^L k, -Q_1^L k) = \tau_2 \Psi \\
Q\vec{k} &= Q_1^L \vec{k} + Q_2^L \vec{i}\vec{k} = Q_1^L - Q_2^L \vec{i} & \Leftrightarrow (Q_1^L, -Q_1^2) = \tau_3 \Psi
\end{aligned}$$

in wich one has used the fact that $Q_\alpha^L \vec{k} = Q_\alpha^L$ and also $\vec{j} = -\vec{ki} = \vec{i}k$. So the form of the τ_k matrices is explained.

Multiplying on the left, or the right, by $\underline{i}$, one obtains the three transformations

$$\phi \to \phi\underline{i}, \phi\underline{j}, \phi\underline{k} \tag{7}$$

which correspond to the transformations $\Phi \to i'\tau_k \Phi$ of a doublet of Dirac spinors.

In these transformations, $\underline{i}, \underline{j}, \underline{k}$ or, as well, $i'\tau_1, i'\tau_2, i'\tau_3$, appear as the generators of $SU(2)$.

Now the change

$$Q_\alpha \to Q_\alpha k, \quad (\alpha = 1, 2) \tag{8}$$

gives, because $ku = \vec{i}\vec{k}u = \underline{i}u = u\underline{i}$, $\underline{i}\vec{i} = \vec{i}\underline{i}$,

$$\phi \to Q_1^L \underline{i} + Q_2^L \vec{i}\underline{i} = \phi\underline{i} \tag{9}$$

and corresponds to the change $\Phi \to i'\Phi$.

This fourth transformation commute with each transformation (7). The pseudo-scalar $\underline{i}$ can be considered as the generator of $U(1)$ and one can consider that one have obtained a $SU(2) \times U(1)$ group, acting on a left doublet, whose generators are $(\underline{i}, \underline{j}, \underline{k}, \underline{i})$. This result is in conformity with what is proposed in [17] for the generators of this group, but it has here a quite different meaning. In particular, we emphasize that, here, $U(1)$ corresponds, as in the Dirac theory of the electron, to a rotation in a real plane of space, to the extent that, in the expressions $Q_1 k, Q_2 k$, the bivector k is the generator of a rotation in the (x, y) plane (when in [17] it seems to be associated with the group of the "duality rotations" [10]).

3. The biquaternionic form of the wave function associated with a doublet of left particles, in the Glashow-Salam-Weinberg theory. The proper frame associated with this doublet

1. We recall that the Clifford product is an associative product of the vectors of an euclidean space $E = R^{p,n-p}$, which is, in particular, reduced to the Grassmann product if all the vectors of the product are orthogonal: for example, if $a, b \in E$ are orthogonal, one has $ab = a \wedge b = -ba$. The reversion

THE TAKABAYASI MOVING FRAME

is an operation which inverses the order of the vectors in the Clifford product, and is denoted $X = a_1..a_p \to \tilde{X} = a_p..a_1$ (see [10]).

Let $B_0 = \{e_\mu \in M, \mu = 0,1,2,3\}$ be an orthonormal fixed frame of M (laboratory frame). Using the Clifford product in M, on sees that the bivectors of M, $\vec{i} = e_1 \wedge e_0 = e_1 e_0$, $\vec{j} = e_2 \wedge e_0 = e_2 e_0$, $\vec{k} = e_3 \wedge e_0 = e_3 e_0$, generate a space isomorphic to E^3. We denote $\underline{i} = e_0 e_1 e_2 e_3$. One obtains $\underline{i}a = -a\underline{i}, \forall a \in M$ and $\underline{i} = \vec{i}\vec{j}\vec{k}$ (see [10]).

Because $(\vec{k})\tilde{} = e_0 e_3 = -\vec{k}$ one can write too

$$\check{u} = \frac{1}{2}(1 - \vec{k}) \Rightarrow u + \check{u} = 1, \ u\check{u} = \check{u}u = 0, \ \check{u}^2 = \check{u} \tag{10}$$

It is easy to deduce from the form given in eq. (4) to a biquaternion Q that $e_\mu Q e_0$ corrresponds to $\gamma_\mu \Psi$ [15]. For example, because $e_0 q = q e_0$, $e_0 \vec{k} = -\vec{k} e_0$, one has

$$e_0 Q e_0 = q_1 e_0^2 + q_2'(-\vec{k})e_0^2 = q_1 - q_2'\vec{k} \quad \Leftrightarrow \quad (\xi_1, -\xi_2') = \gamma_0 \Psi$$
$$e_1 Q e_0 = \vec{i} e_0 Q e_0 = (\vec{i}q_1\vec{k})\vec{k} - \vec{i}q_2'\vec{k} \quad \Leftrightarrow \quad (-\sigma_1 \xi_2', \sigma_1 \xi_1) = \gamma_1 \Psi$$

So the form of the γ_μ Dirac matrices may be explicited, and all the formalism of the matrices may be included in the one of the Hamilton quaternions and biquaternions. In particular $\gamma_5 \Psi = i' \gamma_0 \gamma_1 \gamma_2 \gamma_3 \Psi$ corresponds to $\underline{i}Q k = Q\underline{i}^2 \vec{k} = -Q\vec{k}$ and $[(1-\gamma_5)/2]\Psi$ corresponds to Qu. The decomposition $\Psi = [(1-\gamma_5)/2]\Psi + [(1+\gamma_5)/2]\Psi$ corresponds to $Q = Qu + Q\check{u}$.

2. D. Hestenes has shown in [7] that, if ψ is a biquaternion Q, $j = \psi e_0 \tilde{\psi}$ is a spacetime vector, the "current" associated with ψ, whose components j_μ are written in conventional Dirac formalism $j_\mu = \overline{\Psi}\gamma_\mu \Psi$. Furthermore, if $\psi\tilde{\psi} \neq 0$, one can give to ψ the polar form $\psi = (\rho e^{\underline{i}\beta})^{1/2} R$ described in par. 1, and also, the biquaternion R is such that $\tilde{R} = R^{-1}$ and defines a Lorentz rotation. If ψ depends on the point x of M, the set of spacetime vectors $n_\mu = R e_\mu R^{-1}$, whose $n_0 = v$ is the timelike vector, constitutes a moving orthonormal frame that we will call the Takabayasi frame, or the proper frame, associated with the wave function ψ.

3. Now, we consider the independant wave functions associated with the neutrino and the electron (see [18]), as two biquaternions ψ_ν and ψ_e, the two right singlets $\psi_\nu^R = \psi_\nu \check{u}$, $\psi_e^R = \psi_e \check{u}$, and the left doublet

$$\psi^L = \psi_\nu^L + \psi_e^L \vec{i} = \psi_\nu u + \psi_e u \vec{i} \tag{11}$$

If $\psi^L \tilde{\psi}^L \neq 0$ (that could not be the case if one would have $\psi_\nu = \psi_e$), one can associate with ψ^L a proper frame $\{N_\mu = R^L e_\mu (R^L)^{-1}\}$, where R^L defines the Lorentz rotation associated with ψ^L, that we will call the proper frame of the left neutrino-electron doublet.

One can write for example, because $R^L \underline{i} = \underline{i} R^L$, and $N_\mu = R^L e_\mu (R^L)^{-1}$

$$R^L k = R^L \underline{i} e_3 e_0 = R^L e_2 e_1 R^L = K R^L, \quad K = N_2 N_1 = N_2 \wedge N_1$$

and the multiplication on the left of ψ^L by i, j, k, allows one to define three bivectors I, J, K which may be considered as the generators of the rotations in the three-space $E(x)$, generated by the spacelike vectors N_1, N_2, N_3, orthogonal at the point x of M to the timelike vector N_0. The corresponding operations of multiplication on the doublet Ψ^L, associated to the biquaternion ψ^L, are $i'\tau_1 \Psi^L, i'\tau_2 \Psi^L, i'\tau_3 \Psi^L$. So, a $SU(2)$ change of gauge corresponds to the transformation of $E(x)$ into itself, defined by a change $N_k \to S N_k S^{-1}$.

I have shown in [19] that the three spacetime vectors W^1, W^2, W^3 of the weak field are associated with the three spacetime bivectors $N_2 \wedge N_3, N_3 \wedge N_1, N_1 \wedge N_2$, respectively, in the same way that, in the Dirac theory of the electron, the electomagnetic spacetime vector A is associated with the spin bivector $n_1 \wedge n_2$ (see [8]), whose direction defines the "spin plane". Furthermore, I have established that the relation between the bivectors $\vec{W}_\mu = W_\mu^k N_k \wedge N_0$ of the Yang-Mills theory,

$$\vec{W}_{\mu\nu} = \partial_\nu \vec{W}_\mu - \partial_\mu \vec{W}_\nu + ig(\vec{W}_\mu \vec{W}_\nu - \vec{W}_\nu \vec{W}_\mu), \tag{12}$$

is a consequence of the relation, written in Clifford algebra of M,

$$\partial_\nu \hat{\Omega}_\mu - \partial_\mu \hat{\Omega}_\nu + \frac{1}{2}(\hat{\Omega}_\mu \hat{\Omega}_\nu - \hat{\Omega}_\nu \hat{\Omega}_\mu) = 0, \tag{13}$$

The bivectors $\hat{\Omega}_\mu = -2S \partial_\mu S^{-1}$ are the bivectors of the infinitesimal rotation of the frame $\{N_1, N_2, N_3\}$, in a $SU(2)$ transformation of the three-space $E(x)$ into itself. In analogy with the interpretation I have given in [8] of the movement upon itself of the "spin plane" in the electron theory, the energy associated with the doublet of the left particles neutrino-electron, would to be considered as proportional to the absolute infinitesimal rotation of the frame $\{N_1, N_2, N_3\}$, which is defined by the bivectors $\hat{\Omega}_\mu + S \Omega_\mu^L S^{-1}$, where Ω_μ^L are the bivectors which define the infinitesimal rotation of the proper frame $\{N_\mu\}$ associated with the wave function ψ^L.

On the other side, because

$$\exp(i'\varphi/2)\Psi^L \Leftrightarrow \psi_\nu \exp(k\varphi/2)u + \psi_e \exp(k\varphi/2)u\vec{i}$$

a $U(1)$ change of gauge, induces a rotation, of the same angle φ, in the "spin planes" $P_\nu(x)$, $P_e(x)$, of the neutrino and the electron.

4. As a confirmation of the validity of the biquaternionic form we have given to the doublet of the left particles, let us calculate the contribution to the GSW lagrangian density of the neutral current, written, in conventional notation (see [18])

$$L_N = \frac{g}{2}W^{3\mu}\overline{\Psi}^L\gamma_\mu\tau_3\Psi^L - \frac{g'}{2}B^\mu\overline{\Psi}^L\gamma_\mu\Psi^L - g'B^\mu\overline{\Psi}_e^R\gamma_\mu\Psi_e^R \quad (14)$$

(the right part of the neutrino does not contribute). Here,

$$L_N = \frac{g}{2}\vec{W}^3.j_3^L - \frac{g'}{2}B.j_0^L - g'B.j_e^R \quad (15)$$

where $a.b = a^\mu b_\mu$ means the scalar product of two spacetime vectors a, b, and

$$j_3^L = \psi^L\vec{k}e_0\mathring{\psi}^L = \psi^L e_3\mathring{\psi}^L, \quad j_0^L = \psi^L e_0\mathring{\psi}^L, \quad j_e^R = \psi_e^R e_0\mathring{\psi}_e^R \quad (16)$$

We introduce the spacetime vectors

$$j_\alpha = \psi_\alpha e_0\mathring{\psi}_\alpha, \quad j'_\alpha = \psi_\alpha e_3\mathring{\psi}_\alpha, \quad (\alpha = \nu, e) \quad (17)$$

which represent the "probability current" and the "spin current" associated with the neutrino and the electron (these terms are perhaps not convenient for the neutrino). Using the properties (5), (10), of u and $\tilde{u}$, and because $e_3\tilde{u} = ue_3$, $e_0\tilde{u} = ue_0$, $ue_3 = (e_3 + e_0)/2$, $ue_0 = (e_0 + e_3)/2$, $\vec{i}e_3 = e_3\vec{i}$, $\vec{i}e_0 = -e_0\vec{i}$, one obtains

$$j_3^L = (\psi_\nu u + \psi_e u\vec{i})e_3(\tilde{u}\mathring{\psi}_\nu - \vec{i}\tilde{u}\mathring{\psi}_e) = \psi_\nu ue_3\mathring{\psi}_\nu - \psi_e ue_3\mathring{\psi}_e) = \frac{1}{2}(j_\nu + j'_\nu - j_e - j'_e) \quad (18)$$

$$j_0^L = (\psi_\nu u + \psi_e u\vec{i})e_0(\tilde{u}\mathring{\psi}_\nu - \vec{i}\tilde{u}\mathring{\psi}_e) = \psi_\nu ue_0\mathring{\psi}_\nu + \psi_e ue_0\mathring{\psi}_e = \frac{1}{2}(j_\nu + j'_\nu + j_e + j'_e) \quad (19)$$

$$j_e^R = \psi_e\tilde{u}e_0 u\mathring{\psi}_e = \psi_e\tilde{u}e_0\mathring{\psi}_e = \frac{1}{2}(j_e - j'_e) \quad (20)$$

and so

$$L_N = \frac{1}{4}[-(gW^3 + 3g'B).j_e + (gW^3 - g'B).(-j'_e + j_\nu + j'_\nu)] \quad (21)$$

One remarks that the neutrino intervenes by means of an isotropic current vector $j_\nu + j'_\nu$.

Using the relations of Weinberg (see [18])

$$W^3 = \sin\theta\ A + \cos\theta\ Z,\quad B = \cos\theta\ A - \sin\theta\ Z,\quad g\sin\theta = g'\cos\theta = e, \qquad (22)$$

one obtains the contribution to the lagrangian density of the electromagnetic potential A

$$L_N^A = -eA.j_e \qquad (23)$$

in agreement with the electromagnetic properties of the electron.

4. Conclusion

By the change of a doublet of Dirac spinors into a single entity, the left biquaternion, we have transformed the mixing matrices τ_k acting to the doublet as spacetime bivectors acting on a Hamilton biquaternion. That has allowed us to introduce a new physical concept, the proper moving frame associated, at each point x of the Minkowski space time M, with the doublet of the left particles neutrino and electron of the GSW theory. We have established that a $SU(2)$ change of gauge corresponds to a rotation upon itself of the three-space $E(x)$ orthogonal at x to the timelike vector of the frame. A $U(1)$ change of gauge is a rotation, of the same angle, in the "spin planes" $P_\nu(x)$, $P_e(x)$ associated at each point x to the neutrino and the electron.

We have verified that our geometrical construction is in complete conformity with the GSW gauge model. As an important complement, it allows one to interprete the energy as associated with infinitesimal rotations of moving frame of spacetime, and is a first, but imposed, step to the unification of the theory of elementary particles with the General Relativity.

5. References

1. Costa de Beauregard O., *La Théorie de la Relativité Restreinte*, Masson, Paris, 1949.
2. Halbwachs F., *Théorie relativiste de fluides à spin*, Gauthier-Villars, Paris 1960.
3. Jakobi G. and Lochak G., *C.R. Ac. Sc. (Paris)* **243**, 234 (1956), **243**, 357 (1956).
4. Takabayasi T., *Supp. of the Prog. Theor. Phys.*, **4**, 1 (1957).

5. Vigier J. P., *Structure des Microobjets dans l'Interprétation Causale de la Théorie des Quanta*, Paris, 1956.
6. Halbwachs F., Souriau J.M. and Vigier J. P. , *J. Phys. et le Radium* **22**, 293 (1961).
7. Hestenes D., *J. Math. Phys.* **8**, 798 (1967).
8. Boudet R., *C.R. Ac. Sc. (Paris)* **272** A, 767 (1971).
9. Boudet R., *C.R. Ac. Sc. (Paris)* **278** A, 1063 (1974).
10. Hestenes D. *Space-Time Agebra*, Gordon and Breach, New-York, 1966.
11. Argand R., *Essai sur une manière de représenter les quantités imaginaires dans les constructions géométriques* (1806), Librairie Scientifique et Technique A. Blanchard, Paris, 1971.
12. Dieudonné J., *Linear and Multinear Algebra*, **8**, 1 (1979).
13. Tait P.G., *Traité élémentaire des Quaternions*, Gauthier-Villars, Paris, 1882.
14. Battey-Pratt E. P. and Racey T. J., *Int. Jour. of Theor. Phys.*, **19**, 437 (1980).
15. Lasenby A., Doran C., Gull S., in *Spinors, Twistors, Clifford Algebras*, Z. Oziewics, B. Jancewics, A. Borowiecs eds., Kluwer Ac. Pub., Dordrecht, 1993, p. 233.
16. Cartan E., *Leçons sur la théorie des spineurs*, Gauthier-Villard, Hermann, Paris, 1938.
17. Hestenes D., *Found. of Phys.* **12**, 153 (1982).
18. Elbaz E., *De l'électromagnétisme à l'électrofaible*, (Ed. Marketing, Paris, 1989).
19. Boudet R., in *Clifford algebras and their applications in mathematical physics*, F. Brackx, R. Delanghe, H. Serras, (Kluwer Ac. Pub., Dordrecht, 1992), p. 361

THE NATURE OF THE COSMOLOGICAL REDSHIFT

MARIANO MOLES
Instituto de Astrofísica de Andalucía
Apdo. 3004, 10080 Granada, Spain

Abstract

The standard cosmological view rests primarily on our interpretation of the Hubble law (the phenomenon) as evidence for the expansion of the Universe (the hypothesis). In this frame the redshift of the lines in the spectra of the extragalactic objects is considered as of a purely geometrical nature. It should only be present at sufficiently large cosmological scales where the hypothesis of homogeneity is satisfied.

In spite of the success of the evolutionary models at incorporating other important observational data, the detailed agreement is not without problems. At the same time, the basic, global tests on the nature of the metric have not yet given clear and definitive answers. Finally, the need for ad hoc assumptions about the initial conditions and scale of homogeneity, where the redshift mechanism operates, are also a major concern for that model. This situation is reviewed in the present contribution.

Based on these and other arguments, alternatives to the standard view were proposed after the discovery of the Hubble law. Their common point is to consider the redshift as a physical phenomenon operating at all scales and therefore testable, in principle, even in the laboratory. It is interpreted as a real loss of energy by the photons. That mechanism, as proposed by Vigier and collaborators, is reconsidered here and the lines along which a possible alternative to the standard model could be constructed are discussed.

1. Introduction

Cosmology is one of the domains where Prof. Vigier has applied his ideas about the physical nature of the light and of the space-time. It is true that the views presently dominant in Cosmology rest upon the interpretation of the cosmological redshift as a pure geometrical effect, i.e., in terms of the expansion of the Universe. But Prof. Vigier, following the attitude already manifested by some eminent scientist from the moment of the discovery of the Hubble law, has challenged that view. An important amount of work was thus conducted and promoted by him for the analysis of the theoretical grounds of the modern Cosmology and the relevant observational data.

The standard theory about the redshift and the Universe has proven to be successful to accommodate important observational facts, starting with the Hubble law, and even to predict some new ones. But these, important as they are, constitute consistency arguments and not a positive proof of the nature of the cosmological redshift, which should be addressed in itself and univocally established. Moreover, the situation at present is, contrarily to what is generally advocated, not completely satisfactory, since the data do not always conform, at their face value, to the predictions (Jaakkola, Moles and Vigier 1979; Sandage and Perelmuter 1990; Moles 1991a, b).

The corner stone of the Big Bang theory is indeed the interpretation of the redshift phenomenon (the z-phenomenon) in terms of the time evolution of the space-time metric. Therefore, any attempt to provide a scientific alternative to the Big Bang models should offer, as a starting point, a different explanation for the observed cosmological redshift. In that respect, the first question to elucidate is *whether the redshift is of geometrical or physical origin*. In the first case, as in the standard theory, metric considerations suffice to produce a redshift law. In the second case a specific physical mechanism has to be considered.

This question was, to some extent, already addressed by Zwicky (1929), and explicitly considered by Hubble and Tolman (1935) when trying to test the form of the space-time metric. Schrödinger (1955) would argue later that, in fact, both approaches are not so different since the geometrical explanation does not correspond to a Doppler effect but rather involves some kind of exchange between the geometry - the matter - and the photons, as for any physical explanation. We think that the difference is still there since, contrary to the second case, no specific mechanism has to be invoked in the geometrical approach to produce a redshift.

Moreover, there is in our opinion a second and more fundamental difference between the philosophies underlying both approaches, that of *the scale of the redshift phenomenon*. Indeed, it is obvious that it has to be present at all levels, including the laboratory, in the case it has a physical origin. Then, it could be tested, at least in principle, as any other physical phenomenon (Moles 1978; Vigier 1990). In the geometrical approach the redshift phenomenon, the Hubble law, should only be present at large, cosmological scales, where the hypothesis of homogeneity can be considered as valid and the Friedman -Robertson - Walker metric is of application.

On the observational domain two main ways to test that theory can be devised. The first is just to seek for counterexamples, i.e., situations where the redshift could be found to depend not only on the distance but also on physical properties of the astronomical objects under consideration. The second way is more fundamental and tries to unveil the nature of the metric itself through the so called global tests. Both approaches have produced interesting and, to say the least, controverted results that are still there to remember the complexity of the problem and the importance of the issue (Arp 1987; Sandage 1987, 1988; Sandage and Perelmuter 1990; Moles 1991a).

The question of the nature of the redshift is, in our opinion, not yet solved. It constitutes one of the main issues in the domain of the physical sciences since it would address the fundamental properties of the light and matter. In the present contribution we review some of the salient aspects selected to show the alternatives of a debate that cannot yet be considered as closed.

2. The Geometrical Approach to the z-Phenomenon

A.- The interpretation

It was obvious just after its discovery that the Hubble law can be accommodated within the frame of the evolutionary world models. Just some general considerations about the metric corresponding to an homogeneous and isotropic space-time suffice to produce a redshift law (Robertson 1955),

$$(1+z) = \frac{R_0}{R} \qquad (1)$$

where R is the scale parameter and the subscript 0 refers to the present epoch, provided that $R_0 > R$. It is at once clear from (1) that z is not a Doppler effect since it is only a function of the scale parameter and its evolution with time, as argued by Schrödinger (1955). In that sense it can be said that *the redshift drives the expansion*.

The same metric considerations serve to formulate the global tests, that are not dependent on the gravitational theory used, but just on the form of the metric. At the same time, it follows from (1) that z cannot depend on the properties of the astronomical sources under consideration. These considerations show the way to test the starting hypothesis about the metric of the space-time.

Before reviewing the situation on the observational side, we would like, at the conceptual level we are considering here, to focus the attention on two aspects which appear rather confuse in the standard frame, namely the question of the scale of the phenomenon and that of the initial conditions.

2.1. THE SCALE QUESTION

Since z is only dependent on the metric of the model, it can only be expected to appear at the level where that metric is a good representation of the reality. In other words, a genuine Hubble-like law should be found only at the scale where the Universe starts to be homogeneous and isotropic. In that sense it should be a pure cosmological effect, only present at large scales. This question was explicitly and lively posed by Misner, Thorne and Wheeler (1973). The point is that the phenomenon cannot be present at all scales since then it would not be observable. Quoting their words, ".... only distances between clusters of galaxies and greater distances are subject to the expansion".

Indeed, the observations indicate that the observed Universe cannot be considered as homogeneous at scales below 100 Mpc. *Why then the redshift law is already explicit and neat within 2 Mpc* (Sandage 1988) *where the Universe is grossly non-homogeneous?* It is a difficult problem the large scale behaviour of an Universe that is highly inhomogeneous at more local scales, but it seems clear in any case that the redshift law should be extremely distorted at so small scales, or even non existing.

This is an aspect that has not yet received an adequate answer. That kind of *omnipresence of the redshift law* is in our opinion a problem for the geometrical approach and would plaid for a physical origin of the cosmological redshift.

2.2. THE INITIAL CONDITIONS QUESTION

We would only mention here the well known difficulties met by the standard evolutionary theory to justify the assumption of homogeneity at large (the horizon problem) and the non-trivial value of the density parameter (the flatness problem). The efforts made in the context of the inflationary ideas are there to confirm the importance of a problem that was ignored for a long time. On the other hand, the lack of satisfactory answers until now (see for example, Padmanabhan 1993) explicitly shows the depth of the questions so naively posed.

It seems that evolution and large scale homogeneity together, constitute a too strong condition, difficult to reconcile with the observations. It is however not the homogeneity condition which is in question, but the evolutionary character of the model since it keeps signature of the unphysical character of the initial state in the later phases.

3. The Observations

We feel that theoretical considerations could be enough to pose the problem of the nature of the redshift. But it would be of no more than academic interest if the observations were unanimous in supporting the standard ideas. The last word is to be given, here too, to the well established observational data.

In the frame of the Big Bang model one should expect that the values obtained for the so called cosmological parameters would offer a consistent frame. An earlier analysis by Jaakkola, Moles and Vigier (1979) did show that it was not the case at that time. And the situation today is found to be still the same. The problem of the consistency between the Universe age scale and the age of some stellar systems is an old one, not yet well solved. In fact, acceptation of the high value of the Hubble constant implies the need for a large, positive cosmological constant to avoid a too young Universe (Moles 1991a, b).

Regarding the global cosmological tests, the picture offered by the observational results is that the data, at their face values, do not confirm the standard predictions. Even more, they fit rather well the predictions for a static metric (Sandage 1988; Moles 1991a). It is as if the invoked evolutionary effects were to mislead the astronomers producing a false impression corresponding to a rival metric.

We have recently proposed (Kjaergaard, Jörgensen and Moles 1993) a new method to carry out the surface brightness test without being affected by any evolutionary effect. We hope the results will produce a clear answer in the next years.

We said before that the geometrical, standard case would be in serious danger if any dependence of the observed redshift on properties of the objects were observed. This way of testing has produced many results along the last 30 years (see Arp 1987 and Moles 1991a, for references). In our opinion the accumulated results would indicate an enhancement of the redshift magnitude with the matter density, in some unknown way. But the situation is not easy to decide since it is always difficult and uncertain to ascertain the distance to an astronomical object (see for example Moles et al, 1994, for a discussion along these lines). The situation is somewhat more comfortable when the properties of the objects belonging to a well defined physical association are analyzed. But even in that case the arguments are not completely unambiguous and therefore not enough to convince the scientist otherwise sure of the correctness of the standard model. It is true that the consideration of all those cases is somewhat irritating for the standard views, but not much more can be extracted in our opinion. Of course, was a new paradigm incorporating those possibilities discovered, they would

4. The Physical Approach to the z-Phenomenon

In this case the metric is supposed to be neutral with respect to the observed cosmological phenomena, that have to be explained not on geometrical but on physical terms. For the redshift, the starting consideration is that it represents a real loss of energy by the photons. This energy can go either to other particles or be dissipated in the vacuum.

In some cases the physical mechanism refers to some explicit interaction between the geometry and the photons (Crawford 1979; La Violette 1986). Most cases however put the problem at the photon level to produce a redshift law.

Direct observations are enough to impose some general conditions that any proposed redshift mechanism should satisfy in order to be acceptable in the cosmological domain. These are (Zwicky 1929; Moles 1978):

(i) It is a redshift, i.e., most generally the process implies a loss of energy for the photons. Moreover, when spectral lines are considered, the shift has to be much larger than the broadening caused by the same process.

(ii) The fractional energy loss, $\delta\nu/\nu$, is independent of the energy of the primary photon, in correspondence with the observed independence of the redshift magnitude on the frequency.

(iii) The mechanism produces a negligible deviation of the light rays. This is necessary to avoid the blurring out of the astronomical objects at relatively short distances, and also to avoid the objection raised by Schrödinger (1955).

(iv) Finally, the mechanism is aimed to produced a Hubble law, independently of any metric consideration.

We note that the law can be formulated in differential terms,

$$\frac{\delta\nu}{\nu} = -K\delta r \qquad (2)$$

what reflects the existence of the effect at all levels.

Different mechanisms have been proposed to explain the Hubble law. Among them we would like to consider here those not related with the geometry at any level, the Vacuum induced mechanisms and the Collisional mechanisms. It is important to note that all proposed mechanisms involve new hypothesis not always easy to justify. In this respect the situation is still similar to that depicted by Hubble and Tolman (1935) when they stated that ".... if the redshift is not due to recessional motion, its explanation will probably involve some quite new physical principles". Most of the mechanisms have to introduce, among others, the hypothesis of a non zero rest mass of the photon. This has been shown to be fully compatible with the observational and experimental data (Bass and Schrödinger 1955; Moles and Vigier 1974), and is one of the driving lines of de Broglie and Vigier ideas.

5. A Vacuum induced mechanisms

Several heuristic proposals have been advanced on the possibility for the photons to feel a drag force exerted by the vacuum, that could produce a redshift. Recently Vigier (1990) has taken advantage of his conceptions on a stochastic Dirac-like vacuum to work out one of such possibilities. This vacuum would have a non zero conductivity, σ_0, that could cause the velocity of the plane waves to depend on the frequency. This is equivalent to consider a photon with a non-zero rest mass, with

$$m_\gamma = \frac{\sigma_0 h}{2\sqrt{2}\pi c^2 \epsilon_0 \chi_e} \qquad (3)$$

where h is Planck's constant, ϵ_0 the vacuum dielectric constant and χ_e the dielectric relative constant.

The analysis of the propagation of such a photon in the vacuum shows that there is an energy loss that only depends on the travelled distance, with an essentially zero angular deviation, i.e., a Hubble law. Application of the uncertainty principle allows the connection between the Hubble constant, i.e., the propagation properties of the vacuum, and the photon rest mass,

$$m_\gamma \approx \frac{hH}{2c^2} \qquad (4)$$

5.1. B COLLISIONAL MECHANISMS

In that case the energy lost by the photons is carried out by other particles. The approach was started with the heuristic proposal made by Finlay-Freundlich (1953, 1954a, b), where some kind of photon-photon interaction (see also Born 1953, 1954) was invoked. The specific formula proposed by Finlay-Freundlich was found to be not appropriate, but it was used to connect the redshift phenomenon with the existence of a (then hypothetical) background radiation.

Pecker, Roberts and Vigier (1973) re-elaborated the proposal but some important problems still remained. Later Moles and Vigier (see also Marič, Moles and Vigier 1976, and Moles 1978) made a specific proposal, where the photons transfer energy to some hypothetical, ϕ, particles,

$$\gamma + \phi = \gamma + \phi \qquad (5)$$

In order to satisfy the necessary conditions it was shown that this particle has to be much lighter than the electron, and much heavier than the photon. In the frame of de Broglie's theory of light, the photon and the ϕ particle would correspond to the vectorial and pseudoscalar bosons obtained by the fusion of two spin 1/2 particles. The proposed coupling produces a cross section given by

$$\frac{d\sigma}{dE} = K \frac{E - T}{ET} \qquad (6)$$

where E is the energy of the incident photon and T the transferred energy to the ϕ particle. Expression (6) corresponds to a forward peak scattering, thus avoiding the blurring problems.

The mean fractional energy loss is given by

$$\frac{T}{E} = \frac{K}{2\sigma_t} \qquad (7)$$

which shows that the redshift is independent of the frequency of the incident photon. For a photon traversing a "ϕ-bath" with density $\rho_\phi(r)$, the energy loss is given by

$$\frac{dE}{E} = \frac{\delta\nu}{\nu} = -\frac{K}{2}\rho_\phi(r)dr \qquad (8)$$

Integration along the path yields the redshift law,

$$ln(1 + z) = \frac{K}{2} \int_0^D \rho_\phi dr \qquad (9)$$

which is usually approximated by its linear expansion, i.e., the Hubble law.

Some interesting aspects of such collisional mechanisms are:

(i) It directly depends on the distribution and density of the redshifting particles. Thus, apart from the general Hubble effect, it can account for local effects related either with the light source itself or with the ϕ-density irregularities along the photon path. In general we'll have

$$(1 + z_t) = (1 + z_i)(1 + z_p)(1 + z_H) \qquad (10)$$

The different terms are integrations along the distance travelled by photons in a local bath (z_i), or in localized enhancements along their path (z_p), or all along the distance from the source to the observer, through an smoothed general bath (z_H). In general, the pure distance effect, the Hubble term z_H will dominate, but it could be that the local contribution could be relevant for not too far away objects. On the other hand, light arriving through matter enhancements seem to present a higher redshift than objects situated at the same distance but without such enhancements along its path (Jaakkola et al 1975). This result would support the idea of the existence of the z_p term in (10), as far as the density of ϕ-particles would be directly related with the matter density. In any case, this is a property that can be addressed observationally in a rather unambiguous way.

(ii) The energy carried out by the redshifting particles could reappear in the form of a background radiation. Note that at moderately high redshift, the energy density lost by the redshift effect is equal to the radiated energy, and not far from the energy density of the background radiation (Sandage 1961; Moles 1978). Thus it is possible to conceive rather simple mechanisms to explain the background radiation as a manifestation of a kind of large scale equilibrium.

(iii) Most important, the proposal could be tested at the laboratory level. Such a proposition was made by Moles (1978) using a radiosource that gets occulted by the Sun. A systematic increasing of the observed redshift of the source is predicted as the projected distance between the source and the Sun diminishes. The form of the effect is simply predicted assuming that the hypothetical ϕ particles are distributed as a function of the solid angle around the Sun, supposed to be their source. For the magnitude of the effect, it can be also predicted if the idea that the mechanism is the same operating at cosmological scales is accepted. In that case it would be observable in the proposed experience.

6. Some Considerations About Alternative Approaches

It seems difficult for the moment to produce a change in the dominant cosmological views on observational grounds only. It is as if the data had always the last word but not always the first one. Thus, it seems necessary to introduce a new paradigm, general enough to incorporate new possibilities an explanations. After all, it is often not too difficult to dismiss the importance of results not well fitted by the standard predictions if there are no alternative explanations.

The point is indeed that of the nature of the redshift. And, as for Hubble and Tolman, a new way to explain it has to involve new physical ideas, most probably touching the nature of the light. In this sense, the theories invented by de Broglie, Vigier and others offer a useful frame to new approaches. Those we have reviewed are examples of such kind of work.

Standard Cosmology intends to offer a global answer to the problem of the final nature of the Universe at large scales. The starting point is always some kind of cosmological principle, upon which the whole theory is built up. For the standard theory this is indeed the (Imperfect) Cosmological Principle, stating that the Universe is the same for an observer placed at any place, but not necessarily at any time. In fact, it is the evolutionary character of the Universe what makes possible a geometrical explanation to the redshift law, upon which the whole cosmological theory is built.

In the standard approach, the structure of the whole dominates over and explains the properties of the subsystems. The redshift is but a manifestation of that underlying structure. The background radiation or the (cosmic) He abundance are also considered as properties of the whole, manifested at all scales. But it is obvious that the universality of those and other properties is an intrinsically unverifiable claim. This could confer to the Cosmology, in the way it is now developed, a high degree of arbitrariness and some mythical flavour .... (see Alfvén 1977, for a radical view on that question).

Instead, the reinterpretation of the redshift in terms of some physical mechanism would allow a new way to approach the Cosmology. Now the observed properties should be explained through mechanisms that could be tested at the laboratory. The structure of the whole would be permanent and neutral with respect to the observed phenomena and would serve as some kind of consistency frame. The Perfect Cosmological Principle, which sounds more satisfactory in many respects (Bondi & Gold 1948) could be adopted. This would mean the strict equivalence of all the observers at any time.

The space time metric we propose is indeed that of the Einstein static model. It is the only compatible with the Perfect Cosmological Principle and without offering a geometrical explanation to the z-phenomenon. The model should be implemented with a physical redshift law similar to that discussed in the previous paragraph. The explanation of other cosmological facts like the background radiation or the He abundance should be found on pure physical grounds, probably reflecting the equilibrium mechanism operating at large scale. In that frame the problem of the origin is changed precisely into the problem of those large scale equilibrium mechanisms.

Some different arguments have often been used to discard the Einstein static world model, namely the Olbers paradox and its instability against gravitational perturbations. Both arguments are however inadequate.

Indeed, it has been shown (Harrison 1981; Wesson 1991) that the sky is dark at night just because the stars have finite lifetimes and the speed of the light is finite. The expansion of the Universe is not necessary to solve the paradox and in fact plays a very minor role in reducing the sky background brightness. The night sky can be dark even in a non-expanding Universe. As pointed out by Harrison, to illustrate that point it suffices to think that even if all the matter in the Universe was suddenly converted into radiation, the equilibrium temperature would hardly reach 20K!

On the other hand, Bonnor (1955, and references) has shown that a static Universe can accommodate inhomogeneities at any scale. It is true that the growing of those irregularities poses some problems but, as for the Olbers paradox, the claim that the Einstein metric should necessarily evolve into a expanding or a contracting Universe is not necessarily true (see also Moles 1991b).

Consequently, a Cosmology based upon physical (i.e., non geometrical) explanations of the relevant facts, and erected upon Einstein's static metric is well possible. The central, new, element would indeed be the interpretation of the redshift phenomenon in new terms, most probably involving new views and ideas about the Quantum Nature of the Light. We feel that the pioneering work of de Broglie, Vigier and others is still full of promises even in the cosmological realm.

References

Alfvèn, H.O. 1977, in Cosmology, History and Theology, eds. W. Yourgrau & A.D. Breck, Plenum Press, N.Y., USA
Arp, H.C. 1987, Quasars, Redshifts and Controversies, Interstellar Media, California, USA
Bass, L. & Schrödinger, E. 1955, Proc. Roy. Soc., A232, 1
Bondi, H. & Gold, T. 1948, Mont. Not. Royal Astron. Soc., 108, 252
Bonnor, W,B. 1955, Mont. Not. Royal Astron. Soc., 115, 310
Born, M. 1953, Göt. Nachr., 7, 102
Born, M. 1954, Proc. Phys. Soc., A67, 193
Crawford, D.F. 1979, Nature, 277, 633
Finlay-Freundlich, E. 1953, Göt. Nachr., 7, 95
Finlay-Freundlich, E. 1954a, Proc. Phys. Soc., A67, 192
Finlay-Freundlich, E. 1954b, Philosophical Magazin, 45, 303
Harrison, E.P. 1981, Cosmology, Yhe Science of the Universe, Cambridge University Press, Cambridge, USA
Hubble, E. & Tolman, R. 1935, Astrophys. J., 82, 302
Jaakkola, T., Karoji, H., Le Denmat, G., Moles, M., Nottale, L., Vigier, J.P. & Pecker, J.C. 1975, Mont. Not. Royal Astron. Soc., 177, 191
Jaakkola, T., Moles, M. & Vigier, J.P. 1979, Astron. Nach., 300, 229
Kjaergaard, P., Jörgensen, I. & Moles, M. 1993, Astrophys. J., 418, 617
La Violette, P.A. 1986, Astrophys. J., 301, 544
Marič, Z., Moles, M. & Vigier, J.P. 1976, Astron. Astrophys., 53, 191

Misner Ch.W., Thorne, K.S. & Wheeler, J.A. 1973, Gravitation, Freeman & Co., USA
Moles, M. 1978, Thèse d'Etat, Univ. Paris VI, France
Moles, M. 1991a, in The Physical Universe: The Interface between Cosmology, Astrophysics and Particle Physics, ed. J.D. Barrow, A.B. Henriques, M.T.V.T. Lago and M.S. Longair, Springer-Verlag, Heidelberg, p. 197
Moles, M. 1991b, Astrophys. J., 382, 369
Moles, M. & Vigier, J.P. 1974, C. R. Acad. Sci. Paris, 278B, 969
Moles, M., Márquez, I., Masegosa, J., del Olmo, A., Perea, J. & Arp, H.C. 1994, Astrophys. J., 432, 135
Padmanabhan 1993, Structure Formation in the Universe, Cambridge University Press, Cambridge, U.K.
Pecker, J.C., Roberts, A.P. & Vigier, J.P. 1973, Nature, 241, 338
Robertson, H.P. 1955, Pub. A. S. P., 67, 82
Sandage, A.R. 1961, Astrophys. J., 133, 335
Sandage, A.R. 1987, in IAU Symp. No. 124, Observational Cosmology, ed. A. Hewitt, G. Burbidge and L.A. Fang, Reidel, p. 1
Sandage, A.R. 1988, Ann. Rev. Astron. Astrophys., 26, 561
Sandage, A.R & Perelmuter, J.-M. 1990, Astrophys. J., 350, 481
Schrödinger, E. 1955, Nuovo Cimento, Vol. 1, N. 1, 63
Vigier, J.P. 1990, IEE Trans. on Plasma Science, 18, 64
Wesson, P. 1991, Astrophys. J., 367, 399
Zwicky, F. 1929, Astrophys. J., 15, 773

A QUANTUM *DIGITAL* THEORY OF LIGHT

N. V. POPE
*Department of Mathematics, Keele University,
Keele, Staffordshire. ST5 5BG, U.K.*

Abstract

In the Theory of Relativity, anything travelling at the speed of light registers no intrinsic (i.e., proper) time nor distance in getting from A to B. In that same theory a photon is a quantum of energetic interaction between one atom and another, travelling at precisely that speed, c. So, logically, as Gilbert Lewis pointed out in 1926, according to Relativity, a photon, in getting from A to B takes no time and covers no distance.

In Quantum Theory, also, there can be no intrinsic time nor distance in a quantum interaction. That interaction has to be instantaneous, because any pause between its emission and absorption would break the law of conservation of action, which forbids that the interaction can be held in suspension for the merest instant, let alone for astronomical periods of time. Moreover, if the quantum is thought of as truly integral and irreducible, then it cannot be supposed to have energy to spare for *en route* detection between an emitting and an absorbing atom. In that case, putting a detector between the emitter-atom and the would-be first absorber merely creates an alternative absorber ... and so on. So in that theoretical context there can be no question of our being able to detect or measure the progress of a quantum from A to B *in vacuo*.

Logically, then, Relativity and Quantum Theory are in perfect accord, not only with each other but also with the law of conservation and with Newton's third law of action and reaction (which, like the conservation law, brooks no time-delay). It would therefore appear that for the first time in our history we have the situation in science where everything 'gels'. Yet our accustomed way of thinking about space somehow manages to make this all-attested spatially immediate and instantaneous quantum interaction between distance-separated atoms seem a mystery.

The answer we propose is to switch, at the quantum level of physics, from the usual 'analog' or 'realistic' mode of conceptualisation into a purely 'digital' mode. In other words, we argue that at that ultimate datum-level of all observational/instrumental perception there can be no real, universally self-extended 'vacuum' in which light travels, as waves, or particles, 'real', 'virtual' or whatever. Empirically speaking, all we have at that datum-level of physics are those purely stochastic or probabilistic (in any case, informational) quantum *events* which in optical contexts we rename *phota* (singular: *photum*). We argue that it is out of purely informational patterns and sequences of these data that the dimensions of physical space and time (space-time) and all that is described in those terms are relativistically projected – including the so-called 'propagation' and 'bending' of light.

In this way we see no real conflict between the 'Realist' and the 'Operationalist' (or 'Positivist') approaches to theoretical physics as followed, respectively, by Einstein, Bohm and Vigier on the one hand and the Bohr, or Copenhagen, group on the other. That is, instead of seeing 'Realism' and 'Positivism' as competing world-views we see the need for a proper distinction to be maintained between two radically different uses of language applied to two radically different yet complementary concept-categories. To think of 'photons' as objects 'travelling' in the space that is relativistically projected out of those very same bits of data confuses those categories – like thinking of the letters in the word MOTION as themselves having motion. That logical category mistake, we believe, remains the central source of so many unnecessary partisan conflicts, conceptual conundrums and paradoxes which, for far too long, have inhibited progress in our understanding of nature.

1. Introduction

Some physicists will always believe that philosophy has nothing to do with physics. Yet among physicists there are currently to be distinguished two clearly opposed *philosophical* approaches to the subject. On the one hand is the 'Realist' approach, as followed, for instance, by Einstein, Vigier and Bohm, and on the other hand there is the 'Positivist' approach, of the Copenhagen group, after Niels Bohr. Nature, of course, knows no such distinction. So we at Keele University, in England feel free to propose a *third* philosophical alternative, called Normal Realism,[1] which dispenses with both 'Realism' and 'Positivism' as competing world-views and explores an entirely neutral course of theoretical enquiry, the essence of whose contribution to physics is explained as follows.

2. The *Relative* Aspect of Light

Some see *relativity* as no more than an addendum to classical physics. Even Einstein did! He talked about relativity but took it as axiomatic that light travels in a vacuum in the classically conceived way at a finite and constant, *absolute* speed c. One may, however, quite logically and legitimately dispense with that 'travelling' idea of light and settle for thinking of light as simply *what we see*. In other words, we may quite sensibly think of *all space and all travelling as in the light* rather than of light as travelling in space. In this different way of thinking, space, with all that happens in it, is a *phenomenon* projected by the observer out of informational patterns and sequences of perceived bits of illumination, similarly to the way in which a video drama is projected by the viewer out of patterns and sequences of screen pixels.

Now the 'pixels' out of which we project physical phenomena are customarily called *photons*. However, since the word 'photon' suggests a space-travelling particle, we prefer to call it a *photum*, which is more suggestive of a pure optical *event*. A photum, then, is the ultimate irreducible component of physical phenomena, a quantum of luminosity which is the informational element of the light-patterns we ordinarily see with our eyes and/or by means of optical instruments. The energy a photum manifests is, in the usual way, $h\nu$, where ν is the reciprocal of the photum's characteristic and irreducible spectral period. Patterns of *co-occurring* phota are the objects of ordinary perception (phenomena). The comparative sizes and intensities of these objects, plus their angular or directional features such as perspective and parallax provide the optical information out of which we 'video-project', as it were, the three dimensions of ordinary space, as shown in Figure 1.

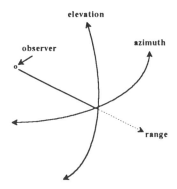

Figure 1. The observocentric (polar) dimensions of space.

Phenomena, then, are essentially *three*-dimensional, and we know that from any location we may actually or instrumentally adopt in those three dimensions the world is distributed in this same *observocentric* or spherical-polar way. Let us call those other material centres *observicles*. By both commonsense and standard mathematical procedures, therefore, an observer-*de*centralised space is interprojected between these observicles with the *non*-polar, or Cartesian, rectangular frame-type dimensions shown in Figure 2

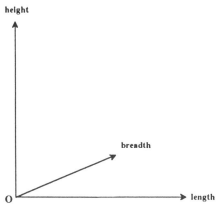

Figure 2. The observer-decentralised (Cartesian) dimensions of space.

For an object to change its position relatively to a centre (observicle) in this communally extrapolated phenomenal space, some measure of *time* has to pass relatively to that centre. We can therefore reinterpret Einstein's 'constant speed of light *in vacuo*' in Bondi's more economical and empirical, radar-based way,[2] simply as a *conversion-factor* interrelating observational distance-measure and time-measure, such that every metre an object travels away from or towards an observicle adds or subtracts 3.3 nanoseconds of time to its duration relatively to that centre – commonly known as the Doppler effect.

We may now see just how easily and naturally *relativity* follows from this. To begin with, the *interdependence* of distance and time signified by the constant c means that time-measure is related to distance-measure as another (fourth) geometrical dimension, so that the distance-measures and time-measures defining the *motion* of a body are *geometrically* related as in Figure 3.

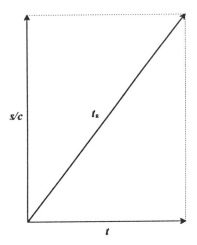

Figure 3. The observational dimensions of motion.

Here, s/c is the distance-time, *in seconds*, moved by the object and t is the time (also *in seconds*) registered by the object itself – say, a clock – in traversing that distance. The resultant t_R of these orthogonal time-components is obtained by Pythagoras, as shown in the diagram.

What we have here then, is a perfectly simple expression of relativistic time-dilation in *purely observational* terms of the distance-time *seen* to be travelled by the object and the proper time the object is *seen* to register in travelling that distance with *no need whatsoever to think of light travelling invisibly in a vacuum*.

That this really *is* relativistic time-dilation can be seen by taking the Pythagorean distance-time-resultant t_R in Figure 3, viz.:

$$t_R = [(s/c)^2 + t^2]^{1/2} \qquad (1)$$

and substituting for s, in that formula the standard equivalent in terms of relative *velocity*, i.e., $s = vt_R$, which produces

$$t_R = [(vt_R/c)^2 + t^2]^{1/2} \qquad (2)$$

By ordinary simplification this becomes

$$t_R = t\,[1 - (v^2/c^2)]^{-1/2} \qquad (3)$$

which is, of course, Einstein's famous time-dilation formula, now derived from ordinary observation and plain Pythagoras.

3. The Quantum Aspect of Light

In 1926, a chemist called Gilbert Lewis[3] discovered that Einstein's assumption that light *travels* creates a strange paradox. For, if v in (3) is put equal to c and the equation is solved for the proper-time t, then we have $t = 0$. So the *proper-time* of any light-interaction, over any distance whatsoever, is *nil* – which means that far from travelling at 'the finite speed c', light-interaction is intrinsically *instantaneous*. As Lewis put it, in a one-way quantum light-interaction the emitter and absorber are physically *touching*. So, what *vacuum* can there possibly be for light-quanta to 'travel' in? The emission and absorption of a quantum are not two events separated by a vacuum; they are *one and the same event*. Space, therefore, is not something that divides a quantum. It is a relativistic information-projection out of observational sequences of whole, indivisible quanta. (As already stated, it is the space that is in the light, not the light that is in space.)

The old dualistic distinction, then, between physical space and observational space, which makes the one 'real' and the other 'subjective' disappears, leaving the two to be regarded as equally real aspects of the same uniformly *phenomenal* space. In this space, all quantum interactions between objects, be they called electromagnetic, gravitational, nuclear or whatever, take place instantaneously – that is, *contiguously*, in accordance with Newton's third law of equal and opposite reaction. For example, in observing a distant star, that object and the observer (or observing instrument) share the same immediate and reciprocal quantum 'jolt'. Of course, this contact is not continuous. That would mean, absurdly, that everything was perpetually jammed together, as in what the Greeks conceived as the 'Eleatic One'. There is no such absurdity in our thesis, in which these contacts are piecemeal and sporadic; that is, discrete and indeterminate – in a word, *probabilistic*. This does not mean that atoms randomly skip across space to touch one another and then return to their places, as in some superluminal barn-dance. To avoid that further absurdity is to see it all consistently the other way around; that is, to think of the distances between bodies in the way that has been described, as *informationally projected* out of sequences of whole quantum events, not as some primordial, pre-extended 'void' that impossibly splits a quantum in two and separates the one half from the other.

Such, then, is space, not a continuous vacuum, field or plenum, nor even a four-dimensional space-time continuum of the sort Einstein and Minkowski envisaged, but a four-dimensional *discretum*, a pure *probability-potential* for instantaneous quantum contact. This answers to what has been called, or, rather, miscalled, 'action-at-a-distance' – miscalled because since quantum interaction is contiguous there is no *distance* involved at that quantum level. In this quantum discretum, as in the Random Access Memory (RAM) of a computer, all informational elements are 'equally close'. This gives us a *phenomenalistic* paraphrase of what David Bohm called 'quantum potential' but minus Bohm's 'hidden variables' and 'implicate' continuum' – and (it has to be said) minus his philosophically out-dated distinction between 'Realism' and 'Positivism'.

Nor, in our alternative paradigm, is there any continuous, overall synchrony of existence. Other than in those sporadic and indeterminate, instantaneous quantum touchings, an 'elementary particle' exists (if that is the word) in isolation so profound that it can scarcely even be called 'locality'. So any question of what a fundamental particle *is* or *does* in between any two of its phenomenal manifestations is meaningless. This explains why those particles are undetectable between source and screen in 'interference' experiments of the sort discussed by people like Franco Selleri and Dipanka Home.[4]

Reality, then, as we interpret it, is no behind-the-scenes classical mechanism. It is a communally interprojected, four-dimensional *phenomenon*. The quantum *discretum* into which this analyses-out, at the microphenomenal level has the same dimensional characteristics as Einstein's space-time continuum but, of course, without the continuity and without the 'implicate' suggestion of a substratum. It has three *statistical* Cartesian dimensions and a purely *statistical* GMT. And since it dispenses with the 'separability' entailed by Einstein's 'velocity' interpretation of light, and since it is couched in terms of proper-time, not relative time, it has built-in 'action-at-a-distance'. In that information-projected, super-fine-grained, interactive 'hologram', the concept of waves or virtual particles *in the vacuum*, as the conveyor of information from one place to another, is replaced by that of *pure probability-functions* of Lewisian 'quantum touchings'. These propagate, among normally distributed observicles, *kinematically* (as in cinematography) with the mathematical wave-like characteristics described by de Broglie and Schrödinger. And since the distances separating these observicles are also times in Bondi's ratio of units c, those probability-waves *travel* in the communally information-projected space at the finite and constant speed c, with all those features of refraction, diffraction, reflection and so on known to the science of optics. So light *does* travel, but *kinematically*, not mechanically – that is, as the purely sequential propagation of instantaneous information-transactions between observicles, not as streams of space-travelling photons – and never, *never in* a void.

Now there is no contradiction in saying that quanta are instantaneous and that *waves* of them travel at the finite speed c. In any wave-phenomenon the way the *elements* of the wave move and the way the *wave* moves are never the same. The molecules of a water-wave, for instance, circle about a fixed point, while the wave travels along the surface with a uniform longitudinal speed. The fact, then, that the elements of a *light*-wave are instantaneous quantum touchings and that *informational sequences* of these *propagate* kinematically at the finite speed *c* presents no paradox whatsoever. Nor is there any so-called 'EPR paradox' in the fact that there are 'spooky superluminal influences' balancing the actions of bodies at a distance. Of *course* there are superluminal influences. There *have* to be, by the laws of physics. This balance is maintained by the instantaneous *quantum potential* linking things together *holistically* – not continuously, in the manner of the classically conceived 'void', 'field' or 'æther', but *dis*-continuously; that is to say, *discretely*, or *digitally*, on a strictly conserved and immediate, one-for-one basis of quantum exchange.

In the space of this article it is not possible to explain all the details of this new and radical *phenomenological* synthesis of relativity and quantum theory.[5] The theory may, however, be summarised by saying that it incorporates the four following design-elements: i) *phenomenalism* in the *anti-Positivistic* works of Austin,[6] Ryle[7] and (the later) Wittgenstein;[8] ii) Lewis's description of *quantum touching*; iii) Bondi's interpretation of *c* as a distance-time *conversion-factor* instead of Einstein's 'velocity of light' and, iv) Occam's Razor, which bids us, regardless of convention, to select the simpler of any two theoretical alternatives. We should always be wary of convention, anyway. Once it saw angels as travelling intermediaries between heaven and earth. Lately it has seen photons as travelling intermediaries between things and our perceptions of them. John Locke, when presented with this view of light as what he called 'travelling globules', declared that it was as though we had been told that light was 'nothing but a company of little tennis-balls which fairies all day long struck with rackets against men's foreheads'. Like Locke, we at Keele are all for abandoning these mystical intermediaries and bringing reality right 'down to earth'. For us this means seeing and feeling things directly, as informational complexes like the patterns of light and shade out of which we communally and holographically (in any event, relativistically) 'video-project' the ordinary everyday world. To think of these complexes analysing-out, ultimately, to quantum light-pixels or *phota* is surely not difficult nowadays, seeing that we know how the space, time and action of a holographic video drama can be projected out of informational sequences of optical blips.

Besides, what alternative is there but to embrace the current view of light as made up of 'photons' travelling inscrutably and telepathically *at the same finite speed with respect to all differently moving observers in an infinitely extended and self-sufficient nothing, or 'void'*. Surely, angels and fairies are easier to believe in than that!

Notes and References

1. Pope, N.V., 1994 'Normal Realism: A Challenge to Physicists', *Proceedings* of the 15th Annual International Meeting of the Alternative Natural Philosophy Association (Sept. 1993), pp.111-115
2. Lewis, G.N., 1926, *Nature* 117, 2937, p.236.
3. Home, D., 1989, *Proc. of the Third Internl. Symp. on the Foundations of Quantum Mechanics*, p. 43
4. Bondi, H. 1965, *Assumption and Myth in Physical Theory*. Cambridge University Press, p. 28.
5. More detailed accounts of this thesis may be found in:
    Pope, N.V., April 1987, 'The Overdue Revolution. *MENSA, pp 28-29.*
    Pope, N.V. and Osborne, A.D., 1987, 'A New Approach to Special Relativity', *International Jnl. Mathematical Education in. Science and. Technology.* **18** (2) 191.
    Pope, N.V. 1989, 'The New World Synthesis', *Philosophia Mathematica (II),* **4**, (1) pp.23-28
    Pope, N.V. and Osborne, A.D., 1992, 'Instantaneous Relativistic Action-at-a-Distance,' *Physics Essays* **5**

(3) 409-421.

Pope, N.V. and Osborne, A.D., 1995, 'Instantaneous Gravitational and Inertial Action-at-a-Distance.' *Physics Essays*, **8**, 3. (in press).

5. Austin, J.L., 1962, *Sense and Sensibilia*, Oxford University Press.
6. Ryle, G. 1963, *The Concept of Mind*, Penguin, London
7. Wittgenstein, L.. 1967. *Philosophical Investigations*, Blackwell, Oxford.

TOWARD A COMPREHENSIBLE PHYSICAL THEORY: GRAVITY AND QUANTUM MECHANICS

A Modern Synthesis

HÜSEYİN YILMAZ
Hamamatsu Photonics K.K.
Hamamatsu City, 430 Japan
Electro-Optics Technology Center, Tufts University
Medford, MA 02155 USA

It is indicated that, of the two fundamental theories of physics, namely, the quantum theory of fields and the spacetime theory of gravitation, the former is mathematically *underdetermined* and the latter *overdetermined*. When properly revised, the two theories seem to merge into a geometrically based general framework theory free of the usual difficulties. In particular, a finite and viable quantum theory of gravity satisfying all necessary correspondence requirements seems to be possible. The result, however, still suffers from the usual lack of intuitive comprehension of quantum processes. In this article an effort is made to remedy the latter situation via a unitary concept of a field quantum we call *wave-particle unity*.

1. Introduction: Three Major Problems

Three major problems of contemporary physical theory seem to be:

 (1) Lack of a general computable framework theory,
 (2) Lack of a quantizable gravitational field theory,
 (3) Lack of comprehensibility in quantum mechanics.

Specifically:

1. The existing framework of quantum field theory seems to lead to infinities which, in general, prevent computability (ability to predict). Renormalization, taken as a principle, is said to restore computability but there exist fundamental physical theories (for example, gravity) which are known to be unrenormalizable. Physical theory seems to be divided into two disjointed camps—renormalizable and unrenormalizable. This is symptomatic of a fundamental crisis in physical theory. Renormalizability cannot possibly be maintained as a physical principle if the price is to give up the physical theory of gravity.

2. A more urgent problem in gravity seems to be its quantizability. Suppose one has found a way of making all quantizable relativistic field theories finite. Still the following questions would remain: Is the conventional theory of gravity a genuine field theory? Does it consistently radiate energy? Is it quantizable? These questions do not seems to be definitely affirmed in the existing literature. Diligent efforts in this respect seem to point to the other direction. Studies of various correspondence limits seem to show that the conventional theory is not a genuine field theory. It seems to be lacking a basic physical element (field stress-energy $t_\mu{}^\nu$) to make it into a genuine field theory that is quantizable.

3. Quantum theory (including quantum field theory) is said to eschew comprehensibility by denying a local reality. The latter problem pervades practically all areas of the quantum phenomena. At present there does not seem to exist a theory which is quantizable, computable and comprehensible at the same time. This seems to be due to the use of the classical concept of a point particle to implement locality. The classical concept of a point particle is too primitive (lacks sufficient structure) to represent a field quantum. A field quantum, [a quasistable (normalized) finite energy wave packet], can behave like a particle and still satisfy locality (local commutativity) in terms of amplitudes. Yet we seem to base our thinking mostly on the classical conception of a particle. This unattended mismatch seems to create the problem of comprehensibility.

In this article we shall briefly review some recent progress on the problems of quantizability, and computability and begin a preliminary study toward the problem of comprehensibility of quantum mechanics and in its more general form, the quantum field theory. We wish to find out why and where our native intuition is defeated by the usual formulations of the theory.

2. A Possible Relativistic and Quantal Framework
   (Relativity and Quantum Mechanics)

Recently it has been suggested that a general and finite (that is computable) relativistic quantum field theory that includes a quantizable theory of gravity may be formulable by amending the usual Feynman path integral method with the following three simple rules: [1]

Rule 1. The Lagrangian of a dynamical system of fields is the sum of the Lagrangians of its individual members plus their mutual interactions:

$$\mathcal{L} = \mathcal{L}_A + \mathcal{L}_B + \mathcal{L}_{AB}$$

The rule is utterly obvious but it must never be forgotten. Some theories, for example, the conventional Einstein theory of gravity, do not satisfy Rule 1 because Einstein's theory does *not* have a genuine *field Lagrangian*. A field theory without a field Lagrangian cannot be quantized because the *Feynman propagator* by which the quantization is implemented, is obtained from the field Lagrangian. (Via the wave equation resulting from the field Lagrangian.) The reason Einstein's theory has no genuine field Lagrangian is that it does not include in its field equations a genuine *field stress-energy tensor* $t_\mu{}^\nu$ for the gravitational field. (It has only the matter stress-energy on the right-hand side of the Einstein-Hilbert tensor, that is, Einstein-Hilbert tensor is deprived of the field stress-energy of gravity by equating it to the matter stress-energy alone.) But without its field stress-energy tensor $t_\mu{}^\nu$ a field cannot have a field Lagrangian because the field Lagrangian and the field stress-energy imply each other. If we have the field Lagrangian we can obtain the field stress-energy tensor from its variation, and, if we have the field stress-energy tensor, we can construct the field Lagrangian from its trace. If we do not have one we cannot have the other. (Can one imagine a Maxwell stress-energy without a Maxwell Lagrangian and vice versa?) Consequently, Einstein's theory cannot be quantized meaningfully by the Feynman method, hence by implication, it cannot be quantized by any known method. In the past the question of field stress-energy tensor in Einstein's theory was confused with a pseudotensor which may appear in some coordinates. (This is usually denoted by $t_\mu{}^\nu$ but we will denote it as $z_\mu{}^\nu$ to avoid confusion.) [2] It is recently shown that $z_\mu{}^\nu$ is an artifact of the use of background noncovariant derivatives. It can be eliminated by use of covariant derivatives with respect to the background metric. (Section 5.1)

Rule 2. The mutual interactions of a dynamical system of fields are absorbable into local gauge-covariant derivatives in terms of which the system appears formally free

$$\mathcal{L} = \mathcal{L}_A + \mathcal{L}_B + \mathcal{L}_{AB} \rightarrow \mathcal{L}_A(D) + \mathcal{L}_B(D')$$

where D, D' are local gauge-covariant derivatives with respect to appropriate curvature and fiber connections. Here we do not detail the variety of gauge relations but simply state the conjectured end result, namely, they turn the interaction into a geometrical statement, hence afford a geometric interpretation of the interaction as in the case of principle of equivalence. Note that the interaction term is inherent in the gauge formalism and cannot be turned off, although at some limits it may be infinitesimally small. Note also that we are not introducing Rule 2 to make the renormalization of some theories possible (according to Rule 3 below there are no infinities) but to restrict possible field theories to a small interrelated size with a common background so that a unified approach may be possible.

Rule 3. The usual *divergent* (therefore also ambiguous) Feynman 1-loop integrals

$$I_F(s, a) = \int_F d^4l\, (l^2)^z (l^2 - a + i\varepsilon)^{-n}$$

$$s = 2 + z - n \geq 0, \quad n > 0, \quad z \geq 0$$

shall be assumed incorrect (possibly due to the inappropriate classical upper limit) and be postulated to have the (finite and unambiguous) physical value

$$I_{Ph}(s, a, a_0) = i\pi^2 \frac{(z+1)!}{(n-1)!} \frac{a^s}{s!} \ln(\frac{a}{a_0})$$

where the further integrations with Feynman parameters are omitted (or postponed) for simplicity. Note that this is the only amendment to be made in the evaluation of Feynman amplitudes, all other results (for example, integrals that are finite) being the same as before. Note also that n, z, s, are strictly integers, $2s$ being the superficial degree of divergence (although here there is no divergence) as in the usual theories upon power counting. In *Figure 1* the positions of the integrals that are considered are depicted by heavy dots. These are not the most general cases one can consider but the general case seems to retain the properties in the $s \geq 0$ sector to be discussed. In other words, the $s \geq 0$ sector seems to be special. For example, it always seems to have a logarithmic term, etc.

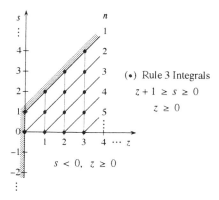

Figure 1. The Existence Region of Rule 3 Integrals (See Section 5.2)
$(s = 2 + z - n \geq 0, \quad n > 0, \quad z \geq 0)$

Some preliminary remarks may be made on Rule 3 : **a)** It seems to be derivable from the principle of causality plus correspondence to its finite counterparts [3] as is indicated in Section 5.2. **b)** The inverse loop propagator a turns out to be of the form

$$a = m^2{}_{av.} - p^2{}_{av.} + p_{av.}{}^2$$

where the l integration seems to induce an average $p_{av.}$ and the loop seems to operate relative to it, namely, $p \to p - p_{av.}$, whereby we get $a = [m^2 - (p - p_{av.})^2]_{av.} = m^2{}_{av.} - p^2{}_{av.} + p_{av.}{}^2$, the averages being over the *normalized* Feynman parameters. [4] For example, in vacuum polarization $a = m^2 - p^2x + p^2x^2$. **c)** In the usual theory infinities occur which are removed by a questionable process of renormalization. But this works only if the infinities are the same in form as the terms of the original Lagrangian. If not, the theory is unrenormalizable and is useless since one cannot calculate with it. In our case there are no infinities and for $a \to a_0$ the original choice of the Lagrangian is recovered. The a_0 induces a kind of group property where $a_0 = m^2$ would be the simplest choice.

As mentioned above, the conventional theory of gravity is not quantizable because it has no field stress-energy $t_\mu{}^\nu$ (it is not included in the field equations) hence also has no field Lagrangian. In other words, the conventional theory does not satisfy Rule 1. This omission seems to have happened because of the existence, in some coordinates, of a pseudotensor $z_\mu{}^\nu$ which was inadvertently taken in place of a genuine field stress-energy tensor $t_\mu{}^\nu$. However, the problem of pseudotensors is now resolved (Section 5.1) and the way to the introduction of a true field stress-energy tensor $t_\mu{}^\nu$ is clear.

Sometimes a linearized approximation is taken and the quadratic part $\Gamma\Gamma$ alone in the Einstein-Hilbert tensor is taken to give the field stress-energy of Einstein's theory. But in second order the other two terms $\partial_\alpha \Gamma$ contain quadratic terms that *exactly cancel* the previous ones in Einstein' theory. This is not surprising, because the field equations of Einstein's theory do not contain the field stress-energy $t_\mu{}^\nu$. The solution is simply confirming the original equations where there is no $t_\mu{}^\nu$.[5] Einstein's theory is therefore modified by explicitly adding the gravitational field stress-energy $t_\mu{}^\nu$ to the matter stress-energy. The theory is then quantizable, and is, by the Rule 3, finite. The new equations are

$$\tfrac{1}{2} G_\mu{}^\nu = \tau_\mu{}^\nu + t_\mu{}^\nu$$

$$\Box^2 \phi_\mu{}^\nu = \tau_\mu{}^\nu$$

$$t_\mu{}^\nu(grav.) = -2[\partial_\mu\phi_\beta{}^\alpha \partial^\nu\phi_\alpha{}^\beta - \tfrac{1}{2}\delta_\mu{}^\nu \partial^\lambda\phi_\alpha{}^\beta \partial_\lambda\phi_\beta{}^\alpha] + \partial_\mu\phi\partial^\nu\phi - \tfrac{1}{2}\delta_\mu{}^\nu \partial^\lambda\phi\partial_\lambda\phi$$

where $\Box^2$ is covariant d'Alembertian, $\phi = \text{trace}(\phi_\mu{}^\nu)$ and, in source-free space $\partial_\nu\phi_\mu{}^\nu = \partial^\mu\phi_\mu{}^\nu = 0$. (In field theory the latter may be written $k_\nu\phi_\mu{}^\nu = k^\mu\phi_\mu{}^\nu = 0$ and corresponds to a gauge fixing.) In order to have a field Lagrangian one must have a field stress-energy (and vice versa). In fact the field Lagrangian of a massless field is a multiple (here $-1/2$) of the trace of the field stress-energy, hence the field Lagrangian is

$$\mathcal{L}_g = -\tfrac{1}{2} t_\mu{}^\mu(grav.) = -\partial^\lambda\phi_\alpha{}^\beta \partial_\lambda\phi_\beta{}^\alpha + \tfrac{1}{2}\partial^\lambda\phi\partial_\lambda\phi$$

where a divergence $\partial_\nu J^\nu$ of a vector is omitted. These equations seem to be exact in many special cases and valid at least in first and second order in general. The modified theory has N-body and multimode gravitational wave solutions of exactly desired kinds (these are the only ones used in practice to calculate experimental predictions.). By the above amendments the usual difficulties in quantum field theory and in the theory of gravity seem to be overcome. It must be emphasized that here we are not criticizing Einstein's theory in order to introduce this new theory. Irrespective of what the new theory does, Einstein's theory has the problems discussed above and it must, in any case, be modified. Note that these developments may be interpreted as:

1. The usual quantum field theory is mathematically *underdetermined*, hence it needs at least one extra statement to render it determinate. This is clear from the fact that after the Feynman integration is carried out the conventional theory still needs to be regulated, renormalized, and is subject to renormalization prescriptions. The very need to renormalize indicates that the theory is *underdetermined* since renormalization implies *extra statements* to make the theory determinate.

2. The usual general relativity is mathematically *overdetermined*, hence it needs at least one extra quantity to absorb (or release) it from the overdetermination. This is clear from the fact that in general relativity

$$\tfrac{1}{2} G_\mu{}^\nu = \tau_\mu{}^\nu + (0)$$

that is, $t_\mu{}^\nu = 0$. But we have the two geometric identities to contend with, namely, Bianchi and Freud identities (here D_ν is the usual covariant derivative)

$$D_\nu G_\mu{}^\nu \equiv 0$$

$$\partial_\nu(\sqrt{-g}\ \tau_\mu{}^\nu) \equiv 0$$

We have *three* equations but only *two* unknowns, $G_\mu{}^\nu$ and $\tau_\mu{}^\nu$. A *third quantity* $t_\mu{}^\nu$ is needed to remove the *overdetermination*. Note that such a stress-energy term is also required for quantization. The extra quantity needed is $t_\mu{}^\nu(grav.)$. As is expected $t_\mu{}^0$ is the gravitational analog of Poynting vector in electromagnetic theory which represents the energy flow. A With the above two modifications we seem to have a well-defined and finite framework to treat any relativistic field theory relevant to physics (in flat or in curved spacetime) including gravity itself. Note that as in the electromagnetic case, the energy radiation is not directly connected to the classical solutions but rather to quantum transitions from less stable states of the bound system to its more stable lower states.

3. Problem of Incomprehensibility of Quantum Mechanics

Although brief, the above seems to highlight the present problems in quantum field theory and the theory of gravity. As to the problem of comprehensibility of quantum mechanics (or, more specifically, quantum field theory), consider the set of five equations below, which may be looked upon as a common denominator of quantum field theories: [6]

$$p\lambda = h \tag{1}$$

$$\frac{\partial \omega}{\partial k} = v_g = v = \frac{\partial E}{\partial p} \tag{2}$$

$$\frac{\omega}{k} = u_\varphi = u = \frac{E}{p} \tag{3}$$

$$uv = c^2 \tag{4}$$

$$p = mv \tag{5}$$

(+) Total energy, (+) Probability

They exhibit strong analogies and relationships between particle and wave concepts. The set is not complete since it is only a common denominator to be used to study general features but it is also made slightly redundant to convey the main features more clearly. For example, of the five equations the first can essentially be inferred from the other four up to a numerical factor h. These equations refer to a quantum in transit, the states being $E = nh\nu$ where $n \geq 0$ is an integer. They imply a strange mix of wave and particle attributes which are, from a classical point of view, confusing and unintelligible.

For example, from a classical point of view $u = E/p$ is unintelligible [7] because:

a) It implies *two* velocities for the same particle!, and,
b) $u \geq c^2/v$ can be larger than the velocity of light!

This is usually a stumbling block in comprehensibility because one usually thinks that something is wrong or mysterious with u. Nevertheless it is consistent and useful as it immediately leads to the two of the most fundamental equations of physics

1. $E = pu = m\underline{uv} = mc^2$

2. $E = pu = \underline{p\lambda}\nu = h\nu$

In this article we shall not be afraid of considering the phase velocity u. In fact we shall try to interpret and use it whenever possible. The list of successful derivations by explicit or implicit use of these equations is very large. They include

3. Increase of mass with speed [8]
4. Uncertainty relations
5. Orbital quantizations
6. Lorentz transformations

7. Phase correlations

.

.

.

and conceivably all of relativity and quantum mechanics. Even conservation laws are derivable (after the Lorentz transformations are derived). Note that the non-relativistic quantum mechanics is not derivable because there $u = v/2$ which is not consistent with $uv = c^2$. It is often believed that $p = h/\lambda$ can be derived from the classical $u = v/2$, $u = \lambda v$ plus $E = p^2/2m = hv$ without special relativity. Let us examine such a derivation:

$$\frac{p^2}{2m} = hv = \frac{hu}{\lambda} = \frac{h}{\lambda}\frac{v}{2} = \frac{h}{\lambda}\frac{p}{2m}$$

$$p^2 = (\frac{h}{\lambda})p$$

This will indeed yield $p = h/\lambda$ but only under the condition that $p \neq 0$, that is, it may not always be able to describe a massive particle at rest consistently. The relativistic derivation does not have this problem. In relativity the relation $uv = c^2$ expresses the orthogonality of rays and wave-fronts and the relativity of simultaneity. These are missing in the nonrelativistic theory. [9]

4. Toward Comprehensibility: Wave-Particle Unity

The above considerations lead us to a kind of thinking where the wave and particle attributes are *united* in the description of quanta instead of being mutually exclusive as in the usual *dualistic* view. Duality connotes *tension, dichotomy* and *mutual exclusivity*, whereas what seems to be inherent in the phenomena is *co-operation, unity* and *mutual inclusiveness*. Although customarily duality itself is often considered as a kind of unity via the concept of complementarity, we shall try to elucidate the distinction by actual examples and try to formulate it physically. We conceive of a *finite-energy, quasistable* solution, say, $\phi = A(ae^{-i(wt-kx)} + \bar{a} e^{i(wt-kx)})$ of a wave equation *box normalized* to have the energy $E = hv$ and momentum $p = h/\lambda = hk/2\pi$. (A is a normalization factor and a, $\bar{a}$ are analogues of complex number operations to be determined later.) Box normalization leads to states similar to standing waves with harmonics n which are interpreted as number of quasistable quanta. In its rest frame a quantum with rest mass appears as if a standing wave. With *quasistable* we mean the wave packet (quantum) behaves as a stable lump in the absence of interaction but in the presence of interaction it can be emitted, absorbed or scattered. A wave packet is where all possible waves interfere constructively to produce the packet and everywhere else destructively to zero. In flight it may roughly be pictured as in *Figure 2*:

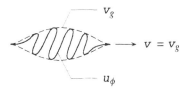

Figure 2. Pictorial Representation. Wave-Particle Unity

where the phase waves propagating with (phase) velocity $u_\varphi = u = E/p$ are operative *inside* of a finite wave envelope, while the energy moves with group (envelope) velocity $v_g = v = \partial E/\partial p$, which is equal to the mechanical velocity v. *(Figure 2.)* The wave packet has not only a *longitudinal* extension but has also a *lateral* extension (not shown in the figure). They account for wave properties (call it a wave aspect) while the lump of energy represented by the normalized wave packet behaves like some stable mechanical object (call it a particle aspect) moving with the group velocity $v_g = v = \partial E/\partial p$. [B] Pictorially speaking, the rays represent things that go on serially in succession, and wavefronts things that happen parallel, that is, simultaneously. Thus inside a wave packet the wavefronts define loci of simultaneity but outside it has no physical reality other than a projection. However, this projection has a physical meaning in the sense of *wave correlations*. For example, if the waves are coherent, they delineate ahead of time where the interference maxima and minima will occur since they propagate in a regular way and maintain the correlations.

Stability can be interpreted as issuing from the normalization of the wave packet (to energy and momentum $h\nu$ and $hk/2\pi$). However, in the presence of interaction the packet is not absolutely stable. It may get absorbed, scattered, or converted to wave packets of different kinds. This quasistable conception is a condition imposed on the essentially classical wave theory. We may call it the *quantum condition (or quantum conditions)*, implying that it is a restrictive condition on the classical wave theory. If we remove it, the theory would revert back to the classical wave theory.

This idea can be understood as follows: Consider the classical, free field of a massive scalar field ϕ and its Lagrangian (simplest for our purpose)

$$\mathcal{L}_\phi = \tfrac{1}{2} \partial^\mu \phi \partial_\mu \phi - \tfrac{1}{2} m^2 \phi^2 \qquad (I)$$

leading to the wave equation $(\partial^\mu \partial_\mu + m^2)\phi = 0$ (where m is rest-mass) with familiar plane wave solutions $\phi = A(a e^{-ikx} + \bar{a} e^{ikx})$. If we impinge a classical plane wave on a 50-50 beam splitter, half of the energy would be transmitted and the other half reflected. This is because the classical wave energy has *no stability,* hence the energy divides into two when the wave is confronted (that is, when it is interacting) with the beam splitter. If now we replace the wave with a normalized wave packet solution, it will be either transmitted or reflected *as a whole* because we are requiring stability both *before* and *after* the interaction. However, the wave packet is not absolutely stable, hence under suitable interactions it may get absorbed or emitted (as a whole) in which cases we say the field made transitions from a state of energy $E = nh\nu$ to a state of energy $E = mh\nu$ or vice versa. Obviously none of this will happen if the wave energy is absolutely stable or if the if the Lagrangian is completely free. In order that they can happen the Lagrangian must have an interaction term $V(\phi)$ and there must be conditions (quantum conditions) to express the rules of transitions.

Experiment (Planck's formula) shows that the rule of transitions is $m \rightarrow n \pm 1$, that is, one quantum at a time is emitted or absorbed in a given direction and this is found to lead to the quantum conditions on a_k, $\bar{a}_k$ as

$$\bar{a}_k a_{k'} - a_{k'} \bar{a}_k = 1_{kk'} \qquad \textbf{(II)}$$

Those solutions that do not obey this condition are not observed, hence they are considered unphysical or forbidden. The net effect of quantum conditions is that a finite energy wave packet is made quasistable and forced to behave like a quasistable particle in transit. This is a kind of synthesis or unification where the original classical concept of a wave and of a particle are being reconciled. [$1_{kk'}$ stands for $\delta_{kk'}$]

The condition **(II)** may look mysterious (so it did even to its discoverers). However, its meaning may be very simple. Let the energy state $E_n = nh\nu$ be represented as $|n> = A_n a^n$, $A_n = 1/\sqrt{n!}$. To accommodate the possibility of transitions and the rule $m \rightarrow n \pm 1$ one must be able to raise and lower the state by one. Let a and $\bar{a}$ be the raising and lowering operations. To raise the state by one we may simply multiply by a. To lower it, however, we cannot just divide by a since this will not give the rule correctly. But if we interpret $\bar{a}$ as a differentiation with respect to a, we get the correct rule. Thus (omitting the indices k, k' for simplicity) one has

$$\bar{a} = \frac{\partial}{\partial a} \qquad \textbf{(II')}$$

Then it is obvious that $\bar{a}a |n> = (n + 1) |n>$, $a\bar{a} |n> = n |n>$, hence

$$(\bar{a}a - a\bar{a})|n> = |n>$$

$$\bar{a}a - a\bar{a} = 1$$

as applied to $|n>$. C We are mentioning these simple points here because they are the most basic and we need an intuitive understanding of them. For example, at thermal equilibrium one would have $<n> + 1 = <n> e^{\beta\omega}$. Solving for $<n>$, the average number of photons in a mode, we have, $<n> = 1/(e^{\beta\omega} - 1)$ which is the Planck law of cavity radiation. [10]

These considerations are not equivalent to the usual Bohrian concept of *wave-particle duality* since in all these cases the wave and particle attributes are always together and nowhere are they mutually exclusive. We may call this togetherness *Wave-Particle Unity*.

This concept is the result of our willingness to consider and use the phase waves and the phase velocity $u = E/p$. The usual abstract concept of duality would seem to imply that while thinking of the wave we should not be able to define their envelope and while thinking of their envelope we should not be able to define the wave, or, physically, while manifesting the wave we should not be able to manifest the envelope and while manifesting the envelope we should not be able to manifest the wave. But this cannot be. A wave and its envelope are inseparable attributes. In the present way of thinking they are both there by definition and the object so formed is nonlocal. This type of nonlocality does not violate relativity. Nowadays it is fashionable to say that there is a conflict between locality and reality. The above analysis seems to imply that the quasistable wave packet is nonlocal and nevertheless the amplitude satisfies local commutativity. It seems that the door is open to a *realistic* view but via a wave-particle unity.

We can try to analyze these conceptual issues in the classical limit where our intuition is more at home. The classical limit of the de Broglie relation may be taken as: [11]

$$p\lambda = h \to 0$$

$$p = \text{finite}, \quad \lambda = 0 \quad (\underline{\text{particle}}, \, not \, \text{wave})$$

$$\lambda = \text{finite}, \quad p = 0 \quad (\underline{\text{wave}}, \, not \, \text{particle})$$

Thus there are actually *two* distinct classical limits, not *one*. Classically (that is, when $h = 0$), waves and particles are mutually exclusive, but quantum mechanically (that is, when $h = h \neq 0$), they are always intimately connected. Bohr's statement that in quantum mechanics wave and particle aspects cannot be simultaneously manifested in the same experiment is therefore questionable—it seems in fact wrong. (See experiment, *Figure 3*.)

Note that the only constraint on the classical wave is its normalization, hence there are an infinity of ways of structuring the shape of the envelope of a wave packet, by using different constituents of phase waves, we shall have an unlimited number of ways to shape the wave packet. A few of them will soon be mentioned (namely, single states, coherent states, squeezed states, thermal states, entangled states, etc.). These states can be produced by controlling certain defining parameters via interaction with appropriate environments. A vast field of *photon shape* engineering can here be envisaged. Various photon shapes for various applications (for example for special chemical reactions or biomedical applications) could, in principle, be manufactured and used.

This idea of the existence of a variety of spectral photons is made possible by the concept of a *wave-particle unity* in the sense of normalized wave packets, whereas the usual *wave-particle duality* connotes the existence of only one kind corresponding to a structureless mechanical particle. Roughly speaking, the wave packet asserts itself more definitely near objects comparable to its wavelength but in flight or near objects much larger than the envelope its effect is like a particle.

One may describe the comprehensibility problem in quantum mechanics as follows:

1) Calculate as in waves,
2) Interpret as in particles,
3) Don't ask reasonable questions.

It seems that comprehensibility suffers because, within quantum mechanics, the classical concepts of waves and particles are not applicable. One can see from the above that one may consider the theory as a classical field theory where only the stable finite energy solutions of the above kind are admitted. Such waves will be emitted, absorbed and scattered, as wholes like particles but they are neither waves nor particles in the usual sense.

The power of such a unitary thinking can be illustrated by many examples. We shall give two that are most illuminating, one experimental (a quantum tunneling experiment) and the other theoretical (a derivation of the Lorentz transformations).

4.1. Quantum Tunneling Experiment Showing Wave-Particle Unity

Single photons coming from a parametric down converter [12] are sent to two counters via a double-prism beam splitter. *(See Figure 3.)* The double prism with an air gap is a tunneling device and classically only waves can tunnel. Classical particles cannot tunnel. Yet the two counters click in perfect anticoincidence, indicating that the waves in question form *finite energy* quasistable wave packets that do not split into two at the beam splitter. On the whole they behave as quasistable mechanical objects possessing energy momentum. [13]

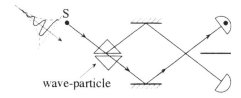

Figure 3. Wave-Particle Unity: Hamamatsu Experiment

The question is what happens at the beam splitter where the single photons are confronted with the air gap. Our pictorial representation says that the phase waves forming the wave packet act just like waves, satisfy wave boundary conditions etc., and via these, try to split as in the classical wave. But the normalization condition (quantum condition) does not allow splitting, hence, as a whole, the wave packet goes one way or the other (that is, either it gets transmitted or reflected but with the same proportion of energy as in the classical wave on the average). Thus at the air gap and vicinity a normalized wave packet manifests its wave structure and its envelope at the same time. This experiment seems to disprove Bohr's contention that in quantum mechanics the wave and particle aspects are mutually exclusive, and that they cannot both be manifested simultaneously. Bohr's statement amounts to a mystery whereas with the *wave-particle unity* there seems to be no mystery.

In his complementarity idea Bohr seems to give special emphasis to interference as a primary aspect of the wave phenomenon. Yet the primary wave aspect is diffraction (tendency to spread as in the Huygens principle). Interference is a result of diffraction plus coherence, hence is indirect. If we consider diffraction (instead) it becomes immediately obvious that wave and particle aspects implicit in the normalized wave packet are always together. Single pinholes in dielectrics and velocities through them may thus be used to test pure particle theories (analogously to Foucault's experiment of wave velocity in refractive media which our *wave-particle unity* can imitate but pure particle theories cannot).

Note that it is not the first time that Bohr has made a misstatement. In the late 1920's Bohr seems to have concluded that the electron's magnetic moment can never be observed outside an actual microscopic atom. This seems to be related to the circumstance that spin, and therefore the intrinsic magnetic moment, does not have a classical analog, hence not amenable to measurement since according to him the measuring device has to be classical. Yet in the last decade by a trapped electron outside the atom the intrinsic magnetic moment is measured. Ironically, not only this measurement is shown possible, but its accuracy being 11 significant digits, it is virtually the best measured quantity in physics. [14] Clearly, the wave-particle duality in the sense of Bohr, that is, the intention of building up quantum mechanics $(h \neq 0)$ from two mutually exclusive classical limits $(h = 0)$, is not a reliable guide. (Note that here we are not blaming Niels Bohr for making a mistake. It is only those who do not do anything who do not make a mistake.)

We now give also a theoretical example of the power of wave-particle unity where the Lorentz transformations are derived from it. This seems to show that wave-particle unity as conceived in the foregoing cannot conflict with the special theory of relativity.

4.2. Derivation of the Lorentz Transformations Via Wave-Particle Unity

As a second example of the power of wave-particle unity we shall derive the Lorentz transformations from it. (That is, we shall use $uv = c^2$, hence implicitly $p = h/\lambda$.) Suppose, in a reference frame (call it S) rain drops (particles with mass m) are coming down vertically. In a frame moving horizontally relative to the first (call it S') with relative velocity V the rain drops will appear slanted by an angle θ and will have the velocity v. *(See Figure 4.)* But according to the wave-particle unity, the particle has also a phase velocity $u = c^2/v$ within the envelope. Assuming the wave packet is sufficiently large so that we can use the phase velocity in the sense of finding the wavefront. The first thing we learn is that $uv = c^2$, which troubles most people is the condition that rays and wavefronts be orthogonal. It will be surprising that this simple setting allows one to derive the Lorentz

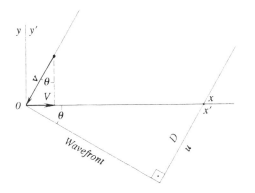

Figure 4. Derivation of Lorentz Transformations

transformations from wave-particle unity. To see this let us first work out the *linear* (first order) approximation. From *Figure 4* we can immediately write

$$\sin \theta = \frac{V}{v} = \frac{D}{x}$$

$$x' = x + \delta x$$
$$t' = t + \delta t$$

Using the phase velocity u and the orthogonality of rays and wave fronts $uv = c^2$ we find

$$\delta x = -Vt \quad , \quad \delta t = -\frac{D}{u}$$

$$-\delta t = \frac{D}{u} = \frac{x}{u}\sin\theta = \frac{Vx}{uv} = \frac{Vx}{c^2}, \quad \text{hence}$$

$$x' = x - Vt \qquad x = x' + Vt'$$

$$t' = t - \frac{Vx}{c^2} \qquad t = t' + \frac{Vx'}{c^2}$$

These formulas already satisfy the principle of relativity [that is, the symmetry of the two frames via $p_x = mv_x = mV$, $p'_x = mv'_x = -mV$, or $p_x = h/\lambda_x$, $p'_x = h/\lambda'_x = -h/\lambda_x$] and the universality of the velocity of light $c' = c$ in x direction [that is, $c = x/t$ leads to $c' = x'/t' = (x - Vt)/(t - Vx/c^2) = c$] but they are not consistent because they do not satisfy the consistency condition that $x = x$. In order to achieve consistency let us multiply the right-hand sides with a normalizing factor A. We have

$$x' = A(x - Vt) \qquad x = A(x' + Vt)$$

$$t' = A(t - \frac{Vx}{c^2}) \qquad t = A(t' + \frac{Vx'}{c^2})$$

The factor A is the same everywhere because this is the only way one can retain the relativistic symmetry between the two frames S and S' and the invariance of c within each frame (Einstein's two postulates). Then, treating A as an unknown and using the consistency condition $x = x$ one finds

$$x = A(x' + Vt') = A^2(x - Vt + Vt - \frac{V^2 x}{c^2}) = A^2(1 - \frac{V^2}{c^2})x$$

$$A = \frac{1}{\sqrt{1 - \frac{V^2}{c^2}}}$$

Substituting the latter back into the previous formulas we have the Lorentz transformations. This derivation of the Lorentz transformations disproves the contention that every kind of nonlocality would conflict with relativity. It does not conflict with our nonlocal concept of wave-particle unity. Not only is there not any conflict, but there is a strong dependence, on each other so that one can be derived from the other. It is interesting that mechanical relativity embedded in $v'_x \rightarrow -v_x$ can also be stated as $\lambda'_x \rightarrow -\lambda_x$.

An earlier derivation of Lorentz transformations [15] from wave-particle unity was made in the degenerate limit $u = v = c$. (See *Figure 5)* where α is the stellar aberration angle. The more general derivation above, incorporating the fields with nonzero mass by using both the group velocity v and the much feared phase velocity u, removes the degeneracy and illustrates the power and the general nature of wave-particle unity. The phase velocity is only inside the wave envelope, hence physically it does not exist outside it. However, if two beams with a common origin come together at a later time and place, the correlations (regularities) are nevertheless maintained (carried) via the synchronous propagation of the phase waves *as if* the correlations were prearranged.

In this way the EPR type experiment, for example, seems to be understandable and does not seem to suggest any conflict with relativity. For, we have a nonlocal object that does not violate local commutativity, hence an element of reality in some nonlocal sense seems to be retainable without violating relativity. The wave function collapse seems to be the greatest stumbling block especially because the roles of the wave function (an operator) and the state function (an operand) are sometimes confusing. [D] It seems that closest attention must be devoted to this distinction so that a *reconciliation* between the resonant transfer of energy, quantum jumps and the collapse may be achieved. As is well known, this is also at the heart of the measurement problem of quantum mechanics.

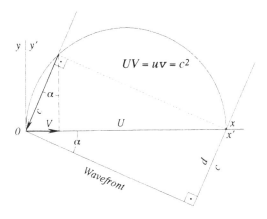

Figure 5. The limit $u = v = c$. Case of zero rest-mass (photon)

The phase waves inside the wave packets are able to take many forms and shapes subject to the normalization conditions enforcing quasistability. Among these are

- Single particle states
- Entangled states
- Mixed states
- Thermal states
- Coherent states
- Squeezed states
-
-

These properties may be better understood if they are studied from a unitary point of view, that is, via a kind of *wave-particle unity* with distribution of energy within the waves but particle-like properties along the rays. In this sense we may well consider the quantum problem as a revision of Huygens' principle whereby the Huygens' waves are replaced by their normalized wave packets. In this way, the mystery in the EPR type experiment seems to be resolvable by noting that the two cascade photons have wave correlations when emitted together that persist in flight and exhibit these correlations upon later measurements. The impression that the measurements seem causally connected in violation of relativity is created because of our insistence on a classical particle concept which cannot carry wave-like correlations.

Classical waves can have correlations of the observed kind, and they indeed have them. If purely classical waves were substituted, the EPR-like correlations would be expected. However, there would be a difference. The classical wave correlations, although

TOWARD A COMPREHENSIBLE PHYSICAL THEORY

qualitatively similar, would be quantitatively different. For example, in a second order two photon coincidence experiment quantum mechanics and the classical wave theory predict qualitatively the same pattern but quantum mechanically the visibility is 1, whereas classically it would be diluted to $\frac{1}{2}$. The difference is that quantum mechanically the detectors are responding to superpositions of wave packets (amplitudes), whereas classically one assumes that they would respond to intensities (individual amplitude squares). The latter would dilute the modulation depth because in squaring the individual amplitudes some of the phase information is lost. It seems that the discussions of the violation of Bell inequalities, etc. would be more fruitful and less mysterious if they were considered from the point of view of waves and wave-particle unity.

With the EPR type cascade and parametric down conversion experiments [12] the Einstein reality is said to be defeated because Einstein nonlocality was incompatible with relativity. It is however, seen that there are objects (normalized wave packets) which are nonlocal as far as the size of the packets is concerned, yet would not violate relativity. The usual thinking is in terms of classical mechanical objects and their locality. It seems that a real, albeit nonlocal, understanding of quantum mechanics should be possible in terms of a wave-particle unity which defines a new kind object for quantum mechanics. The difficulty seems to be in our naive insistence on a mechanical object of a classical kind. This creates a misunderstanding which later turns into a lack of comprehensibility.

5. Some Recent Developments

In this section we would like to review some recent developments which seem to be of relevance to our topic. One of these is the solution of the pseudotensor problem of space-time theory that defied understanding and impeded progress for the last eighty years. The other is a possible derivation of Rule 3. There are other improvements in our understanding which, for lack of space, will be left to a different communication.

5.1 Resolution of Pseudotensor Problem

It is found that the pseudotensor problem arises from the use of noncovariant derivative with respect to the metric of the background geometry. For example, one normally uses a flat background (Minkowski space) for correspondence purposes. If some coordinates of this background space are used to express a solution of the curved space field equations, there arises the question of what happens if we choose Cartesian, spherical or cylindrical, etc. coordinates? It turns out that, in general, anything other than in Cartesian coordinates we have a pseudotensor $z_\mu{}^\nu$. This was noticed as early as 1918 and has been very confusing ever since. It is recently found that the problem is not in the choice of the coordinates, but having chosen them, what kind of derivative will be adopted. It turns out that if the derivative adopted is the covariant derivative with respect to the chosen background, then there is no pseudotensor $z_\mu{}^\nu$. This was actually pointed out by N. Rosen [16] as early as 1939 but it was forgotten. Pseudotensor is thus an artifact of the choice of the kind of derivative and not the effect of the choice of coordinates. It is eliminated if one uses the covariant derivative with respect to the background metric.

This immediately explains why there are no pseudotensors in the Cartesian coordinates. Because the ordinary derivative in Cartesian coordinates of flat space is already the covariant derivative. Therefore, it seems that in order to be free of pseudotensor problem all one has to do is to interpret the *ordinary* derivative symbol ∂_μ with respect to the background metric as the *covariant* derivative with respect to the background metric (which we assume done). The equations of the new theory are then given by

$$\tfrac{1}{2} G_\mu{}^\nu = \tau_\mu{}^\nu + t_\mu{}^\nu$$

$$D_\nu G_\mu{}^\nu \equiv 0$$

$$\partial_\nu(\sqrt{-g}\, \tau_\mu{}^\nu) \equiv 0$$

where all quantities entering are true tensors with respect to the curved metric of the space. Note that, with respect to the flat background metric one has $\sqrt{-g} \rightarrow \sqrt{-g/-\eta}$. This process may be called *derivative transformation*. It transforms away the pseudotensors but not the true tensors. We consider this development extremely important because, as Rosen also pointed out, the Lagrangian process itself can be carried through with covariant derivative with respect to the background coordinates and the pseudotensor problem never arises. Note that with this understanding one can prove that, **a)** General relativity does not have an interior solution for the incompressible fluid because the $t_\mu{}^\nu$ which general relativity disowns is necessary to form the pressure, **b)** These points also explain and answer why some colleagues, for example, S. Antochi [17] erroneously conclude that the pseudotensor problem is a matter of a choice of coordinates alone. **c)** If anybody wants it the expression of the pseudotensor $z_\mu{}^\nu$ is: $U_\mu{}^\nu$ or $u_\mu{}^\nu$ (defined in Reference [2]) evaluated by ordinary derivative with respect to the background metric, minus the same evaluated by the covariant derivative with respect to the background metric. E

5.2. Possible Derivation of Rule 3

Because in its present state the quantum theory of fields is underdetermined, one needs one or more extra statement to render it determinate. For this purpose we have proposed Rule 3 in Section 2. Although Rule 3 is stated there just as a *rule* or *postulate* to be judged by its experimental consequences, it seems to be derivable by using correspondence to the finite counterparts plus the principle of causality. Consider the standard Feynman integral [18]

TOWARD A COMPREHENSIBLE PHYSICAL THEORY

with
$$I_F(s, a) = \int_F d^4l \, (l^2)^z \, (l^2 - a + i\varepsilon)^{-n}$$

$$s = 2 + z - n, \quad n > 0, \quad z \geq 0$$

As is well known, all Feynman amplitudes (divergent ones after renormalization) can ultimately be expressed as a linear combinations of similar integrals (via reduction formulas such as $l^\mu l^\nu \to l^2 g^{\mu\nu}/4$, etc.). Integrated over l from zero to infinity it gives

$$I_F(s, a) = i\pi^2 \, \frac{(z+1)!}{(n-1)!} \, (-a)^s \, \Gamma(-s)$$

which is *finite* and well defined for *negative* integers $s < 0$ (we shall call them *proper*), but *diverges* for zero and positive integers $s \geq 0$ (we shall call them *improper*). Since the latter (as well as the former) regularly occur in the evaluation of Feynman amplitudes, one has the problem of what to do with (or, more properly, how to understand) the cases with $s \geq 0$. [It may be simpler to label the integral as $I_{z,n,F}$, $n = n_1 + n_2$, etc., but here s is a more important label, hence we denote the integral as $I_F(s,a)$ by temporarily omitting z and n.]

Jauch and Rohrlich [19] noted that the case of finite integrals may be used as a model for handling divergent ones. After all, something from the form or structure of the finite cases must survive into the divergent ones. For example, the factor $i\pi^2$ will certainly survive because it comes from the angular parts where the dimensionality is fixed at 4. The form $(z+1)!/(n-1)!$ will also survive because it is a finite and well defined combination of integers. The factor $\Gamma(-s)$ cannot possibly survive because it is the one that diverges. The case of a^s is not so clear but noting that all finite cases are monomials in a^s we may assume that this is generally so by correspondence. [20]

We now note that in all finite (proper) cases, higher values of s correspond to integrals of lower values of s. If we continue this process to $s = -1$ to $s = 0$ and up, we find

$$\int da \, (1/a) = \ln a + C$$

$$\int da \int da \, (1/a) = a \ln a - a + Ca + C_1$$

$$\vdots$$

$$\int da \, \ldots \int da \, (1/a) = (1/s!) a^s [\ln a - Q + C + s! \sum_{r=1}^{S} C_r a^{-r}/(s-r)!]$$

$$I_F(s, a) = i\pi^2 (z+1)!/(n-1)!(1/s!) a^s [\ln a - Q + C + s! \sum_{r=1}^{S} C_r a^{-r}/(s-r)!]$$

where $Q = 1 + 1/2 + \ldots + 1/s$ is a finite number and the C's are integration constants. [Note that all $s \geq 0$ (divergent in the usual theory) integrals have a $\ln a$ term in them.] If now we require that, as in the finite cases, the a^s will survive as a single factor, all integration constants except C must vanish, hence

$$I_F(s, a) = i\pi^2 (z+1)!/(n-1)! (1/s!) a^s[\ln a - Q + C]$$

[As expected, this is no longer the Feynman integral but we shall continue to denote it as $I_F(s, a)$ for continuity.] For the rest of the determination we use the following three arguments: **a)** From a causal point of view only the $\ln a$ should be present in the bracket because a is the (inverse) *causal* loop propagator. **b)** The argument of the logarithm must be dimensionless, hence the logarithm must be of the form, $\ln(a/a_0)$. **c)** The value of a_0 must correspond to the initial choice of the Lagrangian so that when $a \to a_0$ the original Lagrangian is recovered, a_0 being the boundary value of a. The latter allows a group property in the choice of the original Lagrangian. All three steps can be combined into a single operation, namely,

$$I_{Ph}(s, a, a_0) = a^s \int_{a_0}^{a} da\, \partial_a[a^{-s}I_F(s, a)] = a^s[a^{-s}I_F(s, a)]\Big|_{a_0}^{a}$$

$$I_{Ph}(s, a, a_0) = i\pi^2 \frac{(z+1)!}{(n-1)!} \frac{a^s}{s!} \ln(\frac{a}{a_0})$$

which yields our Rule 3. The general form of a 1-loop Feynman amplitudes can then be written down (after use of reduction formulae, etc. and applied order by order in s) as

$$\mathcal{A}_{Ph}(s, a, a_0) = \sum_s \int_0^1 dx\, a^s \int_{a_0}^{a} da\, \partial_a[a^{-s}I_F(s, a)] = \sum_s \int_0^1 dx\, a^s[a^{-s}I_F(s, a)]\Big|_{a_0}^{a}$$

where the x's are Feynman parameters. All it says is that, in each function $I_F(s, a)$ above, we keep only the logarithm of a and write this logarithm as $\ln(a/a_0)$. The complicated looking expression $\mathcal{A}_{Ph}(s, a, a_0)$ will be needed only if one wishes to automate on the computer. Note however, that it seems to be extendible to more general situations than here indicated, for example, when the dimensionality is different from 4 or when the denominator of the integral is of the form $A^{n_1}B^{n_2}$, etc. In other words, the $s \geq 0$ sector seems to be special so as to always accommodate the three statements we have made.

The procedure seems to be simple and general. It seems to be considerable progress compared to the usual renormalization process which is totally unacceptable. [21] Here there are no infinities, nor there seems to be any need for regularizations, the variety of renormalizations (mass, charge, various wave functions, etc.), finite remainders or renormalization prescriptions. Particularly important is the fact that the result is always finite, hence the procedure seems to be applicable to the usually unrenormalizable theories such as gravity. Note also that, in conformity with much physical and mathematical evidence, we hold $d = 4$ dimensional spacetime unique and do not deviate from it even infinitesimally. Thus our approach seems to be new and independent of the usual cut-off and/or dimensional procedures. Moreover, with its evident connection to the principle of causality, the approach seems to be physically sound and mathematically defensible.

SUMMARY

1. A general and calculable (finite) quantum field theory for all quantizable fields seems to be formulable,

2. A quantizable spacetime theory of gravity seems to be obtainable by modifying the usual Einstein theory,

3. Quantum theory seems to be comprehensible if one adheres to a concept of wave-particle unity.

Acknowledgments:

Gratitudes are due to Professor Carroll O. Alley and Mr. Kennett Aschan of the University if Maryland for their genuine interest and many helpful suggestions. Ken Aschan also developed the computer program with which the method of background covariant derivative is incorporated to isolate the pseudotensor and the true stress-energy tensor of the gravitational field.

References:

[1]. Yilmaz, H., *Relativity and Quantum Field Theory*, Annals of New York Academy of Sciences, New York, **75**, (1995), 476-493. (New improvements occurred which are described in Sections 5.1 and 5.2.)

[2]. The pseudotensor z_μ^ν is more general than originally treated in H. Y., *Toward a Field Theory of Gravitation*, Nu. Cim., **107B**, 8, (1992), 941-960. See Section 5.1 and Reference 17.

[3]. In Ref. 1. a version of Rule 3 was introduced via the principle of causality but without the correspondence to the finite integrals. This led to a K-function only the first term of which satisfies the correspondence and which is here reproduced. A possible derivation is given in Section 5.2.

[4]. See Ref. 1, pp. 477-75 and . 488.

[5]. Yilmaz, H., Relativity and Quantum Mechanics, Int. J. Theor. Physics, 21, (Dirac Issue) (1982), 899.

[6]. Yilmaz, H., *Note on Wave-Particle Unity*, Frontiers of Fundamental Physics Eds. M. Barone and F. Selleri, Plenum Press (1994), 526-28.

[7]. Even Schrodinger had a fear of talking about u. Upon deriving u he says: "We are taken aback. We find all our good intentions foiled. ... *horribile dictu*, $u > c$, since $c \geq v$ *The Interpretation of Quantum Mechanics* Edited by M. Bitbol, Ox Bow Press, (1995), 59.

[8]. Start with $uv = EdE/pdp = c^2$ and integrate it into, $E^2 = c^2 p^2 + K = m^2 c^2$. Since $p = mv$, and for $p = 0 \rightarrow K = m_0^2 c^4$, one immediately gets $m = m_0/\sqrt{(1 - v^2/c^2)}$.

[9]. Lochak, G. and R.,Dutheil, *Wave Mechanics and Relativity*, Wave Particle Duality, Ed. F. Selleri, Plenum Press, (1992), 157-67. Also in the same volume P. Guerret, *Dualism Within Dualism*, p. 102-3.

[10]. Note that if a is a Grassmann number $a^2 = 0$, then the $\overline{a} = \partial/\partial a$ leads to the anti-commutation

rules $\bar{a}a + a\bar{a} = 1$, hence $-<n> + 1 = <n> e^{\beta\omega}$, that is, to the Fermi distribution $<n> = 1/(e^{\beta\omega} +1)$.

[11]. Park, D., *Classical dynamics and its Quantum analogues*, Springer-Verlag, Second edition, (1990), 320-21. Park gives to these correspondences a somewhat different interpretation than we pursue here but I believe the tenor is generally in the same direction.

[12]. Alley, C. O., and Shih Y. H., *A New Type of EPR Experiment Using Light Quanta produced by a Nonlinear Optical Process*, ISQM, Physical Society of Japan (1986), 47-52. Alley and Shih pioneered the use of parametric down conversion to make entangled states, now used in many experiments.

[13]. Mizobuchi, Y., *A Test of the Complementarity Principle In Single-Photon States of Light*. Frontiers of Fundamental Physics Eds. M. Barone and F. Selleri, Plenum Press (1994), 511-18.

[14]. Wick., D., *The Infamous Boundary*, Birkhauser, (1995), 128-29.

[15]. Yılmaz, H., *Perception and Philosophy of Science*, Boston Studies on the Philosophy of Science (Seminar was given October 28, 1969.), Vol. **13**, (1974), 1-91. (See pp. 31-34).

[16]. Rosen, N., General relativity and Flat Space I, Phys. Rev. **57**, (1940), 147-50.

[17]. Antochi, S. *On the Asserted Clash Between....* Nu. Cim., **110B**, (1995), 1025-28. Antochi claims that $T_{ik} = \Lambda g_{ik}$ and its solution $g_{00} = 1 - r^2/3\Lambda$, $g_{\theta\theta} = r^2$, $g_{\varphi\varphi} = r^2 \sin\theta$, $g_{rr} = (1 - r^2/3\Lambda)^{-1}$ would be a counter example to our statement, namely, $\partial_\nu(\sqrt{-g}T_\mu{}^\nu) \neq 0$. His mistake is that he did not take into account the transformation $\sqrt{-g} \to \sqrt{-g}/\sqrt{-\eta}$ by which one gets $\partial_\nu(\sqrt{-g}T_\mu{}^\nu) \to \partial_\nu(\Lambda\delta_\mu{}^\nu) \equiv 0$. To his credit, elsewhere in his article Antochi shows the existence of a true stress-energy tensor $t_\mu{}^\nu$ but, instead of concentrating on this, he engages in a futile revival of the pseudotensor.

[18]. Gross, F., *Relativistic Quantum Mechanics and Field Theory*, John Wiley, (1993), 608.

[19]. Jauch, J. M. and Rohrlich, F., *The theory of Photons and Electrons*, Addison-Wesley, (1955), 457.

[20]. Note that in parametric integration lna appears in $\Gamma(n)$ whereas in dimensional regularization, in a^s. In our case in neither; here it comes from integration over $1/a$ via the finite $s < 0$ cases.. For analytical purposes the logarithm and its argument are dimensionless quantities. (See, for example, Veltman, M., *Diagrammatica*, Cambridge, (1994), 145.) Analytically the logarithm behaves like a constant so that $a^s lna$ is counted as a monomial (a circumstance sometimes called self-similarity) in a correspondence argument.

[21]. Feynman, R. P., *QED*, Princeton University Press, (1985), 128-29. Here Feynman expressed his dissatisfaction with the usual renormalization five different ways in a single paragraph. And nothing can match Dirac's, gentle yet devastating statement when he said:"This is just not sensible mathematics. Sensible mathematics involves neglecting a quantity when it turns out to be small—not neglecting it because it is infinitely great and you do not want it." With our proposal the self consistency of quantum electrodynamics seems now quite obvious.

Notes

A. In the case of bound state the classical interaction energy is, of course, negative. During radiation it become more negative and the difference which is positive, partly speeds up the particles of the system and partly radiates. This causes an increase in the frequency of revolutions, that is, a decrease of the periods (period decay). It is essentially the same in gravity as in the atom. However, the mechanism is more subtle because in quantum mechanics it is not the amplitudes but their eigenvalues that represent energy-momenta. In particular, the radiation field is not directly connected to the magnitudes of sources but to the higher derivatives (accelerations) of the sources (here third derivative of the quadrupole) which can be related to the positive outflow of energy into space.

B. This resolves a well-known paradox: According to the geodesic (particle) equations of motion a weightless box filled with radiation (electromagnetic or gravitational) would fall with twice the terrestrial acceleration $(1 + v^2/c^2)\gamma \to 2\gamma$ in violation of the principle of equivalence. However, if we consider the radiation in the box as standing waves, then $v \to v_g \to 0$, and the paradox disappears. This example shows a deep connection between quantum mechanics and the space-time theory of gravity. (If one wishes one can make the interaction energy positive by letting $t_\mu{}^\nu \to \lambda t_\mu{}^\nu$, $\lambda = -1$. This preserves the PPN parameters $\gamma = 1$, and $\beta = 1$; but then the needed N-body solutions will be absent.)

C. The behavior of the classical energy (I) at, say, a 50-50 beam splitter is modified by the quantum condition (II) such that the energy goes one way or the other as if in one direction the waves are reinforced and in the other compensated by a reaction wave from the environment including the vacuum. The classical wave and the quantum reactions seem to be such that in one direction there is a constructive and in the other a destructive interference. (In other words the empty path seems not empty but in a dynamic state of compensation.) Such a construction would seem to explore the geometry of the environment and to accommodate an apparent wave function collapse. It seems also to guarantee that a photon is emitted and absorbed only once. (Similar constructions are sometimes attempted in terms of advanced waves.) Interaction seems to take place through resonance. Since the relativistic frequency is very high, the resonant transfer of energy is very rapid—so rapid that it may appear like a jump to the detector which is of a lower time resolution. In all these effects the quantum appears primarily as a wave whereas the particle aspects enter indirectly through normalisation. If such a construction ever survives it may be saying that quantum mechanics is essentially a revision of the principle of action and reaction of classical physics.

D. If one takes the wave aspect alone seriously one is confronted with the wave-function collapse. (What happens to the wave after the detection?) If on the other hand one takes the particle aspect alone seriously one is confronted with the Hamamatsu experiment. (What happens to the particle in the air gap?) Copenhagen interpretation seems to accommodate the first and Paris-Dublin (de Broglie-Schrodinger) interpretation the latter. The wave-particle unity based on a normalized wave packet seems to accommodate both. However, at present the model seems to be still incomplete and further reconciliations needed.

E. Our exponential solution in gravity is sometimes also unjustly criticized. (Will, C. M., *Theory and Experiment in Gravitational Physics*, Cambridge, (1991), 19.) The exponential is a *shorthand for an iteration* from the linearized solution and, as such, contains noncommutative terms to which the criticism is directed. However, these terms are well defined via Baker-Haussdorff formula. They represent real physics because, in a general interaction process it makes a difference whether something happens before another or vice versa.. A general enough iteration solution of general relativity would have the same feature. It would seem that gravity theories which do not have this feature (for example, the pure scalar theories) that should be criticized. Other criticisms such as by C.W. Misner (unpublished) and E. D. Fackerell (unpublished) are equally unjust and will be answered if they are published. Our gravity field $\phi_\alpha{}^\beta$ is not an arbitrary 16 component construct. Its structure is such that it satisfies the requirement $(\partial_\mu \partial_\nu - \partial_\nu \partial_\mu)\phi_\alpha{}^\beta = 0$. Easiest way to see this is via its definition $\phi_\mu{}^\nu = \int (\sqrt{-g}\sigma u_\mu u^\nu)' dV'/r'$. First, there are 10 and not 16 components due to symmetry. Second, $\sqrt{-g}'$ and $1/r'$ satisfy the condition since they are scalars. On the other hand $(\sigma u_\mu u^\nu)'$ is as in the usual theory and satisfies the condition by definition as in the usual theory.

HIDDEN BACKGROUND FIELD AND QUANTUM NON-LOCALITY

R. RAMANATHAN
Department of Physics
University of Delhi, Delhi-110007 India

Abstract

It is argued that a non-local gravitation-like stochastic background field endowed with certain properties has the possibility of resolving many of the conceptual problems of quantum mechanics in an extremely simple and elegant manner. This proposal has a large overlap and similarity with the ideas of De Broglie, Vigier, Nelson and Penrose, though the mechanism for implementing it is totally novel. The non-linear quantum mechanics of Weinberg also has a counterpart in the present proposal. Incidentally such a background can also endow a very small mass to the photon.

Introduction

The twentieth century has not had the last word in physics, just as the nineteenth century did not. Physics also has to evolve with the evolution of technology which in turn will enable humankind access to yet unknown and more refined data about the phenomenal world. Therefore, if one acknowledges the truth about the evolutionary nature of physics, then it will be natural to seek redress for the shortcomings of present day physics in possible broader schemes, models and theories, which apart from subsuming existing theories gives testable prediction for future experiments. This is the broad motivation for the presentation that follows.

    The crowning achievement of twentieth century theoretical physics in terms of its wide applicability and precision agreement with experimental data is undoubtedly the development of quantum mechanics. Its success is so awesome that any doubt about its correctness is to be set aside. But deep discomfort with the conventional interpretation of quantum mechanics, especially with the superposition principle for the wave function and the Collapse postulate to reconcile it with the actualisation of an observational event, is quite widespread. This discomfort is not only at the philosophical and espistemological levels but also at the more mundane level of physical measurement. All attempts at an alternative formulation of quantum mechanics is born of this intellectual discomfort and one of the most attractive alternatives which has richness in addition to simplicity is the Stochastic formulation of quantum mechanics.

There have been numerous [1] efforts to generate quantum mechanics from a stochastic process ever since the pioneering effort of Schrödinger [2] to develop quantum mechanics, both relativistic and non-relatavistic as well as its field extensions on such a foundation. However, it must be conceded that if a quantum process is indeed a certain form of a stochastic process, it is a very peculiar kind of stochastic process in configuration space. The introduction of the complex functions to describe quantum phenomenon is rather intriguing in the context of a converntional theory of stochastic processes.

There is also an outstanding problem in the formulation of a Markovian stochastic process in a relativistically covariant form. There are general nogo theorems which show that the only relativistically covariant trajectory satisfying the Markovian property is of the form $x^\mu(\tau) = iC^\mu(\tau-\tau_0)$ where τ is the proper time and C^μ the velocity of light and τ_0 is the initial proper time [3,4]. Therefore no stochastic process is possible in ordinary Minkowski space satisfying the Markov condition. Again there have been numerous suggestions to overcome this restriction and for detail the interested reader is referred to Reference 1.

I have over the years been urging attention to a heuristic model of space time which mimics all the characteristic features of relativistic quantum mechanics and as a bonus also offers ways for testing its validity experimentally in future experiments. This proposal [1,5] for an objective formulation of quantum mechanics as a stochastic process in a complex five dimensional space-time which would take care of the hidden background noise field which I postulate as the source of the peculiar properties of quantum phenomenon viewed as a stochastic process. The background noise may probably be of gravitational origin [6], or may be something totally new. Without bothering to go into the details of this noise, I shall present a plausible model of space-time with a hindsight in order that the peculiar properties of quantum phenomenon follow naturally from the model. This kind of approach, which although in its present state of formulation has a great deal of heuristics involved, has the attraction of being a unified scheme which resolves, at least conceptually, the long standing problems of Covariant Markov process and the necessity for complex number in quantum mechanics. Even if the resolution is not completely satisfactory it at least attributes these to an unknown physical source whose manifestations, especially the non-linearities in the evolution equation to be derived, over and above that of the quantum phenomenon, should be sought in future experiments.

It will be in order to suggest a modification of Einstein's "Elevator experiment" to illustrate the effect of the universal background noise. When the elevator is being uniformly accelerated, the observer inside will interpret his motion as evidence for the existence of a uniform gravitational field in action. But supposing the elevator to be subject to a randomly varying acceleration, the observer will attribute the motions inside as due to some probabilistic Brownian-type of forces inside the elevator. He may even develop quantum mechanics to describe the dynamics of particles within such an elevator!

The metric of the complex space-time

The Central idea in our approach is the modelling of the space-time in the presence of a classical noise background which is gravitation-like and therefore affects the space-time geometry and hence the metric of the space-time. We begin by visualising space-time at short distances as a collection of points vibrating with a randomly varying amplitude

$$y_\mu(y,\tau_5) = x_\mu F(y,\tau_5)[\cos(\tan^{-1}\tau_5) + i\theta_\epsilon(\tan^{-1}\tau_5)\sin(\tan^{-1}\tau_5 a)]$$

where $F(y,\tau_5)$ is a random variable and $\theta_\epsilon(\tan^{-1}\tau_5)$ is a smeared heaviside function which has the property that:

$$\theta_\epsilon(\tan^{-1}\tau_5) = 1 \quad \text{for} \quad \tan^{-1}\tau_5 \geq \pi/2 - \epsilon$$

and

$$\theta_\epsilon(\tan^{-1}\tau_5) = 0 \quad \text{for} \quad \tan^{-1}\tau_5 < \pi/2 - \epsilon \quad \text{with} \quad \epsilon \longrightarrow 0_+$$

The variable y is the average value of the space-time point trajectory under the influence of the noise background. We further postulate the existence of a positive definite probability density function $p(\bar{y},\tau_5)$ with the property that:

$$p(y,\tau_5) = p'(x,\tau_5)$$

The line element in this stochastic space will be of the form [5]:

$$ds^2 = g_{\mu\nu}(\tau_5)dy^{\mu*}dy^{\nu}$$

where the * indicates complex conjugation, where the metric is of the form:

$$g_{ij}(\tau_5) = [\cos(\tan^{-1}\tau_5) - i\theta_\epsilon(\tan^{-1}\tau_5)\sin(\tan^{-1}\tau_5)]$$

$$g_{00}(\tau_5) = [\cos(\tan^{-1}\tau_5) + \sin(\tan^{-1}\tau_5)]$$

$$i, j = 1,2,3.$$

This space-time has the property that in the limit of compactification of the global dimension, τ_5, the average of the line element goes over to a Minkowski line element in flat space-time with the Minkowski metric $g_{\mu\nu}$ given by the signatures (+, -, -, -), provided we impose the condition

$$\lim_{\tau_5 \to \alpha} F^*(y,\tau_5) F(\bar{y},\tau_5) = 1$$

The formal expression for the average of a function $A(\bar{y},\tau_5)$ is defined by

$$\bar{A} = \int A(\bar{y},\tau_5) \, p(\bar{y},\tau_5) d^4\bar{y}$$

where $d^4\bar{y}$ is the volume element which is an invariant:

$$\lim_{\tau_5 \to \alpha} d^4\bar{y} = d^4x$$

the volume element in Minkowski space.
Also

$$\lim_{\tau_5 \to \alpha} \bar{y}_\mu = ix_\mu$$

If we endow space-time at quantum mechanical short distance with the above stated formal mathematical properties which have their basis from the postulated noise background in which all microscopic phenomena are supposed to occur, then the dynamics of a point particle in such a space could readily be seen to be similar to relativistic quantum dynamics.

The non-linear Klein-Gordon equations

In the complex five-dimensional space-time we have constructed, we can envisage the motion of a point particle of mass 'm' under the restrictive Markovian condition by generalising the Chapman-Kolmogorov equation:

$$p(\bar{y},\tau_5; \bar{y}',\tau_5') = \int \rho(\bar{y}'',\tau_5'') \, p(\bar{y}'',\tau_5''; \bar{y}',\tau_5') d^4\bar{y}'' \quad \text{for } \tau_5'' > \tau_5 > 0$$

where $p(\bar{y},\tau_5; \bar{y}',\tau_5')$ is the probability of transition from a position $\bar{y}$ at the global parametric value of the fifth dimension equal to τ_5 to another set labelled by $\bar{y}'$, τ_5'.

With this condition we may derive the evolution equation for the probability amplitudes (for details see References 1, 5).

$$-\partial/\partial\tau_5 \phi'(\bar{y},\tau_5) = H^{KFP} \phi'(\bar{y},\tau_5)$$

where H^{KFP} is a positive semi-definite non-linear Hamiltonian given by:

$$H^{KFP} = \frac{d^2}{2} \frac{\partial^2}{\partial \bar{y}^2} \frac{d^2}{2} \Lambda^2 - \frac{d^2}{2} \left[\frac{\partial}{\partial \bar{y}} \ln\phi'(\bar{y},\tau_5) - \Lambda_\mu(\bar{y},\tau_5) \right]^2$$

where

$$\phi'(\bar{y},\tau_5) = \phi(\bar{y},\tau_5) \exp[+ \oint \Lambda_\mu(\bar{y},\tau_5) dy^\mu]$$

and

$$\rho(\bar{y}\tau_5) = [\phi(\bar{y},\tau_5)]^2$$

In the special linear case when the non-linear term in the Hamiltonian vanishes (which is related to the time reversibility of the evolution equation)

$$[\partial/\partial y^\mu \ln\phi'(\bar{y},\tau_5) - \Lambda_\mu]^2 = 0$$

We may write the solution of the linearised equation as:

$$\phi'(\bar{y},\tau_5) = \sum_{n=a}^{\alpha} \phi_n'(\bar{y},\infty) \exp(-E_n, \tau_5)$$

Since all $E_n > 0$ and $E_0 = 0$ in the limit $\tau_5 \to \infty$ we have:

$$\lim_{\tau_5 \to \infty} \phi'(\bar{y},\tau_5) = \phi'_0(x,\infty).$$

Thus only the ground state of the stochastic process survives the limit which serves as the compactification limit, thus eliminating the role of the unobservable hidden dimension τ_5 from the physical evolution equation.

Thus the asymptotic evolution equation for the probability amplitudes is of the form:

$$\left[\frac{\partial^2}{\partial x^2} + \Lambda^2\right]\psi(x) = 0$$

where $\psi(x) = \phi'_0(ix,\infty)$.

This indeed is the Klein-Gordon equation with the Minkowski metric $g_{\mu\nu} = (+, -, -, -)$ if we identify Λ as the inverse Compton wavelength $\Lambda = mc/h$. Thus we see that the relativistic quantum mechanics arises as the ground state equation of a probability amplitude of a generalised Markovian Brownian process in a stochastic space-time with a few specific properties which endow the process with complexity and breakdown of euclidicity of the metric in the compactification limit. Had we retained the non-linear equation without a linearising assumption, and taken the compactification limit, we would have ended up with the non-linear evolution equation (non-linear Klein-Gordon equation):

$$\left[\frac{\partial^2}{\partial x^2} + \frac{m^2c^2}{h^2}\right]\psi(x) + \left[-i\frac{\partial}{\partial x^\mu}\ln\psi(x) + \frac{mc}{h}\left(\frac{h\Lambda_\mu}{mc}\right)\right]^2 \psi(x) = 0$$

This equation would violate time reversibility but has the possibility of opening up a whole new field of non-linear quantum mechanics which encompasses the recent non-linear quantum mechanics proposed by Weinberg [8]. In order not to violate existing experimental data the non-linear term has to be kept very small as in Weinberg's proposal. The non-linear equation proposed by Weinberg for a free particle is

$$i\frac{\partial\psi}{\partial t} = -\frac{1}{2m}\Delta^2\psi + \frac{\partial F(\psi^*\psi)}{\partial\psi^*}$$

with $F(\psi^*\psi) = \sum_{n=0} a_n(\psi^*\psi)^n$, the non-linearity of the non-linear Klein-Gordon equation would be of a different kind at the non-relativistic level but nevertheless will have similar physical implications.

As in all stochastic formulations the probability interpretation of the state vector is built into the theory and the bilinear product of the state vector is automatically the probability density

$$\psi^*(x)\psi(x) = \rho(x,\infty)$$

As in the case of Weinberg's [8] non-linear quantum mechanics, the existence of non-linear terms in the non-linear Klein-Gordon equation of the present proposal would imply breakdown of the superposition principle at some level of observational prevision and would simultaneously imply violation of time reversal invariance. Any minute violation of these two principles in future experiments may be indicative of the existence of the universal background noise which is intrinsically non-local, thus demystifying the reduction postulate of conventional quantum mechanics.

References

1. Ramanathan, R.: *Physica Scripta* 34 (1986), 365; *Phys.Rev.Lett.* 54 (1985), 495; *Int.J.Mod.Phyiscs* A7 (1992), 3035.

2. Zambrini, J.C.: *Phys.Rev.* A 33 (1986), 1532.

3. Hakim, R.: *J.Math.Phys.* 8 (1968), 1805.

4. Mountain, R.: *J.Math.* 4 (1974), 401.

5. Ramanathan, R.: *Physica Scripta* 47 (1993), 484.

6. Penrose, R.: *"Quantum Concepts in Space-time"*, (Oxford Science Publication, 1986), 129; Smolin, L.: *ibid*, 147; Nelson, E.: *"Quantum Fluctuation"*, (Princeton University Press, 1986); Roy, S.: *Phys.Lett. A* 115 (1986), 256; Kyprianidis, A.: *Physics Reports* 115 (1987).

7. Blokhintsev, D.I.: *Sov.J.Pat.Vad.* 5 (1975), 242.

8. Weinberg, S.: *Ann.Physics* 194 (1989), 336.

EINSTEIN–PODOLSKY–ROSEN CORRELATIONS WITHIN BELL REGION

RYSZARD HORODECKI
Instytut of Theoretical Physics and Astrophysics
University of Gdańsk, 80-952 Gdańsk, Poland

Abstract. Direct measurement method of revealing "hidden" EPR correlations for two spin-$\frac{1}{2}$ mixed states satisfying Bell inequalities is presented.

Einstein–Podolsky–Rosen (EPR) correlations have been widely considered in the context of the Bell inequalities for pure entangled states. Those correlations lie beyond Bell region thus they can not be described by any local hidden variable model. However they can be described by nonlocal theories involving the stochastic action–at–a–distance [1, 2, 3]. In the case of mixed states there is difficulty due to the fact that the latter "mixe" the classical and purely quantum information, leading to the correlations which can not be easily classified as purely quantum or classical ones [4, 5]. The purpose of this paper is to show that there is a direct nonlocal measurement effect, which can only occur, if the state is EPR correlated, although Bell inequalities do not need to be violated. By EPR correlated state we mean here a state which, contrary to all classical distributions, is not a mixture of product states [6].

To see it consider the unknown 2×2 state ϱ in the context of the most popular Bell inequality – the one due to Clauser–Horne–Shimony–Holt (Bell–CHSH inequality):

$$|\langle \mathcal{B}_{CHSH} \rangle_\varrho| \leq 2 \qquad (1)$$

To find experimetally, whether given unknown state ϱ violates the inequality one has to perform the EPR type experiment to find the maximal possible value of the respective Bell operator:

$$\mathcal{B}_{CHSH} = \hat{\mathbf{a}}\sigma \otimes (\hat{\mathbf{b}} + \hat{\mathbf{b}}')\sigma + \hat{\mathbf{a}}'\sigma \otimes (\hat{\mathbf{b}} - \hat{\mathbf{b}}')\sigma, \qquad (2)$$

where $\{\sigma_n\}_{n=1}^3$ are the standard Pauli matrices. The correlations needed in the context of Bell inequality

$$E(\hat{\mathbf{a}}, \hat{\mathbf{b}}) = \sum_{m,m'=\pm} mm' p_{\hat{\mathbf{a}},\hat{\mathbf{b}}}(m, m') \qquad (3)$$

are obtained from the joint distriubtions $p_{\hat{\mathbf{a}},\hat{\mathbf{b}}}(m, m')$ i. e. the probabilities of measuring quantum alternatives m and m' ($m, m' = \pm 1$) associated with arbitrary unit vectors $\hat{\mathbf{a}}, \hat{\mathbf{b}}$, respectively. Given an unknown state one has to perform a sequence of measurements with groups of four axes chosen. Then for any chosen group of four axes the probabilities $p_{\hat{\mathbf{a}},\hat{\mathbf{b}}}(m, m')$ are measured according to the usual procedure (i. e. involving random choice of pairs and spatial interval between measurements). Then the results should be substituted into the formula:

$$\langle \mathcal{B} \rangle_\varrho = E(\hat{\mathbf{a}}, \hat{\mathbf{b}}) + E(\hat{\mathbf{a}}, \hat{\mathbf{b}}') + E(\hat{\mathbf{a}}', \hat{\mathbf{b}}) - E(\hat{\mathbf{a}}', \hat{\mathbf{b}}') \qquad (4)$$

Then the needed maximal value $\langle \mathcal{B}_{\max} \rangle_\varrho$ corresponding to the maximal mean value of the operator (2) can be found. If it is strictly greater than

2 one knows that the system in the state ϱ is manifestly nonclassical as it does not admit any local–realistic theories. In some cases the results of such experiments will yield $|\langle \mathcal{B}_{max}\rangle_\varrho| < 2$ and then one can be sure to do with mixed state. As the obtained result lies within Bell region it is natural to make a conjecture that such cases are irrelevant from the point of view of the just performed experiment. Suprisingly, as we shall see, this conjecture is missleading. In fact there is some additional information which can be extracted from the experiment. Namely performing the experiment sketched above one gets all possible joint probabilities which, of course, determine the marginal distributions. The existence of the latter and, especially, their relationship to the correlations has not been customarily taken into account in considering of quantum nonseparability. However it can inform us about new quantum properties. Namely it can show us that the state of the system is EPR–correlated i.e. that in quantum formalism it can not be approximated by manifestly local convex combinations of product states. The mixed EPR–correlated states have young and promising history. They have been recently shown to be able to posess strictly nonlocal statistical properties despite admitting local hidden variable (LHV) model satisfying therby all possible Bell type inequalities [7, 8]. There is a suggestion that *any* system in EPR correlated state, even in the one satisfying Bell inequalities, posesses some nonlocal, so far hidden properties leading to new physical effects. We shall show how in some cases one can easily convince of the existence of "hidden" correlations within the state which lead to violation of classical entropy inequality, though the Bell–CHSH inequality is satisfied.

It can be shown strightforwardly that both classical discrete states [1] and quantum classically correlated states described before satisfy the following entropy inequalities [9]:

$$S_2(\varrho) \geq \max_{i=1,2} S_2(\varrho_i), \qquad (5)$$

$S_2(\varrho)$ stands here for either the quantum Reney entropy $S_2(\varrho) = -\ln Tr\varrho^2$ or its classical counterpart. In quantum case the state ϱ is a density matrix corresponding to the quantum mechanical description of the system, while the states ϱ_i, i=1, 2 are reductions of ϱ. The quntum version of the inequality (5) can be then rewritten as a condition [10]

$$Tr\varrho^2 \leq \min_{i=1,2} Tr\varrho_i^2 \qquad (6)$$

describing manifestly classical relations between purities of the state of the system and its reductions. Furthermore one can show that if we restrict to

---
[1] By the classical discrete states we mean here the distributions $p_{ij} \geq 0$, $\sum_{ij} p_{ij} = 1$

the 2×2 system then the condition (6) involves new physical constraints on results of the EPR type experiment within Bell region. Namely it imposes that:

$$|\langle \mathcal{B}_{CHSH}\rangle_\varrho| \leq 2\sqrt{1 - |||\mathbf{r}||^2 - ||\mathbf{s}||^2|} \qquad (7)$$

for any Bell observable $\mathcal{B}_{CHSH}$, corresponding to any possible choices of pairs of axes in EPR type experiment. The parameters $||\mathbf{r}||, ||\mathbf{s}||$ are lenghts of Bloch vectors describing the spectra of reductions ϱ_1 and ϱ_2 respectively.

One can show that there are infinitely many 2 × 2 states violating the inequality (7) within Bell region. Simple examples of such states are the following [10]:

$$\varrho = \sum_{i=1}^{2} p_i |\psi_i\rangle\langle\psi_i|, \quad \sum_i p_i = 1 \qquad (8)$$

where

$$|\psi_1\rangle = ae_1 \otimes e_1 + be_2 \otimes e_2 \qquad (9)$$
$$|\psi_2\rangle = ae_1 \otimes e_2 + be_2 \otimes e_1 \qquad (10)$$

with $a, b > 0$, $\{e_i\}$ being standard basis in C^2, $0 < (p_1 - p_2)^2 \leq (a^2 - b^2)^2$. Using the condition on violating Bell–CHSH inequality by an arbitrary 2 × 2 state [11] it is straightforward to prove that the states (8) violate the inequality (7). Thus, under the consideration they violate the inequality (6) being then EPR correlated.

Now let us turn back to the EPR type experiment described at the begining. So far one has maximal possible value of the Bell observable $|\langle \mathcal{B}_{\max}\rangle_\varrho|$. However it was obtained from correlation dictributions $p_{\hat{\mathbf{a}},\hat{\mathbf{b}}}(m, m')$ measured in the experiment. Taking their marginals:

$$p_{\hat{\mathbf{a}}}^1(m), \quad p_{\hat{\mathbf{b}}}^2(m'), \quad m, m' = \pm 1 \qquad (11)$$

one can easily find lenghts of Bloch vectors. To do it for the first subsystem, one needs only to find the marginals for which the Klein entropy

$$S_K(\hat{\mathbf{a}}, \varrho_1) := -\sum_{m=\pm} p_{\hat{\mathbf{a}}}^1(m) \ln p_{\hat{\mathbf{a}}}^1(m) \qquad (12)$$

attains its minimum (i. e. when the distance $|p_{\hat{\mathbf{a}}}^1(+) - p_{\hat{\mathbf{a}}}^1(-)|$ is the greatest). Then it is equal to the von Neumann entropy [2]:

$$S(\varrho_1) = -Tr\varrho_1 \ln \varrho_1 \qquad (13)$$

[2] Note that von Neumann measurement is mixing–enhancement [12].

and the marginals corresponding to this minimum coincide with the spectrum ϱ_1 which has the spectral decomposition :

$$\varrho_1 = p_+ P_+ + p_- P_- \qquad (14)$$

with projectors $P_\pm = \frac{1}{2}(I \pm \hat{\mathbf{r}}\sigma)$ and probabilities $p_\pm = \frac{1}{2}(1 \pm ||\mathbf{r}||)$ simply related to seeked parameter $||\mathbf{r}||$. The analogous reasoning leads to finding the parameter $||\mathbf{s}||$ of the second subsystem. Thus if the system has been prepared in the state of the form (8) it can be found during EPR type experiment via violation the inequality (7). The above prescription is the first direct measurement proposal of seeking the EPR correlated states satisfying Bell inequality. In fact, other proposals due to Popescu [7, 8] leading to manifestly nonlocal effects in the case of Werner states either involve nonstandard measurements and can be applied only for higher spins, or require the apparatus measuring nonlocal observables.

Finally we see that the fact that the state is EPR correlated can manifest itself as an effect of violation the inequality (7) within Bell's region. The latter is always satisfied by classicaly correlated states. It is remarkable that the above effect involves strictly nonlocal aspect of the mixed state (correlations) as well as its local ones (marginal distributions), both of them in the same experiment. It is in contrast with the usual EPR–Bell experiment with pure states (see, for instance [13]) where only correlations were needed. Finally one can speculate that if ϱ is interpreted as a reduction of a pure state of some larger system then its nonlocality (nonseparebility) may be viewed as a quantum potential associated with the pure state.

The author would like to acknowledge usefull comments by Nicolas Gisin.

References

1. J.P. Vigier, Non–locality, causality and aether in quantum mechanics, Astron. Nachr. 303 (1982) 55–80
2. N. Cufaro–Petroni and J. P. Vigier, Causal action-at-a-distance interpretation of the Aspect–Rapisarda experiments, Phys. Let. 93A (1983) 383–387.
3. D. Bohm, B. J. Hiley and P. N. Kaloyerou, An ontological basis for the quantum theory, II. A causal interpretation of quantum fields, Phys. Rep. 144 (1987) 349–375
4. R. Horodecki and P. Horodecki, Quantum redundancies and local realism, Phys. Lett. A 194 (1994) 147–152.
5. P. K. Aravind, To what extent do mixed states violate the Bell inequalities?, Phys. Lett. A 200 (1995) 345–349.
6. R. F. Werner, Phys. Rev. A 40 (1989) 4277–4281.
7. S. Popescu, Bell's inequalities versus teleportation: what is nonlocality?, Phys. Rev. Lett. 72 (1994) 797–799.
8. S. Popescu, Bell's inequalities and density matrices: reveling "hidden" nonlocality, Phys. Rev. Lett. 74 (1995) 2619–2622.

9. R. Horodecki, P. Horodecki and M. Horodecki, Quantum α-entropy inequalities: independent condition for local realism?, to be published.
10. R. Horodecki, Two spin-$\frac{1}{2}$ mixtures and Bell inequalities, to be published.
11. R. Horodecki, P. Horodecki and M. Horodecki, Violating Bell inequality by mixed spin-$\frac{1}{2}$ states: necessary and sufficient condition, Phys. Lett. A 200 (1995) 340–344.
12. A. Wehrl, General properties of entropy, Rev. Mod. Phys. 50 (1978) 221–259.
13. A. Aspect and P. Grangier, Tests of Bell's inequalities with pairs of low energy correlated photons: an experimental realization of Einstein-Rosen-Podolski type correlations, *Symposium on the foundations of modern physics*, Joensu, Finland (1985), ed. P. Mittelstaedt (Univ. of Cologne, Köln) pp. 51–71

ELECTROMAGNETIC GAUGE AS INTEGRATION CONDITION
AND SELECTION OF THE SOURCE ADHERING GAUGE

O. COSTA DE BEAUREGARD
76 RUE MURGER 77780 BOURRON-MARLOTTE (FRANCE)

ABSTRACT. Sequel to previous papers with analogous titles. QA^i as potential extra-inertia of a point charge Q moving inside a source adhering 4-potential A^i. Ampère tension as recoil force on a generator and a motor connected by straight parallel conductors.

1. INTRODUCTION

This is a sequel to my contribution[1] in the volume *Advanced Electrodynamics* edited by T.W. Barrett and D.M. Grimes. The aim in it was to show, via Einstein's mass-energy equivalence law and action-reaction opposition between sources of the field, how the electromagnetic gauge is selected as an integration condition[2]; as gauge invariance is a *differential* principle this is quite acceptable. Indeed, the Coulomb gauge is selected in the expression c^{-2} eV of the classical electron mass, or also of the Sommerfeld atomic mass defect.

A *general statement* (G.S.) adduced in this paper is this one: A^i *denoting the source adhering 4-potential* (say, the half-retarded-half-advanced Liénard-Wiechert one used by Wheeler and Feynman[3]), *the "potential 4-momentum"* $P^i = QA^i$ *of a point charge expresses the reaction on the charge's motion due to the sources of the field.* This we interpret as *the "absorber theory"*[4] *version of the reactive force.*

Ad hoc thought experiments exemplify this in Sections 2 to 5 included; finally we derive our G.S. from the Wheeler-Feybman electrodynamics in Section 6.

In some cases not $P^i = QA^i$, but its projection $W^i = c^{-2} QV_k V^k V^i$ on the 4-velocity V^i ($V_i V^i = -c^2$) is involved. As $P_i dx^i = W_i dx^i$, both 4-vectors are equivalent in the variational derivation of the equation of motion. As is well kown[5] non-collinearity of 4-velocity and 4-momentum entails existence of the 6-component spin-and-boost tensor[5] $Q[A^i V^j - A^j V^i]$ (see reference 1, Section 11).

Action-reaction opposition between the vector-potential's sources we have used[1,6] in the magnetostatics of currents to show that $i\mathbf{A}$, with $d\mathbf{A} = r^{-1} i'\mathbf{dl'}$, is the physical, testable stress-tension along a current carrying wire. This "Ampère tension" we revisit here in Sections 4 and 5 in the form of opposite recoils imparted to a generator and a motor connected by straight parallel conductors, or to the anode and cathode of an electron gun -two more proofs of the G.S.

2. INERTIAL MOTION OF A POINT CHARGE
IN THE FIELDLESS VECTOR POTENTIAL OF A TOROIDAL MAGNET

As a preamble consider the electro-magneto-static system consisting of a point charge Q and a toroidal magnet trapping its flux O both at rest in some inertial frame. The magnet can be any topological torus idealized as a wire carrying a magnetic moment O $\mathbf{dl}$ per line element.

Neither in the vacuum nor in the charge nor in the magnet is there mutual enrgy.

As for the *a priori zero mutual momentum, it is entirely located in the charge and the magnet,* in the form of the two opposite *potential* momenta

$$P_c = Q A , \quad P_m = \oint \oint B \times dl ; \quad (1)$$

A and E denote the magnet's vector potential of the charge's electric field, of respective expressions

$$A = \oint \oint r^{-3} r \times dl , \quad E = -r^{-3} Q r . \quad (2)$$

Provided that the Ampère gauge be selected, as expressed in (1), there is action-reaction opposition:

$$P_c + P_m = 0 . \quad (3)$$

Let us make sure that the Poyinting momentum P_m is operational. Suppose that initially the magnet was in a metastable unmagnetized state; when it spontaneously magnetizes a magnetic current appears in it, and the Lorentz style force applied to it by the electric field is just the one needed. As for the point charge, it feels the induced electric field, and gets the right momentum. This is proof of physicality of the vector potential in the line of J.J. Thomson[6] and Konopinski[7].

Now we assume that the magnet and the charge are in relative inertial motion; as we shall see this requires the magnet to be very much heavier than the charge because, while the charge moving inside the magnet's *identically* fieldless vector-potential feels no force, *the magnet does feel the Stern-Gerlach style force generated by the moving charge's magnetic field.*

To simplify the proof we assume that the magnet's loop lies in a plane, and that the axis z along which the charge moves at a velocity v<<c is perpendicular to that plane. Then $dP_m/dt = v \, dP_m/dz$ exhibits the force applied to the magnet as $2 \pi r \oint dB/dz$, where $B = v \times E$ denotes the magnetidc field of the flying charge.

The paradox "no force felt by the charge but a force felt by the magnet" evidences QA as potential momentum -the G.S. statement.

Let us discuss energy conservation. No Maxwell energy builds up in the vacuum, but a Maxwell energy

$$W = Q \oint \oint B.dl = Q \oint \oint [E \times v].dl \quad (4)$$

appears in the magnet; according to formulas (2) it is compensated in the charge by an energy

$$-W = Q A.v . \quad (5)$$

This is the energy of conducting electrons generating the energy $dW = i \, A.dl$ *per line element of a current.* If the circuit is interlaced with a toroidal magnet, this energy produces the mutually induced electromotive force $(di/i \, dt) \oint A.dl$. We show in Section 4 below that this phenomenon does occur along a resistive conducting wire, the so-called Ampère[1,2,8] stress-tension $T = iA$ being thus transferred to the lattice.

Finally we remark that the P and W defined via (1) and (4) do *not* combine into a spacetime vector: P is the space projection of the (here spacelike) P^i, and W the time projection of the (timelike) W^i defined in the Introduction.

3. ARCHIMEDIAN-STYLE ELECTROSTATIC LIFT OR LEST

The lift exerted by a toy helium balloon sitting at the ceiling of a parked caravan is substracted from the weight the caravan would have if air were not "displaced" by helium.

If someone inside the caravan pushes the balloon laterally, barycenter conservation entails anti-recoil of the system; thus, viewed against the background, the person is projected forward (and indeed one has a queer feeling of no-resistance when pushing a toy helium balloon).

So, *Archimedian lift mimicks anti-gravity and anti-inertia.*

Now we show that *Einstein's inertia-gravity equivalence entails an analogous electrostatic phenomenology.*

The way an electromagnetic 6-field B^{ij} accelerates a point charge Q is isomorphic to a space-time angular velocity: QB acts à la Coriolis (whence "Larmor's theorem") and QE acts as a boost.

Uniform Born-style acceleration of a uniformly charged hollow sphere generates inside an electric field, because uniform acceleration is a timelike instanteneous rotation. The covariant expression of the electromagnetic field is c^{-2} V $[U^i U'^j - U^j U'^i]$, V=Q/R denoting the enclosed potential, U^i the 4-velocity and U'^i the 4-acceleration ($U_i U^i = -c^2$, $U_i U'^i = 0$).

Einstein's *inertia-gravity equivalence principle* then entails that a uniform gravity field **g** induces inside a uniformly charged sphere an electric field $E = c^{-2}$ V**g**; as Mach's background gravity potential is $U = c^{-2}$ GM/R = 1, this formula amounts to UE = V**g**.

An alternative form of the statement is: *A point charge Q immersed in a uniform electric potential V is endowed with an extra inertial mass (not rest-mass!)* c^{-2} VQ.

Thus, inside a fieldless potential V ≥ +511 000 volts, an electron should levitate. An experimental test would be difficult due to smallness of the electric field c^{-2} V**g**. But an analogous, more tractable, test using angular momentum conservation has been proposed elsewhere[1].

4. RECOIL FORCES ON GENERATOR AND MOTOR CONNECTED BY STRAIGHT PARALLEL CONDUCTORS

Tension is a potential concept. Dormant along a stretched thread it awakes where the thread breaks or is cut; there, it shows up as *repulsive*. Like pressure is a volume energy density, so the tangential component T.t of a tension T is a linear energy density: think for example of a weight held by a coil spring in a uniform gravity field. Thus the tangential projection T.t of the tension T is the potential energy stored per line element of a filament[1,6].

In the ideal case of zero resistivity denote as ±E the electromotive and counter-electromotive forces of a generator and a motor connected by straight parallel conductors; as ±V=±E/2 the opposite voltages on the lines, and as ±q the linear charge densities they carry [compacted charge carriers move slower, rarefied ones faster]. The V's are Coulomb gauged electric potentials proportional to ±q: V=kq, k denoting a (largely phenomenological) dimensionless constant equal (as will be shown) to the self-induction coefficient per length unit.

Assuming that the transverse Biot-Savart forces repelling the parallel conductors are cancelled by an appropriate binding (as in commercial paired conductors) there remains a repulsive tension T induced along the lines, by emission from the generator and reception by the motor of the power EI at velocity $v=i/q$, i denoting the current's intensity. As opposite momentum densities $\pm Viv$ flow along the lines, these are stressed by a tension $T=Viv=ki^2$. And as $A=Vv$ is, at first order in v/c, the local, self-induced, Ampère gauged vector potential $A=ki$, we get $T=iA$ as value of the repulsive tension along both lines -a result derived in full generality elsewhere [1,6].

This dormant tension is actualizable[9] and measurable[10] anywhere along each line, for example by inserting a mercury loaded junction; the value of the self-induction coefficient then comes out as a by-product.

If there is resistivity o, the tension T is locally transferred to the lattice according to the formula $dT/dl=ivE=oi^2$, $E\equiv dV/dl$ denoting the longitudinal electric field. This is analogous to transfer of the Biot-Savart force to the lattice of a resistive Barlow wheel.

5. RECOIL FORCES ON ANODE AND CATHODE OF AN ELECTRON GUN. CASE OF GUN SHOOTING ALONG THE AXIS OF A CYLINDRICAL CAPACITOR

An electron gun emitting N electrons per second at velocity v and energy -eV confers to each a momentum -eVv, and thus feels a recoil force $T=iVv$, with $i=-Ne$; thus we recover the formula $T=iA$.

As emphasized by de Broglie[11] *the gauge is fixed in this problem as an integration condition*: expressing the dynamic mass $m=m_0\{1+(v/c)^2/2\}-c^{-2}eV$, and thus the momentum as $m_0 v-c^{-2}eA$, fixes the gauge via $V=kq$ and $A=ki$.

Experimental selection of the gauge is then possible. Suppose that the gun shoots along the axis of a cylindrical capacitor enclosing a (Coulomb gauged) fieldless potential $V_0 = q_0 Ln(a/b)$, its cathode being at that potential. Centered upon each flying electron there is a co-moving Maxwellian mutual energy cloud trapped between the capacitor's plates. Each electron, of relativistic mass m, is thus endowed with a renormalized mass $M=m-c^{-2}eV_0$ and momentum Mv (which is easily verified by likening the string of beads to a continuous thread and integrating the mutual Maxwell energy and Poyinting vector between the plates). The same conclusion is drawn by locating the mutual energy on the plates with the surface density qV_0.

The recoil force T=NMv is in principle measurable by collecting the flying electrons on an elastically supported second cathode at potential V_0, acting as a dynamometer.

So, *in analogy with the Aharonov-Bohm effect, here is an example of far-action of an expelled field, equivalent to direct action of a fieldless potential.*

But here *the potential itself, not the potential difference, is measurable*, as V_0 must be expressed in the Coulomb gauge; thus *the electromagnetic gauge is fixed as an integration condition.*

6. LORENTZ CONDITION AND ACTION-REACTION OPPOSITION

As is well known, the Lorentz condition entails that the arbitrary superpotential propagates according to d'Alembert's equation, and is thus likened to a sourceless field magnitude. The general 4-potential solving Maxwell's equations then is the sum of a source-adhering one, say the half-retarded-half-advanced Liénard-Wiechert one used by Wheeler and Feynman[2], plus the general solution of d'Alembert's equation.

Let us show that *discarding the ghostlike solution and retaining the source adhering one yields direct action-reaction opposition between the 4-potential's sources.*

Writing the Lorentz equation of motion of a point charge Q in the form displaying the 4-potential

$$dP^i = d(p^i + QA^i) = Q(\ ^iA_j)\ dx^j \qquad (6)$$

and substituting in it the Wheeler-Feynman[2] 4-potential generated by an other charge Q'

$$dA^i = Q'\ (r^2)dx'^i \qquad (7)$$

we get the W.F. equation of motion

$$dP^i = -dP'^i = F^i ds \qquad (8)$$

displaying direct action-reaction opposition; the 4-force

$$d^2 F^i = QQ'\{d\ (r^2)/dr\}\ dx_j dx'^j\ r^i \qquad (9)$$

is isomorphic to the "shortened Ampère force density"

$$d^2 f = r^{-3}\ ii'(dl.dl')r \qquad (10)$$

of standard use in the magnetostatics of currents.

As the inertial force on a point-particle equals minus the total ponderomotive force we reach this *General Statement: A^i being expressed in the source adhering gauge, the reaction of the other electromagnetic sources on a point charge Q moved by whatever force is the time derivative of the potential 4-momentum QA^i at Q.*

7. A WEBER 1848 FORMULA DERIVED FROM THE LORENTZ FORCE FORMULA

Two point charges following each other along an axis x at the same velocity v (say electrons in a cathodic beam, or charge carriers inside a thin straight wire) each feel the electric field E, but not the magnetic field B generated by the other; E's value, according to the Lorentz transform formulas, is the same as in the comoving frame.

The rest distance r_0 between the charges is Lorentz contracted; also, retarded propagation of the fields must be taken care of which, neglecting radiation, is best done by assuming half-retardation-half-advance. As, at first order in v/c, r_0^{-2} goes into $\{1+(v/c)^2\}\ r^{-2}$, we derive from the Lorentz force formula the Weber 1848 formula[12]

$$F = \pm\ r^{-3}\ Q_a Q_b\ \{1+(v/c)^2\}\ r\ , \qquad (11)$$

a generalized Coulomb force formula exhibiting direct action-reaction opposition.

As for two charge carriers in a thin straight wire, the first term in the r.h.s. (the Coulomb force proper) is cancelled by the charges of the lattice; the second term generates the Amperian repulsive force formula (11) between two current elements along a straight conductor. Transference of this force from the charge carriers to the lattice has been explained in Section 4.

8. ON GRANEAU'S, SAUMONT'S AND ROBSON-SETHIAN'S EXPERIMENTS

Varying at constant intensity i (with the help of an electromotive force) (twice) the self-energy $W = 1/2\,IO$ of a current carrying loop we get (A^l denoting the 4-potential and B^{lj} the 6-field) the work of the Biot-Savart or Grassmann force in the form

$$i\,O = i\quad A.dl = i\,[B \times dl].\,l = i\quad B.[dl \times l] \; ; \qquad (12)$$

integral equivalence of the Ampère tension iA and the Biot-Savart force i B x dl is thus displayed.

Formula (11) has relevance in the Ampère tension tests performed by Graneau[9], Saumont[10] and Robson-Sethian[13]. The latter, implying a zero magnetic flux variation, yields a zero result. The two former ones, where there is a flux variation, do evidence the Ampère tension T (in Saumont's experiments there exists a locally strong magnetic field where the wires bend to dip into the mercury cups).

9. CONCLUSION:

Various thought experiments, plus a general derivation from the Wheeler-Feynman electrodynamics, do evidence that QA^l, with A^l expressed in the source adhering gauge, is an extra potential momentum applied to a moving point charge by the sources of the field; this we interpret as the absorber theory version of the reactive force.

The Aharonov-Bohm effect is not the only one displaying far-action of an expelled electromagnetic field or, equivalently, direct action of a fieldless potential; various examples have been produced.

REFERENCES

1. O. Costa de Beauregard, in *Advanced Electrodynamics*, T.W. Barrett and D.M. Grimes eds, World Scientific, Singapore (in press).
2. O. Costa de Beauregard, *Found. Phys.* **22**, 1485 (1992).
3. J.A. Wheeler and R.P. Feynman, *Rev. Mod. Phys.* **21**, 425 (1949).
4. J.A. Wheeler and R.P. Feynman, *Rev. Mod. Phys.* **17**, 157 (1945).
5. H.C. Corben, *Classical and Quantum Theories of Spinning Particles*, Holden Day, San Francisco (1968).
6. J.J. Thomson, *Phil. Mag.* **8**, 331 (1904); see pp. 347-349.
7. E.J. Konopinski, *Amer J. Phys.* **46** (1978).
8. O. Costa de Beauregard, *Phys. Lett.* A **183**, 41, (1993).
9. P. Graneau, *J. Appl. Phys.* **53**, 6648 (1982).
10. R. Saumont, *Phys. Lett.* A **165**, 307 (1992).
11. L. de Broglie, *C.R. Acad. Sci.* (Paris) **225**, 163 (1947).
12. W.E. Weber, *Ann. der Physik* **73**, 229 (1848).
13. A.E. Robson and J.D. Sethian, *Am. J. Phys.* **60**, 1111 (1992).

AUTHOR INDEX

Ahluwalia, D.V., 443
Andrews, T.B., 181
Badurek, G., 281
Bashkov, V., 159
Baublitz, M. Jr., 193
Beil, R. J., 9
Bergia, S., 259
Bitsakis, E., 333
Borzeszkowski, H.H.von, 395
Boudet, R., 471
Božić, M., 205
Chekhov, A.Yu., 97
Clarkson, M., xiii, 349
Costa de Beauregard, O., 541
Croca, J.R., 305
Dalton, B.J., 235
Desroches, J., 151
Dubovik, V.M., 141
Efremov, G.F., 97, 103
Evans, M.W., 117
Garuccio, A., 373
Gavrilin, A.T., 217
Hathaway, G., 127
Horodecki, R., 535
Hunter, G., 37
Ignatovich, V.K., 293
Jeffers, S., 127, 151
Kajamaa, J., 57
Kälbermann, G., 45
Kraft, D.W., 405
MacGregor, M.H., 17
Marmet, P., 383
Marshall, T.W., 67
Martsenuyk, M.A., 141

Moles, M., 483
Motz, L., 405
Mourokh, L.G., 97, 103
Novikov, M.A., 97, 103, 127
Ord, G.N., 165, 169
Papini, G., 247
Pavičić, M., 311
Pope, N.V., 495
Provost, D., 269
Prosser, R.D., 151
Ramanathan, R., 527
Rowlands, P., 361
Roy, M., 107
Roy, S., 107
Ryff, L.C., 323
Sachs, M., 79
Saha, B., 141
Santos, E., 67
Selleri, F., 413
Sternglass, E.J., 459
Surdin, M., 437
Tchernomorov, A., 159
Vatsya, S.R., 223
Vigier, J.-P., xiii
Whitney, C.K., 1
Wood, W.R.., 247
Yilmaz, H., 503

SUBJECT INDEX

action, 405
action function, 408
action-at-a-distance, 88
Aharanov-Bohm effect, optical, 124
Ampère gauge, 542
asymmetrical beam splitter, 311
$B^{(3)}$ field, 107, 117
B-cyclic equations, 123
B-cyclics, 118
Bell experiment, loophole-free, 317
Bell inequalities, 535
Bell's theorem, 323
black-body radiation, 406
Bohm deterministic theory, 194
Bohm-Vigier stochastic theory, 193, 199
Bohm's interpretation of quantum mechanics, 259
Brittingham solutions, 9
Brownian motion, 169
circularly polarized light, 127
classical trajectories, 411
classical light field, 69
coherent pumping, 323
collapse of the wave packet, 305
conjugate product, 117
correspondence principle, 269
cosmological redshift, 483
cyclically symmetric equations, 117
De Broglie,
    relation, 405
    standing wave, 407
    variable mass problem, 255
    wave packets, 295
    wavelength, 58, 63, 186
    waves, 305
delayed-action-at-a-distance, 85
density of paths, 411
diffusion equation, 194
Dirac equation, 169
Dirac's bispinor equation, 82
Doppler effect, 406
down-converted photon pairs, 316
E-cyclics, 118
Einstein locality, 373
Einstein-De Broglie-Proca Theory, 107
Einstein-Podolsky-Rosen correlations, 535
Einstein's relativity, 383
electromagnetic fields, 166
electromagnetic potentials, 1
electromagnetotoroidics, 141

electron field emission, 195
electron-positron excitations, 106
electroweak unified theory, 53
electroweak and gravitational fields, 119
energy operator, 405
enhancement, 73
EPR correlations, 323
extended models of the electron, 459
Fermat's principle, 224
fermion, 119
Field quantization, 122
gauge group symmetry, 123
Gauss-Mainardi-Codazzi formalism, 248
guidance formula, 253
Hamilton-Jacobi equation, 119
Hamiltonian origin of Schrödinger equation, 260
Hamilton's principal function, 409
hidden variables states, 326
hidden background field, 527
hidden variable theories, 311
Hubble law, 483
induction magnetometers, 128
inertial transformations, 413
informational element, 496
informational sequences, 496
intensity interference, 311
Inverse Faraday Effect, 120, 132
Klein-Gordon equations, 531
Lamb shift, 84
Lienhard-Wiechert potentials, 1
local realism, 326
locality, 327
logical positivism, 79
long-range corrrelations, 270
loophole-free Bell experiment, 317
Lorentz invariance, 437
Mach-Zehnder interferometer, 308
magnetization by light, 118
magneto-optic phenomena, 117
magnetoelectronics, 141
mass as wave motion, 39
Maxwell equations, 151, 159
    vacuum, 118
    unquantized, 67
    nonlinear stochastic, 103
Maxwell's electrodynamics, 45
measurement problem, 349
momentum operator, 405
Nadelstrahlung, 9

nature of particles, 38
neutron interferometry, 198, 281, 344
Noether's theorem, 92
non-locality of quantum mechanics, 373
non-zero photon mass, 107
nonenhancement, 330
nonlinear media, 103
nonlinear Fierz-Pauli equations, 398
nonlocal effects, 254
nonlocality, 311
null direction, 9
operator equation, 408
optical Aharanov-Bohm effect, 124
optomagnetic effects, 127
parametric down conversion, 324
particle's internal De Broglie wave, 40
path-integral formalism, 223
periodic orbits, 278
perturbative general relativity, 395
phase matching condition, 327
photon energy, 406
photons, 1, 17
photons as soliton waves, 42
physical paths, 228
Planck formula, 405
Poincaré group symmetry, 121
Poincaré surface, 273
polarization, 312
position-momentum commutator, 405
positivism, 496
Poynting vectors, 152
preselection experiment, 311
psi-wave packets, 17
quantization of action, 408
quantum beat phenomena, 205
quantum electrodynamics, 45, 80, 97, 395
quantum interference, 205
quantum mechanical tunneling, 349
quantum mechanics, 17, 281
    Bohm's interpretation of, 259
    casual interpretation of, 247
    causal theory of, 352
    comprehensibility in, 503
    deterministic explanation of, 235
    geometric, 200
    incompleteness of, 298
quantum nonlocality, 68
quantum potential, 194, 265
quantum theory of measurement, 205
quantum waves, real, 306

radio-frequency pulses, 120
realism, 496
relativistic quantum systems, 97
relativistic electron pair model, 463
relativistic particle models, 61
relict magnetic field, 124
renormalization, 503
Riemann field metric, 235
s-state, 200
Sagnac effect, 419
Schrödinger equation, 181, 193
    Hamiltonian origin of, 260
second quantization, 312
spin entanglement, 311
spin-correlated interferometry, 312
spin-one boson, 445
spontaneous parametric down conversion, 323
stable orbits as limit cycles, 263
standing De Broglie wave, 407
Stern-Gerlach experiments, 243
stochastic quantum mechanics, 217
stochastic non-Markov model, 217
stochastic theory, 193
strong field limit, 119
Takabayasi moving frame, 471
telegraph equations, 165
teleportation of quantum state, 374
toroid multipole moments, 141
toroidomagnetics, 141
toroidomagnetostatics, 141
transition probability, 260
transverse vector potentials, 119
tunneling, 193
ultracold neutrons, 294
unsharp measurement, 107
vacuum-polarization "P-field", 17
wave equation, 9
wave and particle paradigms, 37
wave function, 405
wave system theory, 183
wave-particle duality, 333, 361
wave-particle duality paradox, 43
wave-particle unity, 510
weak field limit, 119
Weyl geometry, 248
Wheeler-Feynman theory, 86
Wigner Little Group, 121
Wigner densities, 67
zeropoint field, 71

Fundamental Theories of Physics

Series Editor: Alwyn van der Merwe, *University of Denver, USA*

1. M. Sachs: *General Relativity and Matter*. A Spinor Field Theory from Fermis to Light-Years. With a Foreword by C. Kilmister. 1982 ISBN 90-277-1381-2
2. G.H. Duffey: *A Development of Quantum Mechanics*. Based on Symmetry Considerations. 1985 ISBN 90-277-1587-4
3. S. Diner, D. Fargue, G. Lochak and F. Selleri (eds.): *The Wave-Particle Dualism*. A Tribute to Louis de Broglie on his 90th Birthday. 1984 ISBN 90-277-1664-1
4. E. Prugovečki: *Stochastic Quantum Mechanics and Quantum Spacetime*. A Consistent Unification of Relativity and Quantum Theory based on Stochastic Spaces. 1984; 2nd printing 1986 ISBN 90-277-1617-X
5. D. Hestenes and G. Sobczyk: *Clifford Algebra to Geometric Calculus*. A Unified Language for Mathematics and Physics. 1984
   ISBN 90-277-1673-0; Pb (1987) 90-277-2561-6
6. P. Exner: *Open Quantum Systems and Feynman Integrals*. 1985 ISBN 90-277-1678-1
7. L. Mayants: *The Enigma of Probability and Physics*. 1984 ISBN 90-277-1674-9
8. E. Tocaci: *Relativistic Mechanics, Time and Inertia*. Translated from Romanian. Edited and with a Foreword by C.W. Kilmister. 1985 ISBN 90-277-1769-9
9. B. Bertotti, F. de Felice and A. Pascolini (eds.): *General Relativity and Gravitation*. Proceedings of the 10th International Conference (Padova, Italy, 1983). 1984
   ISBN 90-277-1819-9
10. G. Tarozzi and A. van der Merwe (eds.): *Open Questions in Quantum Physics*. 1985
    ISBN 90-277-1853-9
11. J.V. Narlikar and T. Padmanabhan: *Gravity, Gauge Theories and Quantum Cosmology*. 1986 ISBN 90-277-1948-9
12. G.S. Asanov: *Finsler Geometry, Relativity and Gauge Theories*. 1985
    ISBN 90-277-1960-8
13. K. Namsrai: *Nonlocal Quantum Field Theory and Stochastic Quantum Mechanics*. 1986 ISBN 90-277-2001-0
14. C. Ray Smith and W.T. Grandy, Jr. (eds.): *Maximum-Entropy and Bayesian Methods in Inverse Problems*. Proceedings of the 1st and 2nd International Workshop (Laramie, Wyoming, USA). 1985 ISBN 90-277-2074-6
15. D. Hestenes: *New Foundations for Classical Mechanics*. 1986
    ISBN 90-277-2090-8; Pb (1987) 90-277-2526-8
16. S.J. Prokhovnik: *Light in Einstein's Universe*. The Role of Energy in Cosmology and Relativity. 1985 ISBN 90-277-2093-2
17. Y.S. Kim and M.E. Noz: *Theory and Applications of the Poincaré Group*. 1986
    ISBN 90-277-2141-6
18. M. Sachs: *Quantum Mechanics from General Relativity*. An Approximation for a Theory of Inertia. 1986 ISBN 90-277-2247-1
19. W.T. Grandy, Jr.: *Foundations of Statistical Mechanics*.
    Vol. I: *Equilibrium Theory*. 1987 ISBN 90-277-2489-X
20. H.-H von Borzeszkowski and H.-J. Treder: *The Meaning of Quantum Gravity*. 1988
    ISBN 90-277-2518-7
21. C. Ray Smith and G.J. Erickson (eds.): *Maximum-Entropy and Bayesian Spectral Analysis and Estimation Problems*. Proceedings of the 3rd International Workshop (Laramie, Wyoming, USA, 1983). 1987 ISBN 90-277-2579-9

Fundamental Theories of Physics

22. A.O. Barut and A. van der Merwe (eds.): *Selected Scientific Papers of Alfred Landé.* [*1888-1975*]. 1988 ISBN 90-277-2594-2
23. W.T. Grandy, Jr.: *Foundations of Statistical Mechanics.*
    Vol. II: *Nonequilibrium Phenomena.* 1988 ISBN 90-277-2649-3
24. E.I. Bitsakis and C.A. Nicolaides (eds.): *The Concept of Probability.* Proceedings of the Delphi Conference (Delphi, Greece, 1987). 1989 ISBN 90-277-2679-5
25. A. van der Merwe, F. Selleri and G. Tarozzi (eds.): *Microphysical Reality and Quantum Formalism, Vol. 1.* Proceedings of the International Conference (Urbino, Italy, 1985). 1988 ISBN 90-277-2683-3
26. A. van der Merwe, F. Selleri and G. Tarozzi (eds.): *Microphysical Reality and Quantum Formalism, Vol. 2.* Proceedings of the International Conference (Urbino, Italy, 1985). 1988 ISBN 90-277-2684-1
27. I.D. Novikov and V.P. Frolov: *Physics of Black Holes.* 1989 ISBN 90-277-2685-X
28. G. Tarozzi and A. van der Merwe (eds.): *The Nature of Quantum Paradoxes.* Italian Studies in the Foundations and Philosophy of Modern Physics. 1988
    ISBN 90-277-2703-1
29. B.R. Iyer, N. Mukunda and C.V. Vishveshwara (eds.): *Gravitation, Gauge Theories and the Early Universe.* 1989 ISBN 90-277-2710-4
30. H. Mark and L. Wood (eds.): *Energy in Physics, War and Peace.* A Festschrift celebrating Edward Teller's 80th Birthday. 1988 ISBN 90-277-2775-9
31. G.J. Erickson and C.R. Smith (eds.): *Maximum-Entropy and Bayesian Methods in Science and Engineering.*
    Vol. I: *Foundations.* 1988 ISBN 90-277-2793-7
32. G.J. Erickson and C.R. Smith (eds.): *Maximum-Entropy and Bayesian Methods in Science and Engineering.*
    Vol. II: *Applications.* 1988 ISBN 90-277-2794-5
33. M.E. Noz and Y.S. Kim (eds.): *Special Relativity and Quantum Theory.* A Collection of Papers on the Poincaré Group. 1988 ISBN 90-277-2799-6
34. I.Yu. Kobzarev and Yu.I. Manin: *Elementary Particles. Mathematics, Physics and Philosophy.* 1989 ISBN 0-7923-0098-X
35. F. Selleri: *Quantum Paradoxes and Physical Reality.* 1990 ISBN 0-7923-0253-2
36. J. Skilling (ed.): *Maximum-Entropy and Bayesian Methods.* Proceedings of the 8th International Workshop (Cambridge, UK, 1988). 1989 ISBN 0-7923-0224-9
37. M. Kafatos (ed.): *Bell's Theorem, Quantum Theory and Conceptions of the Universe.* 1989 ISBN 0-7923-0496-9
38. Yu.A. Izyumov and V.N. Syromyatnikov: *Phase Transitions and Crystal Symmetry.* 1990 ISBN 0-7923-0542-6
39. P.F. Fougère (ed.): *Maximum-Entropy and Bayesian Methods.* Proceedings of the 9th International Workshop (Dartmouth, Massachusetts, USA, 1989). 1990
    ISBN 0-7923-0928-6
40. L. de Broglie: *Heisenberg's Uncertainties and the Probabilistic Interpretation of Wave Mechanics.* With Critical Notes of the Author. 1990 ISBN 0-7923-0929-4
41. W.T. Grandy, Jr.: *Relativistic Quantum Mechanics of Leptons and Fields.* 1991
    ISBN 0-7923-1049-7
42. Yu.L. Klimontovich: *Turbulent Motion and the Structure of Chaos.* A New Approach to the Statistical Theory of Open Systems. 1991 ISBN 0-7923-1114-0

Fundamental Theories of Physics

43. W.T. Grandy, Jr. and L.H. Schick (eds.): *Maximum-Entropy and Bayesian Methods.* Proceedings of the 10th International Workshop (Laramie, Wyoming, USA, 1990). 1991 ISBN 0-7923-1140-X
44. P.Pták and S. Pulmannová: *Orthomodular Structures as Quantum Logics.* Intrinsic Properties, State Space and Probabilistic Topics. 1991 ISBN 0-7923-1207-4
45. D. Hestenes and A. Weingartshofer (eds.): *The Electron.* New Theory and Experiment. 1991 ISBN 0-7923-1356-9
46. P.P.J.M. Schram: *Kinetic Theory of Gases and Plasmas.* 1991 ISBN 0-7923-1392-5
47. A. Micali, R. Boudet and J. Helmstetter (eds.): *Clifford Algebras and their Applications in Mathematical Physics.* 1992 ISBN 0-7923-1623-1
48. E. Prugovečki: *Quantum Geometry.* A Framework for Quantum General Relativity. 1992 ISBN 0-7923-1640-1
49. M.H. Mac Gregor: *The Enigmatic Electron.* 1992 ISBN 0-7923-1982-6
50. C.R. Smith, G.J. Erickson and P.O. Neudorfer (eds.): *Maximum Entropy and Bayesian Methods.* Proceedings of the 11th International Workshop (Seattle, 1991). 1993 ISBN 0-7923-2031-X
51. D.J. Hoekzema: *The Quantum Labyrinth.* 1993 ISBN 0-7923-2066-2
52. Z. Oziewicz, B. Jancewicz and A. Borowiec (eds.): *Spinors, Twistors, Clifford Algebras and Quantum Deformations.* Proceedings of the Second Max Born Symposium (Wrocław, Poland, 1992). 1993 ISBN 0-7923-2251-7
53. A. Mohammad-Djafari and G. Demoment (eds.): *Maximum Entropy and Bayesian Methods.* Proceedings of the 12th International Workshop (Paris, France, 1992). 1993 ISBN 0-7923-2280-0
54. M. Riesz: *Clifford Numbers and Spinors* with Riesz' Private Lectures to E. Folke Bolinder and a Historical Review by Pertti Lounesto. E.F. Bolinder and P. Lounesto (eds.). 1993 ISBN 0-7923-2299-1
55. F. Brackx, R. Delanghe and H. Serras (eds.): *Clifford Algebras and their Applications in Mathematical Physics.* Proceedings of the Third Conference (Deinze, 1993) 1993 ISBN 0-7923-2347-5
56. J.R. Fanchi: *Parametrized Relativistic Quantum Theory.* 1993 ISBN 0-7923-2376-9
57. A. Peres: *Quantum Theory: Concepts and Methods.* 1993 ISBN 0-7923-2549-4
58. P.L. Antonelli, R.S. Ingarden and M. Matsumoto: *The Theory of Sprays and Finsler Spaces with Applications in Physics and Biology.* 1993 ISBN 0-7923-2577-X
59. R. Miron and M. Anastasiei: *The Geometry of Lagrange Spaces: Theory and Applications.* 1994 ISBN 0-7923-2591-5
60. G. Adomian: *Solving Frontier Problems of Physics: The Decomposition Method.* 1994 ISBN 0-7923-2644-X
61. B.S. Kerner and V.V. Osipov: *Autosolitons.* A New Approach to Problems of Self-Organization and Turbulence. 1994 ISBN 0-7923-2816-7
62. G.R. Heidbreder (ed.): *Maximum Entropy and Bayesian Methods.* Proceedings of the 13th International Workshop (Santa Barbara, USA, 1993) 1996 ISBN 0-7923-2851-5
63. J. Peřina, Z. Hradil and B. Jurčo: *Quantum Optics and Fundamentals of Physics.* 1994 ISBN 0-7923-3000-5

Fundamental Theories of Physics

64. M. Evans and J.-P. Vigier: *The Enigmatic Photon.* Volume 1: The Field $B^{(3)}$. 1994
    ISBN 0-7923-3049-8
65. C.K. Raju: *Time: Towards a Constistent Theory.* 1994 ISBN 0-7923-3103-6
66. A.K.T. Assis: *Weber's Electrodynamics.* 1994 ISBN 0-7923-3137-0
67. Yu. L. Klimontovich: *Statistical Theory of Open Systems.* Volume 1: A Unified Approach to Kinetic Description of Processes in Active Systems. 1995
    ISBN 0-7923-3199-0; Pb: ISBN 0-7923-3242-3
68. M. Evans and J.-P. Vigier: *The Enigmatic Photon.* Volume 2: Non-Abelian Electrodynamics. 1995 ISBN 0-7923-3288-1
69. G. Esposito: *Complex General Relativity.* 1995 ISBN 0-7923-3340-3
70. J. Skilling and S. Sibisi (eds.): *Maximum Entropy and Bayesian Methods.* Proceedings of the Fourteenth International Workshop on Maximum Entropy and Bayesian Methods. 1996 ISBN 0-7923-3452-3
71. C. Garola and A. Rossi (eds.): *The Foundations of Quantum Mechanics – Historical Analysis and Open Questions.* 1995 ISBN 0-7923-3480-9
72. A. Peres: *Quantum Theory: Concepts and Methods.* 1995 (see for hardback edition, Vol. 57) ISBN Pb 0-7923-3632-1
73. M. Ferrero and A. van der Merwe (eds.): *Fundamental Problems in Quantum Physics.* 1995 ISBN 0-7923-3670-4
74. F.E. Schroeck, Jr.: *Quantum Mechanics on Phase Space.* 1996 ISBN 0-7923-3794-8
75. L. de la Peña and A.M. Cetto: *The Quantum Dice.* An Introduction to Stochastic Electrodynamics. 1996 ISBN 0-7923-3818-9
76. P.L. Antonelli and R. Miron (eds.): *Lagrange and Finsler Geometry.* Applications to Physics and Biology. 1996 ISBN 0-7923-3873-1
77. M.W. Evans, J.-P. Vigier, S. Roy and S. Jeffers: *The Enigmatic Photon.* Volume 3: Theory and Practice of the $B^{(3)}$ Field. 1996 ISBN 0-7923-4044-2
78. W.G.V. Rosser: *Interpretation of Classical Electromagnetism.* 1996
    ISBN 0-7923-4187-2
79. K.M. Hanson and R.N. Silver (eds.): *Maximum Entropy and Bayesian Methods.* 1996
    ISBN 0-7923-4311-5
80. S. Jeffers, S. Roy, J.-P. Vigier and G. Hunter (eds.): *The Present Status of the Quantum Theory of Light.* Proceedings of a Symposium in Honour of Jean-Pierre Vigier. 1997
    ISBN 0-7923-4337-9

KLUWER ACADEMIC PUBLISHERS – DORDRECHT / BOSTON / LONDON